Ecological Studies, Vol. 167

Analysis and Synthesis

Edited by

I.T. Baldwin, Jena, Germany
M.M. Caldwell, Logan, USA
G. Heldmaier, Marburg, Germany
O.L. Lange, Würzburg, Germany
H.A. Mooney, Stanford, USA
E.-D. Schulze, Jena, Germany
U. Sommer, Kiel, Germany

Ecological Studies

Volumes published since 1997 are listed at the end of this book.

Springer

Berlin
Heidelberg
New York
Hong Kong
London
Milan
Paris
Tokyo

L. Nagy   G. Grabherr   Ch. Körner   D.B.A. Thompson (Eds.)

Alpine Biodiversity in Europe

With 89 Figures, 17 in Color, and 61 Tables

Springer

Dr. Laszlo Nagy
McConnell Associates
41 Eildon Street
Edinburgh, EH3 5JX
Scotland

Prof. Dr. Christian Körner
Institute of Botany
University of Basel
Schönbeinstr. 6
4056 Basel, Switzerland

Prof. Dr. Georg Grabherr
Department of Vegetation Ecology
and Conservation Biology
Institute for Ecology
and Nature Conservation
University of Vienna
Althanstr. 14
1091 Vienna, Austria

Prof. Dr. Desmond B.A. Thompson
Advisory Services
Scottish Natural Heritage
2 Anderson Place
Edinburgh, EH6 5NP
Scotland

Cover illustration: Background: Mutterberger Seespitze, Stubai Alps, Austria (3302 m, L. Nagy). Dark green fritillary butterfly *Argynnis aglaja*, Swiss Alps (M. Raviglione); dotterel *Charadrius morinellus*, Scottish Highlands (S. Austin); Pyrenean frog *Rana pyrenaica*, Pyrenees (J. Serra Cobo); alpine daisy *Leucanthemopsis alpinum*, Mt. Schrankogel, Stubai Alps, Austria (L. Nagy); isard *Rupicapra pyrenaica*, Pyrenees (J.-P. Martinez-Rica); grasshopper *Podismopsis poppiusi*, South Urals (Y. Mikhailov).

ISSN 0070-8356
ISBN 3-540-00108-5 Springer-Verlag Berlin Heidelberg New York

Library of Congress Cataloging-in-Publication Data

Alpine biodiversity in Europe / L. Nagy ... [et al.].
  p. cm. – (Ecological studies, ISSN 0070-8356 ; vol. 167)
  Includes bibliographical references and index.
  ISBN 3-540-00108-5 (alk. paper)
    1. Mountain ecology--Europe. 2. Biological diversity--Europe I. Nagy, L. (Laszlo),
  1961- II. Series.

QH135.A47 2003
577.5'4'094–dc21   2003042503

This work is subject to copyright. All rights are reserved, whether the whole or part of the material is concerned, specifically the rights of translation, reprinting, reuse of illustrations, recitation, broadcasting, reproduction on microfilm or in any other way, and storage in data banks. Duplication of this publication or parts thereof is permitted only under the provisions of the German Copyright Law of September 9, 1965, in its current version, and permissions for use must always be obtained from Springer-Verlag. Violations are liable for prosecution under the German Copyright Law.
Springer-Verlag Berlin Heidelberg New York
a company of BertelsmannSpringer Science+Business Media GmbH

http://www.springer.de

© Springer-Verlag Berlin Heidelberg 2003

Printed in Germany

The use of general descriptive names, registered names, trademarks, etc. in this publication does not imply, even in the absence of a specific statement, that such names are exempt from the relevant protective laws and regulations and therefore free for general use.

Production: Friedmut Kröner, 69115 Heidelberg, Germany
Cover design: *design & production* GmbH, Heidelberg
Typesetting: Kröner, 69115 Heidelberg, Germany

31/3150 YK - 5 4 3 2 1 0 – Printed on acid free paper

Preface

The United Nations Conference on the Environment and Development (UNCED), held in Rio de Janeiro in 1992, spawned a multitude of programmes aimed at assessing, managing and conserving the earth's biological diversity. One important issue addressed at the conference was the mountain environment. A specific feature of high mountains is the so-called alpine zone, i.e. the treeless regions at the uppermost reaches. Though covering only a very small proportion of the land surface, the alpine zone contains a relatively large number of plants, animals, fungi and microbes which are specifically adapted to cold environments. This zone contributes fundamentally to the planet's biodiversity and provides many resources for mountain dwelling as well as lowland people. However, rapid and largely man-made changes are affecting mountain ecosystems, such as soil erosion, losses of habitat and genetic diversity, and climate change, all of which have to be addressed. As stated in the European Community Biodiversity Strategy, "the global scale of biodiversity reduction or losses and the interdependence of different species and ecosystems across national borders demands concerted international action". Managing biodiversity in a rational and sustainable way needs basic knowledge on its qualitative and quantitative aspects at local, regional and global scales. This is particularly true for mountains, which are distributed throughout the world and are indeed hot spots of biodiversity in absolute terms as well as relative to the surrounding lowlands.

At the European scale, ALPNET, a European Science Foundation sponsored Alpine Biodiversity Network (1998–2000), was established to synthesise available knowledge on alpine biodiversity. ALPNET was established following an ESF-ICALPE (International Centre for Alpine Environments) exploratory workshop at the Centro di Ecologia Alpina, Monte Bondone, Italy, in 1995. Priority was given to the cataloguing and synthesis of the available knowledge on: (1) biodiversity of alpine environments, (2) spatial patterns along elevation gradients, and (3) vegetation dynamics. As ALPNET was considered to be the first stage of integrating alpine biodiversity, field research was restricted to characterising the bioclimate of the European alpine areas through the use of soil temperature data loggers. This book is the culmination of ALPNET's work. Three workshops (in Monte Bondone, Italy; Innsbruck, Austria; Cargése, Corsica) between 1998-2000 contributed substantially to much of the material presented here, amounting to our current understanding of alpine biodiversity in the high mountain regions of Europe.

The ALPNET model has since been adopted on a global scale to establish the Global Mountain Biodiversity Assessment programme of DIVERSITAS (http://www.unibas.ch/gmba). Another new initiative, with many participants from ALPNET, is GLORIA-Europe (http://www.gloria.ac.at, launched on 1 January 2001), the European component of a long-term global observation system for recording vegetation and temperature at alpine summits.

Essentially, this book is about the taxonomic richness of plant and animal communities, their pattern and diversity in space, their temporal changes in species richness and community composition, and the underlying ecosystem processes above the treeline in the European high mountains (except the dry Mediterranean high mountains of Crete, the Canary Islands and the Azores).

The book begins with an overview of Europe's alpine areas (Section I, Chap. 1). This is followed by a synthesis of bioclimate of the alpine zone across locations ranging from northern Scandinavia to the Sierra Nevada in a north-south direction and from the Sierra Nevada to the Jakupica Mountains in central Macedonia, FYR, from west to east. Section I concludes with a physiographic and ecological characterisation of the European high mountain regions (Chaps. 3.1-3.10). Section II (Chaps. 4-11) discusses plant taxonomic and community (assemblage) diversity patterns. The scales range from local through the sub-regional (mountain range) to the European continent. The patterns are, in the main, discussed in relation to physiographic factors: latitude, longitude, altitude, substratum and topography. Section III (Chaps. 12-17) presents a collection of accounts on high mountain invertebrate species and assemblages, ordered similarly to Section II on plants. The major factors discussed in relation to processes and changing patterns are global change (primarily climate warming) and land use (grazing). Section IV (Chaps. 18-23) focuses on vertebrate distribution, population ecology and ecosystem impacts, especially grazing. Several accounts in Section V (Chaps. 24-28) examine changes ranging from the timberline in the Urals to those observed at the upper limit of plant life in the Alps. Whilst these reports focus on climate amelioration as the main factor of changes (the reports are from areas little influenced by man), the role of herbivores and their management are at least of equal importance. The closing paper (Chap. 29) is a synthesis of alpine biodiversity in space and time.

Unless stated otherwise, plant nomenclature in this volume follows *Flora Europaea* (Tutin et al. 1964, 1968-1980).

Preface

Acknowledgements. We thank the ESF for funding the ALPNET network and contributing to the cost of colour illustrations in this volume. The Bristol Foundation, Switzerland, and Scottish Natural Heritage, Scotland, contributed support towards financing the editing work.

References

Tutin TG, Heywood VH, Burges NA, Valentine DW, Walters SM, Webb DA (eds) (1964) Flora europaea Vol 1. Cambridge University Press, Cambridge
Tuting TG, Heywood VH, Burges NA, Moore DM, Valentine DW, Walters SM, Webb DA (eds) (1968–1980) Flora Europaea Vols. 2–5. Cambridge University Press, Cambridge

February 2003   The Editors

Punta Minuta, Corsica, 2556 m (L. Nagy)

Fundamental to the establishment of ALPNET was the unstinting enthusiasm, driving commitment and knowledge of Jennifer Nagy (née McConnell), who died shortly before the first workshop in 1998. Our bryologist, Patricia Geissler, was killed in a traffic accident in Geneva in 1999. We shall always remember Jennifer and Patricia.

This book is dedicated to the memory of these two excellent scientists, friends and colleagues. They were very special people.

Contents

I	**Europe's Alpine Areas**	
1	**An Outline of Europe's Alpine Areas**	3
	G. Grabherr, L. Nagy and D.B.A. Thompson	
1.1	The Mountains of Europe	5
1.2	An Ecological Classification of Europe's Alpine Areas	8
1.3	The Regional Accounts (Chaps. 3.1–3.10)	11
References	....	11
2	**A Bioclimatic Characterisation of Europe's Alpine Areas** ..	13
	Ch. Körner, J. Paulsen and S. Pelaez-Riedl	
2.1	Introduction	13
2.2	Site Selection and Methods	13
2.2.1	What Was Measured and Where?	13
2.2.2	Sensors and Sensor Positioning	14
2.2.3	Data Analysis	16
2.3	Results	17
2.3.1	Between-Year and Between-Site Variation	17
2.3.2	Temperature Extremes	19
2.3.3	The Daily Course of Ground Temperature	20
2.3.4	The Seasonal Course of Ground Temperature	20
2.3.5	Seasonal Mean Temperatures	22
2.3.6	Growing Season Length	22
2.3.7	Thermal Sums	23
2.4	Alpine Biota Share a Common Climate Across Europe ...	26
References	....	28

3	The Regional Accounts	
3.1	**The High Mountain Vegetation of the Scandes** R. VIRTANEN	31
3.1.1	Geography, Geology, and Ecological Conditions	31
3.1.2	Climate	31
3.1.3	The Flora	32
3.1.4	The Alpine Zone and the Treeline Ecotone	32
3.1.4.1	Number of Plant Taxa	32
3.1.4.2	Rare Taxa	33
3.1.4.3	Altitude Zones	33
3.1.4.4	Raunkiaer's Biological Types	34
3.1.4.5	Vegetation and Its Responses to a Changing Environment	34
References		36
3.2	**The High Mountain Vegetation of Scotland** L. NAGY	39
3.2.1	Introduction	39
3.2.2	Geology, Soils and Climate	39
3.2.3	The Flora	40
3.2.3.1	The Present-Day Flora	40
3.2.4	Vegetation	41
3.2.5	Species and Habitat Diversity and Causes of Change	41
References		45
3.3	**Vegetation of the Giant Mountains, Central Europe** J. JENÍK and J. ŠTURSA	47
3.3.1 Introduction		47
3.3.2 Physical-Geographical Setting		47
3.3.3 Flora and Vegetation		48
3.3.4 Refugia of the Polygenous Flora		50
References		51
3.4	**The Alpine Vegetation of the Alps** P. OZENDA and J.-L. BOREL	53
3.4.1	Physiography	53
3.4.2	Floristic Diversity	56
3.4.3	Species Richness vs. Altitude	56
3.4.4	Endemism	58

Contents   XIII

3.4.5	Taxonomic Diversity	59
3.4.6	Life Forms	59
3.4.7	Comparison of Different Mountain Ranges – Species Richness, Floristic and Community Similarities	60
References		63

3.5 The Alpine Flora and Vegetation of the South-Eastern Carpathians 65
G. COLDEA

3.5.1	Geography and Geology	65
3.5.2	The Treeline Ecotone (Subalpine) and the Alpine Zone	65
3.5.3	Climate	66
3.5.4	Flora	67
3.5.4.1	History of the Flora	67
3.5.4.2	Number of Taxa	68
3.5.5	Vegetation	68
3.5.6	Human Impacts vs. Diversity	70
References		71

3.6 The High Mountain Flora and Vegetation of the Apennines and the Italian Alps 73
F. PEDROTTI and D. GAFTA

3.6.1	Geography	73
3.6.2	Geology and Climate	74
3.6.3	History of the Flora	76
3.6.4	Today's Flora	76
3.6.5	Altitudinal Vegetation Zonation	80
3.6.6	Floristic Diversity and Distinctiveness of Communities	81
3.6.7	Causes of Change and Conservation	82
References		83

3.7 The Vegetation of the Alpine Zone in the Pyrenees 85
D. GÓMEZ, J.A. SESÉ and L. VILLAR

3.7.1	Geography	85
3.7.2	History, Geology, Soils and Climate	85
3.7.3	History of the Flora	87
3.7.4	Today's Flora	87
3.7.5	Vegetation	89
References		92

3.8	**High Mountain Vegetation of the Caucasus Region**	93
	G. Nakhutsrishvili	
3.8.1	The Caucasus	93
3.8.1.1	Geography and Geology	93
3.8.1.2	Climate	94
3.8.2	The Flora	95
3.8.2.1	History of the Flora	95
3.8.2.2	Human Impacts	95
3.8.2.3	Number of Taxa	95
3.8.3	High Mountain Vegetation	97
3.8.3.1	The Treeline Ecotone	97
3.8.3.1.1	Tall-Herb Vegetation	99
3.8.3.1.2	Meadows, Feather-Grass Steppes and Scrub at the Treeline	99
3.8.3.2	The Alpine Zone Proper	100
3.8.3.3	The Subnival Zone	100
3.8.4	A Comparative Analysis of the High Mountain Vegetation of the Caucasus and the Alps	101
References		102
3.9	**The Vegetation of the Corsican High Mountains**	105
	J. Gamisans	
3.9.1	Geography	105
3.9.2	Climate	105
3.9.3	The Flora	106
3.9.3.1	History of the Flora	106
3.9.3.2	Human Impact	106
3.9.3.3	The Number of Taxa	107
3.9.4	Vegetation Altitude Zones	108
References		110
3.10	**The High Mountain Vegetation of the Balkan Peninsula**	113
	A. Strid, A. Andonoski and V. Andonovski	
3.10.1	Geography	113
3.10.2	Climate	114
3.10.3	Soils	115
3.10.4	The Flora	116
3.10.4.1	Endemism and Speciation	116
3.10.5	Vegetation	117
References		121

II Plant and Vegetation Diversity

4 Overview: Patterns in Diversity 125
U. MOLAU

4.1 Scale and Diversity 125
4.2 Temporal Diversity 127
4.3 Diversity at Landscape and Continent Scales 128
References 131

5 Taxonomic Diversity of Vascular Plants in the European Alpine Areas 133
H. VÄRE, R. LAMPINEN, C. HUMPHRIES and P. WILLIAMS

5.1 Introduction 133
5.2 Methods 134
5.2.1 The Selection of Taxa 134
5.2.1.1 Taxon Richness 134
5.2.1.2 Range-Size Rarity 140
5.2.2 Area Selection 142
5.2.2.1 Near-Minimum Area Set to Represent All 142
5.2.2.2 Hotspot Areas 144
5.3 Results and Discussion 144
5.3.1 Taxon Richness 144
5.3.2 Regional Mountain Range Richness 145
5.3.3 Area Selections 145
5.4 Conclusions 146
References 147

6 Patterns in the Plant Species Richness of European High Mountain Vegetation 149
R. VIRTANEN, T. DIRNBÖCK, S. DULLINGER, G. GRABHERR, H. PAULI, M. STAUDINGER and L. VILLAR

6.1 Introduction 149
6.2 The Data 156
6.3 Geographical and Ecological Trends in Plant Species Richness 157
6.4 Discussion 162
6.4.1 Latitudinal Trends 162
6.4.2 The Size of Mountain Areas 162

6.4.3	Soil Substratum	163
6.4.4	Snow Cover	164
6.4.5	Local and Regional Species Richness	164
6.5	Conclusions	165
References		166

7	**Altitude Ranges and Spatial Patterns of Alpine Plants in Northern Europe** J.I. HOLTEN	173
7.1	Introduction	173
7.1.1	Study Area	174
7.1.2	Methods	174
7.2	Altitude Ranges and Patterns	176
7.2.1	Altitudinal Ranges of Alpine Plant Species	176
7.2.1.1	Boreal-Alpine (Montane) Group	176
7.2.1.2	Widespread (Ubiquitous) Group	177
7.2.1.3	Centric Group	178
7.2.2	Alpine Vascular Plant Species Richness vs. Altitude	179
7.3	What Limits the Altitudinal Ranges and Patterns in the Alpine Zone?	180
7.3.1	Limiting Factors	181
7.3.1.1	High Summer Temperatures and Drought	181
7.3.1.2	Winter Temperatures	181
7.3.1.3	Alpine Permafrost	182
7.4	Conclusions	182
References		183

8	**Vascular Plant and Bryophyte Diversity Along Elevational Gradients in the Alps** J.-P. THEURILLAT, A. SCHLÜSSEL, P. GEISSLER, A. GUISAN, C. VELLUTI and L. WIGET	185
8.1	Introduction	185
8.1.1	Study Areas	185
8.1.2	Methodology	186
8.1.2.1	The Transects	186
8.1.2.2	Analyses	186
8.2	Patterns of Distribution	187
8.2.1	Elevation and Species Richness	187
8.2.2	Life Forms, Growth Forms and Elevation	188

8.2.3	Distribution of Elevational Species Groups	188
8.2.4	Similarity and Elevation	189
8.2.5	Cumulative Species Richness and Elevation	189
8.3	Discussion	190
8.3.1	Theoretical Background	190
8.3.2	The Temperature–Physiography Hypothesis	191
8.4	Conclusions	192
References		192

9	**Assessing the Long-Term Dynamics of Endemic Plants at Summit Habitats**	**195**
	H. Pauli, M. Gottfried, T. Dirnböck, S. Dullinger and G. Grabherr	
9.1	Introduction	195
9.2	Study Areas	196
9.3	Methods	197
9.4	Results	198
9.4.1	Vegetation Cover and Vascular Plant Species Cover	198
9.4.2	Species Richness and Endemic Species	199
9.5	Discussion	202
9.5.1	Orographic Isolation and the Cryophilic Nature of Endemics	202
9.5.2	Climate Warming and the Vulnerability of Cryophilic Endemic Species	204
9.6	Conclusions	205
References		206

10	**Mapping Alpine Vegetation**	**209**
	T. Dirnböck, S. Dullinger, M. Gottfried, C. Ginzler and G. Grabherr	
10.1	Introduction	209
10.2	Study Site	212
10.3	Methods	212
10.4	Results and Discussion	215
10.4.1	The Map	215
10.4.2	Accuracy Assessment	215
10.4.3	Remotely Sensed Maps and Biodiversity Research	216
10.5	Conclusions	217
References		218

11	A GIS Assessment of Alpine Biodiversity at a Range of Scales	221

U. Molau, J. Kling, K. Lindblad, R. Björk, J. Dänhardt and A. Liess

11.1	Introduction	221
11.2	GIS in Ecology	222
11.3	The Latnjajaure Catchment	222
11.4	Biodiversity Assessment by GIS	225
11.4.1	Field Sampling	225
11.4.2	GIS Database and Modelling	226
11.5	Results and Conclusions	226
11.5.1	Plants	227
11.5.2	Animals	227
References		228

III	Invertebrates	
12	Overview: Invertebrate Diversity in Europe's Alpine Regions	233

P. Brandmayr, R. Pizzolotto and S. Scalercio

12.1	Macrolepidoptera	233
12.2	Carabids	234
12.3	A Comparison of the Diversity Patterns of Arachnids, Carabids and Lepidoptera	235
12.4	Conclusions	236
References		237

13	The Geographical Distribution of High Mountain Macrolepidoptera in Europe	239

Z. Varga

13.1	Introduction	239
13.2	Faunal Types in the European Orobiomes	239
13.3	The Alpine Faunal Type	242
13.4	Speciation and Endemism in the Alpine Faunal Type	243
13.5	Biogeographical Characterisation of the Xeromontane Faunal Type	245
13.6	Biogeographical Connections and Similarities of the European High Mountains	246

13.7	Geographical and Phylogenetic Patterns in the European High Mountain Macrolepidoptera	247
13.8	Conclusions	249
References		250

14 High Altitude Invertebrate Diversity in the Ural Mountains 259
Y.E. MIKHAILOV and V.N. OLSCHWANG

14.1	Introduction	259
14.2	Species Diversity: State of Knowledge	261
14.2.1	Historical Origins of Zoological Research	261
14.2.2	A Brief Inventory of the Fauna	261
14.3	The Composition and Origin of the High Altitude Fauna	265
14.4	Dynamics of Species Diversity: Long-Term Observations	269
14.5	Altitudinal Gradients in Species Diversity and Abundance of Soil Fauna	270
14.6	Herbivore–Vegetation Interactions in the Alpine Zone	271
14.7	Phytophagous Insects as Potential Bioindicators in Alpine Ecosystems	272
14.8	Conclusions	273
References		274

15 The Diversity of High Altitude Arachnids (Araneae, Opiliones, Pseudoscorpiones) in the Alps 281
K. THALER

15.1	Introduction	281
15.2	High Altitude Arachnids of the Alps: a Systematic Overview	281
15.2.1	Opiliones and Pseudoscorpiones	282
15.2.2	Araneae	284
15.3	Regional Distribution	285
15.4	Biology	287
15.4.1	Vertical Distribution	287
15.4.2	Spatial Distribution	289
15.4.3	Phenology, Winter Activity, Life Cycles	290
15.4.4	Biotic Interactions	291
15.4.5	Various Aspects	292
15.5	Conclusions	292
References		293

16	Patterns of Butterfly Diversity Above the Timberline in the Italian Alps and Apennines	297
	L. Tontini, S. Castellano, S. Bonelli and E. Balletto	

16.1	Introduction	297
16.2	Materials and Methods	298
16.2.1	Assemblage Description	298
16.2.2	Grazing Impacts	299
16.3	Results	299
16.3.1	Richness and Density of Species	299
16.3.1.1	Species Richness Along Latitude and Altitude	300
16.3.1.2	Population Densities and Assemblage Structure Variation	301
16.3.2	Exclusively Alpine Species Richness and Densities	301
16.3.3	Grazing Impacts	302
16.4	Conclusions	304
References		305

17	Diversity Patterns of Carabids in the Alps and the Apennines	307
	P. Brandmayr, R. Pizzolotto, S. Scalercio, M.C. Algieri and T. Zetto	

17.1	Introduction	307
17.2	Overall Features of Alpine Carabid Communities in the Alps	308
17.3	Species Recruitment in Alpine Habitats	309
17.4	Diversity Patterns in the Alps	311
17.5	Carabid Assemblages of the Central Apennines	312
17.6	Alpine Carabid Assemblages and Climate Change	313
17.7	Conclusions	314
References		314

IV	Vertebrates

18	Overview: Alpine Vertebrates – Some Challenges for Research and Conservation Management	321
	D.B.A. Thompson	

18.1	Species Diversity Variation	321
18.2	Population Cycling and Herbivores	322
18.3	Habitat Use	323

| 18.4 | Conclusions | 324 |
| References | | 325 |

19 Breeding Bird Assemblages and Habitat Use of Alpine Areas in Scotland 327
D.B.A. THOMPSON, D.P. WHITFIELD, H. GALBRAITH, K. DUNCAN, R.D. SMITH, S. MURRAY and S. HOLT

19.1	Introduction	327
19.2	Study Area and Methods	328
19.3	Results	330
19.3.1	The Breeding Bird Assemblages	330
19.3.2	Habitat Use	331
19.4	Discussion	334
19.4.1	Breeding Bird Assemblages	334
19.4.2	Habitat Use	335
19.4.3	Consequences of Habitat Modification	335
19.5	Conclusions	336
References		337

20 Rodents in the European Alps: Population Ecology and Potential Impacts on Ecosystems 339
D. ALLAINÉ and N.G. YOCCOZ

20.1	Introduction	339
20.2	Is There an 'Alpine Rodent'?	339
20.3	Population Dynamics and Demography	342
20.4	Alpine Rodents and Plant Communities	343
20.5	Consequences of Changes in the Management of Alpine Ecosystems and in Climate	345
20.6	Needs for Further Work	346
20.7	Conclusions	346
References		347

21 Large Herbivores in European Alpine Ecosystems: Current Status and Challenges for the Future 351
A. LOISON, C. TOÏGO and J.-M. GAILLARD

| 21.1 | Introduction | 351 |
| 21.2 | The Species Concerned and Their Distribution in Europe | 354 |

21.3	Ecological Features	354
21.3.1	Morphology and Physiology	354
21.3.2	Habitat	355
21.3.3	Feeding Ecology	355
21.3.4	Social and Mating System	355
21.3.5	Spatial Features	356
21.4	Dynamics in Space and Time	356
21.4.1	Demographic Patterns	356
21.4.2	Density Dependence and Effects of Climate	357
21.5	The Role of Large Herbivores on the Mountain Landscape and Plant Communities	358
21.6	Changes in Mountain Ecosystems and New Challenges for Large Herbivores	359
21.6.1	Interaction with Domestic Ungulates, Diseases and Parasites	359
21.6.2	The Increase of Lowland Ungulates and Inter-specific Competition	359
21.6.3	The Return of Large Carnivores	359
21.6.4	Habitat Fragmentation	360
21.6.5	Recreational Use of the Mountains and Disturbance	361
21.7	Conclusions and Future Research Needs	361
References		362

22	**Diversity of Alpine Vertebrates in the Pyrenees and Sierra Nevada, Spain**	367
	J.P. Martínez Rica	
22.1	Introduction	367
22.2	Alpine Vertebrate Species	368
22.3	Species Richness vs. Diversity Indices	370
22.4	Altitudinal Gradients and Biodiversity	371
22.5	Conclusions	374
References		374

23	**The Impacts of Vertebrate Grazers on Vegetation in European High Mountains**	377
	B. Erschbamer, R. Virtanen and L. Nagy	
23.1	Introduction	377
23.2	The Historical Perspective	377
23.3	The Grazers and Their Diet	378
23.4	Grazing Impacts	379

23.5	The Effects of Grazing Exclusion on the Vegetation	380
23.5.1	Biomass and Species Composition	380
23.5.2	Dead Standing Plant Material, Litter, and Soil Microbes	382
23.6	Evidence from Outside Exclosures	384
23.6.1	Canopy Structure, Growth Form, Flowering and Seed Production	384
23.6.2	Succession After the Abandonment of Pasturing	385
23.6.3	Nutrient Enrichment and Community Changes	385
23.7	Grazing in the Alpine Zone: Ecological Problems and Prospect	386
23.8	Conclusions	388
References		389

V Long-Term Vegetation Dynamics

24 Overview: Alpine Vegetation Dynamics and Climate Change – a Synthesis of Long-Term Studies and Observations 399
G. Grabherr

24.1	Introduction	399
24.2	Treeline Ecotone Trees and Forest Line Dynamics	399
24.3	Vegetation Dynamics Within the Alpine Zone	
24.4	Dynamic Processes at the Limits of Plant Life	406
24.5	Conclusions	406
References		407

25 Long-Term Changes in Alpine Plant Communities in Norway and Finland 411
R. Virtanen, A. Eskelinen and E. Gaare

25.1	Introduction	411
25.2	Study Areas	412
25.3	Data Collection and Analysis	412
25.4	Changes in Species Composition and Abundance	413
25.4.1	Heath Vegetation	413
25.4.2	Snowbed Vegetation	417
25.4.3	Alpine Soligenous Mires	418
25.5	Discussion	419
25.6	Conclusions	420
References		421

26	Vegetation Dynamics at the Treeline Ecotone in the Ural Highlands, Russia	423
	P.A. MOISEEV and S.G. SHIYATOV	

26.1	Introduction	423
26.2	Materials and Methods	424
26.2.1	Study Areas	424
26.2.2	Long-Term Weather Data	425
26.2.3	Repeat Photography and Image Analysis	426
26.3	Four Examples of Photo Pairs	427
26.4	Discussion	432
26.5	Conclusions	433
References		433

27	Recent Increases in Summit Flora Caused by Warming in the Alps	437
	M. BAHN and CH. KÖRNER	

27.1	Introduction	437
27.2	The Mount Glungezer Summit	437
27.3	Floristic Changes Between 1986–2000	439
27.4	Conclusions	441
References		441

28	The Piz Linard (3411 m), the Grisons, Switzerland – Europe's Oldest Mountain Vegetation Study Site	443
	H. PAULI, M. GOTTFRIED and G. GRABHERR	

28.1	Introduction	443
28.2	Method	443
28.3	Results	444
28.3.1	Changes in Species Richness	444
28.3.2	Changes in Species Abundance	446
28.4	Discussion	446
28.5	Conclusions	447
References		448

VI	**Synthesis**	
29	**Alpine Biodiversity in Space and Time: a Synthesis**	453
	L. Nagy, G. Grabherr, Ch. Körner and D.B.A. Thompson	
29.1	Spatial Trends in Animal and Plant Species Richness	453
29.1.1	Latitudinal Trends	453
29.1.2	Altitude Trends	456
29.1.3	Animals	458
29.2	Temporal Changes	459
29.2.1	Environment	459
29.2.2	Grazers	460
29.3	Outlook	461
References		463

Subject Index 465

Contributors

ALGIERI, M.C., Dipartimento di Ecologia, Università della Calabria, Arcavacata di Rende, 87036 Cosenza, Italy

ALLAINÉ, D., UMR-CNRS 5558 "Biométrie et Biologie Evolutive", Université LYON 1, Boulevard du 11 novembre 1918, 69622 Villeurbanne Cedex, France

ANDONOSKI, A., Skopje University, Faculty of Forestry, P.O. Box 235, 91000 Skopje, Republic of Macedonia

ANDONOVSKI, V., Skopje University, Faculty of Forestry, P.O. Box 235, 91000 Skopje, Republic of Macedonia

BAHN, M., Institut für Botanik, Universität Innsbruck, Sternwartestraße 15, 6020 Innsbruck, Austria

BALLETTO, E., Dipartimento di Biologia Animale e Dell'uomo, Via Accademia Albertina 13–17, 10123 Torino, Italy

BJÖRK, R., Botanical Institute, Gothenburg University, P.O. Box 461, 405 30 Gothenburg, Sweden

BONELLI, S., Dipartimento di Biologia Animale e Dell'uomo, Via Accademia Albertina 13–17, 10123 Torino, Italy

BOREL, J.-L., Laboratoire 'Ecosystèmes et Changements Environnementaux', Centre de Biologie Alpine, Université Joseph Fourier, B.P. 53, 38041 Grenoble Cedex 9, France

BRANDMAYR, P., Dipartimento di Ecologia, Università della Calabria, Arcavacata di Rende, 87036 Cosenza, Italy

CASTELLANO, S., Dipartimento di Biologia Animale e Dell'uomo, Via Accademia Albertina 13–17, 10123 Torino, Italy

COLDEA, G., Institute of Biological Research, 48 Republicii Str., 3400 Cluj-Napoca, Romania

DÄNHARDT, J., Department of Ecology, Lund University, 223 62 Lund, Sweden

DIRNBÖCK, T., Institute for Ecology and Nature Conservation, Department of Vegetation Ecology and Conservation Biology, University of Vienna, Althanstraße 14, 1091 Vienna, Austria

DULLINGER, S., Institute for Ecology and Nature Conservation, Department of Vegetation Ecology and Conservation Biology, University of Vienna, Althanstraße 14, 1091 Vienna, Austria

DUNCAN, K., Advisory Services, Scottish Natural Heritage, 2 Anderson Place, Edinburgh EH6 5NP, Scotland

ERSCHBAMER, B., Institut für Botanik, Universität Innsbruck, Sternwartestraße 15, 6020 Innsbruck, Austria

ESKELINEN, A., Department of Biology, University of Oulu, P.O. Box 3000, 90014 Oulu, Finland

GAARE, E., NINA, Tungasletta 2, 7485 Trondheim, Norway

GAFTA, D., Department of Plant Biology, Babes-Bolyai University, Republic Street 42, 3400 Cluj-Napoca, Romania

GAILLARD, J.-M., CNRS-UMR5558, Laboratoire de Biométrie et d'Ecologie Evolutive, University Lyon 1, Boulevard du 11 novembre 1918, 69622 Villeurbanne Cedex, France

GALBRAITH, H., Advisory Services, Scottish Natural Heritage, 2 Anderson Place, Edinburgh EH6 5NP, Scotland

GAMISANS, J., Laboratoire d'Écologie Terrestre, Université Paul Sabatier, 39 allées Jules Guesde 31062 Toulouse Cedex, France

GEISSLER, P., Conservatoire et Jardin botaniques de la Ville de Genève, CP 1, 1292 Chambésy, Switzerland (deceased)

GINZLER, C., Swiss Federal Institute for Forest, Snow and Landscape Research, Landscape Inventories, Advisory Service for Mire Conservation, 8903 Birmensdorf, Switzerland

GÓMEZ, D., Instituto Pirenaico de Ecología, CSIC, Apartado 64, 22700 Jaca (Huesca), Spain

GOTTFRIED, M., Institute for Ecology and Nature Conservation, Department of Vegetation Ecology and Conservation Biology, University of Vienna, Althanstraße 14, 1091 Vienna, Austria

GRABHERR G., Institute for Ecology and Nature Conservation, Department of Vegetation Ecology and Conservation Biology, University of Vienna, Althanstraße 14, 1091 Vienna, Austria

Contributors

GUISAN, A., Institut d'Écologie et de Géobotanique, Bâtiment de Biologie, Université de Lausanne, 1015 Lausanne, Switzerland

HOLT, S., Advisory Services, Scottish Natural Heritage, 2 Anderson Place, Edinburgh EH6 5NP, Scotland

HOLTEN. J.I., Terrestrial Ecology Research (TerM), Skogaromveien 19, 7350 Buvika, Norway

HUMPHRIES, C., The Biogeography and Conservation Laboratory, The Natural History Museum, London SW7 5BD, UK

JENÍK, J., Charles University, Faculty of Science, Benátská 2, 128 00 Praha 2, Czech Republic

KLING, J., Swedish Society for Nature Protection, P.O. Box 7005, 402 31 Gothenburg, Sweden

KÖRNER, C., Institute of Botany, University of Basel, Schönbeinstrasse 6, 4056 Basel, Switzerland

LAMPINEN, R., Botanical Museum, Finnish Museum of Natural History, Box 7, 00014 University of Helsinki, Finland

LIESS, A., Botanical Institute, Gothenburg University, P.O. Box 461, 405 30 Gothenburg, Sweden
(Present address: Steinickeweg 7, 80798 Munich, Germany)

LINDBLAD, K., Botanical Institute, Gothenburg University, P.O. Box 461, 405 30 Gothenburg, Sweden

LOISON, A., CNRS-UMR5558, Laboratoire de Biométrie et d'Ecologie Evolutive, University Lyon 1, Boulevard du 11 novembre 1918, 69622 Villeurbanne Cedex, France

MARTÍNEZ RICA, J.P., Instituto Pirenaico de Ecología, Avda. Montañana, 177, 50080 Zaragoza, Spain

MIKHAILOV, Y.E., Department of Botany and Forest Protection, Urals State Forestry-Engineering University, Sibirsky trakt 37, 620100 Yekaterinburg, Russia

MOISEEV P.A., Department of Botany and Forest Protection, Urals State Forestry-Engineering University, Sibirsky trakt 37, 620100 Yekaterinburg, Russia

MOLAU, U., Botanical Institute, Gothenburg University, P.O. Box 461, 405 30 Gothenburg, Sweden

MURRAY S., Advisory Services, Scottish Natural Heritage, 2 Anderson Place, Edinburgh EH6 5NP, Scotland

NAGY, L., McConnell Associates, 41 Eildon Street, Edinburgh EH3 5JX, Scotland

NAKHUTSRISHVILI, G., Institute of Botany, Georgian Academy of Sciences, Kojori Road 1, 380007 Tbilisi, Georgia

OLSCHWANG V.N., Institute of Plant and Animal Ecology, Urals Branch of Russian Academy of Sciences, 8 Marta St., 202, 620144 Yekaterinburg, Russia

OZENDA, P., Laboratoire 'Ecosystèmes et Changements Environnementaux', Centre de Biologie Alpine, Université Joseph Fourier, B.P. 53, 38041 Grenoble Cedex 9, France

PAULI, H., Institute for Ecology and Nature Conservation, Department of Vegetation Ecology and Conservation Biology, University of Vienna, Althanstraße 14, 1091 Vienna, Austria

PAULSEN, J., Institute of Botany, University of Basel, Schönbeinstrasse 6, 4056 Basel, Switzerland

PEDROTTI, F., Department of Botany and Ecology, University of Camerino, Via Pontoni 5, 62032 Camerino, Italy

PELAEZ-RIEDL, S., Institute of Botany, University of Basel, Schönbeinstrasse 6, 4056 Basel, Switzerland

PIZZOLOTTO, R., Dipartimento di Ecologia, Università della Calabria, Arcavacata di Rende, 87036 Cosenza, Italy

SCALERCIO, S., Dipartimento di Ecologia, Università della Calabria, Arcavacata di Rende, 87036 Cosenza, Italy

SCHLÜSSEL, A., Conservatoire et Jardin botaniques de la Ville de Genève, CP 1, 1292 Chambésy, Switzerland

SESÉ J.A., Instituto Pirenaico de Ecología, CSIC, Apartado 64, 22700 Jaca (Huesca), Spain

SHIYATOV, S.G., Laboratory of Dendrochronology, Institute of Plant and Animal Ecology of the Ural Division of the Russian Academy of Sciences, 8 Marta Street, 202, Ekaterinburg, 620144, Russia

SMITH, R.D., Advisory Services, Scottish Natural Heritage, 2 Anderson Place, Edinburgh EH6 5NP, Scotland

STAUDINGER, M., Institute for Ecology and Nature Conservation, Department of Vegetation Ecology and Conservation Biology, University of Vienna, Althanstraße 14, 1091 Vienna, Austria

STRID, A., Botanic Garden, Carl Skottsbergs gata 22, 41319, Gothenburg, Sweden

Štursa, J., Krkonoše National Park, Dobrovského 3, 543 11 Vrchlabí, Czech Republic

Thaler, K., Institute of Zoology and Limnology, University of Innsbruck, Technikerstrasse 25, 6020 Innsbruck, Austria

Theurillat, J.-P., Centre alpien de Phytogéographie, Fondation J.-M. Aubert, CP 42, 1938 Champex, Switzerland

Thompson, D.B.A., Advisory Services, Scottish Natural Heritage, 2 Anderson Place, Edinburgh EH6 5NP, Scotland

Toïgo, C., Office National de la Chasse, 8 Impasse Champ-Fila, 38000 Eybens, France

Tontini, L., Dipartimento di Biologia Animale e Dell'uomo, Via Accademia Albertina 13–17, 10123 Torino, Italy

Väre, H., Botanical Museum, Finnish Museum of Natural History, Box 7, 00014 University of Helsinki, Finland

Varga, Z., Department of Zoology and Evolution, Faculty of Sciences, Debrecen University, Egyetem tér 1, 4010 Debrecen, Hungary

Velluti, C., Conservatoire et Jardin botaniques de la Ville de Genève, CP 1, 1292 Chambésy, Switzerland

Villar L., Instituto Pirenaico de Ecología, CSIC, Apartado 64, 22700 Jaca (Huesca), Espana

Virtanen, R., Department of Biology, University of Oulu, P.O. Box 3000, 90014 Oulu, Finland

Whitfield, D.P., Advisory Services, Scottish Natural Heritage, 2 Anderson Place, Edinburgh EH6 5NP, Scotland

Wiget, L., Laboratoire d'Ecologie végétale et de Phytosociologie, rue Emile-Argand 11, 2000 Neuchâtel 7, Switzerland

Williams, P., The Biogeography and Conservation Laboratory, The Natural History Museum, London SW7 5BD, UK

Yoccoz, N.G., Department of Arctic Ecology, Norwegian Institute for Nature Research, Polar Environmental Centre, 9296 Tromsø, Norway

Zetto, T., Dipartimento di Ecologia, Università della Calabria, Arcavacata di Rende, 87036 Cosenza, Italy

I Europe's Alpine Areas

1 An Outline of Europe's Alpine Areas

G. Grabherr, L. Nagy and D.B.A. Thompson

Geographers, biologists, meteorologists, mountaineers and a large range of other groups of people each have their views on what is 'alpine'. Some think that the alpine region encompasses only the 960-km-long Alps separating southern and central Europe, whilst others think of specific high snowy ranges such as the Tyrol, Pyrenees, or the Dinaric Alps to be alpine. This book considers all of the 'alpine' areas of Europe, and defines qualifying ecosystems as those lying at or above the 'treeline' ecotone. We define the treeline as the connection between the highest elevation groups of trees [which form distinct patches and are at least of 3-m height, Körner (1999); Table 1.1]. This connecting line falls within the treeline ecotone, which ranges from the timberline (the upper limit of the montane forest) and the tree species line (the upper limit of isolated individual trees). The treeline, as defined here, is a climate-driven boundary, often modified or displaced by land use activities. Given that climate-defined treelines are sometimes absent, some authors have tried to define vegetation zones or belts by using climatic parameters alone (e.g. Holdridge 1947; Rivas-Martinez 1995). This should, however, be supported by true climate measurements such as those reported by ALPNET, the Alpine Biodiversity Network (Chap. 2).

Mountain ranges have a variety of climatic conditions and ecosystems. Variation in topography and altitude create steep gradients in temperature, patchiness of moisture and nutrient availability, variable degrees of wind exposure, and uneven, seasonal snow cover. In addition, variation in geological substrata, soil and cryological processes are among the major natural sources of habitat diversification. The biological richness of high mountain ecosystems is a result of various combinations of these physiographic factors and human management effects. The biological richness of mountain biota exceeds that of many in the lowlands (Grabherr et al. 1995; Grabherr 1997). At a global scale, the alpine life zone covers ca. 3% of the land area and accounts for about 4% of all higher plant species (Körner 1995) whilst, on the European scale, the latter figure is ca. 20% (Chap. 5). The associated macro- and micro-fauna also make a considerable contribution to the biological diversity of this life zone (Chaps. 12–17); for

Table 1.1. Working definitions adopted for alpine areas in Europe and formally used by ALPNET

Feature	Definition adopted for ALPNET
Linear	
Line of closed arborescent vegetation	The line where the closed forest (cf. timber line or forest line) or abutting scrub (formed by e.g. *Pinus mugo, Alnus viridis, Genista* spp.) ends, as seen from adistance
Treeline	The line where closed groups of trees taller than 3 m end. This is readily visible on many high mountains in N Europe
Tree species line	The line beyond which no individuals of a tree species occur
Altitudinal zone	
Treeline ecotone	The zone between the forest line and tree species line
Alpine	The zone between the treeline and the upper limit of closed vegetation (cover >20–40%); vegetation is an important part of the landscape and its physiognomy
Lower alpine	The zone where dwarf-shrub communities are an important part of the vegetation mosaic (incl. the thorny cushion formations of the Mediterranean mountains)
Upper alpine	The zone where grassland, steppe-like and meadow communities are a significant part of the vegetation mosaic
Nival	The zone of open vegetation above the upper alpine zone; no predominating life form with frequent cushion and small rosette plants; vegetation is a rather small part of the landscape and its physiognomy
Alpine-nival ecotone (subnival)	The transition between the upper alpine zone and the nival zone; coincides with the permafrost limit

a review of microorganisms, see Broll (1998). As we shall see later in the book, much of this richness derives from particular facets of the alpine environment.

The definitions of elevation zones and boundaries adopted for ALPNET are listed in Table 1.1. These definitions aid the synthesis on a continent-wide scale by reducing the ambiguities associated with the different terms relating to altitude belts or zones (especially the sub-alpine). The treeline ecotone is used synonymously with what is often called 'subalpine'. As the term sub-alpine has been used for a wider altitude range than just the treeline ecotone its use is avoided in this volume as far as possible. (For nomenclature traditionally used by researchers in the Alps, see, e.g., Löwe 1970; Reisigl and Keller 1987; Ellenberg 1988, 1996 or Ozenda 1994; and for north west Europe, see Horsfield and Thompson 1996).

1.1 The Mountains of Europe

The mountains of Europe are of old, worn-down or young, rugged formations (e.g. Hubbard 1937). The oldest are those of the Caledonian range: the Kiolen range of Norway and Sweden, and the Scottish Highlands, which date back to the Precambrian (~500 million years). The other old mountains of the Variscan or Hercynian system (Meseta, Massif Central, Vosges, the Black, Thuringian and Bohemian Forests, Harz and Erzgebirge, and the Sudetes) and of the Urals are of younger Palaeozoic (c. 355–290 million years) origin. The young, rugged mountains, such as the Sierra Nevada, Pyrenees, the Balearic mountains, the Alps, Apennines, Dinaric Alps, Carpathians, the mountains of Greece (incl. Crete) and the Caucasus, were largely shaped in the Cenozoic (ca. 1–12 million years), in most cases, by several uplifts. Some of the Hercynian massifs (parts of the N and S Carpathians, Corsica and Sardinia) were also involved in the Alpine folding. Past (e.g. Caucasus) and recent volcanism (Mt. Etna) was – and is – locally important.

The largest ranges include the Scandes, Urals, Caucasus, Carpathian mountains, Alps, Dinaric Alps together with the Hellenides, Apennines, and the Pyrenees-Cantabrian mountain complex (Ozenda 1994). The largest alpine areas are found in the highest ranges of the Alps and the Caucasus. The smaller ranges include the Sierra Nevada and the Baetic mountains, the central Iberian mountains, the Jura, Rhodope, and Balkan. Small alpine areas are also present in the central European Hercynian mountains.

Glaciation has played an important role in shaping landforms and watercourses, especially in the Highlands of Scotland, and in the Scandes, Pyrenees and Alps. Glaciation occurred also in the Mediterranean mountains, although mostly restricted to the highest peaks. There is clear evidence of glaciation in, for example, Corsica (Gauthier 2000) and the Sierra Nevada (Gomez Ortiz and Salvador Franch 1998).

Elevation and relief have important implications for the amount of rainfall, the proportion of rain to snow, persistence of snow, temperature and exposure to sunshine. Geomorphology or relief may modify and locally overrule climatic factors in determining the ecology of an area. For example, steep slopes may impose topographic limits on the distribution of trees or closed grasslands at altitudes below their climatic limits. The main features of the individual mountain ranges are listed in Table 1.2.

Table 1.2. Some of the key characteristics of high mountain systems throughout Europe

Mountain system	Geology	Origin	Geomorphology	Extent	Elevation (max.) in metres
Scandinavian Highlands (Scandes)	Crystalline (gneiss, granite, mica schist); sandstone, occasional limestone	Caledonian	West – steep, dissected glacial; east – less steep and less dissected with rounded plateaux	58.5 N–70.5 N	Numerous peaks over 2000 (2469)
Scottish Highlands	Crystalline (gneiss, granite, gabbro); sandstone	Caledonian	Glacial with remnant pre-glacial land forms; west – steep, dissected glacial; east – less steep and less dissected with rounded plateaux	56 N–58.5 N	Numerous peaks above 1000 (1343)
Urals	Crystalline (diverse)	Precambrian, Caledonian, Hercynian	Elongated structural forms divided by deep-seated faults; active glacial processes in the north	52 N–69 N	Numerous peaks above 1000 (1894)
Hercynian Mountains of central Europe	Crystalline (gneiss, granite sediments); mica schists	Hercynian	Block mountains with plateau-type relief and some glacial features	48.5 N–51 N	Some peaks above 1000 (1602)
Alps	Crystalline, schists, calcareous	Alpine	Crystalline: glacial, massive summits and ridges; limestone: sharp crests, ridges, rock walls; others: eroded	44 N–48 N	Numerous peaks above 4000 in the Western, and above 3000 in the Eastern Alps (4810)
Carpathian mountains	Crystalline/shales, limited schists, metamorphic and limestone	Alpine	Locally glacial; predominantly fluvial	45 N–50 N	Several peaks above 2000 (2663)

An Outline of Europe's Alpine Areas

Mountain system	Geology	Origin	Geomorphology	Extent	Elevation (max.) in metres
Pyrenees	Crystalline (axial zone); limestones, sandstones	Alpine	Several parallel ridges E-W, Central sector highest; east – high, steep, fractured; west – less steep; glacial features mainly in eastern and central, planation surfaces	42 N–43 N	Over 100 peaks above 3000 (3404)
Caucasus	Limestones, crystalline, schists, metamorphic (e.g. serpentinite), igneous	Alpine	North (Greater Caucasus) – high, steep, glacial features, volcanic; south (Minor Caucasus) – less high and steep, volcanic	39 N–47 N	Numerous peaks above 3000 Greater Caucasus (5595); Minor Caucasus (4090)
Corsica	Granite, crystalline schists	Hercynian	Alpine-type relief with glacial features above 2000 m	42 N–43 N	Over 100 peaks above 2000 (2710)
Dinaric Alps and Mts. of Greece	Limestones, metamorphic (ophiolites or serpentine), crystalline schists	Alpine	Glacial in the highest mountains, karst, fluvial	35 N–45 N	Numerous peaks above 2000 (2917)
Rila-Rodopi Massif	Crystalline	Hercynian	Dissected with glacial features at the highest peaks	42 N–45 N	Numerous peaks above 2000 (2925)
Apennines	Sedimentary; limestone	Alpine	Karstic; ridges with glacial features at the highest peaks (Central Apennines)	38 N–45 N	Several summits above 2000 (2912)
Baetic mountains (incl. Sierra Nevada)	Metamorphic, igneous, sedimentary	Alpine	More or less isolated convex mountains with broad summits, limestone and dolomite cliffs	37 N–38 N	Several peaks above 2000 (3478)
Sicily (Mount Etna)	Igneous	Alpine	Volcanic	38 N	(3263)

Data from Embleton (1984); Chaps. 3.1–3.10; C. Coldea, D. Gomez, G. Grabherr, J. Jeník, and G. Nakhutsrishvili (pers. comm.)

1.2 An Ecological Classification of Europe's Alpine Areas

For an ecological classification of Europe's high mountains two criteria are of particular relevance: (1) the geographical position of the different life zones, and (2) their elevation (Figs. 1.1 and 1.2). Four major types of alpine life zone can be distinguished: Mediterranean, temperate, boreal, and arctic (Fig. 1.1). The mountains of the Canary Islands and the Azores deviate in many respects from mainland Europe and are not considered in detail here. The regional accounts (Chaps. 3.1–3.10) provide comprehensive details on the alpine areas, and additional information is found in Ozenda (1985, 1994), Ellenberg (1988, 1996) and Wielgolaski (1997). Below we synthesise some of the over-arching points that help define common and contrasting features of Europe's alpine areas.

Boreal and arctic alpine environments receive moderate snow in winter and are characterised by severe frosts. In the summer, long days result in an extended light period that may selectively favour certain adapted plant species. The treeline in the European boreal mountains, in contrast to the conifer forest of the mountains of continental Siberia, is formed by birch (*Betula pubescens* spp. *czerepanovii*). Mixed dwarf-shrub heath with dwarf birch (*Betula nana*) and shrubby willows cover large areas within the treeline ecotone. Above this, ericaceous dwarf-shrub heath (*Vaccinium* spp., *Empetrum nigrum* ssp. *hermaphroditum*) appears, replaced at higher elevations by fellfields with small cushion plants and creeping dwarf shrubs (e.g. *Diapensia lapponica*, *Loiseleuria procumbens*), sedges (e.g. *Carex bigelowii*)

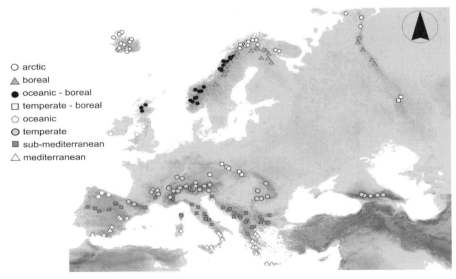

Fig. 1.1. Life zone characterisation of Europe's high mountains

and rushes (e.g. *Juncus trifidus*). The lack of deep snow cover allows cryoturbation, and solifluction and gelifluction on slopes, all of which are active on the fellfields. Patterned ground expands over large areas. Huge desert-like, block fields, known as gol'tsy, are typical for the northern Ural Mountains. The smooth relief of most boreal mountains gives rise to glacier cover almost throughout the nival zone. A few nival plant assemblages can occupy sunny niches on some nunataks. The alpine zone in the arctic (e.g. the mountains of Spitsbergen) starts at sea level, where it is almost identical to the zonal tundra, which gradually grades into polar mountain desert. The mountains of Iceland are unique because of their volcanic nature, but show typical alpine elevation zones where not affected by human disturbance.

Temperate mountains are often characterised by heavy snow accumulation which provides protection from deep soil frost. A distinctive series of plant communities can be related to snow cover and snow lie. Avalanche pathways interrupt the treeline at many sites. The treeline trees are coniferous species of alpine-boreal origin (*Larix decidua*, *Picea abies*, *Pinus cembra*). Some species are specific to particular parts, e.g. in the Pyrenees, where *Pinus uncinata* is the only conifer at the treeline; in the Caucasus, with the endemic *Betula litwinowii* or *Picea orientalis* at the treeline; and the *Pinus mugo* scrub in the Eastern Alps, the Carpathians and the Dinarids (with *Alnus* scrub on snow-rich slopes). Temperate alpine vegetation consists of dwarf-shrub communities at and immediately above the treeline. Most areas of the alpine zone are occupied by a variety of graminoid-dominated communities that are commonly called alpine grasslands, although the predominance of sedges (*Carex* spp.) or rushes (*Kobresia*=*Elyna myosuroides*, *Juncus* spp.) indicates that these

Fig. 1.2. Elevation classification of the European high mountains

may be better termed alpine sedge or rush heaths. Floristically, the many dominating sedges (e.g. *Carex curvula*, *C. ferruginea*, *C. firma*, *C. sempervirens* and *C. tristis*), rushes (e.g. *Juncus jacquinii*), grasses (e.g. *Festuca halleri* group or *F. varia* group) and their associates belong to a specific group of plants which is endemic to the mountains of the Alps system, and which may have originated from the old Tertiary mountain flora of southern Europe. Nival summits are clustered in the Alps and the Caucasus (Fig. 1.2), along with a few peaks in the Pyrenees. In the nival zone, scattered assemblages of cushion plants, small rosette plants, and small grasses occupy favourable sites, but, floristically, clearly defined communities (associations) of species can still be distinguished in different habitats.

Mediterranean high-mountain environments are characterised by relatively dry and warm summers. Mediterranean conifer forests (e.g. *Abies cephalonica*, *A. nebrodensis*, *Cupressus sempervirens*, *Pinus* spp.) form the treeline or the treeline is replaced by scrub (e.g. *Genista aetnensis*). Treelines are rarely natural because of human disturbance since prehistoric times. The most characteristic feature of the Mediterranean alpine (or cryoro-Mediterranean) zone is the presence of xerophytic thorny cushion communities (which contain many endemics). A nival zone above the permanent snow line (cryo-Mediterranean) occurs only on Mt. Etna and the Sierra Nevada in Spain (Fig. 1.2). At Mt. Etna the uppermost reaches are affected by volcanic activity. Nival plant assemblages of endemic cushion plants and low stature grasses are typically developed in the Sierra Nevada above 2800 m.

South of the Alpic system (sensu Ozenda 1985) or mid-latitude high mountains, further extensive mountain systems occur. These include the mountains of the Central Meseta in Spain, the southern Pyrenees, most of the Apennines, the mountains of Corsica, and the mountains of the Balkan Peninsula (such as parts of the Dinarids, and the Rhodopi and Rila mountains), which are neither temperate nor typically Mediterranean (Fig. 1.1). Each of these sub-Mediterranean mountains has its specific character. The forests at the treeline ecotone are often deciduous (*Fagus sylvatica*, *Alnus viridis* ssp. *suaveolens*) or dominated by endemic pines (e.g. *Pinus nigra* ssp. *laricio* in Corsica, *Pinus peuce* in Macedonia). The alpine zone consists of tussock-like grasslands (*Festuca* spp., *Stipa* spp., *Poa* spp.) and dwarf-shrub heath is of minor importance. Many of these mountains have been refugia for old Tertiary elements. This specific endemic mountain flora is not only concentrated in habitats such as rocks and screes, but also in the grassland communities. No sub-Mediterranean mountain reaches above the permanent snow line and, with the exception of a small area on Mt. Olympus, a specific nival belt does not exist.

From west to east a gradient of increasing continentality (or decreasing oceanity) exists. The Cantabrian mountains of northern Spain, the mountains of Scotland, NW England, Wales and Ireland, and the Norwegian west coast are wet and snow-rich mountains (Fig. 1.1). Extensive heathland in the lower alpine zone, and lichen- or moss-rich fellfields in the upper alpine zone, are the major components of the vegetation mosaic. The treeline is composed of *Genista* species in the Cantabrian mountains, *Betula pubescens* spp. *czerepanovii* in Norway, and *Pinus sylvestris* in Scotland (though here it is largely absent owing to human influences). There are many examples where a clear altitude zonation is absent. One of them is the carbonatic rocky Picos de Europa, Cantabrian mountains, where the rugged relief masks a clear zonation (the uppermost trees are beeches).

Many of Europe's mountains are so-called middle mountains (Fig. 1.2) where only the highest reaches are treeless (e.g. Black Forest, Giant Mountains, Southern Urals, Southern Apennines). The vegetation of these pseudoalpine zones is in many cases depauperate derivates from former extensive alpine grasslands, dwarf shrub-heath, fellfields, or Mediterranean hemispherical thorny dwarf-shrub formations. These act as outposts of zonal alpine floras and vegetation, and are of particular scientific interest and importance for detecting the effects of climate change.

1.3 The Regional Accounts (Chaps. 3.1–3.10)

There are ten regional accounts of alpine regions ranging from the Scandes to the Caucasus. These deal with the climate, physiography, flora (historical and extant) and vegetation (altitude zonation of vegetation types and life form analysis) of each geographical region. The taxonomic diversity of local floras (flowering plants, ferns, bryophytes, and lichens) at and above the treeline ecotone are reported. Plant communities are discussed in relation to upward altitudinal sequences of functional communities (treeline ecotone, dwarf-shrub heath, grasslands, tall-herb meadows, mires, snowbeds, springs and flushes, rock, scree and open cryptogamic communities). Each chapter also considers additional ecological factors, including the potential impacts of increased temperature, modified precipitation patterns, pollution and land management.

Acknowledgements. We thank Ch. Körner, R. Virtanen and M. Wrightham for helpful comments.

References

Broll G (1998) Diversity of soil organisms in alpine and arctic soils in Europe. Review and research needs. Pirineos 151/152:43–72

Ellenberg H. (1988) Vegetation ecology of central Europe, 4th edn. Cambridge University Press, Cambridge

Ellenberg H (1996) Vegetation Mitteleuropas mit den Alpen. 5. Aufl. Ulmer, Stuttgart
Embleton C (ed) (1984) Geomorphology of Europe. Macmillan, London
Gauthier A (2000) Corse des sommets. Albiana, Ajaccio
Gomez Ortiz A, Salvador Franch F (1998) El glaciarismo de Sierra Nevada, el mas meridional de Europa. In: Gomez Ortiz A, Perez Alberti C (eds) Las huellas glaciares de la montañas espanolas. Servicio de Publicationes da Universidade de Santiago de Compostela, Santiago, pp 385–430
Grabherr G (1995) Alpine vegetation in a global perspective. In: Box EO, Peet RK, Masuzawa T, Yamada I, Fujiwara K, Maycock PF (eds) Vegetation science in forestry. Kluwer, Dordrecht, pp 441–451
Grabherr G (1997) The high mountain ecosystems of the Alps. In: Wielgolaski FE (ed) Ecosytems of the world, part 3. Polar and alpine tundra. Elsevier, Amsterdam, pp 97–121
Grabherr G, Gottfried M, Gruber A, Pauli H (1995) Patterns and current changes in alpine plant diversity. In: Chapin FS III, Körner C (eds) Arctic and alpine biodiversity: patterns, causes and ecosystem consequences. Springer, Berlin Heidelberg New York, pp 167–181
Holdridge LR (1947) Determination of world plant formations from simple climatic data. Science 105:267–268
Horsfield D, Thompson DBA (1996) The uplands: guidance on terminology regarding altitudinal zonation and related terms. Information and advisory note no. 26. Scottish Natural Heritage, Battleby
Hubbard GD (1937) The geography of Europe. D Appleton-Century, New York
Körner C (1995) Alpine plant diversity: a global survey and functional interpretations. In: Chapin FS III, Körner C (eds) Arctic and alpine biodiversity: patterns, causes and ecosystem consequences. Springer, Berlin Heidelberg New York, pp 45–62
Körner C (1999) Alpine plant life. Springer, Berlin Heidelberg New York
Löwe D (1970) Subarctic and subalpine: where and what? Arct Alp Res 2:63–73
Nakhutsrishvili G, Ozenda P (1998) Aspects géobotaniques de la haute montagne dans le Caucase essai de comparaison avec les Alpes. Ecologie 29:139–144
Ozenda P (1985) La vegetation de la Chaine Alpine. Masson, Paris
Ozenda P (1994) Végétation du continent européen. Delachaux et Niestlé, Lausanne
Reisigl H, Keller R (1987) Alpenpflanzen im Lebensraum. Alpine Rasen, Schutt- und Felsvegetation. G Fischer, Stuttgart
Rivas-Martinez S (1995) Classification bioclimatica de la terra. Fol Bot Matrit 16:1–29
Wielgolaski F (ed) (1997) Polar and alpine tundra. In: Goodall DW (ed) Ecosystems of the world, part 3. Elsevier, Amsterdam

2 A Bioclimatic Characterisation of Europe's Alpine Areas

Ch. Körner, J. Paulsen and S. Pelaez-Riedl

2.1 Introduction

The natural high altitude treeline, the sole bio-reference for defining the alpine zone, integrates local thermal conditions in such a way that it occurs at equal temperatures, at both European and global scales (Körner 1998). This is seen by its occurrence at progressively lower elevations along a northward latitudinal gradient. Does this mean in practice that, for example, a Norwegian fellfield offers comparable life conditions to that of a Macedonian alpine heath? For our study, it was of interest if what in various localities in Europe has been referred to as alpine did indeed represent a quantitatively comparable environment.

We undertook a concerted programme of bioclimatological data collection spanning 30 degrees in latitude across Europe between 1998 and 2000. We report the results of soil temperatures measured 10 cm belowground at 23 sites in the alpine zone of European mountain systems. As will be discussed below, ground temperature is the single best surrogate for the alpine bioclimate in general.

2.2 Site Selection and Methods

2.2.1 What Was Measured and Where?

A comparative study such as this depends on using standardised methods. Root zone temperature was decided to be the most useful and feasible measure of alpine life conditions. Air temperature is more difficult to measure (radiation errors) and is known to deviate substantially from temperature near to ground surface or belowground and hence its relevance for life processes in low stature vegetation is questionable (Grace 1988; Körner 1999). Canopy temperature is influenced by plant morphology, stature, vegetation density, and local wind regime, and therefore a Europe-wide standardisation of probe positioning would have been difficult. Other climatic factors have been found to either be correlated with ground temperature (e.g. the radiation regime) or vary at the micro-scale (e.g. air humidity within a stand), not permitting standardisation of data collection. Empirical evidence suggested that, except for the most southern peaks in the Mediterranean, precipitation does not limit alpine plant life, but controls season length via snow cover duration, which can be detected by measuring ground temperature. The above considerations led to an agreement to measure ground temperatures at 10 cm below the soil surface under closed vegetation in the alpine zone, 200–250 m above the climatic treeline (the line connecting the highest elevation patches of trees of at least 3 m height). In the absence of a natural treeline, the logger positions were based on best expert knowledge (Fig. 2.1). The sites in 18 regions ranged from north of the Polar Circle to the mid-Mediterranean; from Europe's most westerly mountains to the eastern central European mountains. Elevations ranged from 900 m in the north to 2530 m in the south (Fig. 2.2).

2.2.2 Sensors and Sensor Positioning

We used self-contained sealed automatic one-channel data loggers (Tidbit StowAway temperature logger; Onset Corporation, Bourne, Massachusetts, USA) programmed for a 1-h measurement interval. These sensors have an accuracy of 0.2 K. Before distribution, all sensors were calibrated in a common ice water bath at 0 °C. As none of them deviated by more than 0.2 K from 0 °C no correction was applied. The loggers were positioned, using a common protocol, under closed vegetation (preferably grassland) on level ground, unscreened by protruding geomorphological structures such as rocks.

The sensors were buried at the end of the 1998 (snow permitting) or at the beginning of the 1999 growing season. All loggers were retrieved between the last week of August 2000 and mid-September 2000. Hence, for each site, we had at least one full year of data; for most sites, there were nearly two years of data. Some sites had two loggers so that local variability (effects of slope exposure) could be recorded. Data from a reference site in the Swiss Alps, operated over 4 years, illustrates temporal variation (Fig. 2.3).

The site positions selected showed that some of them were outside the target elevation. The two sites in the Sierra Nevada, Spain, and the one on Mount Etna, Sicily, were about 300 m too low. Hence, the correct alpine logger position would have been at 2800–2850 m. The too low an elevation combined with sparse vegetation cover may explain why the data for these sites fell outside the range obtained for all the other sites. In the following, we treat these

A Bioclimatic Characterisation of Europe's Alpine Areas

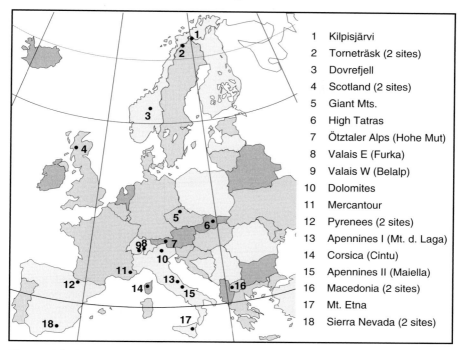

Fig. 2.1. The location of soil temperature measurement sites across Europe, 1998–2000

1 Kilpisjärvi
2 Torneträsk (2 sites)
3 Dovrefjell
4 Scotland (2 sites)
5 Giant Mts.
6 High Tatras
7 Ötztaler Alps (Hohe Mut)
8 Valais E (Furka)
9 Valais W (Belalp)
10 Dolomites
11 Mercantour
12 Pyrenees (2 sites)
13 Apennines I (Mt. d. Laga)
14 Corsica (Cintu)
15 Apennines II (Maiella)
16 Macedonia (2 sites)
17 Mt. Etna
18 Sierra Nevada (2 sites)

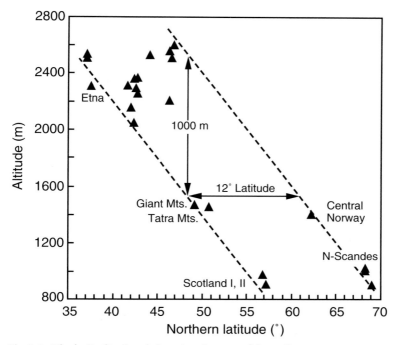

Fig. 2.2. The latitudinal and elevational range of the soil temperature measurement sites

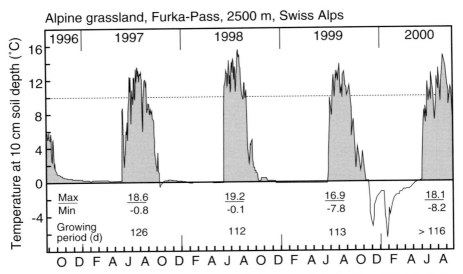

Fig. 2.3. Soil temperature measured at 10 cm belowground, Furka Pass, E Valais, Swiss Alps, 1996–2000. The site is at an exposed plateau with a short and thin graminoid cover of the *Carex curvula* type, typical of the central Alps. Winter snow cover, and thus soil temperature, is largely determined by wind direction during heavy storms

two Mediterranean regions separately. Although they are not comparable with the other sites, they illustrate how different the thermal regime at such treeless, lower than alpine elevations, might be, as compared with those sites that conform to our above definition of alpine and the associated desired data logger position. Hence, the absence of trees alone at high elevation is an insufficient zonation criterion.

Two other sites also deviated from the protocol; the site in the Swiss Central Alps (Furka Pass, Valais East) was about 300 m above the treeline (ca. 100 m too high) and the Maiella, Apennines site was in the krummholz belt (about 100 m too low). These two sites were retained in the comparison because statistically the two elevation deviations almost cancel each other; however, they may add to the variance of the data.

2.2.3 Data Analysis

Absolute minimum and maximum temperature, daily mean temperature (24-h means) and their seasonal course were used to characterise the thermal regime at each site. The beginning of the growing season was defined by the usually obvious and sharp warming of the ground after snowmelt (first incidence of a temperature above +2 °C). The first cold period in the autumn, when ground temperature at 10 cm belowground dropped to +2 °C (indicating severe frost aboveground), signalled the end of the season. Warmer periods later were not included in calculating the season length. The median day temperatures for the season and thermal sums (day degrees) above 0, 5, and 7 °C were calculated. Where data were available for >1 year, means of corresponding days were used. One sensor in the Spanish Pyrenees was excavated by an animal in July 2000 and all data recorded thereafter were disregarded. The site on Mount Etna became covered by several cm of volcanic ash in August 2000 that caused a slight lowering of temperature.

The data are presented graphically in such a way that they are centred on mid-summer and start on January 1, irrespective of when the actual recording at a site commenced. Therefore, one annual course of data may be composed of data actually collected during the first half of the second year and the second half of the first year. Since the time overlap of recording across sites was >80 % this procedure does not affect the comparative value and permits visualisation.

2.3 Results

2.3.1 Between-Year and Between-Site Variation

The data for the Swiss Alps show similar growing season temperatures after snowmelt over 4 years (Fig. 2.3). However, winter temperatures showed a large year-to-year variability. There were some winters without ground freezing (Fig. 2.3), and one (1999/2000) with deep ground freezing, due to mid-winter snow removal by strong winds. This demonstrates the well-known importance of snow cover on winter ground temperature.

July and August temperatures across all sites for which data were available for two seasons differed by <1 K between years (Table 2.1). The mean for all sites indicated no difference of the temperature for July and August 1999 and 2000. This provides some confidence that the comparison of mountain sites is relatively robust, but we cannot exclude that some of the sites had an exceptional summer. From their experience, none of the participants found that the seasons were exceptional at their sites. The only exception was perhaps the unusually large amount of snow at one of the Swiss sites (Furka Pass) in 1999 that lasted until July.

Ch. Körner, J. Paulsen and S. Pelaez-Riedl

Table 2.1. 1999 versus 2000 comparison of July and August ground temperatures

Site	Northern latitude (°)	July 1999 (min.) Mean (max.)	July 2000 (min.) Mean (max.)	Diff.	August 1999 (min.) Mean (max.)	August 2000 (min.) Mean (max.)	Diff.	Diff. July+August
N-Finland	69.0	(4.8) 9.2 (14.6)	(3.7) 9.5 (16.4)	+0.3	(2.8) 5.4 (9.4)	(6.3) 8.6 (12.0)	+3.2	+1.8
N-Sweden I	68.4	(2.5) 8.2 (14.7)	(1.6) 9.5 (17.6)	+1.3	(1.3) 5.1 (10.0)	(4.3) 8.0 (13.8)	+2.9	+2.1
N-Sweden II	68.4	(2.2) 6.9 (11.7)	(2.2) 7.4 (12.7)	+0.5	(2.5) 5.2 (9.2)	(4.0) 7.0 (10.1)	+1.9	+1.2
Central Norway	62.3	(5.7) 9.5 (14.9)	(3.3) 8.4 (14.6)	−1.0	(4.8) 8.9 (14.9)	(3.9) 7.8 (12.0)	−1.1	−1.1
Scotland II	57.2	(6.3) 9.8 (13.5)	(6.3) 9.9 (13.8)	+0.1	(5.2) 8.8 (13.8)	(8.6) 10.7 (12.4)	+1.9	+1.0
Giant Mts.	50.8	(6.5) 9.9 (15.2)	(4.2) 6.6 (12.6)	−3.3	(5.7) 8.6 (14.0)	(4.8) 9.9 (15.8)	+1.3	−1.0
Ötztaler Alps	46.8	(6.3) 10.0 (14.7)	(2.6) 8.9 (15.9)	−1.2	(6.3) 10.7 (14.4)	(4.6) 11.1 (16.2)	+0.4	−0.4
Alps Valais I (E)	46.6	(−0.1) 10.6 (16.3)	(4.2) 10.4 (17.2)	−0.2	(7.9) 11.9 (16.9)	(7.0) 12.7 (18.1)	+0.7	+0.3
Alps Valais II (W)	46.4	(6.6) 10.1 (13.5)	(4.8) 8.8 (13.2)	−1.3	(7.4) 10.0 (13.5)	(7.4) 10.3 (13.5)	+0.3	−0.5
Dolomites	46.3	(5.1) 12.6 (16.7)	(8.9) 12.9 (17.3)	+0.2	(10.3) 12.9 (16.4)	(9.2) 14.7 (19.1)	+1.8	+1.0
Mercantour	44.2	(5.2) 11.4 (16.2)	(5.2) 11.3 (16.7)	−0.2	(7.8) 11.6 (15.3)	(6.6) 11.5 (16.5)	−0.1	−0.1
Pyrenees I	42.7	(7.3) 14.6 (19.4)	(4.1) 12.0 (17.5)	−2.6	(10.5) 14.0 (18.9)	(6.0) 11.7 (17.1)	−2.3	−2.4
Corsica	42.4	(5.1) 10.3 (16.1)	(3.3) 9.9 (16.7)	−0.4	(4.2) 10.5 (15.8)	(4.2) 9.7 (14.6)	−0.8	−0.6
Central Apennines II	42.2	(7.8) 13.3 (19.7)	(7.5) 13.9 (16.7)	+0.6	(8.9) 14.9 (19.7)	(8.9) 14.9 (17.0)	0.0	0.3
Macedonia I	42.0	(8.5) 12.1 (16.7)	(7.1) 12.9 (19.3)	+0.8	(8.8) 13.1 (18.1)	(8.0) 13.3 (18.4)	+0.2	+0.5
Macedonia II	41.7	(9.2) 12.4 (16.4)	(9.2) 13.1 (17.9)	+0.8	(10.1) 13.3 (17.3)	(10.1) 12.7 (15.9)	−0.7	+0.1
Across regions mean[a]		(5.6) 10.8 (15.8)	(4.9) 10.3 (15.9)	−0.5	(6.7) 10.5 (15.0)	(6.5) 11.0 (15.2)	+0.5	0.0

[a] Means across regions are calculated in such a way that regions with two sites are considered only once, namely by the mean for those two sites

2.3.2 Temperature Extremes

Ground temperature minima and maxima showed a substantial variation from site to site with no latitudinal trend (Fig. 2.4). The variation in minima was likely to be associated with variable snow cover. From these data we can conclude that any of the sites (perhaps with the exception of the Scottish Highlands because of their maritime climate) may at one time experience ground temperatures as low as −13 °C at 10 cm belowground. However, any of these sites may experience a winter without any ground freezing. The same holds for maximum temperatures, which showed no obvious latitudinal trend except for the Sierra Nevada and Mount Etna sites, which however are discussed separately. Most sites had a peak temperature of about 20 °C except for the oceanic mountains of Scotland where the maxima were about 15 °C. The overall temperature amplitude at 10 cm belowground was 12 K in Scotland,

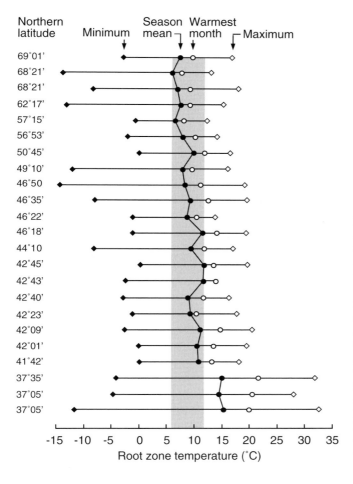

Fig. 2.4. A latitudinal comparison of mean, minimum and maximum soil temperature at 10 cm belowground from sites at typical alpine elevations covered by short vegetation. *Grey shading* indicates the 7–12 °C seasonal mean for all sites excluding the Sierra Nevada and Mount Etna sites, which were placed 300 m too low

28 K in the Ötztal Alps and 45 K in the Sierra Nevada. At the ground surface, the extremes as well as the amplitude would have been substantially larger as surface temperatures in closed alpine vegetation may exceed 30 K on bright summer days (Körner 1999). However, the overall spatial trends across Europe would be similar.

2.3.3 The Daily Course of Ground Temperature

Figure 2.5 illustrates that the means discussed here represent quite different thermal regimes during the day. Ground temperatures at 10-cm depth on bright days showed a 24-h cycle of 7 K in Corsica and 2–3 K in Scotland, with most other sites around 5 K (not shown). The ground surface and low stature alpine vegetation undergoes a much less buffered temperature regime with diurnal amplitudes of 30 °C quite common (Körner 1999).

2.3.4 The Seasonal Course of Ground Temperature

Overall, the seasonal ground temperatures showed a similar trend across Europe (Fig. 2.6). The three sites in northern Scandinavia had lower thermal sums (shaded areas in Fig. 2.6) than some of the most southern sites. However, most of the sites showed similar temperature patterns. For instance, the data for Corsica did not differ from those measured in the Alps, nor did the data from the Apennine I site or Macedonia differ from those in the Pyrenees or in central Norway. The temperatures in the Tatra and in Macedonia were similar.

Perhaps only a coincidence was a common peak of temperature at nearly all sites around 1 July 2000, indicated by an arrow in Fig. 2.6. The duration of ground freezing was longer and it occurred more regularly in some of the higher latitude sites than in the Alps. However, severe ground freezing may also occur in the alpine zone in the Mediterranean (Fig. 2.7).

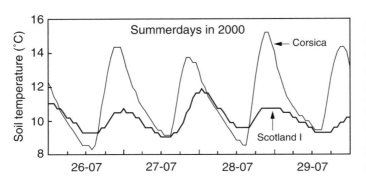

Fig. 2.5. An example of the diurnal variation of the soil temperature 10 cm below typical alpine vegetation from Corsica and Scotland on bright days in midsummer

A Bioclimatic Characterisation of Europe's Alpine Areas

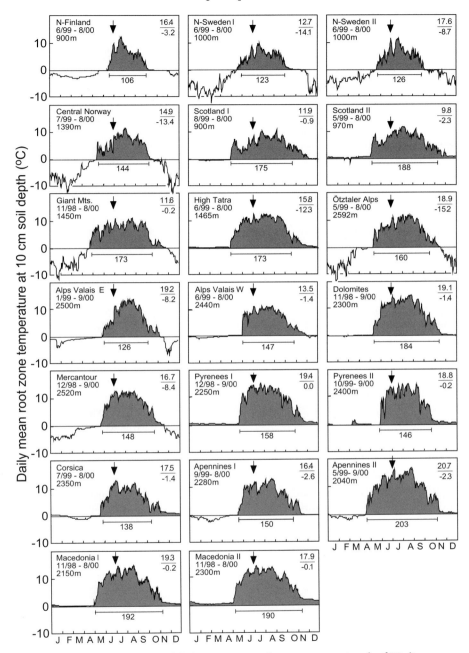

Fig. 2.6. The seasonal course of daily mean ground temperature at each of 20 sites across Europe. The duration of the data collection is indicated in the *top left corner of each graph*. Absolute maxima over minima are given in the *top right corner*. The growing season length in days is given at the bottom

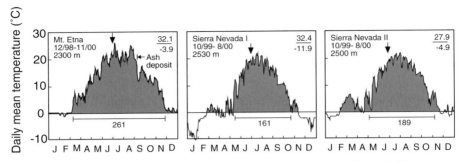

Fig. 2.7. The seasonal course of daily mean ground temperature at three Mediterranean locations, 300 m below the target elevation of the network's site comparison under open vegetation. The duration of the data collection is indicated in the *top left corner of each graph*. Absolute maxima over minima are given in the *top right corner*. Bottom figure is growing season lengths in days

2.3.5 Seasonal Mean Temperatures

The mean ground temperature during the growing season ranged between 5.7 and 11.6 °C (Fig. 2.8, Table 2.2). The three southernmost sites, which were below the project's target elevations and had scant vegetation cover, were between 14.2 and 15 °C. Excluding these sites, the mean for the other sites was 8.8 °C. There was a weak trend of mean seasonal ground temperatures increasing from the northernmost sites to the Mediterranean with little change as far south as 46°N latitude. The between-site variation within one region was 0.2 K in Macedonia, 0.8 K in the Sierra Nevada, 1.4 K in N Scandinavia, Scotland and the Pyrenees and 3.2 K in the Alps (Ötztal vs. Dolomites). The two sites in the Apennines are not at comparable elevations. All except two of the N Scandinavian sites within a region were between 1 and 100 km apart. This intra- vs. interregional comparison illustrates that local climatic variation may be larger than the effect of a 20–30° difference in latitude, even when the sites were carefully selected for similarity in life zone, exposure and plant cover.

2.3.6 Growing Season Length

Growing season length ranged from 106 to 203 days. There was a slight trend towards longer seasons at lower latitudes, but this was not very pronounced and shorter seasons could be found in the central Alps, the Apennines, and the Scandes (Fig. 2.8). Once more, it is likely that season length data reflect the local and temporal variability in snow cover. Although it may vary from year to year, 4 months appear to be average for the alpine season length in Europe. The shortest growing season was at the Furka Pass, Central Swiss Alps, in 1999 with 105 days as opposed to the 126 day 4-year mean at the same site. The winter

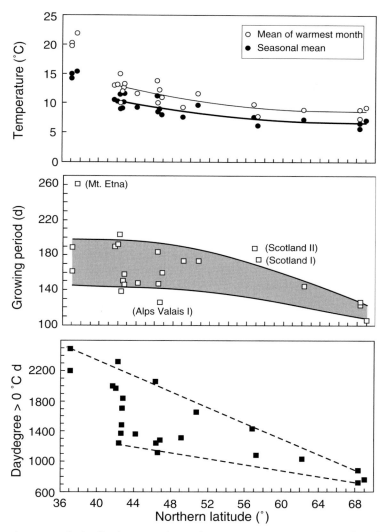

Fig. 2.8. A latitudinal comparison of mean temperatures, season length and thermal sums above 0 °C. Each point represents one of the 23 data logger sites

snow load exceeded several metres in that year and, 150 m away from the logger site, closed alpine grassland remained under snow throughout the season.

2.3.7 Thermal Sums

The shaded areas in Figs. 2.6 and 2.7 illustrate the sum of all degree hours >0 °C. In addition, Table 2.2 shows thermal sums of >5 and >7 °C. Sites dif-

Table 2.2. Summary of climatological data

Sites	Geographical coordinates	Elevation of logger (m)	Date of start
N-Finland, Kilpisjärvi	69°01'N, 20°50'E	900	15-06-99
N-Sweden I, Torneträsk	68°21'N, 18°29'E	1000	05-06-99
N-Sweden II, Torneträsk	68°21'N, 18°29'E	1000	05-06-99
Central Norway, Dovrefjell	62°17'N, 09°21'E	1390	01-07-99
Scotland I, An Cabar, NW Highlands	57°15'N, 04°23'W	900	12-08-99
Scotland II, Glas Maol, E-Grampians	56°53'N, 03°22'W	970	16-05-99
Giant Mts., Certova Louka, Czech Republic	50°45'N, 15°40'E	1450	13-11-98
High Tatras, Slovak Republic	49°10'N, 20°10'E	1465	20-05-99
Ötztaler Alps, Hohe Mut, Austria	46°50'N, 11°02'E	2592	25-04-99
Alps Valais I, Furka-Pass, Switzerland	46°35'N, 08°23'E	2500	15-09-96
Alps Valais II, Belalp, Switzerland	46°22'N, 07°57'E	2450	16-06-99
Dolomites, Passo Rolle, Italy	46°18'N, 11°48'E	2200	15-11-98
Mercantour, France	44°10'N, 07°10'E	2520	17-11-98
Pyrenees I, Collarada N, Spain	42°45'N, 00°28'W	2250	21-11-98
Pyrenees II, Collarada S, Spain	42°43'N, 00°27'W	2360	17-09-99
Central Apennines I, Monti d. Laga, Italy	42°40'N, 13°21'E	2280	21-08-99
Corsica, Mt. Cintu	42°23'N, 08°56'W	2350	18-06-99
Central Apennines II, Maiella, Italy	42°09'N, 14°07'E	2040	10-05-99
Macedonia I, Ceripasina Rudoka	42°01'N, 20°52'E	2150	04-11-98
Macedonia II, Preslap Jakupica	41°42'N, 21°22'E	2300	05-11-98
Sicily, Mt. Etna	37°35'N, 15°00'E	2300	22-11-98
Sierra Nevada II, Spain	37°05'N, 03°19'W	2500	28-09-99
Sierra Nevada I, Spain	37°05'N, 03°22'W	2530	29-09-99
Mean across all sites		1935	
Mean without Sierra Nevada and Etna		1858	

Vegetation period (d)	Abs Min	Abs Max	Mean warmest month	Seasonal mean	Median value	Degree hours >0 °C/ day	>5 °C/ day	>7 °C/ day
106	−3.2	16.4	(J) 9.3	7.1	6.6	755	264	137
123	−8.7	17.6	(J) 8.9	6.6	6.2	877	278	143
126	−14.1	12.7	(J) 7.3	5.7	5.6	720	172	61
144	−13.4	14.9	(J) 8.9	7.2	6.9	1034	366	172
175	−0.9	11.9	(S) 7.7	6.2	6.3	1085	289	94
188	−2.3	13.8	(J) 9.8	7.6	7.5	1429	526	265
173	−0.2	16.2	(A) 11.6	9.6	9.7	1653	807	501
173	−12.3	15.8	(A) 9.3	7.6	7.5	1309	492	237
160	−15.2	18.9	(A) 10.9	8.0	8.1	1281	543	316
126	−8.2	19.2	(A) 12.3	9.0	9.6	1111	529	341
147	−1.4	13.5	(A) 10.1	8.5	8.7	1243	538	291
184	−1.4	19.1	(A) 13.8	11.2	11.6	2063	1159	826
148	−8.4	16.7	(A) 11.6	9.2	9.4	1361	645	401
158	0.0	19.4	(J) 13.3	11.6	11.9	1832	1050	753
146	−0.2	18.8	(A) 12.6	10.2	10.3	1482	777	528
150	−2.6	16.4	(A) 12.0	9.1	9.3	1368	645	398
138	−1.4	17.5	(A) 10.1	9.0	9.1	1244	578	350
203	−2.3	20.7	(A) 15.0	11.4	11.6	2318	1320	957
192	−0.2	19.3	(A) 13.2	10.3	10.4	1972	1044	710
190	−0.1	17.9	(A) 13.0	10.5	10.9	2000	1068	730
261	−3.9	32.1	(A) 21.9	15.3	15.4	3988	2700	2213
189	−4.9	27.9	(J) 20.2	14.2	15.0	2493	1664	1361
161	−11.9	32.4	(J) 19.7	15.0	14.4	2200	1483	1219
164	−5.1	18.7	12.3	9.6	9.7	1601	823	565
158	−4.8	16.8	11.0	8.8	8.9	1407	655	411

fered far more in thermal sums than they did in terms of mean temperature. The weak trend towards warmer mean temperatures plus the similarly weak trend towards longer snow-free periods in the south added up to substantially higher thermal sums in the south (Fig. 2.8). This was so despite the fact that the day length in summer (July/August) is 24 h near the Polar Circle as opposed to ca. 16 h in the Alps. Although the likelihood for radiative nighttime cooling is greater in the south, steeper solar radiation and perhaps less cloudiness during daylight hours cause significantly greater thermal sums (1300–2000 degree hours south of 44°N compared with 800–900 degree hours in the far north). The higher the thermal threshold is set (e.g. 5 and 7 °C) the steeper this N-S gradient is because higher temperatures occur more frequently at low latitudes.

2.4 Alpine Biota Share a Common Climate Across Europe

The major finding of this continent-wide comparison of alpine bio-temperatures is the similarity of a number of measures describing the temperature regime. The fact that season length, mean temperatures or minima and maxima did not show a strong latitudinal trend and the means were similar (with the exception of the data from the Sierra Nevada and Mt. Etna) adds to our confidence that what experts call alpine does indeed reflect life conditions which stand the physical test of similarity. Oceanity and regional climate seem to be at least as important in determining the regional temperature regime as latitude.

The latitudinal thermal gradients were smaller than expected. One explanation is the greater day lengths at high latitudes which allow longer positive soil heat flux and reduce nighttime cooling. Longer day length appears to compensate for reduced solar inclination. Locally occurring plants have evolved an adaptation to a 24-h photoperiod in the summer, as was shown by a continent-wide transplant experiment (Prock and Körner 1996). Mid-latitude (47°N) alpine plants thrived in the north; however, plants from high latitudes (68°N) performed badly in the Alps, presumably because of the short day length.

The importance of solar radiation for vegetation and ground warming is illustrated by concurrently collected data at the treeline at three of the sites (Fig. 2.9). Ground temperatures under treeline trees 200–250 m below the alpine site were generally colder. This underlines the positive influence of solar radiation on bio-temperature and explains, in part, why trees do not grow at elevations higher than they do. In addition to the fact that their apical meristems experience much cooler temperatures then low stature alpine vegetation (Grace 1988), trees create a cold root zone by screening solar radiation. Crawford (1997) presented a similar explanation for the position of the

A Bioclimatic Characterisation of Europe's Alpine Areas

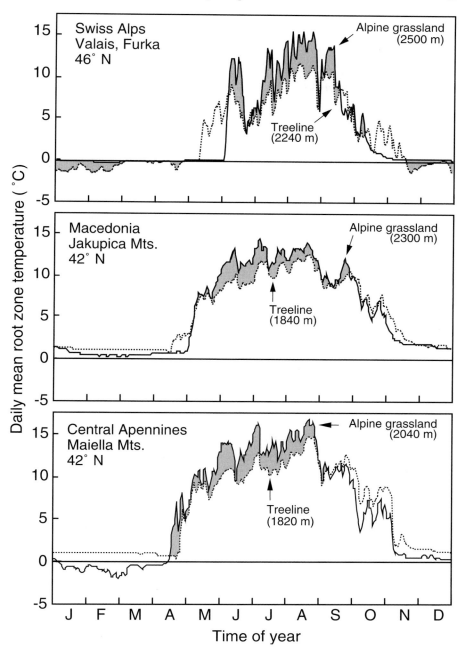

Fig. 2.9. A comparison of −10 cm ground temperatures under trees at the treeline and alpine vegetation (at 200–250 m higher elevations). Note that the alpine sites are usually warmer than the treeline sites during the growing season (for treeline, Ch. Körner, unpubl.)

Siberian latitudinal treeline. As climate warms, trees establish on former permafrost ground. However, as the forest closes, permafrost returns and the trees die. Alpine vegetation at 200–250 m above the treeline not only experiences similar season length, but also experiences equal or warmer canopy and root zone temperatures than trees at the treeline do. The alpine zone is thus an environment that is not colder than the upper montane forest and low plant stature, and rapid ground warming creates a microclimate that allows alpine plant communities to thrive far higher than trees.

Acknowledgements. The work was carried out as part of the European Science Foundation Network on Alpine biodiversity (ALPNET). The purchase of the data loggers was financed by the Swiss Academy of Sciences. We thank A. Andonoski, E. Balletto, G. Blanca, J.-L. Borel, D. Caruso, J. Holten, J. Jeník, R. Kaufmann, U. Molau, L. Nagy, F. Pedrotti, J.-P. Theurillat, L. Villar, R. Virtanen, and V. Zingerle for placing and recovering the temperature loggers.

References

Crawford RMM (1997) Oceanity and the ecological disadvantages of warm winters. Bot J Scotl 49:205–222
Grace J (1988) The functional significance of short stature in montane vegetation. In: Werger MJA, Van der Aart PJM, During HJ, Verhoeven JTA (eds) Plant form and vegetation structure. SPB Academic Publishers, The Hague, pp 201–209
Körner C (1998) A reassessment of high altitude treeline positions and their explanation. Oecologia 115:445–459
Körner C (1999) Alpine plant life. Springer, Berlin Heidelberg New York
Prock S, Körner C (1996) A cross-continental comparison of phenology, leaf dynamics and dry matter allocation in arctic and temperate zone herbaceous plants from contrasting altitudes. Ecol Bull 45:93–103

3 The Regional Accounts

3.1 The High Mountain Vegetation of the Scandes

R. Virtanen

3.1.1 Geography, Geology and Ecological Conditions

The Scandinavian mountain chain (Scandes) extends from 58°N to 70°N. Its highest peak is Galdhöppigen (2469 m above sea level, a.s.l.), Jotunheimen massif, Norway. The Scandes belong to the Caledonides and were formed as a result of a collision between the Baltic and Laurentian shields ca. 550–400 million years ago. Their main landforms were strongly modified by later tectonic movements, weathering and glaciations (e.g. Binns 1978; Bax 1989; Svenningsen 1995; Eide et al. 1999). The geology of the Scandes is variable and includes Precambrian, igneous, schistose metamorphic and sedimentary rocks. The latter, containing high concentrations of calcium and magnesium carbonate, are of great botanical importance. The upper elevation of the mountain birch (*Betula pubescens* ssp. *czerepanovii*) treeline is at ca. 1300 m a.s.l. in the mountains of SE Norway, descending almost to sea level in the coastal mountains of northern Norway (Aas and Faarlund 2001). A recent overview of the ecological conditions in the Scandes can be found in Väisänen (1998), Sonesson and Molau (1998) and Moen (1999).

3.1.2 Climate

The western or Atlantic side of the Scandes is characterised by an oceanic climate with high precipitation, relatively mild winters with abundant snow. The eastern side has a relatively continental climate with less precipitation, less snow cover and a higher amplitude in temperature than in the west. For further details on climatic conditions, see, e.g., Moen (1999) and Chap. 2.

3.1.3 The Flora

According to current palaeogeographic knowledge, the Scandes were fully glaciated during the Pleistocene, but, during the Early Weichselian, some coastal areas of Norway were probably ice free (Forsström and Punkari 1997). This suggests that no chance existed for glacial survival and the Scandes were re-vegetated through plant immigration after deglaciation. The vegetation underwent marked changes in the late glacial and postglacial era (Birks et al. 1994) and the elevation of the timberline has fluctuated (Birks et al. 1996).

A number of recent studies have brought controversial evidence to explain modern plant distributions. Most studies seem to suggest that the glacial survival or nunatak theory is superfluous and the essential pattern of distribution of the mountain plants of the Scandes can be explained by today's ecological conditions (Birks 1993, 1996). Model-based analyses of land uplift do not suggest that mountain areas with thin ice were common (Fjeldskaar 2000).

Genetic studies may also be interpreted as indications that re-colonisation after the retreat of the ice may have been effective itself. The findings of Tollefsrud et al. (1998), which showed relatively little regional genetic differentiation among populations of *Saxifraga cespitosa*, suggest that long-distance dispersal (by seeds) may have been relatively efficient (see also Brochmann et al. 1996). *Papaver radicatum* combines several reproductive and demographic traits that may lead to potentially rapid evolutionary differentiation (Nordal et al. 1997). Moreover, the studies of Selin (2000) showed that some north Scandinavian races of *P. radicatum* are not differentiated so markedly as suggested earlier and that the more distant the populations are the more differentiated they are in morphological traits.

The potential role of climate change impacts was recently reviewed by Sonesson and Molau (1998) and Hofgaard (1999) gave an account of human impacts at the northern Swedish timberline ecotones.

3.1.4 The Alpine Zone and the Treeline Ecotone

3.1.4.1 Number of Plant Taxa

According to a database collected from Nilsson (1987), Gjærevoll (1990), Lid and Lid (1994) and Jonsell (2000), 463 vascular plant taxa (species or subspecies) occur in the alpine zone of the Scandes. The number excludes lowland plants only incidentally found in the alpine zone. Of the 463, 233 are arctic-alpine which mainly occur above the treeline or are widespread elsewhere in arctic and/or alpine areas. There are 28 endemic taxa, 11 species and 17 subspecies, respectively. In relative terms, 12.1 % of the arctic-alpine vascular plants and 6.1 % of all the vascular plants growing in the alpine zone are endemic. About 558 bryophyte species (156 liverworts and 402 mosses) occur in the alpine zone of the Scandes (R. Virtanen, unpubl.). In the mainland area of Norway, 1066 bryophyte species have been reported (Frisvoll et al. 1995). For other plant groups, the numbers are poorly known.

3.1.4.2 Rare Taxa

There are no reliable frequency data to assess the rarity of the alpine vascular plants in the Scandes. About 50 plant species or subspecies are rare and have a small distribution area and/or a small population size. Typical rare species include *Papaver* spp., *Gentiana purpurea* and *Nigritella nigra*. One recent study has elucidated the origin of the rare *Saxifraga opdalensis* (Steen et al. 2000). One species, *Crepis multicaulis*, became extinct in the 1940s (Lid and Lid 1994), probably as a result of fencing its site to exclude grazing by sheep (Alm and Often 1997).

3.1.4.3 Altitude Zones

The vegetation zonation of the Scandes comprises four zones (e.g. Haapasaari 1988; Virtanen et al. 1999b): orohemiarctic (upper treeline ecotone), lower oroarctic (low alpine), middle oroarctic (middle alpine) and upper oroarctic (high alpine). An additional subnival zone is present on some high peaks. The orohemiarctic zone is characterised by scattered mountain birch and dwarf birch (*Betula nana*) heaths, scattered individual trees, grey willow scrub (*Salix lapponum*, *S. glauca* and *S. lanata*) and abundant *Vaccinium myrtillus* heath. In the lower oroarctic zone, *V. myrtillus* starts to grow patchily confined to snow-protected sheltered slopes. Various *Empetrum* heaths dominate; low sedge-grass-herb meadows are common in sites with late snowmelt. In the middle oroarctic zone snowbeds are characterised by dwarf willows (*Salix herbacea* and *S. polaris*) and grass heaths (*Festuca ovina* and *Juncus trifidus*) are typical. *Empetrum* heaths are fragmented, *V. myrtillus* is absent and in the north *Cassiope tetragona* is common. Stone and polygon fields are common, together with snow patches melting in July–August. The upper oroarctic zone is characterised by late lying, or even permanent, patches of snow, stone and polygon fields. A few vascular plants, such as *Luzula arcuata* ssp. *confusa* and *Ranunculus glacialis*, form a discontinuous vegetation cover.

3.1.4.4 Raunkiaer's Biological Types

In the alpine zone of the Scandes, hemicryptophytes predominate while chamaephytes and geophytes make up ca. 33% of the flora (Table 3.1.1). Aquatic plant types (helophytes and hydrophytes), therophytes and phanerophytes constitute only a minor part of the flora. These proportions are comparable with those reported from, e.g., the Giant Mts. (Chap. 3.3), Corsica (Chap. 3.9), and the Pyrenees (Sesé et al. 1999).

3.1.4.5 Vegetation and Its Responses to a Changing Environment

The alpine vegetation of the Scandes is well documented in over 100 studies. Påhlsson (1994) gave a summary of the 48 alpine vegetation types with literature references. The number of described vegetation units is much larger. If some rare types are considered, about 80 types of vegetation in total can be recognised in the Scandes. In the middle oroarctic zone of the NW Finnish Lapland alone, Virtanen and Eurola (1997) described 31 vegetation types. The alpine vegetation types show marked variation in species richness (Table 3.1.2). The oligotrophic wind-swept heaths and snowbeds are poorest in vascular plants with a species-rich cryptogam flora. The vegetation types on eutrophic soil substrata are normally richer in vascular plants and poorer in cryptogams than the vegetation on nutrient-poor sites.

The results from many long-term experiments in the alpine zone tundra in northern Sweden have been summarised by Shaver and Jonasson (1999). They concluded that the alpine ecosystems of the northern Scandes were more responsive to temperature manipulations than zonal tundra ecosystems. Kudo et al. (1999) studied the effects of snowmelt on the growth of alpine dwarf shrubs. They found that an extension of the growing season decreased

Table 3.1.1. Raunkiaer's life forms of all flowering plants in the alpine zone of the Scandes with arctic-alpine species shown in parentheses

Life form	Number of taxa	(%)
Chamaephytes	84 (65)	18 (28)
Geophytes	59 (20)	13 (9)
Helophytes	6 (1)	1 (<1)
Hemicryptophytes	273 (139)	59 (60)
Hydrophytes	8 (0)	2 (0)
Phanerophytes	23 (5)	5 (2)
Therophytes	10 (3)	2 (1)
Total	463 (233)	100

Table 3.1.2. A generalised scheme of the main categories of alpine vegetation on the Scandes (Påhlsson 1994) and species richness of vascular plants, bryophytes and lichens for some characteristic plant community types

Vegetation and plant community types (reference for species richness: locality)	Number of communities described	Mean number of vascular plant species and (cryptogams) for 1 m² [*, 4 m²]
Open cryptogamic communities	51	
Rock and scree communities	7	
Snowbed vegetation	17	
Cassiope hypnoides–Salix herbacea type (Nordhagen 1943:Sikilsdalen)		10 (15)
Salix polaris type (Gjærevoll 1956: S Scandes)		19 (10)
Alpine mire vegetation	24	
Mesotrophic sedge fen (Nordhagen 1928: Sylene)		9 (8)
Eutrophic sedge fen (Nordhagen 1928: Sylene)		22 (4)
Tall-herb vegetation	9	
Cicerbita alpina–Geranium sylvaticum type (Nordhagen 1943: Sikilsdalen)		33 (4)*
Grasslands	6	
Ranunculus acris–Poa alpina type (Gjærevoll 1956: N Scandes)		15 (5)
Dwarf-shrub heath and heath-like vegetation on calcareous soil substrata	25	
(a) Wind-exposed heaths		6 (17)
Loiseleuria procumbens–Arctostaphylos alpina–Empetrum type (Nordhagen 1943: Sikilsdalen)		24 (16)*
Dryas octopetala type (Nordhagen 1955: Solvaagtind)		16 (17)
(b) Snow-protected heaths		
Vaccinium myrtillus–Phyllodoce caerulea type (Nordhagen 1943: Sikilsdalen)		13 (8)
Salix reticulata–Poa alpina type (Gjærevoll 1956: N Scandes)		
Treeline ecotone incl. scrub	11	

leaf nitrogen concentration and enhanced leaf turnover. Nyléhn and Totland (1999) showed that warmer temperature increased the growth and reproduction of individual *Euphrasia frigida* plants. However, their population densities were likely to decrease in a denser vegetation resulting from warmer temperatures. Sandvik and Totland (2000) showed that *Saxifraga stellaris* L. would have more seeds set under warmer climatic conditions that would enhance its chance to establish at new sites. Based on factorial field experiments, Michelsen et al. (1999) demonstrated that soil microbes could affect plant growth through nutrient immobilisation. *Festuca ovina* reacted strongly to changes in soil nutrient supply whereas *Vaccinium uliginosum* responded little to the same treatments.

The Scandes is one of the northernmost mountain chains of Europe and this is reflected in their species composition and ecosystem structure. The circumpolar arctic element is relatively well represented in flora. Similarly, a characteristic element is the presence of west arctic species which are also found in Greenland and in the northern parts of Canada. There are few endemic taxa, in contrast to the situation in the Alps, for instance. The vegetation differs from that of the Ural Mts. in species composition owing to differences in bedrock composition and climatic conditions (Virtanen et al. 1999a). Some high-altitude vegetation formations of the Scottish Highlands resemble those found in some parts of the Scandes.

Acknowledgements. The work was financially supported by the Finnish Research Council of Natural Resources and Environment. I thank Laszlo Nagy for useful comments on the manuscript.

References

Aas B, Faarlund T (2001) The holocene history of the Nordic mountain birch belt. In: Wielgolaski FE (ed) Nordic mountain birch ecosystems. Parthenon Publishing Group, New York, pp 5–22

Alm T, Often A (1997) Species conservation and local people in E Finnmark, Norway. Plant Talk 11:30–31

Bax G (1989) Caledonian structural evolution and techtonostratigraphy in the Rombak-Sjangeli window and its covering sequences, northern Scandinavian Caledonides. Nor Geol Unders Bull 415:87–104

Binns RE (1978) Caledonian nappe correlation and orogenic history in Scandinavia north of lat 67°N. Geol Soc Am Bull 89:1475–1490

Birks HH, Paus A, Svendsen JI, Alm T, Mangerud J, Landvik JY (1994) Late Weichselian environmental change in Norway, including Svalbard. J Quat Sci 9:133–145

Birks HH, Alm T, Vorren K (eds) (1996) Holocene treeline oscillations, dendrochronology and palaeoclimate. Paläoklimaforschung/Palaeoclimatic research 20. Gustav Fischer, Stuttgart

Birks HJB (1993) Is the hypothesis of survival on glacial nunataks necessary to explain the present-day distributions of Norwegian mountain plants? Phytocoenologia 23:399–426

Birks HJB (1996) Statistical approaches to interpreting diversity patterns in the Norwegian mountain flora. Ecography 19:332–340

Brochmann C, Gabrielsen TM, Hagen A, Tollefsrud MM (1996) Seed dispersal and molecular phylogeography: glacial survival, tabula rasa, or does it really matter? Det Norske Videnskaps-Akademi. I. Mat Nat Kl Avh NS 18:54–68

Dahl E (1990) History of the Scandinavian alpine flora. In: Gjaerevoll O (ed) Maps of distribution of Norwegian vascular plants, vol II. Alpine plants. Tapir Publishers, Trondheim, pp 16–21

Eide EA, Torsvik TH, Andersen TB, Arnaud NO (1999) Early carboniferous unroofing in western Norway: a tale of alkali feldspar thermochronology. J Geol 107:353-374

Fjeldskaar W (2000) An isostatic test of the hypothesis of ice-free mountain areas during the last glaciation. Norsk Geol Tidskr 80:51-56

Forsström L, Punkari M (1997) Initiation of the last glaciation in northern Europe. Quat Sci Rev 16:1197-1215

Frisvoll AA, Elvebakk A, Flatberg KI, Økland RH (1995) Sjekkeliste øver norske mosar. NINA Temahefte 4:1-104

Gjærevoll O (1956) The plant communities of the Scandinavian alpine snowbeds. Kongel Nor Vidensk Selsk Skr 1956:1-405

Gjærevoll O (ed) (1990) Maps of distribution of Norwegian vascular plants. II. Alpine plants. Tapir Publishers, Trondheim

Haapasaari M (1988) The oligotrophic heath vegetation of northern Fennoscandia and its zonation. Acta Bot Fenn 135:1-219

Hofgaard A (1999) The role of 'natural' landscapes influenced by man in predicting responses to climate change. Ecol Bull 47:160-167

Jonsell B (ed) (2000) Flora Nordica 1. Bergius Foundation, Stockholm

Kudo G, Nordenhäll U, Molau U (1999) Effects of snowmelt timing on leaf traits, leaf production, and shoot growth of alpine plants: comparisons along a snowmelt gradient in northern Sweden. Écoscience 6:439-450

Lid J, Lid DT (1994) Norsk flora. 6th edn. (by R Elven). Det Norske Samlaget, Oslo

Michelsen A, Graglia E, Schmidt IK, Jonasson S, Sleep D, Quarmby C (1999) Differential responses of grass and a dwarf shrub to long-term changes in soil microbial biomass C, N and P following factorial addition of NPK fertilizer, fungicide and labile carbon to a heath. New Phytol 143: 523-538

Moen A (1999) National atlas of Norway: vegetation. Norwegian Mapping Authority, Hønefoss

Nilsson Ö (1987) Nordisk Fjällflora, 2nd edn. Bonniers, Stockholm

Nordal I, Hestmark G, Solstad H (1997) Reproductive biology and demography of Papaver radicatum - a key species in Nordic plant geography. Opera Bot 132:33-41

Nordhagen R (1928) Die Vegetation und Flora des Sylenegebietes. I. Die Vegetation. Skr Nor Vidensk Akad Oslo. I. Mat Nat Kl 1927:1-612

Nordhagen R (1943) Sikilsdalen og Norges fjellbeiter. En plantesosiologisk monografi. Bergens Mus Skr 22:1-607

Nordhagen R (1955) Kobresieto-Dryadion in northern Scandinavia. Sv Bot Tidskr 49:63-87

Nyléhn J, Totland Ø (1999) Effects of temperature and natural disturbance on growth, reproduction, and population density in the alpine annual hemiparasite *Euphrasia frigida*. Arct Antarct Alp Res 31:259-263

Påhlsson L (ed) (1994) Vegetationstyper i Norden. TemaNord 1994:665, Nordiska ministerrådet, Copenhagen

Sandvik SM, Totland Ø (2000) Short-term effects of simulated environmental changes on phenology, reproduction, and growth in the late-flowering snowbed herb *Saxifraga stellaris* L. Écoscience 7:201-213

Selin E (2000) Morphometric differentiantion between populations of Papaver radicatum (Papaveraceae) in northern Scandinavia. Bot J Linn Soc 133:263-284

Sesé JA, Ferrández JV, Villar L (1999) La flora alpina de los Pirineos. Un patrimonio singular. In: Villar L (ed) Espacios Naturales Protegidos del Pirineo. Ecología y cartografía. Consejo de Protección de la Naturaleza de Aragón, Zaragoza

Shaver GR, Jonasson S (1999) Response of Arctic ecosystems to climate change: results of long-term field experiments in Sweden and Alaska. Polar Res 18:245-252

Sonesson M, Molau U (1998) The Caledonian mountains, northern Europe, and their changing ecosystems. Pirineos 151/152:111-130

Steen SW, Gielly L, Taberlet P, Brochmann C (2000) Same parental species, but different taxa: molecular evidence for hybrid origins of the rare endemics *Saxifraga opdalensis* and *S. svalbardensis* (Saxifragaceae). Bot J Linn Soc 132:153–164

Svenningsen OM (1995) Extensional formation along the Late-Precambrian-Cambrian Baltoscandian passive margin: the Sarektjåkkå Nappe, Swedish Caledonides. Geol Rundsch 84:649–664

Tollefsrud MM, Bachman K, Jakobsen KS, Brochmann C (1998) Glacial survival does not matter – II: RAPD phylogeography of Nordic Saxifraga cespitosa. Mol Ecol 7:1217–1232

Väisänen R (1998) Current research trends in mountain biodiversity in NW Europe. Pirineos 151/152:131–156

Virtanen R, Eurola S (1997) Middle oroarctic vegetation in Finland and middle-northern arctic vegetation on Svalbard. Acta Phytogeogr Suec 82:1–60

Virtanen R, Oksanen L, Razzhivin V (1999a) Topographic and regional patterns of tundra heath vegetation from northern Fennoscandia to the Taimyr Peninsula. Acta Bot Fenn 167:29–83

Virtanen R, Pöyhtäri P, Oksanen L (1999b) Topographic and altitudinal patterns of heath vegetation on Vannøya and the northern Varanger Peninsula, northern Norway. Acta Bot Fenn 167:3–28

3.2 The High Mountain Vegetation of Scotland

L. Nagy

3.2.1 Introduction

Scotland lies between 54°N and 61°N 8°W and 1°W and its total land area, including the islands, is about 78,829 km². Estimates vary between 2% (Miles et al. 1997) and 4% (Thompson and Brown 1992) about the extent of the area above the potential treeline. The only existing natural treeline is at 640 m in the eastern Highlands. Much of the treeline is suppressed by fire and grazing (Crawford 1989). It is difficult to provide an overall estimate of the elevation of the potential treeline, as vegetation similar to that found above the potential treeline in the eastern Highlands occurs at lower altitudes in the far west or in the far north (McVean and Ratcliffe 1962; Nagy 1997). The lower boundary of the alpine zone is often delineated at 750 m a.s.l., a realistic elevation in the light of experimental growth of plantation conifers. The alpine vegetation at elevations above ca. 900 m consists largely of prostrate dwarf shrub and grass-moss formations of short stature.

3.2.2 Geology, Soils and Climate

The Scottish mountains are part of the Caledonian mountain belt and as such are of Ordovician origin (ca. 450 million years). They are characterised by weathered and eroded igneous and metamorphic rocks and by glacial geomorphology. The geology of Scotland is diverse with most of the alpine zone being acidic granites, quartzite and schists. The soils overlying them are base-poor and leached (podsolised). There are localised occurrences of calcareous parent material such as calcareous mica schists (predominantly along the line from Beinn Luigh in the west via Ben Lawers to Caenlochan in the east). With less acidic soil conditions, these hills host a large proportion of the rare high mountain species of Scotland. Peat blankets some plateaux and gentle sloping hillsides.

Scotland's climate is oceanic; the average annual temperature range for 1985–1988 was –5.0 °C (January) to + 6.9 °C (July) at the Cairngorm summit (1245 m a.s.l.) and the annual precipitation at 1000 m in the Cairngorms was in the range 1700–2000 mm (McClatchey 1996). Snow lie at 1000 m is dependent on topography and is estimated to be 50–200 days (Barry 1992). Growing season was estimated to be about 180 days based on continuous measurement of soil temperatures at 10 cm between June 1999 and August 2000 at Glas Maol (970 m), Eastern Grampians and Ben Wyvis (900 m; see Chap. 2). The sum of thermal time, defined as the sum of temperatures on days when soil temperature was >5 °C, was 526 day degrees in chionophilous *Vaccinium* heath (Glas Maol) and 289 day degrees in *Carex bigelowii–Racomitrium lanuginosum* vegetation (Ben Wyvis). Interestingly, the accumulated air temperatures above 5.6 °C on sites at about 1000 m are >500 day degrees year^{-1}.

3.2.3 The Flora

An arctic-alpine flora with similar vegetation to today's Scandinavian mountains and tundra characterised Scotland ca. 12,000 years ago (Walker 1984). The extreme cold of the Younger Dryas (ca. 11,000 years ago) left behind an impoverished flora. The Holocene brought the advance of forests, possibly to about 200 m higher elevation than today's only natural treeline at 640 m at Creag Fhiachlach, Cairngorms (Birks 1997). The latter, however, has been stable for the past 100–1000 years (McConnell 1996). The Holocene advance of forest resulted in the extinction of around 19 % of the arctic-alpine taxa (Birks 1997). Today's potential vegetation of Scotland is possibly an extreme oceanic variant of the boreal zone of Europe (Ozenda 1994; Huntley et al. 1997).

3.2.3.1 The Present-Day Flora

Scotland has a small flora (Table 3.2.1) and this is particularly true for its high mountains. For example, the complete flora of the Cairngorm Mountains, which have the largest area in the alpine zone of all mountains in Scotland, comprises about 280 vascular plant species. Sydes (1997) estimated that, in addition to the native species, about 40 % of today's vascular plants in Scotland were introduced taxa. He also estimated that native subspecies contributed an additional 56–78 and native hybrids another 89–125 taxa. There are nine Scottish endemics excluding microspecies.

The life-form spectrum (following the scheme of Pignatti 1982) of the 193 vascular plants recorded by McVean and Ratcliffe (1962) above 750 m is 66.9 % hemicryptophytes, 15.5 % chamaephytes (11.6 % without nanophytes),

Table 3.2.1. The number of various plant taxa in Scotland

Plant group	Total	Above the treeline	Exclusive to the alpine
Lichens	1486[a]	ca. 700[d]; 55[f]	275[d]
Bryophytes	928[b]	212[f]	89[e]
Native vascular plant species	1117[c]	193[f]+31[g]	58[g]

[a] O'Dare and Coppins (1993).
[b] Usher (1997).
[c] Sydes (1997) (excluding 442 apomictic microspecies of *Hieracium* and *Taraxacum*).
[d] Frydey (1997) commented that 275 species were 'mountain specialist', most of them occurring in open lichen communities. McVean and Ratcliffe (1962) recorded 55 species from vegetation above 750 m.
[e] Horsfield and Hobbs (1993) gave 443 as the number of species associated with the uplands, i.e. above enclosed farmland. Some of these do not reach above the treeline; 89 were quoted as most abundant or confined to the alpine zone.
[f] McVean and Ratcliffe (1962), figure from all 399 vegetation samples above 750 m.
[g] Ratcliffe (1991) listed 58 species that are found exclusively above the treeline, 27 of these were also listed in McVean and Ratcliffe (1962).

12.2% geophytes including Juncaceae, 4.4% therophytes, and 1.1% phanerophytes (5% if nanophytes and phanerophytes are taken together).

3.2.4 Vegetation

Table 3.2.2 shows the major vegetation types in Scotland that occur at and above the treeline. The number of communities distinguished may vary by the area considered and is always somewhat subjective. For example, the quoted figures in Table 3.2.2 are based on McVean and Ratcliffe (1962), who distinguished 42 communities in the Scottish Highlands. Later work by Birks and Ratcliffe (1981) and Rodwell (1991, 1992) reduced the numbers to 38 and 28, respectively, although Rodwell distinguished numerous subcommunities.

3.2.5 Species and Habitat Diversity and Causes of Change

The diversity of habitats arising from climatic, topographical and geological variation gives rise to moss and lichen heaths, snowbeds, high-altitude blanket bog and dwarf-shrub heath. The number of communities increases with altitude and from east to west; with more rare communities in the west and fewer widespread ones in the east (Brown et al. 1993).

Table 3.2.2. Community types and their species richness at and above the treeline in the Scottish Highlands

Community type	Elevation range (m)	Number of communities	Principal species	Number of all species[a] and (vascular species only) per community[b]; range of all species and (of vascular species) 4 m^{-2}
Open cryptogamic communities[c]		?	*Lecanora* spp., *Miriquidica* spp., *Rhizocarpon* spp.	40 (17); 5–19 (2–10)
Springs and flushes	–910	6	*Philonotis fontana* *Saxifraga stellaris, Pohlia glacialis, Anthelia julacea*	45 (34); 13–29 (9–21)
Scree communities	820–1100	1	*Cryptogramma crispa, Athyrium alpestre*	102 (41); 17–46 (8–21)
Late snowbed communities	910–1220	3	*Polytrichum alpinum, Kiaeria starkei, Salix herbacea*	64 (21); 12–26 (3–9)
Peat bogs and mires	–1070	7	*Eriophorum vaginatum, Calluna vulgaris, Empetrum nigrum* ssp. *hermaphroditum*, *Sphagnum* spp., *Carex* spp.	61 (23); 19–33 (8–15)
				99 (53); 19–41 (9–24)
Tall-herb meadows	–850	1	*Deschampsia cespitosa, Luzula sylvatica, Trollius europaeus*	93 (58); 19–54 (12–36)
Grasslands[d] (grass and moss heaths)	610–1220	10	*Nardus stricta, Carex bigelowii, Juncus trifidus*	35 (9); 15–19 (2–4)
				47 (19); 10–22 (3–13)
Dwarf-shrub heath	–1130	12	*Calluna vulgaris, Vaccinium myrtillus* *Empetrum nigrum* ssp. *hermaphroditum, Dryas octopetala*	53 (24); 9–26 (5–18)
				118 (78); 18–65 (10–44)
Scrub	670–910	1	*Salix lapponum*	92 (54); 37–42 (15–26)
Treeline ecotone	640–910	1	*Pinus sylvestris, Calluna vulgaris, Juniperus communis* ssp. *nana*	60 (17); 15–29 (5–12)

[a] Including vascular plants, bryophytes and lichens.
[b] Where there is more than one community values are given for the most species-poor and the most species-rich community.
[c] Epilithic communities on rocks, boulders, screes, and crags have not been classified to date (but see Frydey 1997).
[d] Excludes anthropogenic grasslands and those on calcareous substrata.

Apart from habitat characteristics, vegetation is directly influenced by domestic and wild grazers [sheep (*Ovis*, domestic), feral goats (*Capra*, domestic), red deer (*Cervus elaphus* L.), hare (*Lepus timidus* L.), ptarmigan (*Lagopus mutus* Hartert)], and airborne pollution and global climate change play a yet unquantified role.

The high mountain areas of Scotland have been least affected by human impact compared with lowlands; however, impacts of grazing locally may have caused significant vegetation change. Grazing impacts through the removal of biomass, nutrient enrichment through defecation and urination, and mechanical damage to plants caused by trampling may vary over scales from local population dynamics (Miller et al. 1999) to landscapes (MacDonald et al. 1998).

There are instances where long-term grazing is likely to have resulted in a change in vegetation. It is recognised that grassland communities on calcareous substrata are all influenced by grazing. It has been suggested that the ungrazed equivalent of the *Festuca-Alchemilla-Silene* community is the *Dryas octopetala-Silene acaulis* community. A comparison of the component species shows that there are a number of grazing-sensitive species (*Dryas, Salix reticulata, S. arbuscula, Angelica sylvestris, Geranium sylvaticum, Geum rivale, Rhodiola rosea, Saussurea alpina, Hieracium* spp., cushion plants and bryophytes) in the *Dryas-Silene* community that are absent from the *Festuca-Alchemilla-Silene* community (McVean and Ratcliffe 1962; Rodwell 1992). It is thought that the *Festuca-Agrostis* community was derived and maintained by grazing and McVean and Ratcliffe (1962) suggested that some stands of it might be the derivative of former *Dryas* communities. On moist ledges, it is likely that a tall herb community type was the original vegetation. *Festuca ovina-Agrostis capillaris-Alchemilla alpina* grass heath derived from *Dryas-Silene* ledge vegetation (McVean and Ratcliffe 1962; Rodwell 1992). The *Festuca ovina-Agrostis capillaris-Silene acaulis* dwarf herb community either derived from *Dryas-Silene* community through grazing or snow-lie is more important (McVean and Ratcliffe 1962).

The potential impacts of nitrogen (N) deposition on high altitude bryophytes have been reported. Baddeley et al. (1994) suggested that N input from atmospheric deposition stimulated grass growth at the expense of *Racomitrium lanuginosum* in moss heaths dominated by that species. High N concentrations have been found in late snowbed bryophyte species, also (Woolgrove and Woodin 1996a). Extreme concentrations of NO_3^- and SO_4^{2-} (pH 3.2, NO_3^- 447 µeq l^{-1}, SO_4^{2-} 706 µeq l^{-1}) based maximum values reported for the Cairngorms had significant transient effects on the photosynthetic system of *Kiaeria starkei*. The length of recovery indicated that the species' long-term survival might be threatened if extreme loadings became the norm.

The addition of N alone on skeletal soil on serpentine with low initial plant cover (about 5%) induced no vegetation changes (Nagy and Proctor 1997).

Only after the addition of NPK was a significant cover increase through an increase in reproduction and recruitment of herbs and through increased grass growth. In the exclosures, quadrats that initially had grasses tended to be dominated by them at the expense of other herbs. As it is phosphorus (P) that is limiting in many sites, the implications of an increased N input may not be directly extrapolatable to predict vegetation changes without examining detailed nutrient budgets. In closed vegetation, an increase in soil nutrient status as a result of excrement deposition may favour the growth of fine-leaved grasses and herbaceous species such as *Potentilla erecta* and *Galium saxatile* (MacDonald et al. 1998). As increasing grass cover is likely to attract more grazers, as was found on Meall Mór an Chaorainn, Monadhliath Mountains (McConnell Associates 1999), it may become a self-perpetuating process.

Recreational trampling is increasing and impacts on vegetation have been observed. Some plants are more susceptible to trampling than others depending on their morphology and their susceptibility to damage (resistance) and ability to recover from damage (resilience) will determine their response. The least tolerant (combination of resistance and resilience) species tend to decrease in abundance most rapidly. The more tolerant species may increase in abundance with low levels of trampling due to reduction in competition but will also decline at high impact levels (Legg 2000). Along footpaths, a reduction in species richness of about 30 % may occur (L. Nagy and C.J. Legg, unpubl.). Jonsdottir (1991) found in Iceland that trampling reduced *Racomitrium lanuginosum* cover in *Carex bigelowii–Racomitrium lanuginosum* heath. An increase in lichen cover in exclosures has been reported by Frydey (1997) and *Cladonia* spp. were found lower in sheep-grazed parts of Ben Wyvis as opposed to those without sheep grazing; a significantly higher cover in *Carex bigelowii* was recorded in the sheep-grazed area (McConnell Associates 2000).

Increased temperature is likely to further reduce the number of arctic-alpine species in the Scottish high mountains. Species which reach their southern distribution in Scotland with isolated stations and habitat specialists are particularly susceptible (Ratcliffe 1991; Birks 1997; Frydey 1997). Elmes and Free (1994) estimated that 14 % of rare species (Red Data Book listed species in all habitats in the whole of Britain) would half their range on a 2 K increase in temperature and about 15 % would become extinct. This, however, should be seen in the light of evidence that there was no change in temperature in NW-Scotland over the twentieth century (Schönwiese and Rapp 1997) with an increase of 0.2–0.4 K in other parts of lowland Scotland (Harrison 1997) as opposed to observed increases of up to 2.5 K in more continental parts of Europe.

Acknowledgements. I thank D. Horsfield, J. Jeník, J. Proctor and R. Virtanen for their comments.

References

Baddeley JA, Thompson DBA, Lee JA (1994) Regional and historical variation in the N-content of *Racomitrium lanuginosum* in Britain in relation to atmospheric N deposition. Environ Pollut 84:189–196

Barry RG (1992) Mountain weather and climate. Routledge, London

Birks HJB (1997) Scottish biodiversity in a historical context. In: Fleming LV, Newton AC, Usher MB (eds) Biodiversity in Scotland: status, trends and initiatives. The Stationery Office, Edinburgh, pp 21–35

Birks HJB, Ratcliffe DA (1981) Classification of upland vegetation types. Nature Conservancy Council, Edinburgh

Brown A, Horsfield D, Thompson DBA (1993) A new biogeographical classification of the Scottish uplands. I. Descriptions of vegetation blocks and their spatial variation. J Ecol 81:207–229

Crawford RMM (1989) Studies in plant survival. Blackwell, Oxford

Elmes GW, Free A (1994) Climate change and rare species in Britain. The Stationery Office, London

Frydey A (1997) Montane lichens in Scotland. Bot J Scotl 49:367–374

Grace J (1997) The oceanic tree-line and the limit for tree growth in Scotland. Bot J Scotl 49:223–236

Harrison SJ (1997) Changes in the Scottish climate. Bot J Scotl 49:287–300

Horsfield D, Hobbs A (1993) North-west region. Report on upland habitats and flora. Scottish Natural Heritage, Edinburgh

Huntley B, Daniell RG, Allen JRM (1997) Scottish vegetation history: the highlands. Bot J Scotl 49:163–176

Jonsdottir IS (1991) Effects of grazing on tiller size and population dynamics in a clonal sedge (*Carex bigelowii*). Oikos 62:177–188

Legg CJ (2000) Review of published work in relation to monitoring of trampling impacts and change in montane vegetation. Scottish Natural Heritage Review No. 131. Scottish Natural Heritage, Battleby

MacDonald A, Stevens P, Armstrong H, Immirzi P, Reynolds P (1998) A guide to upland habitats. Surveying land management impacts, vols 1 and 2. Scottish Natural Heritage, Battleby

McClatchey J (1996) Spatial and altitudinal gradients of climate in the Cairngorms – observations from climatological and automatic weather stations. Bot J Scotl 48:31–49

McConnell J (1996) The history of the *Pinus sylvestris* L. treeline in the Cairngorms, Inverness-shire. PhD Thesis, University of Edinburgh, Edinburgh

McConnell Associates (1999) Testing the consistency of field assessments made using the SNH field guide to upland habitats. Scottish Natural Heritage, Edinburgh, Scotland

McConnell Associates (2000) Ben Wyvis grazing project. Scottish Natural Heritage, Edinburgh, Scotland

McVean DN, Ratcliffe DA (1962) Plant communities of the Scottish highlands. Nature conservancy monograph no 1. The Stationery Office, Edinburgh

Miles J, Tudor G, Easton C, Mackey EC (1997) Habitat diversity in Scotland. In: Fleming LV, Newton AC, Usher MB (eds) Biodiversity in Scotland: status, trends and initiatives. The Stationery Office, Edinburgh, pp 43–56

Miller GR, Geddes C, Mardon DK (1999) Response of the alpine gentian *Gentiana nivalis* L. to protection from grazing by sheep. Biol Conserv 87:311–318

Nagy J (1997) Plant biodiversity in Scottish alpine vegetation: a review and research needs. Bot J Scotl 49:469–477

Nagy L, Proctor J (1997) Plant growth and reproduction on a toxic alpine ultramafic soil: adaptation to nutrient limitation. New Phytol 137:267–274

O'Dare AM, Coppins B (1993) Biodiversity inventory – lichens. Scottish Natural Heritage, Edinburgh

Ozenda P (1994) Végétation du continent européen. Delachaux et Niestlé, Lausanne

Pignatti S (1982) Flora d'Italia. Edagricole Calderini, Bologna

Ratcliffe RD (1991) The mountain flora of Britain and Ireland. Br Wildl 3:1–21

Rodwell JS (ed) (1991) British plant communities. vol 2. Mires and heaths. Cambridge University Press, Cambridge

Rodwell JS (ed) (1992) British plant communities, vol 3. Grasslands and montane communities. Cambridge University Press, Cambridge

Schönwiese C-D, Rapp J (1997) Climate trend atlas of Europe based on observations 1891–1990. Kluwer, Dordrecht

Sydes C (1997) Vascular plant biodiversity in Scotland. In: Fleming LV, Newton AC, Usher MB (eds) Biodiversity in Scotland: status, trends and initiatives. The Stationery Office, Edinburgh, pp 89–103

Thompson DBA, Brown A (1992) Biodiversity in montane Britain – habitat variation, vegetation diversity and some objectives for conservation. Biol Conserv 1:179–208

Usher MB (1997) Scotland's biodiversity: an overview. In: Fleming LV, Newton, AC, Usher MB (eds) Biodiversity in Scotland: status, trends and initiatives. The Stationery Office, Edinburgh, pp 5–20

Walker MJ (1984) A pollen diagram from St-Kilda, Outer Hebrides, Scotland. New Phytol 97:99

Woolgrove CE, Woodin SJ (1996a) Current and historical relationships between the tissue nitrogen content of a snowbed bryophyte and nitrogenous air pollution. Environ Pollut 91:283–288

Woolgrove CE, Woodin SJ (1996b) Effects of pollutants in snowmelt on *Kiaeria starkei*, a characteristic species of late snowbed bryophyte dominated vegetation. New Phytol 133: 519–529

3.3 Vegetation of the Giant Mountains, Central Europe

J. Jeník and J. Štursa

3.3.1 Introduction

The flora of the Giant Mountains, the highest range of the Hercynian middle-mountains in central Europe, consists of a broad spectrum of species of diverse chorological and ecological affinity. Many of the species distributed in Europe's alpine, Nordic and lowland areas occur together in a relatively small area (with about 10 % of the area above the treeline ecotone), and at a comparatively low latitude in the Giant Mts. A combination of relief, topoclimate, rock type, soil fertility and selective disturbance of avalanches have enabled the co-existence and preservation of multiple relict organisms. Treeless summits and plateaux with periglacial scree, grassy tundra and mires, and the corries with cliffs and avalanche paths, have served as habitats for disjunct and endemic populations of vascular plants. A highly polygenous flora generates a pattern of unique plant communities that creates difficulties in vegetation classification.

3.3.2 Physical-Geographical Setting

The Giant Mts. – Krkonoše and Karkonosze in Czech and Polish, respectively – lie at 50°37'N latitude and 15°45'E longitude, at the eastern limit of the Hercynian (or Variscan) system (sensu Holmes 1965), along the boundary between Czechia and Poland. The summit areas of this range culminate in the Sněžka Peak at 1603 m. They are influenced by severe climatic factors (Migala et al. 1995). The average annual temperature is +1 °C with frequent freeze-and-thaw cycles, a high proportion of the precipitation falling as snow, and there are strong westerly winds. These factors induce prominent cryogenic, niveo-glaciogenic and eolian processes that affect the soil, plants and animal life. Dominant rocks are gneiss, phyllite, and granite; only local small basalt, porphyrite and limestone outcrops create nutrient-rich parent rocks. Two parallel ridges are connected by two ancient Tertiary etchplains, but the emerging peaks and some valley heads are considerably sculpted by Pleistocene glaciers and recent nivation (Soukupová et al. 1995). Most of the area above the forest line (1250 m a.s.l.) is in the treeline ecotone of the upper montane coniferous forest and krummholz zones (ca. 9%) and about 1% of the range is covered by alpine heaths and grasslands. Avalanche frequency has far-reaching impact on habitat fragmentation, biodiversity and vegetation pattern. Between 1962 and 1999, 735 avalanche events were recorded (Spusta and Kocianová 1998).

3.3.3 Flora and Vegetation

A complete enumeration of the flora of the Giant Mts. has been available since 1969 (Šourek 1969), but new taxonomical methods have extended the species list even by new endemic species (*Minuartia corcontica*). The flora of the Giant Mts. consists of 1226 vascular species (within 424 genera and 105 families). Of this number, 437 species occur above the treeline. According to Raunkiaer's system of life forms, the hemicryptophytes prevail (83%), while chamaephytes (5.7%), geophytes (4.6%), and therophytes (0.2%) account for about 10%. Twenty-eight species of phanerophytes (6.4%) found their refuge mainly on the avalanche tracks in the corries. Sixty-four species (14.5%) of the total are non-native.

Though less definitive in their enumeration, the bryologists estimate the occurrence of 500 species (150 moss and 350 liverwort species), about half of which are distributed above the treeline. Of the 250 lichen species, over two-thirds are components of the upper montane, lower alpine and alpine zones.

There are 27 species and subspecies and four varieties of vascular plants which are endemic to the Giant Mts. They grow mostly in glacial cirques (corries) that also harbour many small populations of alpine plants found on the Alps, Scandes, or Scottish mountains. On the avalanche tracks descending down to the montane zone, these endemic and Nordic populations often mix with plants belonging to beech forests and grasslands of the lower montane zone. The biogeographic 'crossroads' position is apparent also in mires where *Pinus mugo* grows at its northernmost boundary, while, in the same plant community, *Rubus chamaemorus* reaches its southernmost limit. Another example, *Saxifraga nivalis*, has its nearest population on Ben Lawers in Scotland, and in Scandinavia.

Quasi-periglacial climate and related exogeodynamic and ecological processes have created a pattern of habitats which harbour – irrespective of altitude – biotic communities described from the Scandes or European high mountains. Examples include the grassy sedge-heaths of the *Nardus*

Vegetation of the Giant Mountains, Central Europe

Table 3.3.1. Altitudinal range of plant communities in the central area of the Giant Mts. (Adapted from Moravec et al. 1995)

Vegetation type	Number of plant communities (alliance/ association)	Altitude range (m)	Community species richness
Open cryptogamic communities	5/5	1000–1600	n.d.
Late snowbed communities	1/1	1450–1600	5–10
Grasslands	5/10		
Species-poor meadows in upper alpine zone		1300–1600	5–20
Species-rich meadows in lower alpine zone		1200–1400	30–75
Rock communities	1/1	1250–1400	18–33
Scree communities	1/1	1000–1450	n.d.
Springs and flushes	2/8	1200–1500	10–27
Ombrogenous and soligenous mires	4/9	1200–1450	13–22
Tall herb meadows	2/5	1200–1450	17–77
Dwarf-shrub heath	1/1	1450–1600	7–26
Treeline ecotone, incl. scrub	3/5	1200–1450	20–29

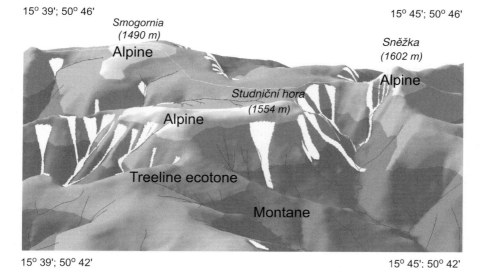

Fig. 3.3.1. View of the Giant Mts. from the south up the glacial valley of Obří Důl. Avalanche tracks, shown in *white*, generate refugia for alpine and arctic-alpine species in the treeline ecotone and in the montane forest

stricta–Carex rigida type, *Salix herbacea* snowbeds, and stony summits of *Juncus trifidus*-dominated vegetation (Matuszkiewicz and Matuszkiewicz 1975; Moravec et al. 1995). The position of the treeline varies according to relief, wind and snow distribution (anemo-orographic systems, sensu Jeník 1997, 1998). Avalanches recurrently disturb the closed-canopy krummholz dominated by dwarf pine (*Pinus mugo*) and taiga of Norway spruce (*Picea abies*), and rejuvenate the leached topsoil along a wide altitudinal gradient. This generates a mixture of interlacing plant communities, some of which are distributed along the broad flanks and valleys, others are limited to particular azonal habitats or to rather specialised ecosystems in the glacial cirques (Hadač and Štursa 1983). If viewed in terms of altitudinal zonation of plant communities, the corries represent a particular 'intercalary zone' of the Giant Mts. (Table 3.3.1, Fig. 3.3.1) and their vegetation is difficult to classify (Moravec et al. 1995).

3.3.4 Refugia of the Polygenous Flora

Species richness in the Giant Mts. is enhanced by the presence of endemic species and small populations of disjunct plants occurring as relicts in the corries. Called by the local idiom as 'gardens', these treeless refugia derive from long-term arrival and survival of immigrants established in the course of the Holocene period. These habitats developed due to a particular combination of primary georelief, topoclimate, parent rock and perturbation by avalanches. This combination sustained a treeless environment and controlled competition by graminoids, shrubs and trees. The species-rich communities seldom occupy north-facing slopes; however, they always grow on relatively base-rich bedrock, in a leeward position, under a deep snow drift, and on avalanche paths. Some of these habitats are irrigated by base-rich springs. (For details on the key factors of refugia, see Jeník et al. 1983.)

Local differences among the refugia are obviously caused by (1) mechanical disturbances (rock-falls, landslide, soil erosion or avalanches), (2) site modification by established vegetation, and (3) succession. Colonisation by tussock grasses or trees has been a permanent threat. The survival of glacial and mid-Holocene relict and disjunct populations for millennia, and veritable speciation of some new taxa (Jeník 1983), suggest that avalanche activity maintained the open-habitats in the intercalary zone. This probably also applies to most of the Postglacial age, including the period of general ascent of the forest-line by an estimated 200–300 m altitude.

The threat of an ascending treeline to the refugia in the Giant Mts. is particularly relevant with regard to the projected warming of the atmosphere in the twenty-first century (Grabherr et al. 1995). Wojtun et al. (1995) have already reported essential floristic changes in plant communities; however, these appear to have been affected also by air pollution. Curiously as it may appear, the fragile polygenous flora of the Giant Mts. might be damaged by the absence of destructive snow avalanches.

References

Grabherr G, Gottfried M, Gruber A, Pauli H (1995) Patterns and current changes in alpine plant diversity. In: Chapin FS, Körner C (eds) Arctic and alpine biodiversity. Springer, Berlin Heidelberg New York, pp 167–181

Hadač E, Štursa J (1983) Syntaxonomic survey of plant communities of the Krkonoše (Giant Mountains). Opera Corcontica 20:79–98

Holmes A (1965) Principles of physical geology. Nelson and Sons Ltd, London

Jeník J (1983) Evolutionary stage of the Sudetic corries. Biol Listy 48:241–248 (in Czech)

Jeník J (1997) Anemo-orographic systems in the Hercynian Mts. and their effects on biodiversity. Acta Univ Wratisl Prace Inst Geogr Ser C Meteorol Klimatol 4:9–21

Jeník J (1998) Biodiversity of the Hercynian Mountains in central Europe. Pirineos 151/152:83–99

Jeník J, Bureš L, Burešová Z (1983) Revised flora of Velká Kotlina cirque, the Sudeten mountains, parts 1 and 2. Preslia 55:25–61, 123–141

Matuszkiewicz W, Matuszkiewicz A (1975) Mapa zbiorowisk roslinnych Karkonoskiego Parku Narodowego. Ochr Przyrody 40:45–112

Migala K, Pereyma J, Sobik M (1995) Current climatic factors and topoclimatic differentiation of the Giant Mts. In: Fischer Z (ed) Ecological issues of the high-mountain part of the Giant Mts. Wydaw. Inst. Ekologii PAN, Warszawa, pp 51–78 (in Polish)

Moravec J, Hadač E, Jeník J, Krahulec F et al. (1995) Plant communities of the Czech Republic and their conservation status, 2nd edn. Okr Vlast Muzeum, Litoměřice (in Czech)

Spusta V, Kociánová M (1998) Avalanche cadastre in the Czech part of the Giant Mts. during winter seasons 1961/62–1997/98. Opera Corcontica 35:3–205

Soukupová L, Kociánová M, Jeník J, Sekyra J (eds) (1995) Arctic-alpine tundra of the Krkonoše, the Sudetes. Opera Corcontica 32:5–88

Šourek J (1969) Flora of the Giant Mountains. Academia, Praha (in Czech)

Wojtun B, Fabiszewski J, Sobierajski Z, Matula J, Zolnierz L (1995) Current changes in the high-mountain plant communities of the Giant Mts. In: Fischer Z (ed) Ecological issues of the high-mountain part of the Giant Mts. Wydaw. Inst. Ekologii PAN, Warszawa, pp 213–245 (in Polish)

3.4 The Alpine Vegetation of the Alps

P. Ozenda and J.-L. Borel

3.4.1 Physiography

The Alps stretch over an arch of more than 1200 km from the Mediterranean Sea to the Vienna basin between latitudes 44°N and 48°N. Its width ranges between 120 and 250 km (Munich-Verona axis). About 15,000 km^2 of its estimated total surface area of 200,000 km^2 lies above the treeline. There are numerous summits reaching 4000 m and over between the Ecrins and the Bernina massif. The differences in elevation are often large over short distances such as between deep valleys and summits.

There is a marked geological and climatic contrast between the Pre-Alps and the Inner Alps. The Pre-Alps are humid and mostly calcareous (except the Piedmontese and the Styrian Pre-Alps), whilst the Inner Alps are relatively dry and largely siliceous (Fig. 3.4.1). This geographic structure is of fundamental importance for the distribution of plant species, including those in the alpine belt.

The mean annual temperature lapse rate with altitude is 0.55 K 100 m^{-1} with a value of about 0.7 K in the summer and about 0.4 K in winter. From an ecological point of view, this thermic lapse rate is a main influencing factor of life conditions and causes a sequence of vegetation belts (Ozenda 1985). An increasing proportion of the annual precipitation falls as snow because of a decrease in temperature with altitude. For example, in the French W Alps, about half the annual precipitation is in the form of snow above 2000 m and above 3600 m it is the only form. At high elevations extended snowpack may limit the duration of the growing season. Gensler (1946) estimated that the reduction in growing season length in the Swiss Alps was 7 days 100 m^{-1} on N slopes and 6 days 100 m^{-1} on S slopes.

The stature and density of the trees decrease from closed montane forest to the treeline ecotone, where contorted growth forms dominate, the so-called *Kümmerformen* (Fig. 3.4.2). Dwarf-shrub heaths cover most of the ground between the crooked trees. The ca. 100-m belt between the continuous woodland and the scattered last trees above is classically called the *Kampfzone*.

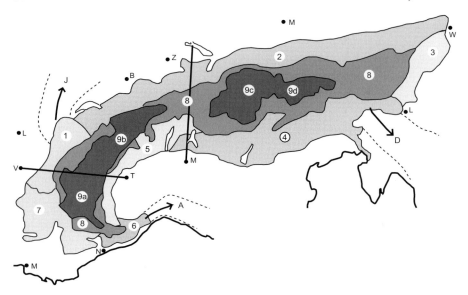

Fig. 3.4.1. Ecological zones in the Alps and biogeographical sectors based on relatedness between the Alps and the peripheral chains: *1–7* Pre-Alps sectors (with a predominance of carbonated rocks, except *3* and *5*; *1* Delphino-Jurassian sector of the southern Jura; *2* north-eastern Pre-Alps; *3* Suprapannonian sector; *4* Illyrian and Gardesan-Dolomitic sector extending into the Dinarid Mountains; *5* Insubrian-Piedmontese sector; *6* Preligurian sector extending into the northern Apennines; *7* High Provençal sector; *8, 9* sectors with a siliceous predominance and a continental climate forming the intra-alpine axis. Around the two poles of continentality (*9*) are the intermediate Alps or *Zwischenalpen* (*8*), uninterrupted in the eastern Alps but largely fragmented in the western Alps. (Ozenda 1985)

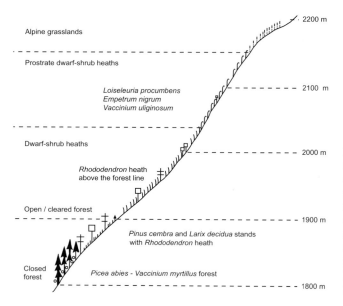

Fig. 3.4.2. Transition from woodland to dwarf-shrub heaths and the grasslands of the alpine belt in the outer crystalline massifs of the French north-western Alps. (Adapted from Richard and Pautou 1982)

The Alpine Vegetation of the Alps

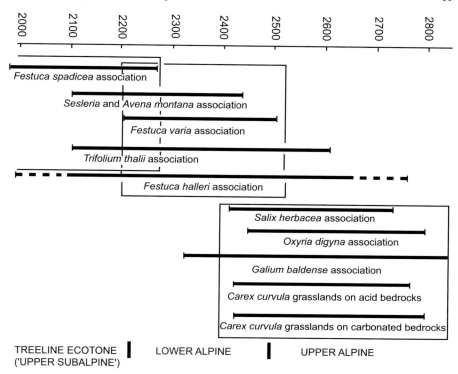

Fig. 3.4.3. Altitudinal distribution of some alpine herbaceous communities, Upper Tinée basin, Mercantour massif (from Guinochet 1938). Note that the altitudinal distribution of these communities often overlap; several of them are common to all three altitude belts indicated

Often, this natural zone is absent today because of human impact. In many massifs of the Alps, the treeline has been depressed by an estimated 200–300 m. Closed alpine grasslands (alpine vegetation belt), of all the European mountain ranges, are best developed in the Alps. They are found between ca. 2200 and 2900 m in the northern part of the French Western Alps and, southwards, between 2300 and 3000 m in the Briançonnais and in the Mercantour massifs. There is no sharp delimitation between the grasslands of the treeline ecotone, where they form a mosaic with dwarf-shrub heaths, and the authentically alpine grasslands. The high altitude herbaceous vegetation occurring across a wide altitude range appears as a kind of continuum (Fig. 3.4.3).

3.4.2 Floristic Diversity

Precise estimates of the alpine flora are difficult because of the varying notion of species and differences in the interpretation of the treeline ecotone and the alpine belt. We consider the mid- and upper-alpine belts, which are above the dwarf-shrub dominated lower-alpine belt and are characterised by closed grasslands (Table 3.4.3). The species we base our estimates on are those listed in the Flora Europaea. This number and those in the classic floras differ by about 5% as micro-species of apomictic genera are largely absent or little represented at high altitudes, except for *Alchemilla* and *Hieracium*.

Chapter 5 estimated that the total flora of the treeline ecotone and the alpine zone comprised ca. 2500 species and subspecies in Europe. Our figures concern only exclusively alpine species. The total vascular plant flora of the Alps comprises 4530 plant species according to D. Aeschimann et al. (pers. comm., cited in Theurillat 1995), a figure equivalent to about 40% of the native flora of Europe (Ozenda 1985). There are two widely differing figures for the alpine flora of the Swiss Alps. Jerosch (1903) reported 420 species as opposed to the 790 of Landolt (1991). As the Swiss flora has been well known for a long time this large discrepancy is probably due to different definitions of the term alpine (both with respect to vegetation belt and species) by these two workers. For the whole of the Alps, Favarger (1972) gave 1050 species and Theurillat (1995), by extrapolating the data of Landolt (1991), estimated it to be 1000–1100. Meusel and Hemmerling (1979) listed 650 species, but their work had considerable omissions. Overall, 750–800 (of which 270 are endemics) is a likely figure.

3.4.3 Species Richness vs. Altitude

Species richness decreases with elevation in the alpine (Fig. 3.4.4) and in the nival belts (Fig. 3.4.5). Overall species richness declines more or less linearly with slight blips at the treeline, the alpine-nival ecotone and at the limits of vascular plant life. Discontinuities at the above transition zones may appear more pronounced in studies conducted at the landscape scale (Grabherr et al. 1995; Theurillat and Schlüssel 1996; Chap. 8) as opposed to species counts from floristic databases. A comprehensive study to evaluate the different approaches using different scales and stratified according to habitats is yet to be carried out.

Species richness in the nival belt has been increasing recently. Braun-Blanquet (1955, 1957) reported that the phanerogam flora of two Swiss summits of 3400-m elevation increased from eight species in 1911 to 15 species in around 1950. Hofer (1992) has shown that on 14 peaks of the Bernina group of moun-

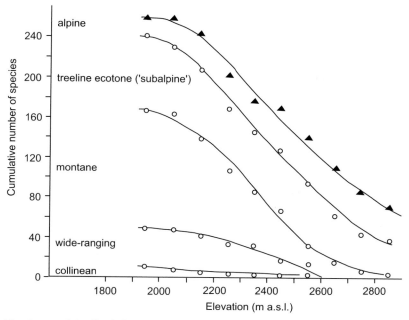

Fig. 3.4.4. Altitudinal distribution (in 100-m altitude belts) of phanerogam species in the Valais, Swiss Alps (modified from Theurillat and Schlüssel 1996, in Körner 1999). Values at each elevation are cumulative values for species groups (as listed along inside of Y-axis). Species groups are based on the classification of species altitude distributions. The number of high mountain plant species is independent of altitude while that of mid and low mountain species regularly decreases with altitude

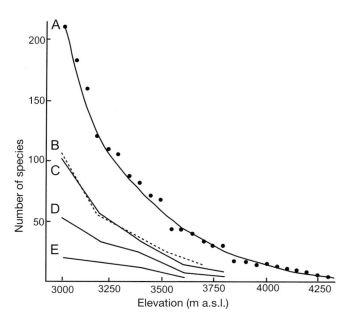

Fig. 3.4.5. Species richness of the nival flora vs. altitude in the eastern (*A*) and Oetztal (*B–E*) Alps. *A* nival phanerogamic flora of the whole eastern Alps (modified from Grabherr et al. 1995); *B* flowering plants; *C* lichens; *D* mosses; *E* hepatics. (After Reisigl and Pitschmann 1954a,b; Pitschman and Reisigl 1958)

tains, the mean number of plant species increased on average from 16 to 28 between 1920 and 1990. The concurrent increase in mean summer temperature was about 0.6 K. Grabherr et al. (1994), based on observations at 26 summits >3000 m of the Eastern Alps, estimated that species established upslope by 0.1–0.4 m year^{-1} over a 90-year period in the twentieth century. Contemporaneous annual temperature increase was estimated to be 0.7 K.

3.4.4 Endemism

The fundamental study of Pawlowski (1970) in the Alps and in the Carpathians remains the essential basis on endemism (Table 3.4.1). About two-thirds of the species in Table 3.4.1 occur in the high mountains.

The *Campanula*, *Gentiana* (s. l.), *Primula*, *Salix* and *Saxifraga* have the highest proportion of endemics of those genera which are of the greatest importance for the composition of the alpine flora in the Alps. *Androsace* is an additional one in the Pyrenees. The siliceous massifs appear poorer than the calcareous ones; however, this difference is probably not linked to edaphic conditions, but rather to the continuity and the homogeneity of the siliceous Inner Alps landscape as opposed to the fragmented nature of the calcareous Pre-Alps.

The pioneer communities are by far the richest in endemics; 35–40 % of the endemic species are on rocks and screes. One-third of the endemic genera and sections (50 % in the Maritime Alps) are rupicolous (Pawlowski 1970).

Table 3.4.1. The total number of the endemic species of the Alps, all vegetation belts put together. (Pawlowski 1970)

	Endemics	Subendemics	Total
Western Alps	148	16	165
Eastern Alps	173	30	203
Central Alps	9	1	10
Panalpine endemics	66	8	74
Total	396	55	452

3.4.5 Taxonomic Diversity

The systematic composition of the alpine flora is obviously quite different from that of the lowland regions. The Brassicaceae and Cyperaceae are relatively over-represented in the alpine belt. In the same way, 80 % of the European Primulaceae are confined to high mountains. Conversely, Lamiaceae and Apiaceae are under-represented. The genera *Androsace, Artemisia, Astragalus, Campanula, Cerastium, Draba, Gentiana* (s. l.), *Pedicularis, Phyteuma, Potentilla, Primula, Ranunculus, Salix, Saxifraga* and *Viola* make up about 30 % of the flora of the alpine belt; 30 % in the Swiss Alps, 34 % in the Valais, 33 % in the Vanoise massif, 27.5 % in the Scandes and 32.4 % in the list of Meusel and Hemmerling (1979) for the whole of the Alps. Forty-six percent of the 102 angiosperms that occur above 3000 m in the Ötztal belong to only four families (Asteraceae, Brassicaceae, Caryophyllaceae and Poaceae); these families have a 25 % share of the flora in the lowlands. The combined contribution of the Fabaceae, Lamiaceae, Liliaceae and Rubiaceae is 3 % above 3000 m as opposed to 20 % in the lowlands. Although the nival belt is physiognomically dominated by the dicotyledonous cushion plants, their proportion to monocotyledons is the same as in the lowlands. The fundamental difference between lowland and nival monocots is that the latter comprise graminoids only (Cyperaceae, Juncaceae and Poaceae).

The number of cryptogams decreases with altitude similarly to phanerogams (Pitchsmann and Reisigl 1954a,b; Reisigl and Pitschmann 1958); however, altitude affects their cover relatively less. The high mountain cryptogamic flora is still incompletely listed. Lichens certainly exceed 200 species in the nival belt of the Alps chain and similarly to other chains (e.g. Scandes, Altai); there is a remarkable development of the genera *Cetraria, Parmelia* (s. l.) and *Umbilicaria*. About 100 species of algae are known from the nival belt of the Alps, mostly unicellular ground or soil-dwelling species.

3.4.6 Life Forms

The distribution of the plant species of the alpine and nival belts according to the life forms of Raunkiaer is very different from that of the lowlands (Table 3.4.2). The majority of the flora is formed by hemicryptophytes. Phanerophytes are absent and the only ligneous species are dwarf or prostrate chamaephytes (*Loiseleuria, Empetrum*). There are few annuals such as *Euphrasia minima* (which may reach 3500 m), *Gentiana tenella, Gentiana nivalis, Sagina saginoides* and *Sedum atratum*. Geophytes are rare.

Table 3.4.2. The percentage distribution of life forms at low and high altitudes in Switzerland. (Raunkiaer 1911, cited in Wielgolaski 1997)

Life form	Low elevation (<1000 m)	High elevation (>2500 m)
Phanerophytes	10	0.0
Chamaephytes	5	24.5
Hemicryptophytes	50	68.0
Geophytes	15	4.0
Therophytes	20	3.5

3.4.7 Comparison of Different Mountain Ranges – Species Richness, Floristic and Community Similarities

In the Pyrenees, there are about 420 vascular plant species in the alpine grassland belt (Ozenda 1997). This number is lower than that usually quoted. Nevertheless, on an area-to-area basis, species richness is higher in the Pyrenees than in the Alps. For the Carpathian Mountains, our estimate is about 300 alpine plant species within an area not exceeding 1000 km^2 (Chap. 3.5). In the Rila Mountains, Bulgaria, there are ca. 400 species (of which 211 are alpine) in the treeline ecotone and the alpine zone, which together extend to about 740 km^2 (Roussakova 2000). Of the 211 alpine species, 79% are present in the Carpathian Mountains, 75% in the Alps and somewhat less than 50% in the Apennines (Stojanov and Stefanov 1922). In the Central Apennines, which has some similarities with the Alps, the alpine flora includes over 200 species, of which about 50 are endemics (Tamaro 1983, 1986; Chap. 3.6).

Over 50% of the species recorded in the alpine belt of the Swiss Alps occur in the Carpathian Mountains and in the Pyrenees (Fig. 3.4.6). There is a lower percentage of species common to the Swiss Alps and the Scandes (46%), despite the marked presence of arctic-alpine elements in both areas. The vegetation is rather different, too. In the Alps, dwarf-shrub heath is restricted mostly to a narrow zone (lower alpine), immediately above the treeline ecotone. In contrast, in the Scandes, dwarf-shrub heath dominated by, e.g., *Arctostaphylos* spp., *Betula nana*, *Empetrum nigrum* ssp. *hermaphroditum*, *Loiseleuria procumbens* and *Vaccinium myrtillus* covers over one-third of the alpine belt (lower alpine). A possible explanation for the differences is the different geographical distribution of the plant species that form the bulk of the vegetation (Ozenda 1993, 2002). The *Carex*- and *Festuca*-dominated alpine grasslands of the Alps and the *Betula* and *Salix* dwarf-shrub heaths of the Scandes were independently formed and have no community types (syntaxa) in common, except in some extreme habitats (such as dominated by *Loiseleuria*, or *Carex bicolor*; Dahl 1987; Dierssen 1992). Grabherr (1995, 1997) distinguished zonal and azonal communities in the alpine belt of the mid-lati-

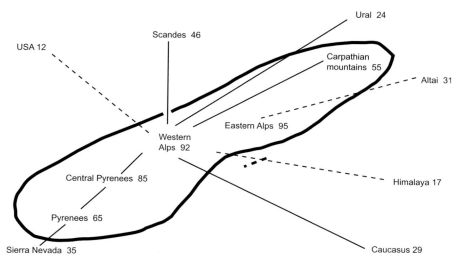

Fig. 3.4.6. Floristic relationship between the main mountain ranges (expressed as the proportion of the Swiss alpine plant species present in the other main mountain ranges). *Solid line* shows the limit of the medio-European orosystem, sensu Ozenda (1985)

tude (nemoral) mountain ranges of Europe. Zonal or climatic climax communities are closed grasslands. Azonal plant communities develop in habitats (rocks, screes and mires) that impose constraints such as space, water limitation or excess, in addition to climate. The above-mentioned differences between the nemoral and the boreal alpine belts make the zonal grasslands a type of vegetation unique to the mid-latitude mountains of Europe.

Endemic vascular plants account for about 15% of the total alpine flora in each of the above chains (sensu Favarger 1972 and dividing the Alps into a western and an eastern part). The figure, however, depends on the geographical reference used. Ozenda (1995) extended the concept of endemic species to some characteristic species of the medio-European orosystem, sensu Ozenda (1985, 2002). According to this extension, ecological similarities among the ranges of the system are reflected by endemic or subendemic supraspecific taxa such as *Soldanella, Cardamine* subgen. *Dentaria, Primula* subgen. *Auriculastrum, Gentiana* sect. *Coelanthe* and *Megalanthe, Saxifraga* sect. *Aizoonia*.

At the level of plant community types (alliances as defined by Braun-Blanquet 1954), the similarities between the alpine belt of the Alps and other ranges are: SE Carpathian Mountains 73%, Dinarids 67%, Pyrenees 59%, Bulgarian high mountains 50% (Ozenda and Borel 1994).

Table 3.4.3. Mean species richness in alpine vegetation formations in the Alps. Data after Grabherr (1999) and Grabherr and Mucina (1993)

Community type	No. of communities	Principal species	Mean number of species across communities[a]
Springs and flushes	15*	*Philonotis fontana, Saxifraga stellaris*	14–17** (6–32)***
Scree communities	40	*Thlaspi rotundifolia, Androsace alpina, Oxyria digyna, Papaver* spp.	15–25 (>5–<40)
Rock communities	35	*Potentilla caulescens, P. clusiana, Androsace helvetica, A. vandelii, Campanula* spp., *Primula* spp., *Asplenium* spp.	12–17 (>5–<30)
Snowbed communities	10	*Salix retusa, S. reticulata, S. herbacea, Cardamine alpina, Polytrichum sexangulare*	10–30 (>5–<55)
Soligenous mires	5	*Carex nigra, Trichophorum cespitosum*	20–25 (3–40)
Tall-herb meadows	15	*Cicerbita alpina, Adenostyles alliaria, Rumex alpinus*	19–24 (12–36)
Grasslands (calcareous)	25	*Carex ferruginea, C. firma, C. sempervirens, Sesleria albicans, Festuca violacea* (s.l.)	29–42 (12–61)
Grasslands (acid)	15	*Carex curvula, Nardus stricta, Festuca violacea* (s.l.), *F. varia, Agrostis schraderiana*	24–38 (12–56)
Wind-blasted	5	*Kobresia* (= *Elyna*) *myosuroides, Carex rupestris*	30–40 (20–50)
Pastures	15	*Nardus stricta, Festuca nigrescens, Poa alpina, P. supina*	10–40 (5–60)
Dwarf-shrub heath	6	*Rhododendron ferrugineum, Empetrum nigrum, Vaccinium uliginosum, Juniperus communis* ssp. *nana, Arctostaphylos alpinus, Loiseleuria procumbens*	30–40 (20–60)
Treeline ecotone scrub	5	*Pinus mugo, Alnus viridis, Salix helvetica*	15–20 (10–30)

[a]* Communities are equated with associations according to the central European classification (e.g. Grabherr and Mucina 1993) with values extended to the whole of the Alps; ** range of mean; *** minimum to maximum.

References

Braun-Blanquet J (1955) Die Vegetation des Piz Languard, ein Masstab für Klimaänderungen. Svensk Bot Tidskr 49:1–9

Braun-Blanquet J (1957) Ein Jahrhundert Florenwandel am Piz Linard (3414 m), Volume Jubilaire W. Robyns. Bull Jard Bot Bruxelles 137:221–232

Dahl E (1987) Alpine-subalpine plant communities of south Scandinavia. Phytocoenologia 15:455–484

Dierssen K (1992) Zur Synsystematik nordeuropäischer Vegetationstypen. 1. Alpine Vegetation und floristisch verwandte Vegetationseinheiten tieferer Lagen sowie der Arktis. Ber Reinh Tüxen Ges 4:191–226

Favarger C (1972) Endemism in the mountain floras of Europe. In: Valentine DH (ed) Taxonomy, phytogeography and evolution. Academic Press, London, pp 191–204

Gensler GA (1946) Der Begriff der Vegetationszeit. Engadin Press, Samedan

Grabherr G (1995) Alpine vegetation in a global perspective. In: Box EO, Peet RK, Masuzawa T, Yamada I, Fujiwara K, Maycock PF (eds) Vegetation science in forestry. Kluwer, Dordrecht, pp 441–451

Grabherr G (1997) The high-mountain ecosystems of the Alps. In: Wilgolaski FE (ed.) Ecosystems of the world, part 3. Polar and alpine tundra. Elsevier, Amsterdam, pp 97–121

Grabherr G, Gottfried M, Gruber A, Pauli H (1995) Patterns and current changes in alpine plant diversity. In: Chapin FS III, Körner C (eds) Arctic and alpine biodiversity. Springer, Berlin Heidelberg New York, pp 167–181

Grabherr G (1999) Vascular plant species richness of plant communities in the eastern Alps in relation to regional diversity. ALPNET News Issue 2, pp 7–9 (http://website.lineone.net/~mccassoc1/alpnet/2plant.PDF)

Grabherr G, Mucina L (1993) Die Pflanzengesellschaften Österreichs. Teil II: Natürliche waldfreie Vegetation. Fischer, Jena

Grabherr G, Gottfried M, Pauli H (1994) Climate effects on mountain plants. Nature 369:448

Guinochet M (1938) Etudes sur la végétation de l'étage alpin dans le bassin supérieur de la Tinée (Alpes Maritimes), Thèse de doctorat és sciences, Station Internationale de Géobotanique Méditerranéenne et Alpine, Montpellier, Communication No. 59. Bosc Frères M et L Riou, Lyon

Hofer HR (1992) Veränderungen in der Vegetation von 14 Gipfeln des Berninagebietes zwischen 1905 und 1985. Ber Geobot Inst Eidgenössisches Tech Hochschule Stiftung Rübel 58:39–54

Jerosch MC (1903) Geschichte und Herkunft der schweizerischen Alpenflora. Eine Übersicht über den gegenwärtigen Stand der Frage. Verlag von Wilhelm Engelman, Leipzig

Körner C (1999) Alpine plant life. Functional plant ecology of high mountain ecosystems. Springer, Berlin Heidelberg New York

Landolt E (1991) Gefährdung der Farn- und Blütenpflanzen in der Schweiz mit gesamtschweizerischerischen und regionalen roten Listen. Selbstverlag, Bern

Meusel W, Hemmerling J (1979) Pflanzen zwischen Schnee und Stein. Verlag Leipzig, Leipzig

Ozenda P (1985) La végétation de la chaîne alpine dans l'espace montagnard européen. Masson, Paris [German edition: Die Vegetation der Alpen im europäischen Gebirgsraum, Gustav Fischer Verlag, Stuttgart, 1987]

Ozenda P (1993) Etage alpin et Toundra de montagne: parenté ou convergence? Fragmenta Floristica Geobot Suppl 2:457–471

Ozenda P (1995) L'endémisme au niveau de l'ensemble du Système alpin. Acta Bot Gall 14:753–762

Ozenda P (1997) Aspects biogéographiques de la végétation des hautes chaînes. Biogeographica 73:145–179

Ozenda P (2002) Perspectives pour une géobiologie des montagnes. Presses Polytechniques et Universitaires Romandes, Lausanne

Ozenda P, Borel J-L (1994) Biocoenotic diversity patterns in the alpine belt of the mountains in western and central Europe. In: Colloques Phytosciologiques n° XXIII 'Large area vegetation surveys'. Cramer, Stuttgart, pp 723–735

Pawlowski B (1970) Remarques sur l'endémisme dans la flore des Alpes et des Carpathes. Vegetatio 21:181–243

Pitschmann H, Reisigl H (1954a) Zur nivalen Mossflora der Ötztaler Alpen (Tirol). Rev Bryol Lichénol 23:123–131

Pitschmann H, Reisigl H (1954b) Beiträge zur nivalen Flechtenflora der Ötztaler und Ortleralpen. Rev Bryol Lichénol 24:138–143

Reisigl H, Pitschmann H (1958) Obere Grenzen von Flora und Vegetation in der Nivalstufe der zentralen Ötztaler Alpen (Tirol). Vegetatio 8:93–129

Richard L, Pautou G (1982) Alpes du Nord et Jura méridional. Notice détaillée des feuilles 48 Annecy – 54 Grenoble, Carte de la Végétation de la France au 200 000. Editions du CNRS, Paris

Roussakova V (2000) Végétation alpine et sous alpine supérieure de la Montagne de la Rila (Bulgarie). Braun-Blanquetia 25:1–132

Stojanov N, Stefanov B (1922) Phytogeographical and floristic characteristics of the Pirin mountains. Ann Univ Sofia 18:1–27

Tamaro F (1983) Compendio sulla flora del Gran Sasso d'Italia. Monografia dei quaderni del Museo speleologia. L'Aquila 2:1–58

Tamaro F (1986) Documenti per la conoscenza naturalistica della Majella. Giunta regionale d'Abruzzo, Chieti

Theurillat J-P (1995) Climate change and the alpine flora: some perspectives. In: Guisan A, Holten JI, Spichiger R, Tessier L (eds) Potential ecological impacts of climate change in the Alps and Fennoscandian Mountains. An annex to the intergovernmental panel on climate change (IPCC). Second Assessment Report, Working Group II-C (Impacts of climate change on mountain regions), Publication hors-série n° 8 des Conservatoires et Jardin botaniques de la Ville de Genève, pp 121–127

Theurillat J-P, Schlüssel A (1996) L'écocline subalpin-alpin : diversité et phénologie des plantes vasculaires. Bull Murithienne 114:163–169

Wielgolaski FE (1997) Fennoscandian tundra. In: Ecosystems of the world, part 3. Polar and alpine tundra. Elsevier, Amsterdam, pp 27–83

3.5 The Alpine Flora and Vegetation of the South-Eastern Carpathians

G. Coldea

3.5.1 Geography and Geology

The Carpathian mountain chain was formed in the Mesozoic (Triassic and Cretaceous). It lies between latitudes 44°30'–48°N and longitudes 21°30'–21°E. The south-eastern Carpathians account for 54%, or 66,303 km², of the whole Carpathian chain. Of this, about 7000 km² (>10%) lies above the treeline.

The East Carpathians are predominantly of crystalline schists in the central part and of volcanic parent material in the western part, where their highest peak is Pietrosu (2305 m). The South Carpathians are of crystalline schists with granite intrusions and ancient limestones. They are less fragmented than the East Carpathians and comprise the highest peaks (Moldoveanu 2544 m and Negoiu 2535 m). The Apuseni Mts. do not have a clear alpine belt (highest peak Vladeasa 1835 m) and are formed from various volcanic and sedimentary rocks, with patchily distributed limestones.

Acid brown, ferriluvial, alpine meadow, lithosols, rendzinas and hydromorphic soils are found in the alpine zone.

3.5.2 The Treeline Ecotone (Subalpine) and the Alpine Zone

The potential timberline is at 1750 m altitude and is formed by *Picea abies*, *Pinus cembra*, *Larix decidua* and *Sorbus aucuparia*. In the alpine zone, arctic-alpine microthermic species such as *Carex curvula*, *Juncus trifidus*, *Kobresia simpliciuscula*, *K. myosuroides*, *Oreochloa disticha*, *Minuartia recurva*, *M. sedoides*, *Silene acaulis*, *Ranunculus glacialis*, *Saxifraga bryoides*, *Gentiana nivalis*, *G. frigida*, *Salix herbacea*, *S. reticulata* and endemic microthermic species such as *Lychnis nivalis*, *Dianthus glacialis* ssp. *gelidus*, *Cardaminopsis neglecta*, *Silene dinarica*, *Thesium kernerianum*, *Thymus pulcherrimus*, *Poa deylii* and *Festuca bucegensii* occur. The treeline ecotone (subalpine belt) lies between 1700 and 2200 m and is characterised by *Rhododendron myrtifolium–Pinus mugo* dwarf pine scrub and secondary grasslands of the *Scorzonera purpurea* ssp. *rosea-Festuca nigricans* and *Viola declinata–Nardus stricta* type. The vegetation of the alpine zone (2200–2544 m) is of primary grasslands (*Primula–Carex curvula* and *Oreochloa disticha–Juncus trifidus* associations), oligothermic dwarf-shrub (*Cetraria–Loiseleuria procumbens* and *Rhododendron myrtifolium–Vaccinium* associations) and late snowbed communities (*Salix herbacea* and *Soldanella pusilla–Ranunculus crenatus* associations).

Palynological studies from peat sediments from the treeline ecotone (Farcas et al. 1999) showed a cold steppic phase with *Artemisia* (60%) and Chenopodiaceae in the Younger Dryas (11,140 B.P.), before the expansion of *Pinus sylvestris* and *Alnus viridis* in the pre-Boreal (10,000 B.P.). Although the last glaciation induced major changes in the climate of the south-eastern Carpathians, species of some tree genera such as *Picea* and *Betula* survived.

3.5.3 Climate

The climate of the south-eastern Carpathians is moderately continental. The annual mean temperature in the treeline ecotone is +2 °C and the annual mean precipitation is 1100 mm year^{-1}. In the alpine zone, the annual mean temperature is –2 °C with a mean precipitation of 1400 mm year^{-1}. On the highest peaks, snow lies on average for 217 days and the length of the growing season is 148 days (Omul, 2504 m, Bucegi Mts., South Carpathians). Growing season precipitation ranges from 575 mm at the Omul peak to 690 mm Iezer,

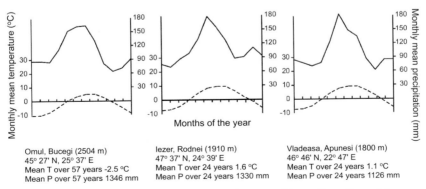

Fig. 3.5.1. Climate diagrams from the South Carpathians, East Carpathians and the Apuseni Mountains. *Top line* precipitation; *bottom line* temperature

Rodnei Mts., East Carpathians. An absolute minimum of air temperature of −38 °C was recorded in 1929 (Omul). There are 5–6 frosty days per month in summer (Fig. 3.5.1).

3.5.4 Flora

3.5.4.1 History of the Flora

The history of botanical research of the south-eastern Carpathians began with Baumgarten (1816) followed by several accounts of flora (Heuffel 1858; Schur 1866; Brandza 1883; Simonkai 1887; Grecescu 1898; Pax 1898, 1909; Savulescu 1952–1972) and vegetation (Borza 1934, 1946, 1959; Pauca 1941; Morariu 1942; Borhidi 1958; Buia 1962; Beldie 1967; Resmerita 1970; Boscaiu 1971; Coldea 1990, 1991).

The flora of the south-eastern Carpathians has, in all its vegetation belts, species of autochthonous Tertiary origin, with a dominance of boreal elements. It is also rich in meridional elements such as *Leontopodium alpinum*, *Saussurea discolor*, *Scutellaria alpina*, *Taraxacum fontanum* and *Taraxacum alpinum*, indicating connections to Mediterranean and Irano-Turanian floras (Boscaiu 1971). In the Carpathian flora, there are also several pan-Carpathic and Dacic paleoendemics, together with relicts of Balkano-Illyric and Moesic origin (Coldea 1991). All pan-Carpathic and Dacic paleoendemics date from the Tertiary. The high florogenetic potential of the Carpatho-Danubian basin is demonstrated by the predominance of active endemisms (e.g. apomictic taxa such as *Hieracium* spp.) over passive endemisms (palaeoendemic taxa). Some of the Carpathian endemic species, such as *Syringa josikaea*, *Hepatica transsilvanica* and *Veronica bachofenii*, show close affinities with species from eastern and central Asia (*Syringa wolfii*, *Hepatica falkoneri* and *Veronica dahurica*). The links with the Asiatic flora were established via the Balkanic massif between the Eocene and the Miocene (Pontian), when the Aegean mainland submerged and the Balkans became a peninsula. These paleogeographic connections explain the presence of Daco-Balkano-Caucasian connective elements (*Carex dacica*, *Scleranthus uncinatus*, *Bruckenthalia spiculifolia*, *Plantago gentianoides*; Boscaiu 1971). The present flora of the south-eastern Carpathians with about 18% alpine and alpine-arctic elements was strongly affected by the Pleistocene Ice Ages when the extreme catatherms drastically reduced the autochthonous arcto-tertiary stock.

3.5.4.2 Number of Taxa

A total of 4190 plant species have been recorded in the south-eastern Carpathians with 1650 native flowering plants, 65 pteridophytes, 600 bryophytes, and 1875 lichens. Above the timberline (1750 m asl), there are 663 flowering plant species, 9 pteridophyte species, 65 bryophyte species and 218 lichen species. The flowering plants belong to 246 genera of 64 families, the pteridophytes 13 genera of eight families, the bryophytes 46 genera of 25 families, and the lichens 53 genera of 25 families. Some of the most species-rich flowering plant families are Asteraceae (99 species), Poaceae (53), Caryophyllaceae (47), Brassicaceae (35), Rosaceae (32), Scrophulariaceae (32), Cyperaceae (31) and Ranunculaceae (29).

The family abundance of Rosaceae and Fabaceae is much reduced with altitude. They are represented by two (Rosaceae) and three (Fabaceae) species in the alpine zone as compared with 30 and 20 species in the treeline ecotone. The Saxifragaceae show less decrease (8 species in the alpine compared with 10 in the treeline ecotone).

The Raunkiaer life forms show a clear dominance of hemicryptophytes (70%), followed by chamaephytes (13.6%), geophytes (7.5%), therophytes (6.5%) and phanerophytes (2.4%). The dominance of hemicryptophytes and chamaephytes in the alpine zone and in the upper part of the treeline ecotone, and the low percentage of therophytes and phanerophytes, mirrors the potential area of primary alpine grasslands. In the alpine zone (2200–2540 m), 130 flowering plant and 34 bryophyte species occur, 50 of the flowering plants being exclusive to this zone.

Of all cormophyte and bryophyte species occurring above the timberline, 33% are common and 27% are abundant; 40% are rare boreal-alpine and endemic species that are present only in some localities in the alpine zone. Of all 127 endemic flowering plant taxa of Romania (Morariu and Beldie 1976), 59% (48 species and 22 subspecies) are found in the plant communities above the timberline. The endemic pan-Carpathian as well as the endemic south-eastern Carpathian species have a special phytogeographical and syntaxonomic importance as they are found in all vegetation types above the timberline.

3.5.5 Vegetation

Eighty communities have been described from the treeline ecotone and the alpine zone of the south-eastern Carpathians. The large majority of them contain Carpathian and Carpatho-Balkanic endemics, which is why they are considered vicariant associations of those described from central and western

Table 3.5.1. Characteristics of the vegetation in the treeline ecotone and in the alpine zone in the south-eastern Carpathians

Vegetation type	No. of plant communities	Altitude range (m asl)	Minimum and maximum community species richness with endemics in parentheses	Shannon index (H)
Scrub				
Acidophilous	4	1450–2200	44 (2)–88 (6)	1.45–2.53
Basiphilous	1	1700–1960	83 (11)	2.89
Dwarf-shrub heath				
Acidophilous	4	1650–2400	23 (3)–58 (5)	1.36–1.65
Grasslands				
Acidophilous	7	1600–2500	31 (7)–67 (13)	1.31–1.74
Basiphilous	12	1600–2500	26 (5)–92 (18)	1.33–2.38
Tall-herb meadows				
Acidophilous	9	1400–2500	39 (8)–87 (16)	1.45–3.62
Ombrogenous and soligenous mires				
Acidophilous	3	1800–2300	25 (1)–50 (2)	1.77–2.71
Basiphilous	3	1500–1800	36 (2)–65 (0)	2.26–3.43
Springs and flushes				
Acidophilous	4	1400–2200	26 (1)–32 (1)	1.65–2.63
Basiphilous	1	1400–1900	55 (8)	2.30
Late snowbed communities				
Acidophilous	7	1650–2500	19 (2)–42 (4)	1.54–2.65
Basiphilous	2	1800–2400	36 (3)–58 (8)	2.26–2.85
Rock communities				
Acidophilous	3	1550–2350	8 (1)–18 (2)	0.59–1.58
Basiphilous	6	1400–2150	21 (4)–49 (8)	0.97–2.93
Scree communities				
Acidophilous	5	1800–2500	34 (1)–44 (3)	1.43–2.07
Basiphilous	5	1600–2200	26 (4)–61 (6)	1.63–2.60
Open cryptogamic communities				
Acidophilous	4	2200–2500	26 (1)–60 (1)	2.46–4.68

Europe (Coldea 1991). The 80 communities fall into 10 community types (Table 3.5.1), each having a distinct physiognomy and ecology (Coldea and Cristea 1998).

Scrub, tall grass meadows and grasslands occupy large areas, have a zonal distribution and have the highest species richness (a mean value of 64). Spring communities and late snowbeds, all of which occupy small areas and have an azonal distribution, have lower species richness (mean 35). Rock communities include a lower number of species (mean 29) than scree communities (mean 38).

Within the same functional type, acidophilous plant communities have a lower number of species (both total and endemic) than basiphilous ones. A larger number of endemic species (mean 10) are found in grasslands and tall-herb meadows, than in spring and meadow communities (1–2 species).

The Shannon diversity index (H) calculated for relevés is dependent on the number of species per relevé and is higher in the communities dominated by 3–4 species than in those with a single dominant. The diversity index in basiphilic communities (i.e. those on calcareous or dolomitic parent material) is always higher than in acidophilic communities, regardless of functional community type.

In common high mountain azonal habitats, such as snowbeds, springs and peat bogs, the plant communities are similar to those described from the Austrian Alps (Grabherr and Mucina 1993) and from the northern Carpathians (Krajina 1933; Mucina and Maglocky 1985). In contrast, the acid alpine grasslands and rock communities from the south-eastern Carpathians are more similar to the communities described from the Balkans (Horvat et al. 1937; Simon 1958).

The xerophytic and chionophobic (harsh winter with little snow cover) saxicolous lichen communities (*Aspicilia cinerea* and *Umbilicaria hirsuta*) are less species-rich and have a lower diversity index value than the hygrophytic, chionophilous communities (*Acorospora sinopica* and *Rhizocarpon alpicola* communities; Ciurchea and Crisan 1993). The primary lichen associations with prevalently crustaceous thalli (*Aspicilia cinerea* and *Acorospora sinopia* communities) evolve naturally into lichen communities of an *Umbilicaria* and *Parmelia* type (*Umbilicaria hirsuta* community) with prevalently foliaceus thalli. They contribute to weathering and eventually the development of lithosols where saxicolous bryophyte and cormophyte associations can establish.

3.5.6 Human Impacts vs. Diversity

Land clearing for pastures in the nineteenth century reduced the area of the potential scrub vegetation (*Rhododendron myrtifolium–Pinus mugo*, *Campanula abietina–Juniperus communis* ssp. *nana* and *Salix silesiaca–Alnus viridis* communities) in the upper part of the treeline ecotone of the south-eastern Carpathians by ca. 50–60%. This initiated soil erosion and increased floods in the lowlands (Coldea 1990).

In the last decade of the twentieth century, a general warming process was recorded. Air temperature in the treeline ecotone and in the alpine zone has risen by 1.4–1.5 K, while precipitation decreased by 200–400 mm as compared with multi-annual means. According to NASA forecasts, the global climate will become warmer by another 1.5 K by 2020 (Petrescu 1990). This may have dramatic impacts on the survival of several narrowly distributed arctic-alpine and endemic orophytic species. Species present in a sole location in the south-eastern Carpathians such as *Andryala levitomentosa*, *Cardaminopsis neglecta*, *Carex bicolor*, *C. lachenalii*, *Juncus castaneus*, and *Ranunculus glacialis* are in particular likely to disappear.

The decrease in the extent of ombrogenous and soligenous mire vegetation recorded since the 1980s indicates that these vegetation types may be sensitive indicators of climate change impacts in the Carpathians and their use for monitoring may be considered.

References

Baumgarten G (1816) Enumeratio Stirpium Magno Transsilvaniae Principatui. Ed. Camesinae, Vindobonae

Beldie A (1967) Flora si vegetatia Muntilor Bucegi. Ed. Academiei, Bucuresti

Borhidi A (1958) Gypsophilion petraeae foed. nova et contributions à la végétation du Mont Ceahlau (Carpathes Orientales). Acta Bot Sci Hung 4:211–231

Borza A (1934) Studii fitosociologice in Muntii Retezatului. Bull Grad Bot Cluj 14:1–84

Borza A (1946) Vegetatia muntelui Semenic din Banat. Studii fitosociologice. Bull Grad Bot Cluj 26:24–53

Borza A (1959) Flora si vegetatia vaii Sebesului. Ed. Academiei, Bucuresti

Boscaiu N (1971) Flora si vegetatia Muntilor Tarcu, Godeanu si Cernei. Ed. Academiei, Bucuresti

Brandza D (1883) Prodromul Florei Romane. Ed. Academie, Bucuresti

Buia A (1962) Pajistile din Masivul Parang si inbunatatirea lor. Ed. Agro-Silvica, Bucuresti

Ciurchea M, Crisan F (1993) Vegetatia lichenologica saxicola din Rezervatia stiintifica a Parcului National Retezat. In: Popovici I (ed) Parcul National Retezat. Studii Ecologice Ed West Side Brasov, Brasov, pp 58–77

Coldea G (1990) Muntii Rodnei. Studiu geobotanic. Ed. Academiei, Bucuresti

Coldea G (1991) Prodrome des associations végétales des Carpates du sud-est (Carpates roumaines). Doc Phytosoc 13:317–539

Coldea G, Cristea V (1998) Floristic and community diversity of subalpine and alpine grasslands and grazed dwarf-shrub heaths in the Romanian Carpathians. Pirineos 151/152:73–82

Farcas SD, Beaulieu J-L, Reille M, Coldea G, Diaconeasa B, Goeury C, Goslar T, Jull T (1999) First ^{14}C datings of late glacial and Holocene pollen sequences from Romanian Carpathes. CR Acad Sci Paris Sci 322:799–807

Grabherr G, Mucina L (1993) Die Pflanzengesellschaften Österreichs. Teil II. Natürliche Waldfreie Vegetation. Fischer Verlag, Jena

Grecescu D (1898) Conspectul Florei Romaniei. Ed. Dretpatea, Bucuresti

Krajina V (1933) Die Pflanzengesellschaften des Mlynica-Tales in den Vysoke Tatry (Hohe Tatra) I, II. Beih Bot Centralbl 50:774–957; 51:1–224

Heuffel J (1858) Enumeratio plantarum in Banatu Temisensis sponte crescentium et frecquentius cultorum. Caroli Ueberreuter Verlag, Vindobonae

Horvat I, Pawlowski B, Walas J (1937) Phytosoziologische Studium uber die Hochgebirgsvegetation der Rila Planina in Bulgarien. Bull Acad Polon Sci Lett Ser B Sci Nat 8:159–189

Morariu I (1942) Vegetatia Muntelui Tibles. Bull Soc Reg Geogr 61:143–180

Morariu I, Beldie A (1976) Endemismele din Flora Romaniei. In: Beldie A, Morariu I (eds) Flora R S Romaniei, vol 13. Ed. Academiei, Bucuresti, pp 97–105

Mucina L, Maglocky S (1985) List of vegetation units of Slovakia. Doc Phytosoc 9:175–220

Pauca A (1941) Studiu fitosociologic in Muntii Codru si Muma. Ed. Academiei, Bucuresti

Pax F (1898, 1909) Grundzüge der Pflanzenverbreitung in den Karpathen. vols I, II. Wilhelm Engelmann Verlag, Leipzig

Petrescu I (1990) Perioadele glaciare ale pamintului. Ed. Tehnica, Bucuresti

Resmerita I (1970) Flora, vegetatia si potentialul productiv pe masivul Vladeasa. Ed. Academiei, Bucuresti

Savulescu T (1952–1972) Flora Republicii Romane. Ed. Academiei, Bucureti

Schur F (1866) Enumeratio Plantarum Transsilvaniae. Wilhelm Braumüller Verlag, Vindobonae

Simon T (1958) Über die alpinen Pflanzegesellschaften des Pirin-Gebirges. Acta Bot Hung 4:159–189

Simonkai L (1887) Enumeratio florae Transsilvanicae vasculosae critica. Királyi Magyar Természettudományi Társulat Kiadó, Budapest

3.6 The High Mountain Flora and Vegetation of the Apennines and the Italian Alps

F. Pedrotti and D. Gafta

3.6.1 Geography

The chain of the Apennines extends from the Cadibona Pass (on the border between Piedmont and Liguria) to the Aspromonte in Calabria. The treeline varies from 1900 to over 2000 m according to exposure. Twenty-one mountains have summits over 1900 m (mostly in the Central Apennines) and the areas above treeline cover about 5% (Fig. 3.6.1).

The Italian Alps (henceforth referred to as Alps) extend from Liguria (Maritime Alps) to Venetia Julia (Julian Alps) and are bordered in the south by the Po Valley. The climatic potential treeline ranges from 1850 to 2100 m depending on exposure, position within the Alpic chain and on geomorphology. About 40% of the total mountain area lies above the treeline (Fig. 3.6.1).

All areas above the treeline belong to the Euro-Siberian subregion, except for Sirino, Pollino, Sila and Aspromonte massifs of the Apennines which are Mediterranean (Rivas-Martinez 1996). The lower alpine belt is characterised by evergreen scrub and heath, while the upper alpine belt is marked by more or less closed grasslands.

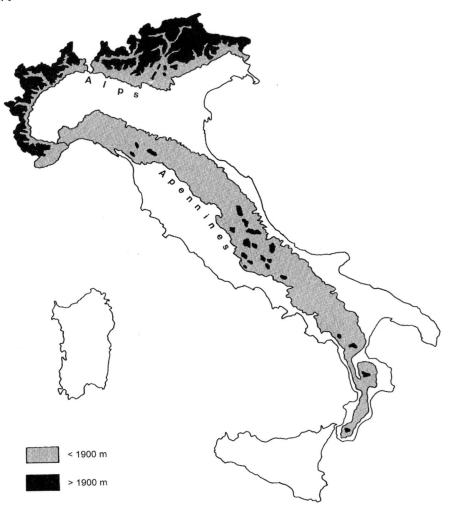

Fig. 3.6.1. Geographic distribution of the mountain areas of the Italian Alps and the Apennines; areas above 1900 m a.s.l. (the approximate elevation of the treeline) are shown in *black*

3.6.2 Geology and Climate

The Apennines are predominantly calcareous (Jurassic and Cretaceous), except for the northern Apennines and the Laga Mountains in the central Apennines, which are marly-arenaceous (Miocene and Eocene). The Alps contain a large variety of sedimentary (limestone, dolomite), metamorphic (gneiss, schist) and igneous (porphyry, granite) parent material.

The climate of the treeline ecotone and alpine belt in the Apennines has a suboceanic character because of a sufficient rainfall during the entire year, except for a pronounced minimum during July (Biondi and Baldoni 1994). The mean duration of the snow cover at Campo Imperatore (2137 m a.s.l.), the only high-altitude meteorological station in the Apennines, is 187 days year^{-1}, between October and May (Baldoni et al. 1999). The extent of Calderone, the only glacier in the Apennines, has nearly halved between 1794 and 1990 (D'Orefice et al. 2000).

The climate of the Alps differs longitudinally, with a higher mean rainfall (3000 mm year^{-1}) in the east (Friulian Alps) than in the west. Latitudinally, there is a strong, complex continentality gradient from the Pre-Alps to the inner Central Alps (Gafta and Pedrotti 1998; Figs. 3.6.2 and 3.6.3).

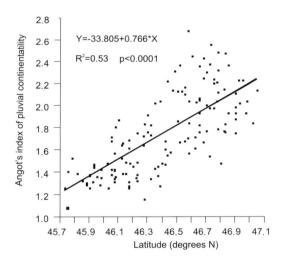

Fig. 3.6.2. Linear fit of the pluvial continentality index (Angot 1906) as a function of latitude in the central Italian Alps (redrawn from Gafta and Pedrotti, 1998). The index is the ratio of the sum of mean rainfall in the warmest 6 months to that in the coldest 6 months in a year. Values over 1.5 indicate continental climates such as those that characterise the inner Alps

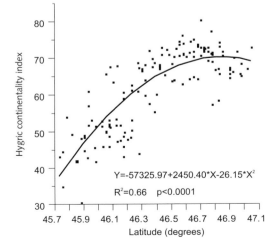

Fig. 3.6.3. Parabolic fit of the modified hygric continentality index (Michalet 1991) in winter as a function of latitude in the central Italian Alps. The index is a function of elevation and winter precipitation. Values larger than 65° are typical for the inner Alps valleys where, unlike in the Pre-Alps, a low amount of precipitation falls in winter

The average duration of the growing season throughout the Italian Central Alps is about 76 days (>+10 °C air temperature) at the treeline, decreasing with elevation at a mean rate of 8.5 days 100 m^{-1} (Gafta and Pedrotti 1998). The average duration of the snow cover on the Ortles-Cevedale mountain group (Central Alps) increases quadratically with altitude, giving an estimate of 3050 m for the lower limit of the permanent snow in the early 1960s (Desio 1967). Today's snow line is at about 3200 m.

A climate warming is evidenced by the dynamics of the glaciers in the Alps (Belloni 1992). We found that those in the Ortles-Cevedale retreated by a mean altitude of 105 m between 1926 and 1961 (t=7.4; $p<0.0001$, paired-t test). The rate of retreat across the whole of the Italian Alps was 106.5 m between 1980 and 1998 ($n=18$, t=17, $p<0.0001$, paired-t test).

3.6.3 History of the Flora

The flora of the Apennines originated in two distinct zones, which were located to the east and west in respect to the current mountain chain ridge. The endemic species of the Apennines, particularly on the massifs of Abruzzo, differentiated during the Tertiary, simultaneously with the uplift of the mountain chain. Glacial relicts such as *Dryas octopetala* and *Salix herbacea* are among the most recent recruitment to the Apennine flora.

The arcto-tertiary flora of the Alps was for the most part replaced by an arctic-alpine flora (e.g. *Carex, Pulsatilla, Loiseleuria*) during the Ice Ages with the survival of some so-called glacial relicts (e.g. *Linnaea borealis, Trientalis europaea*). On the other hand, reproductive isolation of plant populations in unglaciated parts of the Italian Alps, especially in the Pre-Alps, caused a high degree of speciation (conservative endemisms and vicariant species).

3.6.4 Today's Flora

In the Apennines, including all elevations, 3091 vascular plant species occur out of the total of 5599 species-rich Italian flora (Pignatti 1982). In the treeline ecotone and above there are 728 (23.6%) vascular species, including some widespread Mediterranean species that occur from sea level.

Among the Phanerophytes the percentage of endemisms is about 15%, slightly higher than the 13% calculated for the central Apennines by Favarger (1986). The richest massifs are Gran Sasso, Majella and Monte Pollino, where species such as *Adonis distorta, Viola magellensis* occur.

The treeline ecotone flora of the Apennines is characterised by contorted trees and/or dwarf shrubs (12.4% of all Apennine species) that are also found in the Alps, while *Juniperus communis* ssp. *hemisphaerica* occurs only on Monte Pollino. The tree species most representative of the contorted growth form, *Pinus mugo*, has disappeared entirely from the Laga, Gran Sasso and Sibylline Mountains, where its occurrence in the past has been evidenced by pollen analyses (Marchesoni 1959). Under direct or indirect human influence, many species of the montane belt, such as *Bromus erectus*, are now widely found in the treeline ecotone and even in the lower alpine belt.

Of the 497 bryophytes occurring in the Apennines, only 150 (30.2%) are present above the treeline. Among the rarest bryophyte species are *Pohlia andalusica* (Mainarde Group), *Lophozia opacifolia* and *Polytrichum sexangulare* (Laga Mountains), and *Asterella gracilis* (Abruzzo region).

Over half the vascular flora of Italy, that is 3264 species, occurs in the Alps region (Pignatti 1982). Of these, 1122 (34.4%) are found above the treeline. Among the most species-rich genera are *Carex* (55), *Saxifraga* (41), *Gentiana* (25), *Primula* (19), *Pedicularis* (19), *Potentilla* (18), *Campanula* (18), *Moehringia* (17), *Androsace* (13) and *Silene* (12).

About 21% of the vascular species are endemic (Table 3.6.1). This is two-thirds of the percentage calculated by Favarger (1972) for the whole of the Alps (31%), indicating that the highest level of speciation happened on the southern border (Ligurian Alps, Lombardian Pre-Alps, Benacensi Pre-Alps, Dolomites and Julian Alps). Among the rarest species, with limited distribution, are *Saxifraga florulenta* (Maritime Alps), *Sanguisorba dodecandra* (Valtellina), *Linaria tonzigii* (Monte Arera), *Daphne petraea* (Monte Tremalzo), and *Rhizobotrya alpina* (Dolomites).

There are 1032 species of bryophytes in the Italian Alps, which is 81.7% of the Italian bryoflora (Cortini Pedrotti 1992; Aleffi and Schumacker 1995). Above the treeline 439 species occur, that is a number proportional to the area above treeline. Only two bryophytes – *Radula visianica* and *Riccia breidleri* – are considered endemic. Among the rarest species, present only in few stations are *Voitia nivalis*, *Oreas martiana* and *Paludella squarrosa* (Cortini Pedrotti 1979; Schumacker et al. 1999).

The areas above treeline in the Apennines and the Alps have many species in common and there are some evident similarities in species pattern. However, it is important to note that the southern limit of distribution of some common high-mountain species is in the northern Apennines (*Rhododendron ferrugineum*, *Vaccinium vitis-idaea*, *Luzula spicata*). The floristic affinities between the Alps and Apennines are also evidenced by the vicariant species pairs distributed above treeline, e.g. *Juniperus communis* ssp. *nana*–*J. communis* ssp. *hemisphaerica*, *Festuca violacea*–*F. violacea* ssp. *italica* (central Apennines) or ssp. *puccinellii* (N Apennines), and *Sesleria albicans*–*S. apennina* (Pignatti 1994).

The pteridophytes are weakly represented in both mountain chains (about 4% above the treeline), which indicates that they are less capable than the

Table 3.6.1. Taxonomic richness above the treeline in the Italian Alps and the Apennines

	Families	Genera	Species	Endemics (%)
Alps				
Bryophyta	79	223	439	2 (0.5)
Pteridophyta	12	20	45	0
Spermatophyta	61	268	1077	227 (21.1)
Apennines				
Bryophyta	58	179	150	0
Pteridophyta	12	19	31	0
Spermatophyta	53	233	697	106 (15.2)

phanerogams of adapting to microthermic conditions. There are no endemic pteridophytes (Table 3.6.1).

Hemicryptophytes dominate in both the treeline ecotone and the alpine belt, followed by chamaephytes (alpine) and geophytes (treeline ecotone; Table 3.6.2). The higher percentage of the chamaephytes in the alpine belt than in the treeline ecotone is in part due to the numerous cushion species that occur at high altitudes (Table 3.6.3). Some of the therophytes (*Poa annua, Capsella bursa-pastoris, Medicago lupulina, Senecio viscosus*) are synanthropic species that have spread from the montane belt following sheep grazing. For similar reasons many grass species (*Nardus stricta, Agrostis capillaris, Anthoxanthum odoratum, Bromus erectus*) are found today in pastures of the treeline ecotone.

The species exclusively found within the treeline ecotone are more numerous than those confined to the alpine belt (Table 3.6.3). However, almost half the species above the treeline occur in bot the treeline ecotone and in the alpine belt (41.9% for the Apennines and 46.2% for the Alps). This is because of the diffusion of some alpine species into the disturbed habitats of the treeline ecotone. There are numerous cushion-forming species in the alpine belt; however, the specific richness of the graminoids is on the whole the highest (Table 3.6.3).

Irrespective of the taxonomic group considered, there are few abundant species (with a recorded cover >25%; Table 3.6.4). The percentage of rare species is generally higher in the Apennines. For pteridophytes the most common species of the Alps are the rarest in the Apennines. This seems to be consistent with the higher occurrence of mesic (moist) habitats in the Alps.

Table 3.6.2. Life form structure of the vascular flora above the treeline in the Italian Alps and the Apennines. *T* therophytes; *Hc* hemicryptophytes; *G* geophytes; *Ch* chamaephytes; *Ph* phanerophytes; *Hd* hydrophytes

	Total	T	Hc	G	Ch	Ph	Hd
Alps							
Total number of species	1122 (100)	41 (3.6)	763 (68.0)	113 (10.1)	165 (14.7)	31 (2.8)	9 (0.8)
Number of species exclusive to the treeline ecotone (%)	340 (100)	15 (4.4)	215 (63.3)	46 (13.5)	33 (9.7)	27 (7.9)	4 (1.2)
Number of species exclusive to the alpine belt (%)	264 (100)	8 (3.0)	181 (68.6)	17 (6.4)	57 (21.6)	0	1 (0.4)
Apennines							
Total number of species (%)	728 (100)	30 (4.1)	486 (66.8)	80 (11.0)	104 (14.3)	27 (3.7)	1 (0.1)
Number of species exclusive to the treeline ecotone (%)	233 (100)	11 (4.7)	139 (59.7)	33 (14.2)	22 (9.4)	27 (11.6)	1 (0.4)
Number of species exclusive to the alpine belt (%)	190 (100)	4 (2.1)	131 (69.0)	16 (8.4)	39 (20.5)	0	0

Table 3.6.3. Growth form structure of the vascular flora above the treeline in the Italian Alps and Apennines

	Total	Tree	Shrub	Graminoid	Forb	Cushion	Other
Alps							
Treeline ecotone species (%)	340 (100)	5 (1.5)	31 (9.1)	34 (10.0)	21 (6.2)	9 (2.6)	240 (70.6)
Alpine species (%)	264 (100)	0 (0.0)	3 (1.1)	22 (8.3)	2 (0.8)	46 (17.4)	191 (72.4)
Apennines							
Treeline ecotone species (%)	233 (100)	3 (1.3)	29 (12.4)	12 (5.0)	14 (5.8)	2 (0.8)	173 (74.2)
Alpine species (%)	190 (100)	0	3 (1.6)	26 (13.7)	0	25 (13.2)	136 (71.5)

Table 3.6.4. Estimation of species occurrence and abundance in the Italian Alps and Apennines. *Common* rather frequent occurrence; *abundant* constantly dominant in at least one plant community type; *rare* occasional occurrence

	Common	Abundant	Rare
Alps			
Bryophyta (%)	194 (44)	72 (17)	173 (39)
Pteridophyta (%)	26 (58)	6 (13)	13 (29)
Spermatophyta (%)	442 (41)	109 (10)	526 (49)
Apennines			
Bryophyta (%)	89 (59)	30 (20)	31 (21)
Pteridophyta (%)	9 (29)	2 (6)	20 (65)
Spermatophyta (%)	234 (34)	48 (7)	415 (59)

3.6.5 Altitudinal Vegetation Zonation

There are very few plant communities common to the two mountain chains, e.g. *Empetrum–Vaccinium gaultheroides* heath and *Salix herbacea* snowbeds. However, there are some communities that are ecologically and functionally analogous such as, for example, the *Erica–Pinus mugo* (Alps) and *Polygala chamaebuxus–Pinus mugo* (Apennines) scrub or the *Sesleria albicans–Carex sempervirens* and *Sesleria apennina* calcareous grasslands.

The treeline in the Apennines is formed by *Fagus sylvatica* alone or in mixed stands with *Abies alba* (Central Apennines) and by *Pinus leucodermis–Juniperus communis* ssp. *hemisphaericus* open woodlands (Mt. Pollino). The more pronounced arctic-alpine character of the Italian Alps compared with the Apennines is evidenced by the absence of some communities in the latter, e.g. *Larix–Pinus cembra* forest, *Salix helvetica* scrub and *Loiseleuria–Cetraria* prostrate dwarf-shrub. There are conspicuous differences between the snowbed communities of the Apennines and the Alps, the latter having by far more arctic-alpine species.

The large expanse of the anthropogenic grasslands of the *Koeleria splendens–Bromus erectus*, *Bellardiochloa variegata–Nardus stricta* and *Brachypodium–Festuca spadicea* types in the Apennines suggests that the climatic altitudinal limit of closed arborescent vegetation was higher in the past. Paone (1987) estimated that there were about 5.5 million sheep in Abruzzo, Molise and Puglia at the beginning of the seventeenth century.

There are some differences between the Pre-Alps and the Inner Alps. Thus, pure *Larix decidua* stands of the Pre-Alps are replaced by *Juniperus sabina–Larix decidua* stands in the dry inner valleys and by *Larix decidua–Pinus cembra* stands in the remaining part. Where man has destroyed the scrub in the upper part of the treeline ecotone, anthropo-zoogenic grasslands of the *Sieversia montana–Nardus stricta* type are found.

3.6.6 Floristic Diversity and Distinctiveness of Communities

In agreement with the general pattern observed in high mountain areas, vascular species richness decreases with increasing altitude; however, this relationship is not monotonic because of a conspicuous maximum reached in the alpine grasslands (Tables 3.6.5 and 3.6.6). Within the same functional community type, both species richness and the proportion of endemics are generally higher on calciferous bedrocks than on siliceous ones. However, there are some exceptions. Acidophilous scrub, dwarf-shrub heath and snowbed communities from the northern Apennines are always more species rich than their basophilous counterparts in the Central Apennines. Pastures tend to have a higher floristic diversity than the primary grasslands probably due to high spatial variability caused by local trampling and nutrient enrichment effects. The highest proportion of endemics and rare species is found in the alpine belt, especially in open communities (e.g. screes, snowbeds). Within the

Table 3.6.5. Mean species richness of the main vegetation types in the Italian Alps

Vegetation type (number of plant communities)	Elevation range (m)	Mean species richness[a]	Endemics and (rare species) (%)
Siliceous substrata			
Natural grasslands (5)	1900–3200	34	4.6 (5.4)
Snowbed (1)	2400–3000	17	4.3 (8.5)
Scree communities (2)	2000–3000	15	12.3 (16.8)
Rock communities (1)	1900–3000	17	8.5 (36.7)
Dwarf-shrub heath (2)	2000–2600	22	6.5 (5.9)
Anthropogenic grassland (1)	1700–2370	38	1.6 (1.8)
Treeline ecotone incl. scrub (3)	1800–2300	25	3.4 (1.9)
Base-rich substrata			
Snowbed (1)	1920–3000	25	8.7 (16.4)
Natural grasslands (6)	1900–2800	35	8.6 (13.1)
Scree communities (2)	1800–2900	13	5.9 (4.5)
Treeline ecotone incl. scrub (4)	1750–2400	27	4.7 (1.9)

[a] Sampling area varied according to the vegetation type: 4–100 m² (scree and rock communities), 4–50 m² (snowbed communities), 10–100 m² (grasslands), 25–100 m² (dwarf-shrub heaths), and 50–200 m² (scrub).

Table 3.6.6. Mean species richness of the main vegetation types in the Apennines

Vegetation type (number of plant communities)	Elevation range (m)	Mean species richness[a]	Endemics and (rare species) (%)
Siliceous substrata			
Natural grasslands (3) – N Apennines only	1760–2250	22	8.1 (5.1)
Dwarf-shrub heath (2) – N Apennines only	1700–2250	20	0.6 (6.7)
Snowbed (1) – N Apennines only	1750–2040	18	6.9 (33.2)
Anthropogenic grassland (1)	1700–2220	31	7.6 (6.3)
Scrub (1) – N Apennines only	1740–1920	23	0.0 (8.0)
Base-rich substrata			
Snowbed communities (2)	1965–2910	17	19.2 (28.5)
Scree communities (6)	1700–2800	18	26.4 (36.1)
Natural grasslands (4)	1850–2770	26	15.1 (23.5)
Dwarf-shrub heath (1)	1790–2020	13	3.3 (3.8)
Anthropogenic grassland (1)	2000–2215	19	12.4 (4.5)
Scrub (6)	1720–2420	18	9.2 (15.9)
Open woodland (1)[b]	1900–2200	13	20.4 (15.5)

[a] Sampling area varied according to the vegetation type: 4–100 m² (scree and rock communities), 4–50 m² (snowbed communities), 10–100 m² (grasslands), 25–100 m² (dwarf-shrub heaths), and 50–200 m² (scrub).
[b] Mt. Pollino only

same functional community and bedrock type, species richness is higher and the proportion of endemics and rare species is lower in the Alps than in the Apennines (Tables 3.6.5 and 3.6.6).

3.6.7 Causes of Change and Conservation

Although glaciations had less impact on the Apennines than the Alps, the latter have a richer alpine vascular flora. This is probably a result of the larger area above the treeline in the Alps. However, several factors, such as the younger orogenetic age of the Apennines (mostly Pliocenic), their geological homogeneity (almost exclusively calcareous) and the extensive livestock husbandry in the central-southern part of the Italian peninsula, are also contributory factors. The occurrence of montane species in the alpine belt is more marked in the Apennines, where it is largely attributable to sheep grazing. The greater proportion of rare species on the Apennine summits than on the Alpine ones, especially of pteridophytes, makes the Apennines more vulnerable to species extinction and a reduction in biodiversity.

The alpine plant communities of the Alps have a higher alpha-diversity than the ecologically analogous ones in the Apennines, which may be explained at least in part by the higher gamma (regional) diversity in the Alps. However, the alpine communities in the Apennines include a higher proportion of endemics and rare species than their eco-functional counterparts in the Alps. A greater regional diversity and their closeness to the Maritime Alps may in part account for the higher species richness in acidophilous communities of the northern Apennines than that recorded in equivalent basophilous communities from the central Apennines. From a conservation point of view the most valuable are the grasslands because of their high species richness, and the scree and rock communities because they harbour many endemics and rare species.

To date there is no clear evidence of an altitudinal shift of communities to support the theory of global climate change. However, the retreat of glaciers is a proof of a warming of the climate in Europe since the nineteenth century.

References

Aleffi M, Schumacker R (1995) Checklist and red-list of the liverworts (Marchantiophyta) and hornworts (Anthocerotophyta) of Italy. Flora Med 5:73–161

Angot A (1906) Étude sur le régime pluviométrique de la Méditerranée. Congr Soc Sav Paris 120–134

Baldoni M, Biondi E, Frattaroli R (1999) Caratterizzazione bioclimatica del Gran Sasso d'Italia. Braun-Blanquetia 16:7–20

Belloni S (1992) Oscillazioni frontali dei ghiacciai italiani e clima: un sessantennio di ricerche. Geogr Fis Dinam Quat 15:43–57

Biondi E, Baldoni M (1994) The climate and vegetation of peninsular Italy. Coll Phytosoc 23:675–721

Cortini Pedrotti C (1979) La distribuzione di Paludella squarrosa (Hedw.) Brid. in Italia. Stud Trent Sci Nat 56:21–35

Cortini Pedrotti C (1992) Checklist of the mosses of Italy. Flora Med 2:119–221

Desio A (1967) I ghiacciai del Gruppo Ortles-Cevedale (Alpi centrali). Tamburini Editore, Milano

D'Orefice M, Pecci M, Smiraglia C, Ventura R (2000) Retreat of Mediterranean glaciers since the Little Ice Age: case study of Ghiacciaio del Calderone, central Apennines, Italy. Arct Antarct Alpine Res 32:197–201

Favarger C (1972) Endemism in the mountain floras of Europe. In: Valentine DH (ed) Taxonomy, phytogeography and evolution. Academic Press, London, pp 191–204

Favarger C (1986) Endémisme, biosystématique et conservation du patrimoin génétique. Atti Ist Bot Lab Critt 5:5–14

Gafta D, Pedrotti F (1998) Fitoclima del Trentino-Alto Adige. Stud Trent Sci Nat 73:55–111

Marchesoni V (1959) Importanza del fattore storico-climatico e dell'azione antropica nell'evoluzione della vegetazione forestale dell'Appennino Umbro-Marchigiano. Ann Acc Ital Sci For 8:327–343

Michalet R (1991) Nouvelle synthèse bioclimatique des milieux méditerranéens. Application au Maroc septentrional. Rev Ecol Alp 1:45–60

Paone N (1987) La transumanza. Immagini di una civiltà. Cosmo Iannone Editore, Isernia
Pignatti S (1982) Flora d'Italia. Edagricole, Bologna
Pignatti S (1994) The climax vegetation above timberline in the northern and central Apennines. Fitosociologia 26:5–17
Rivas Martinez S (1996) Bioclimatic map of Europe. Cartographic Service, University of León, León, Spain
Schumacker R, Soldan Z, Aleffi M, Miserere L (1999) The bryophyte flora of the Gran Paradiso National Park (Aosta Valley and Piedmont, Italy) and its immediate surroundings: a synthesis. Lejeunia 160:1–107

3.7 The Vegetation of the Alpine Zone in the Pyrenees

D. Gómez, J.A. Sesé and L. Villar

3.7.1 Geography

The 100-km-wide chain of the Pyrenees extends over 400 km in a west-east direction from the Atlantic Ocean to the Mediterranean Sea. About 1760 km^2 (4.5%) of its ca. 40,000 km^2 total area is in the alpine zone, with the treeline being at 2300 m a.s.l. (Sesé et al. 1999). In spite of the small area, 130 peaks exceed 3000 m in altitude (Buyse 1993).

The Pyrenees are bordered by the Aquitaine Plain to the north and the Ebro Valley to the south. There is a sharp contrast between the topography of the northern and southern slopes and the watershed separates the eurosiberian and Mediterranean biogeographic regions. To the west, the hills gradually descend into the Basque Depression from the Balaitus (3144 m) over 150 km; to the east, there is a sharper decline to sea level over 70 km from the Puigmal (2910 m) and Canigó (2784 m).

3.7.2 History, Geology, Soils and Climate

The eastern Pyrenees (from Andorra to the Mediterranean Sea) are predominantly siliceous; between Pallars-Ariège and Ossau-Tena there are siliceous and calcareous rocks and to the west limestones dominate. During the Hercynian orogeny, an underlying axis uplifted and Palaeozoic granites, sandstones and schists formed. The highest peak, Aneto (3404 m), is found among these mountains in the central Pyrenees. During the Alpine folding, at the early Tertiary, Mesozoic rocks, mainly limestone mountains, formed on both sides of the Axial Pyrenees, including Monte Perdido (3355 m), the second highest peak of the range (the highest calcareous mountain in Europe). There are also conglomerates of Oligocene and Miocene origin and Eocene marls. Patchily there are rocks of volcanic origin. All these rocks were subject to glacial activity during the Quaternary, and therefore

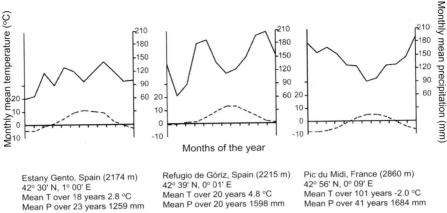

Fig 3.7.1. Climatic diagrams for three weather stations in the treeline ecotone and alpine zone in the Pyrenees. *Solid line* precipitation; *dashed line* temperature

there are till deposits above 800 m together with gravel beds and alluvial terraces.

The northern and western slopes have an oceanic climate with precipitation throughout the year, mild winters and cool summers. The southern slopes are drier and more continental (much insolation, torrential rainfalls at the equinoxes, dry summers, strong temperature variations and very cold winters). Finally, near the Mediterranean, both the northern and southern sides are influenced by a coastal Mediterranean climate.

Snowfalls begin in October and snow lies until June or July, especially above 2000 m, mostly in the western section. Persistent snowbeds and isolated glaciers can be found above 2800 m and are under protection as 'natural monuments'. In the central and eastern area, mainly in Spain, many peaks have long dry periods between snowfalls, so that winter is the driest season (e.g. the Refugio de Góriz, Monte Perdido massif, Spain, at 2215 m, only registers 16% of the total annual precipitation; Fig. 3.7.1).

At 2000 m or above, it may freeze or snow throughout the whole year. The average annual air temperature is 2.4 °C at 2300 m; –0.9 °C at 2900 m; and –3.2 °C at 3300 m. By using the regression equations of Barrio et al. (1990), the length of the growing season in the alpine belt has been estimated to be 22 days at 2300 m; 19 days at 2900 m, and none at 3300 m or above (Gómez et al. 1997).

3.7.3 History of the Flora

Plant names follow Flora Europaea (Tutin et al. 1964, 1968–1980) or, when names of authorities are given, Flora Iberica (Castroviejo et al. 1986, 1993). Together with some of the ancient vascular plants of the world, such as the Selaginellaceae or the Lycopodiaceae, some palaeoendemic species and genera have survived in the Pyrenees. On limestone cliffs and screes, two species of the genus *Borderea* (Dioscoreaceae) testify that a subtropical climate dominated in the central Pyrenees in the Tertiary. *B. pyrenaica* reaches the alpine zone while *B. chouardii* is restricted to the montane zone. The genus *Ramonda* (Gesneriaceae) appears to be ancient, although related genera and species are found on the southern Balkan Peninsula (e.g. Montenegro, Bulgaria, Macedonia). Among the Umbelliferae, there is an endemic and monotypic genus (*Xatardia*), restricted to the eastern Pyrenees, between Andorra and Puigmal, and two species of an endemic genus of the Pyrenean and Cantabrian mountains, *Endressia castellana* (west) and *E. pyrenaica* (east). More recent, but presumably of pre-glacial origin, are many endemics such as, for example, *Minuartia cerastiifolia*, conserved on nunataks, some *Petrocoptis* (Caryophyllaceae), *Androsace* (Primulaceae) and *Cirsium* (Compositae) species. Many alpine species reach their extreme distribution in SW Europe, such as, for example, *Carex bicolor* and *C. curvula*, which perhaps arrived during the last cold period. Others, such as *Hippophaë rhamnoides* spp. *fluviatilis*, reached the Pyrenees during an interglacial period and yet others, such as *Salix daphnoides*, colonised morainic deposits. Post-glacial forests, mainly beech, fir and pines, established 4000–6000 years ago (Jalut 1988). The genera *Saxifraga*, *Festuca*, *Veronica* and *Oxytropis* are much diversified in the alpine zone of the Pyrenees.

3.7.4 Today's Flora

The Pyrenees are one of the European centres of plant diversity (Villar and Dendaletche 1994). According to Dupias (1985), the total vascular plant flora of the Pyrenees is about 3500 species and subspecies. Of bryophytes, about 530 species have been recorded (Casas 1986). J. Etayo (pers. comm.) estimated that there were about 1500 species of lichens in the Pyrenees of which 150–200 could colonise the alpine zone. In addition, many undescribed species of fungi (parasymbionts, parasites, saprophytes or hyperparasymbionts) normally live on other lichens.

The alpine flora of the Pyrenees (above 2300 m a.s.l.) is of 809 species and subspecies (31 ferns, three Gymnospermae and 775 Angiospermae) belonging to 294 genera and 78 families. Sixty-three percent (about 500 species) are abundant (39%) or frequent (24%) and we consider them true alpine and the remaining 37% as accidental. Twenty-two species occur exclusively in the alpine zone.

Not including apomictic genera such as *Alchemilla, Taraxacum* and *Hieracium*, the 10 best-represented families in the alpine zone are Compositae (79 taxa, 9.7% of the total occurring in the alpine zone), Gramineae (66, 8.1%), Caryophyllaceae (48, 5.9%), Cyperaceae (48, 5.9%), Cruciferae (44, 5.4%), Rosaceae (47, 5.7%), Scrophulariaceae (42, 5.2%), Ranunculaceae (38, 4.7%), Leguminosae (35, 4.4%), and Saxifragaceae (29, 3.6%). The best-represented genera are *Carex* (29 taxa), *Saxifraga* (28), *Festuca* (20), *Ranunculus* (20), *Potentilla* (15), *Veronica* (14), *Androsace* (12), *Sedum* (12), and *Poa* (11). Some genera, such as *Androsace, Carex, Draba, Gentiana, Oxytropis, Pedicularis, Primula, Saxifraga*, and *Sedum*, occur mainly in the alpine zone.

About 180 taxa (excluding apomictic microspecies) are confined to the Pyrenees chain, or to one or more of its three sectors (eastern, central, and western). The western Pyrenees are rich in calcicolous endemics, such as *Androsace cylindrica* ssp. *hirtella, Buglossoides gastonii, Cirsium carniolicum* ssp. *rufescens, Saxifraga hariotii* and *Thalictrum macrocarpum*. This sector also harbours some endemic plants such as *Aster pyrenaeus, Dethawia tenuifolia, Euphorbia chamaebuxus, Ranunculus parnassiifolius* ssp. *favargeri* Küpfer and *Rumex cantabricus*, or extending some isolated populations into the Cantabrian Mountains. In the eastern Pyrenees, there exist also some endemic plants such as, for example, *Senecio leucophyllus* and *Salix ceretana* (P. Monts.) Chmelar. In the central section, there are siliceous or calcareous, and Pyrenees-wide endemics, e.g. *Ramonda myconi, Veronica aragonensis, Cirsium glabrum, Silene borderei*, and *Salix pyrenaica*.

In the whole of the Pyrenees, less than 3% of the flora (about 90 species) is endemic whereas, in the alpine zone (above 2300 m), this proportion is 11.8%. The higher value in the alpine zone might be explained in relation to the glacial events and vertical movements of the flora during the Quaternary (Sesé et al. 1999; Villar 1999). Furthermore, the abundance of screes and crevices could have provided habitats to colonise and to survive for several species, such as: *Androsace ciliata, A. hirtella, A. pyrenaica, Buglossoides gastonii, Onosma bubanii, Saxifraga hariotii, S. media, S. pubescens* ssp. *pubescens, Vicia argentea*, and other endemics. In addition to the endemics, orophytes (35%), eurosiberian species (21%), boreo-alpines (14.4%) and others (17.8%) complete the chorological spectrum for the alpine zone.

Many high mountain taxa such as *Leontopodium alpinum, Rhododendron ferrugineum, Saxifraga bryoides, Androsace helvetica*, and *Artemisia umbelliformis* reach their westernmost limit of distribution in the Pyrenees.

Sesé et al. (1999) have shown that in the Pyrenees as a whole, hemicryptophytes (45%) predominate with annuals (20%) and chamaephytes (14%); geophytes, hydrophytes and phanerophytes are scarce. In the alpine zone, hemicryptophytes (64.6%), together with chamaephytes (21.5%), make up about 85% of the flora. Above 2900 m, chamaephytes and hemicryptophytes further increase in dominance and are exclusive above 3200 m.

3.7.5 Vegetation

The lower limit of the alpine zone is 2300 m (Rivas Martínez 1988). The *Pinus uncinata* treeline ranges from 2100 m on the northern slopes in the western sector (Anie massif) to 2600 m in the more continental central sector (e.g. Néouvielle massif and the Encantats area, Boí-Aigües Tortes National Park). The climatic treeline in many places has been lowered by fire, grazing and logging; however, forest recovery has been observed in some localities since about the 1950s (Métailié 1999). The plant cover in the alpine zone is mostly <50% and there are frequent screes, cliffs and outcropping rock. Above 2800–3000 m, there is a discontinuous subnival zone.

About 70 plant communities have been described from the alpine zone of the Pyrenees (Table 3.7.1) and the whole range harbours >300 communities (Rivas Martínez 1988). Some of them, such as the *Androsace ciliata* screes (with *Saxifraga iratiana, Androsace ciliata, Minuartia cerastiifolia*, and *M. sedoides*) or some snowbed communities with *Carex pyrenaica* and *Cardamine alpina* type are exclusive to the Pyrenean high mountains (Vigo 1976; Rivas Martínez 1988).

Species richness decreases by about 58 taxa every 100 m of altitude between 1200 and 3404 m. This decline is steeper above 2300 m and, above 3000 m (Fig. 3.7.2), a possible indication of the beginning of alpine and subnival belts. Species richness per 25 m^2 varies between 2–4 species for aquatic vegetation and 26–30 species for stony open pastures on limestone (Table 3.7.1, last column). Scree communities are species poor (8–14 species) whilst closed grasslands, especially on calcareous bedrock, e.g., the *Kobresia* (= *Elyna*) *myosuroides* or the *Primula intricata* type and snowbeds, are species rich (22–28).

The vegetation of the Pyrenees, especially its northern slopes in the central and western sectors, is similar to that of the Alps. The eastern sector and some isolated areas on southern slopes are more similar to the Iberian mountains and the Sierra Nevada, where the so-called subalpine and alpine zones are replaced by oro-Mediterranean and cryoro-Mediterranean zones. The climatic contrast between the western oceanic side and the Mediterranean and continental southern, central and eastern sectors and the varied bedrock results in a rich mosaic of flora and vegetation (Atlantic, alpine, montane and sub-Mediterranean elements). When compared with the Alps, the Pyrenees have a smaller number of vascular plants and plant communities, but with a slightly higher proportion of endemics in the alpine zone (12% in the Pyrenees vs. 10% in the Alps). The number of endemic or subendemic genera exceeds that in the Alps, where there is a single one.

Table 3.7.1. Mean species richness in alpine plant communities in the Pyrenees. Species richness values are from a vegetation sample (relevé) database

Habitat type	Principal species in community types on siliceous (Si) or calcareous (Ca) substratum	Number of plant associations	Species richness (no. of species/25 m^2)
Snowbed communities	Salix herbacea, Cardamine alpina, Omalotheca supina (Si)	4	10–14
	Salix retusa, S. reticulata, Carex parviflora (Ca)	2	13–17
Rock communities	Androsace vandellii, A. pyrenaica, Primula hirsuta (Si)	6	11–16
	Cystopteris fragilis, Viola biflora, Asplenium viride (Ca)	3	10–14
	Saxifraga iratiana, Valeriana globulariifolia, Potentilla nivalis (Ca)	4	10–14
Scree communities	Senecio leucophyllus, Luzula alpinopilosa, Oxyria digyna (Si)	1	8–12
	Dryopteris oreades, Cryptogramma crispa (Si)	1	8–12
	Androsace ciliata, Minuartia cerastiifolia, Festuca borderei (Ca, Si)	2	10–14
	Iberis spathulata, Veronica aragonensis, Borderea pyrenaica (Ca)	7	12–16
	Saxifraga praetermissa, Arenaria purpurascens, Veronica nummularia (Ca, Si)	2	9–14
Grasslands	Festuca eskia, Campanula ficarioides, Trifolium alpinum (Si)	5	20–24
	Festuca airoidis, Gentiana alpina, Minuartia recurva (Si)	4	21–25
	Nardus stricta, Ranunculus pyrenaeus, Meum athamanticum (Si)	5	15–19
	Kobresia (= Elyna) myosuroides, Oxytropis pyrenaica, Carex curvula (Ca)	2	24–28
	Primula intricata, Trifolium thalii, Horminum pyrenaicum (Ca)	2	22–26
	Festuca scoparia, Sideritis hyssopifolia, Saponaria caespitosa (Ca)	3	26–30
Spring, flushes and aquatic vegetation	Saxifraga aquatica, S. stellaris, Epilobium alsinifolium (Si)	3	5–9
	Potamogeton gramineus, P. alpinus	1	2–4
	Cratoneuron commutatum, Cochlearia pyrenaica, Philonotis calcarea (Ca)	1	4–7
	Carex nigra, Juncus filiformis, Leontodon duboisii (Si)	2	10–15
	Carex davalliana, Juncus alpinus, Primula farinosa (Ca)	3	14–18

The Vegetation of the Alpine Zone in the Pyrenees

Table 3.7.1. (*Continued*)

Habitat type	Principal species in community types on siliceous (Si) or calcareous (Ca) substratum	Number of plant associations	Species richness (no. of species/25 m^2)
Prostrate dwarf shrub heath	*Loiseleuria procumbens, Vaccinium uliginosum, Arctostaphylos alpina* (Si)	2	14–19
Dwarf shrub heath and scrub	*Rhododendron ferrugineum, Vaccinium myrtillus, Sorbus chamaemespilus* (Si)	2	12–16
Tall-forb vegetation in livestock resting areas	*Rumex pseudoalpinus, Chenopodium bonus-henricus, Sisymbrium pyrenaicum* (Ca, Si)	1	11–16
Treeline ecotone	*Pinus uncinata, Cotoneaster integerrima, Juniperus communis* ssp. *alpina* (Ca, Si)	1	12–17

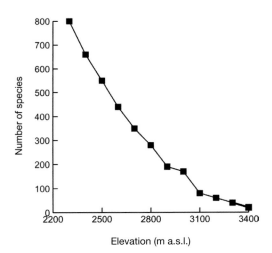

Fig. 3.7.2. The decrease with altitude in vascular plant species numbers from the treeline (2300 m a.s.l.) to the highest peaks in the Pyrenees. Data are from herbarium and field records

References

Barrio G del, Creus J, Puigdefàbregas J (1990) Thermal seasonality of the high mountain belts of the Pyrenees. Mountain Res Dev 10:227–233

Buyse J (1993) Los tresmiles del Pirineo. Ediciones Martín Roca SA, Barcelona

Casas C (1986). Catálogo de los briófitos de la vertiente española del Pirineo central y de Andorra. Collect Bot (Barcelona) 16:255–321

Castroviejo S, Laínz M, López González G, Montserrat P, Muñoz Garmendia F, Paiva J, Villar L (eds) (1986) Flora Iberica, vol I. Real Jardín Botánico, Madrid

Castroviejo S, Aedo C, Cirujano S, Laínz M, Montserrat P, Morales R, Muñoz Garmendia F, Navarro F, Paiva J, Soriano C (eds) (1993) Flora Iberica, vol III. Real Jardín Botánico, Madrid

Dupias G (1985) Végétation des Pyrénées. CNRS, Toulouse

Esteve F, González V, Arenal F (1997) Los Macromicetes de las áreas alpinas y subalpinas del Parque nacional de Ordesa y zonas limítrofes. Instituto de Estudios Altoaragoneses, Huesca

Gómez D, Sesé JA, Ferrández JV, Aldezabal A (1997) Altitudinal variation of floristic features and species richness in the Spanish Pyrenees alpine belt. 36th IAVS Symposium, Ser. Informes no 40. Universidad de La Laguna, Tenerife, pp 113–123

Jalut G (1988) Les principales étapes de l'histoire de la forêt pyrénéenne française depuis 15.000 ans. In: Villar L (ed) Homenaje a Pedro Montserrat. Instituto de Estudios Altoaragoneses e Instituto Pirenaico de Ecología, Huesca y Jaca, pp 609–615

Métailié JP (1999) Le conquérant des estives. Les feuilles du pin à crochets 1:27–37

Rivas Martínez S (1988) La vegetación del piso alpino superior de los Pirineos. In Villar L (ed) Homenaje a Pedro Montserrat, Instituto de Estudios Altoaragoneses e Instituto Pirenaico de Ecología, Huesca y Jaca, pp 719–728

Tutin TG, Heywood VH, Burges NA, Valentine DW, Walters SM, Webb DA (eds) (1964) Flora Europaea, vol 1. Cambridge University Press, Cambridge

Tutin TG, Heywood VH, Burges NA, Moore DM, Valentine DW, Walters SM, Webb DA (eds) (1968–1980) Flora Europaea, vols 2–5. Cambridge University Press, Cambridge

Sesé JA, Ferrández JV, Villar L (1999) La flora alpina de los Pirineos. Un patrimonio singular. In: Villar L (ed) Espacios Naturales Protegidos del Pirineo. Ecología y cartografía. Consejo de Protección de la Naturaleza de Aragón, Zaragoza, pp 57–76

Vigo J (1976) L'alta muntanya catalana. Flora i vegetació. Ed. Montblanch, Barcelona

Villar L (1999) Some notes on the alpine flora of the Pyrenees. ESF ALPNET News 1:9–10

Villar L, Dendaletche C (1994) Pyrenees. France, Spain and Andorra. In: Heywood VH, Davis SD, Hamilton AC (eds) Centres of plant diversity. A guide and strategy for their conservation. WWF and IUCN, Cambridge, pp 61–64

3.8 High Mountain Vegetation of the Caucasus Region

G. Nakhutsrishvili

3.8.1 The Caucasus

The Caucasus is characterised by contrasting environmental conditions and a high diversity of plant communities. The treeline ecotone and alpine vegetation are present almost throughout the Greater and Minor Caucasus. The treeline ecotone is especially rich in a mosaic of communities of contrasting character ranging from hyper-humid relicts (Colchis, Western Caucasus) to xero- and hemixerophytic vegetation types (Eastern and Minor Caucasus).

Today's treeline (2000–2600 m) is 300–400 m lower than it was in the nineteenth century owing to human impact (Dolukhanov 1966a). The alpine zone (2400–3000 m) is characterised by the dominance of short grasslands, dwarf-herb meadows, alternating with *Rhododendron caucasicum* shrubs and rock and scree vegetation. The subnival zone extends up to 3800 (4000) m.

3.8.1.1 Geography and Geology

The Caucasus lies over an area of 441,000 km^2 (of which 15 % is above the treeline), between 39°N and 47°N and 40°E and 50°E. The 3600 m main range of the Greater Caucasus forms the watershed between the North and Transcaucasus. Its highest peaks are Elbrus (5595 m) and Shkhara (5068 m). The Minor Caucasus consists of a number of highlands rising from 2000 to 3500 (4000) m. The highlands are separated from each other by minor lowland ranges. The highest peak of the Minor Caucasus is Mt. Aragaz (4090 m).

The Greater Caucasus (Kavkasioni) is of Pliocene igneous origin and largely overlain by Jurassic and Cretaceous rocks. The western section of the Greater Caucasus, west of Mt. Elbrus, is predominantly limestone. In the central section, Palaeozoic crystalline formations occur, together with igneous rocks. East of Mt. Kazbegi, Central Caucasus, the main divide is formed primarily of Jurassic bedrocks. The Transcaucasian massif (Dzirula) consists of amphibolite and greenschist accompanied by marbles and serpentinite.

3.8.1.2 Climate

The different regions of the Caucasus show a considerable range of contrasting climatic conditions caused by their orographic structure and the presence of the Black and the Caspian Seas. Precipitation is highly variable: the Western Greater Caucasus has a humid climate (>2000 mm year^{-1}) whilst some regions in the Eastern and Southern Caucasus are less so ($\geq$ 500–600 mm year^{-1}). However, even in the driest parts, droughts above the treeline are very rare. Snow cover above the treeline is from November until April and from October until June in the subnival zone. For high altitude climatic data for Kazbegi, see Table 3.8.1.

Table 3.8.1. Climatic data for the Kazbegi region, central Caucasus. The data are means of long-term observations at each of three meteorological stations located at altitudes of 1850, 2396 and 3650 m a.s.l.

Characteristic	Treeline	Alpine	Subnival
Length of growing season (days)	180	120	90
Mean annual air temperature (°C)	5.8	0.3	−6.1
Absolute minimum air temperature (°C)	−35	n.d.	−42
Absolute maximum air temperature (°C)	32	27	16
Mean annual precipitation (mm)	743	1249	961
Mean monthly precipitation (May–August, mm)	100	216	148
Number of frost-free days	124	118	62
Number of days with risk of summer frost	179	175	140

3.8.2 The Flora

3.8.2.1 History of the Flora

The high mountain flora of the Caucasus consists mainly of boreal and arctic-alpine elements which are thought to have colonised during the Pleistocene. However, Fedorov (1952) and Kharadze (1960) proposed that the most typical representatives of the Caucasian high mountain flora were of autochthonous origin, based around a Tertiary nucleus and developed during the Quaternary. In his analysis of the Caucasian endemisms, Grossheim (1936) indicated that the area of the Greater Caucasus has been colonised by both boreal and Anterior Asian elements. Autochthonous hemi-xerophytic elements are well represented in the areas of the Central Caucasus from the montane to the alpine zones. The genera *Edraianthus*, *Nepeta*, *Scabiosa* and *Symphyandra* indicate ancient connections with the Mediterranean flora (Kharadze 1960) whilst *Astragalus* spp., *Betonica nivea*, *Salvia canescens*, *S. daghestanica*, *Scutellaria* spp., *Silene* spp., and *Ziziphora* spp. are of Anterior Asian origin.

3.8.2.2 Human Impacts

Perturbation of the high mountain vegetation by man began in prehistoric times and is likely to have strongly modified it. For reasons including past climate changes and the existence of a rich and diverse fauna of wild herbivores before human times, it is difficult to reconstruct what primary vegetation may have been like. The abundance of poisonous, thorny, or otherwise unpalatable species of the genera *Aconitum*, *Adenostyles*, *Anemone*, *Anthriscus*, *Aquilegia*, *Astrantia*, *Chaerophyllum*, *Cirsium*, *Colchicum*, *Delphinium*, *Digitalis*, *Euphorbia*, *Fritillaria*, *Galega*, *Gentiana*, *Lilium*, *Nardus*, *Pedicularis*, *Pulsatilla*, *Ranunculus*, *Rumex*, *Scrophularia*, *Swertia*, *Trollius* and *Veratrum* attest to a strong prehistoric herbivory pressure.

3.8.2.3 Number of Taxa

The flora of the Caucasus comprises 6350 native species of vascular plants, 900 mosses and 1500 lichens. Of the 6350 vascular plants, 1600 species are endemic (Sakhokia 1958; Dolukhanov 1966a; Gagnidze 2000). Especially high endemism occurs among high mountain plants such as *Astragalus*, *Campanula*, *Cirsium*, *Delphinium*, *Draba*, *Jurinea*, *Onobrychis*, *Psephellus*, *Primula*, *Pyrethrum*, *Pyrus*, *Rosa*, *Saxifraga*, and *Symphyandra*. Nine of the 15 Caucasian endemic genera occur exclusively in high mountains (Dolukhanov

Table 3.8.2. The number of total and endemic species of flowering plants and the percentage of endemic species in altitude zones of the high mountains (> 1900 m) in the Kazbegi region, central Greater Caucasus. (Sakhokia and Khutsishvili 1975)

Species by altitude	Number of species	Number of endemic species (% of total)
Total high mountain	761	203 (26.7)
Treeline ecotone	302	72 (23.8)
Treeline ecotone – alpine	290	78 (26.9)
Alpine	104	32 (30.7)
Treeline ecotone – subnival	17	1 (5.9)
Alpine – subnival	21	6 (28.6)
Subnival	27	14 (52.0)

Table 3.8.3. Number and percentage of species of flowering plants belonging to different life forms in the subnival zone in the Kazbegi, Elbrus, Mamisoni and Svaneti regions, Greater Caucasus. The classification of life forms is according to Raunkiaer, the lignification type of the shoot, life history, phyllotaxis and the structure of root. (Nakhutsrishvili and Gamtsemlidze 1984)

Life forms	Kazbegi	Elbrus	Mamisoni	Svaneti
1. Raunkiaer's system				
a) Chamaephytes	11 (13.1)	9 (10.6)	5 (6.8)	8 (10.8)
b) Hemicryptophytes	70 (83.3)	72 (83.5)	64 (87.7)	63 (85.1)
c) Geophytes	3 (3.6)	4 (5.5)	4 (5.5)	3 (4)
2. Lignification type of the shoot				
a) Herbs	76 (90.5)	79 (92.9)	71 (97.3)	69 (93.2)
b) Shrubs	8 (9.5)	6 (7.1)	2 (2.7)	5 (6.7)
3. Life history				
a) Monocarps	8 (9.5)	8 (9.5)	9 (12.3)	7 (9.5)
b) Polycarps	76 (90.5)	77 (90.5)	64 (87.7)	67 (90.5)
4. Phyllotaxis				
a) Cushion	10 (11.9)	9 (10.6)	6 (8.2)	9 (12.2)
b) Rosette	14 (16.7)	16 (18.8)	13 (17.8)	14 (18.9)
c) Semi-rosette	48 (57.1)	47 (55.3)	45 (61.6)	43 (58.1)
d) Non-rosette	12 (14.3)	13 (15.3)	9 (12.3)	8 (10.8)
5. The structure of root				
a) Long taproot	16 (19)	13 (15.3)	8 (10.9)	10 (13.5)
b) Short taproot	5 (5.9)	5 (5.9)	5 (6.8)	4 (5.4)
c) Filamentous taproot	2 (2.4)	3 (3.5)	3 (4.1)	3 (4)
d) Rhizome – taproot	18 (21.4)	22 (25.9)	18 (24.6)	18 (24.3)
e) Short rhizome	13 (15.5)	12 (14.1)	12 (16.4)	13 (17.5)
f) Long rhizome	12 (14.3)	11 (12.9)	8 (10.9)	8 (10.8)
g) Stoloniferous	1 (1.2)	1 (1.2)	1 (1.4)	1 (1.3)
h) Fibrous root	3 (3.6)	4 (4.7)	4 (5.5)	4 (5.4)
i) Tuberous	2 (2.4)	2 (2.3)	2 (2.7)	1 (1.3)
j) Bulbiferous	1 (1.2)	2 (2.3)	2 (2.7)	2 (2.7)
k) Cespitose	11 (13.1)	9 (11.8)	10 (13.7)	10 (13.5)
Total	84 (100)	85 (100)	73 (100)	74 (100)

1966b). For example, in the Kazbegi region, 203 of the 761 high mountain (>1900 m a.s.l.) vascular plant species are endemic. The highest percentage of endemism is estimated to be in the subnival zone (Table 3.8.2).

Table 3.8.3 shows the number and percentage of species of flowering plants belonging to different life forms according to different classification systems in the Kazbegi, Elbrus, Mamisoni and Svaneti regions of the Greater Caucasus. In the subnival and nival belts, polycarpic herbs with rosette or semirosette leaves and a long taproot, rhizome-taproot or short rhizome are typical. According to Raunkiaer's life form system, hemicryptophytes prevail in this altitude belt.

3.8.3 High Mountain Vegetation

Table 3.8.4 shows the main community types, elevation range, principal species and mean number of species per area in the high mountain areas of the central Caucasus. The richest are the meadows at the treeline ecotone and in the alpine zone. The lowest number of species is found in the plant assemblages (nanocoenoses) of the subnival belt, and in scree, rock and snowbed communities.

3.8.3.1. The Treeline Ecotone

The treeline ecotone consists of open and crooked-stemmed (krummholz) forest, tall herbaceous vegetation, meadows, and often steppe and thorny cushion vegetation. In the treeless areas of the South Georgian Uplands, mountain steppes interdigitate with alpine meadows.

Low stature *Pinus kochiana* (= *P. sosnovskyi*) and *Picea orientalis* forests occur on relatively dry slopes, whereas crooked-stemmed forests (*Betula litwinowii, Acer trautvetteri, Sorbus caucasigena*) are developed under moist conditions. Dark coniferous forests extend to an altitude of 2000–2200 m. Montane pine forests occur from 700 m up to about 2400 m a.s.l. In several regions of the Caucasus pines have been recorded at 2500–2600 m (Tumajanov 1980). Crooked-stemmed forests of *Fagus orientalis* are characteristic of the Colchis and extend to 2300 m. The Colchis endemics, *Betula medwedewii* and *Quercus pontica*, are moderately frost-resistant; however, they are well adapted to cold and moist winters and can reach 2000–2200 m. The upper limit of *Quercus macranthera* varies between 2400 and 2700 m, being higher under dry conditions than moist. Crooked-stemmed forests of *Betula litwinowii, B. medwedewii* and *Acer trautvetteri* are the most typical communities of the treeline. Under relatively continental climatic conditions the upper boundary of *B. litwinowii* varies between 2400 and 2600 m (Dolukhanov 1966a).

Table 3.8.4. Community types, elevation range, principal species and the maximum number of species recorded per 1 m² and 25 m² in the Caucasus

Community type	Elevation range (m)	Principal species	Number of species ($1\,m^{-2}/25\,m^{-2}$)
Subnival nanocoenoses	2900–3800 (4000)	Cerastium kazbek, Lamium tomentosum, Pseudovesicaria digitata, Scrophularia minima, Symphyoloma graveolens	2/6
Springs and flushes	–2500	Batrachium divaricatum, Hyppuris vulgaris, Veronica anagalis-aquatica	?
Snowbed communities	2500–3800	Minuartia aizoides, Pedicularis crassirostris, Ranunculus helenae, Taraxacum porphyranthum	15/17
Scree and rock communities	1800–2600	Asperula albovii, Campanula bellidifolia, Onosma caucasica, Primula bayernii, Scutellaria leptostegia, Sempervivum pumilum	11/20
Alpine meadows	2500–2900	Alchemilla caucasica, Carex tristis, Carum caucasicum, Festuca supina, F. varia, Nardus stricta, Geranium gymnocaulon, Sibbaldia semiglabra	16/29, 20/39
Mires	–2900	Blysmus compressus, Callitriche verna, Carex kotschyana, C. wilnicea, Menyanthes trifoliata, Sphagnum angustifolium	?
Wet meadows	1900–2900	Equisetum palustre, Gladiolus tenuis, Iris sibirica, Parnassia palustris, Phragmites australis	16/23
Dwarf-shrub heath	1900–2900	Daphne glomerata, Dryas caucasica, Empetrum nigrum, Vaccinium myrtillus, V. vitis-idaea	?
Treeline ecotone meadows	1900–2500 (2600)		
a) Grass meadow		Calamagrostis arundinacea, Hordeum violaceum	25/41
b) Herb meadow		Anemone fasciculata, Geranium ibericum	17/22
c) Xero-mesophytic grass meadow		Bromopsis variegeta, Festuca varia	36/47
d) Steppe meadow		Carex buschiorum, Festuca ovina Pulsatilla violacea	26/38
Tragacanthic vegetation	–2500	Agropyron gracillinum, Artemisia splendens, Astragalus aureus, A. denudatus, Berberis vulgaris, Ephedra procera, Juniperus communis ssp. hemisphaerica, Scutellaria leptostegia, Stipa tirsa	23 (100 m²)

Table 3.8.4. (*Continued*)

Community type	Elevation range (m)	Principal species	Number of species ($1\,m^{-2}/25\,m^{-2}$)
Tall-herb vegetation	1900–2400	*Aconitum nasutum, Angelica tatianae, Cephalaria gigantea, Gadellia lactiflora, Heracleum sosnowskyi, Lilium monadelphum*	6/15
Scrub communities	1900–2500 (2800)	*Daphne albowiana, Rhododendron caucasicum, Juniperus communis* ssp. *hemisphaerica, J. Sabina, J. pigmaea*	27 (100 m²)
Treeline ecotone	1900–2600	*Acer trautvetteri, Betula litwinowii, B. medwedewii, B. raddeana, Fagus orientalis, Pinus kochiana, Quercus macranthera, Q. pontica*	?

3.8.3.1.1 Tall-Herb Vegetation

Unlike many other mountain systems (e.g. Alps, Rocky Mts., Pamir Mts., Himalayas), species-rich tall-forb vegetation (megaforbs of 3–4 m height) is well represented at the treeline ecotone in the Caucasus. The total species richness is 90 according to Gagnidze (1974).

3.8.3.1.2 Meadows, Feather-Grass Steppes and Scrub at the Treeline

Meadows are abundant in open forest in areas with a moist temperate climate. Herb meadows typically occupy level sites, depressions and slopes of northern and north-westerly exposure but often slopes of southern exposures at the treeline ecotone. Their dominants are mainly broad-leaved herbaceous plants. Xero-mesophytic grass meadows are predominantly found on southern slopes and on plateaux. They often ascend beyond the treeline ecotone up to 2600 m. Wet meadows occur near springs, streams and rivers. Steppe-like meadows (mountain steppes) develop in more humid and less warm climatic conditions (Dolukhanov et al. 1946; Nakhutsrishvili 1999).

Feather-grass steppes, which in some places reach the alpine zone, are mostly found in the Minor Caucasus (Zangezuri, Daralagez, Alagez, Shakhdag, and Tsalka-Javakheti) and in the eastern part of the Greater Caucasus (e.g. Daghestan and Pirikita Khevsureti). Mountain steppes are more diverse in their floristic composition than plain steppes (Magakian 1941; Grossheim 1948; Table 3.8.4).

Daphne albowiana and *D. woronowii* scrub, the latter on limestone, occurs abundantly in the Colchis. *Juniperus communis* ssp. *hemisphaerica* and *J. sabina* are distributed in many regions.

3.8.3.2 The Alpine Zone Proper

The alpine zone is dominated by short grass meadows, alpine mats, alternating with *Rhododendron caucasicum* Scrub, dwarf-shrub and open rock and scree vegetation.

The alpine meadows include tussock grass and sedge meadows, herb-rich meadows and tussocky grass-herb meadows (Table 3.8.4). Snowbed communities occur in the upper part of the alpine zone usually as patches among large boulders (Table 3.8.4). Short-cropped secondary vegetation mats, caused by heavy grazing and enriched in meadow elements (especially *Alchemilla* species), occupy considerable areas.

Steep northern slopes are covered by shrubs of *Rhododendron caucasicum* ('dekiani' in Georgian) up to 2700–2800 m (Table 3.8.4). Chionophilous communities of *Daphne glomerata* are very characteristic of Caukasus. Prostrate shrub communities of *Dryas caucasica* favour cold stony slopes of northern exposure Greater Caucasus.

The rock and scree vegetation shows a strong similarity to the upper alpine meadows and dwarf-herb meadows. They are characterised by a high degree of endemism (Shetekauri and Gagnidze 2000), which reaches its highest rates in Daghestan, Eastern Caucasus, and the Colchis, Western Caucasus (Kharadze 1960).

3.8.3.3 The Subnival Zone

Despite the harsh environment, over 200 species of vascular plants occur in the subnival zone of the Greater Caucasus with 109 of them exclusively growing there (Kharadze 1965). The upper limit of plant distribution varies; in the highest regions of the Caucasus (Svaneti, Kazbegi) it reaches 3950 m a.s.l., while it is 3600 m in the continental Daghestan (Prima 1974) and 4000 m on Mount Aragats, Minor Caucasus (Voskanian 1976). The total number of phanerogamic plant species above 3200 m is 94 for the whole of the Caucasus; in the Central Caucasus, 34 species grow above 3350 m, whereas in the Minor Caucasus only nine species reach 3800 m (Aragats).

Local endemic species grow in close proximity to the snow line. For example, in the western Caucasus, the Caucasian endemic *Saxifraga scleropoda* reaches 4000 m altitude, with *Cerastium kazbek* at 3950 m in the central Caucasus, and *Draba araratica* at 4000 m on Mt. Aragats. Typical subnival species found at 3600–3800 m are, e.g., *Alopecurus dasyanthus*, *Cerastium pseudokas-*

bek, *Colpodium versicolor, Draba supranivalis, Pseudovesicaria digitata, Saxifraga moschata, S. exarata, S. flagellaris, Senecio karjagini, Tripleurospermum subnivale, Veronica minuta* and *V. telephiifolia.*

A number of endemic species and genera occur in the subnival zone. Among them, monotypic genera include *Pseudovesicaria* (Brassicaceae), *Symphyoloma* (Apiaceae), and *Pseudobetckea* (Valerianaceae); oligotypic genera of Caucasian-Anterior Asian origen include *Coluteocarpus, Didymophysa, Eunomia* (Brassicaceae) and *Vavilovia* (Fabaceae). They are mainly distributed in high mountains of the Minor Caucasus and Asia Anterior (Nakhutsrishvili and Gagnidze 1999).

Mosses (e.g. *Bryum* spp., *Dicranoweisia crispula, Dicranum elongatum, Pohlia elongata, Pogonatum nanum, Tortella tortuosa, Tortula muralis*) and lichens (e.g. *Caloplaca elegans, Cetraria islandica, C. nivalis, Hypogymnia encausta, Lecidea atrobrunea, Parmelia vagans, Placolecanora melanophthalma, P. rubina, P. murilis, Rhizocarpon geographicum, Stereocaulon alpinum, Thamnolia vermicularis, Umbilicaria cylindrica* and *U. virginis*) grow abundantly in all types of vegetation in the subnival zone.

3.8.4 A Comparative Analysis of the High Mountain Vegetation of the Caucasus and the Alps

The characteristic feature of high mountain vegetation of the Caucasus is the development of xero- and hemi-xerophytic vegetation, i.e. feather-grass steppes, steppe meadows, thorny cushion plant communities and meadows of the Lamiaceae family at the treeline and in the lower reaches of the alpine zone. These communities are not found in the high mountains of the Carpathians, the Alps and the Pyrenees (Grabherr and Mucina 1993; Ellenberg 1996; Ozenda 1997; Nakhutsrishvili and Ozenda 1998).

The tall herbaceous vegetation of the Caucasus has no analogue in the Alps. Although some species of tall-herb vegetation are common to the Alps and the Caucasus (e.g. *Aconitum nasutum, Doronicum macrophyllum, Telekia speciosa*), the tall-herb vegetation of the Caucasus is unique in the number of species, level of endemism and stature. The *Betula–Adenostyles* tall-herb community of Ellenberg (1996) has little in common with the tall-herb vegetation of the Caucasus, which ecologically is more of an equivalent of the *Alnus viridis* scrub of the Alps (Nakhutsrishvili and Ozenda 1998).

Meadows of *Calamagrostis arundinacea, Festuca varia, Hordeum violaceum,* and *Trisetum flavescens* occur both in the Caucasus and in the Alps; however, in contrast to the Alps, *Arrhenatherum elatius* is not a characteristic grassland species in the Caucasus.

Dry meadows with *Bromopsis variegata* and diverse herb meadows with *Anemone fasciculata, Betonica macrantha, Geranium ibericum, Inula orien-

talis, Ranunculus caucasicus, R. oreophilus, Scabiosa caucasica, and *Trollius patulus* are widespread in the Caucasus but are absent from the Alps. *Kobresia* (= *Elyna*) *myosuroides* (Alps) and *Kobresia capilliformis* (Caucasus) rush heaths, *Woronowia speciosa–Carex pontica* (Caucasus, on limestones) and *Carex firma* (Alps, on limestones) sedge heaths can be regarded as vicarious communities.

Many subnival plants in the Caucasus are local endemic species, which have phylogenetic relationships with species from Asia Anterior or Asia Minor (Nakhutsrishvili and Gagnidze 1999), but unrelated to species of the (sub)nival zone of the Alps. The vascular plant species of the Alps are widely distributed and occur in most European high mountain areas, while local endemism prevails in the Caucasus.

Acknowledgements. I am grateful to Dr. M. Akhalkatsi for her help during the preparation of the manuscript.

References

Dolukhanov AG (1966a) Zakonomernosti Geograficheskogo Raznoobraziya Rastitel'nosti I Verkhnaya Granitsa Lesa V Gorakh Zakavkaz'ya [Regularities of geographical diversity of vegetation and timberline in the Trans-Caucasian mountains]. Probl Bot 8:196–207

Dolukhanov AG (1966b) Rastitel'nost' [Vegetation]. In: Gerassimov IP (ed) Kavkaz [Caucasus]. Nauka, Moscow, pp 223–251

Dolukhanov AG, Sakhokia MP, Kharadze AL (1946) Osnovnye Osobennosti Rastitel'nogo Pokrova Verkhnei Svanetii [Main features of the vegetation cover of upper Svanetia]. Trudy Tbilissk Bot Inst 9:79–130

Ellenberg H (1996) Vegetation Mitteleuropas mit den Alpen. 5. Aufl. Ulmer, Stuttgart

Fedorov AA (1952) The history of high mountain flora of the Caucasus in the Quaternary period as an example of autochtonous development of the Tertiary floristic basis. In: Lavrenko E (ed) Materiali po Chetvertichnomu Periodu. SSSR, 3, Moscow, pp 49–86

Gagnidze RI (1974) Botanicheskii i Geograficheskii Analiz Florotsenoticheskogo Kompleksa Vysokotrav'ya Kavkaza [Botanical and geographical analysis of the florocoenotic complexes of tall herb vegetation of the Caucasus]. Metsniereba, Tbilisi

Gagnidze R (2000) Diversity of Georgia's flora. In: Berutchashvili N, Kushlin A, Zazanashvili N (eds) Biological and landscape diversity of Georgia. WWF Georgia country Office, Tbilisi, pp 21–33

Grabherr G, Mucina L (eds) (1993) Die Pflanzengesellschaften Österreichs. Band 2. Fischer, Jena

Grossheim AA (1936) Analiz flory Kavkaza. [The analysis of the Caucasian flora]. Trudy Bot In-ta Azerbaidzh Fil AN SSSR. Izd Az Fil AN SSSR, Baku

Grossheim AA (1948) Rastitel'nyi pokrov Kavkaza [The vegetation of the Caucasus]. MOIP, Moscow

Kharadze AL (1960) Ob endemichnom hemixerophil'nom elemente Bolshogo Kavkaza [An endemic hemixerophilous element of the Greater Caucasus uplands]. Probl Bot 5:115–126

Kharadze AL (1965) O subnivalnom poiase Bolshogo Kavkaza [On the subnival belt of the Greater Caucasus]. Zametki Sistem Geogr Rastenii Inst Bot AN GSSR 25:103–104

Kolakovsky AA (1961) Rastitel'nyi Mir Kolkhidy [Plant life of the Colchis]. MOIP, Otd Bot 10th edn. MGU, Moscow

Magakian AK (1941) Rastitel'nost' Armyanskoi SSR [Vegetation of the Armenian SSR]. Izd. AN SSSR, Leningrad

Nakhutsrishvili G (1999) The vegetation of Georgia (Caucasus). Braun Blanquetia 15:5–74

Nakhutsrishvili G, Gagnidze RI (1999) Die subnivale und nivale Hochgebirgsvegetation des Kaukasus. Phytocoenosis 11:173–182

Nakhutsrishvili G, Gamtsemlidze Z (1984) Zhyzn rastenii v ekstremal'nykh usloviakh [Plant life in the extreme environments]. Nauka, Leningrad

Nakhutsrishvili G, Ozenda P (1998) Aspect Geobotaniques de la Haute Montagne dans le Caucase. Essai de Comparaison avec les Alpes. Ecologie 29:139–144

Ozenda P (1997) Aspects Biogéographiques de la Végétation des Hautes Chaînes. Biogeographica 73:145–179

Prima VM (1974) Subnivalnaia flora Vostochnogo Kavkaza, stroenie, ecologicheskii, biologicheskii i geographicheskii analiz [The subnival flora of the eastern Caucasus, its composition, ecological, biological and geographical analysis]. In: Galushko A (ed) Flora i Rastitel'nost Vostochnogo Kavkaza. Orjonikidze, Orjonikidze, pp 46–48

Sakhokia MP (ed) (1958) Botanicheskie ekskursii po Gruzii [Botanical excursions in Georgia]. Izd AN GSSR, Tbilisi

Sakhokia MP, Khutsishvili C (1975) Conspectus Florae Plantarum Vascularium Chewii. Metsniereba, Tbilisi

Shetekauri Sh, Gagnidze R (2000) Diversity of high-mountain endemic flora of the Greater Caucasus. In: Berutchashvili N, Kushlin A, Zazanashvili N (eds) Biological and landscape diversity of Georgia. WWF Georgia Country Office, Tbilisi, pp 135–159

Tumajanov II (1980) Luga I kryvolesie Kavkasa [Caucasian meadows and elfin woodlands]. In: Gribova S, Isashenko T, Lavrenko E (eds) Rastitel'nost' evropeiskoi chasti SSSR [Vegetation of the European part of the USSR]. Nauka, Leningrad, pp 198–202

Voskanian VE (1966) O biologicheskikh osobenostiakh rastenii v Verkhnei chasti alpiiskogo poieasa [On some biological peculiarities of plants in the upper part of the alpine belt of the Aragats Mountain]. Bot Zh 2:257–265

3.9 The Vegetation of the Corsican High Mountains

J. GAMISANS

3.9.1 Geography

A large Mediterranean island of 8748 km², Corsica is situated between 41° and 43°N. Until the lower Miocene, Corsica, together with Sardinia, was adjacent to the coast of Provence and Languedoc (see Fig. 7 in Gamisans 1999) where it was part of a Hercynian mountain arch joining the Pyrenees and the east Iberian massifs with the outer crystalline massifs of the Alps. The alpine zone is estimated to occupy about 100 km² in the dominant crystalline part of Corsica, composed of granite, gneiss and rhyolite (Fig. 15 in Gamisans 1999). The vegetation in the small and fragmented alpine zone is largely composed of often patchy chionophilous grassland on northern slopes, dry grassland in southerly exposure and open scree and rock communities.

3.9.2 Climate

The overall climate in Corsica is Mediterranean, marked by a strong summer drought. However, there are large differences in temperature and precipitation between the lowland coastal areas and the high mountains. The latter are more humid and rather cold with the summer drought being short or absent; there is a regular snow cover of up to 6 months (Gamisans 1999; Gauthier 2000). The absolute minimum air temperature recorded at 2000 m a.s.l. was −18.6 °C in 1986 (Gauthier 2000). The growing season at 2350 m was estimated to be 138 days in 1999–2000. The mean soil temperature 10 cm below surface at 2350 m was 9.0 °C with a range of −1.4 to +17.5. The accumulated thermal day temperature (>5 °C) was 578 day degrees (Chap. 2). Summer frosts (−5 to 0 °C) were regularly observed at altitudes of 1400–2700 m during 1966–1980 (J. Gamisans, pers. observ.).

3.9.3 The Flora

3.9.3.1 History of the Flora

Braun-Blanquet (1926) considered that the Corso-Sard flora was a 'paleogenic flora that had developed in situ', that it is a flora from the middle Tertiary which has since evolved in a closed system with very little contribution from the outside. The relationships of the Corso-Sard flora with those of the Balearic Islands, the Pyrenees, the Hyères Islands and Provence support the ancient origin of at least part of the flora.

According to a model by Bocquet et al. (1978), the Messinian trauma (a cold and dry period in the Mediterranean) caused serious damage to the former Arcto-Tertiary flora which was not then able to develop into the current orophilic flora. The Corsican mountains would then have been populated by a flora of Asian origin that had followed a westward migration. The weak point in this part of the hypothesis is that it does not explain the absence from Corsica of species representative of orophilic flora belonging to the genera *Achillea*, **Androsace*, *Artemisia*, *Campanula*, *Gentiana*, **Oxytropis*, **Pedicularis*, *Primula*, **Rhododendron*, and **Soldanella* (the genera marked with an asterisk are absent from the Corsican flora). Many of these taxa are indeed considered to be of Asian origin. Conversely, Favarger (1975) suggested that the Arcto-Tertiary flora was not destroyed and is at the origin of a large part of the orophilic flora of southern Europe.

Each of these various hypotheses leave some aspects of the Corsican flora unsatisfactorily explained. However, there is a certain link between the percentage of endemics and the duration of isolation of the Corso-Sard block. This block has, in certain instances, acted as a reserve for species and genera that have probably been eliminated from other areas by competition. The flora of Corsica is not purely of relict character, however. Plant species have evolved here also, similarly to that on the continent, but probably with less genetic exchange and much less input from the outside. Consequently, certain endemics have been able to occupy a large number of habitats that they would not have had access to if competition had been stronger. These endemics have thus become widespread, occurring from the meso-Mediterranean level up to the alpine (Gamisans 1981).

3.9.3.2 Human Impact

The presence of man for about 9000 years (Gamisans 1996) has, through agricultural, and particularly pastoral activities, altered the vegetation quite dramatically. For instance, the deciduous oak forest has receded, evergreen oaks have expanded and the timberline has descended (Reille et al. 1999). It is quite likely that man has had an impact on the treeline ecotone. Charcoal evidence shows that the treeline reached above 1900 m. Note that in Corsica the lower ecotones of the alpine zone do not correspond to the limit of trees but to that of *Alnus* scrub (on northern slopes) or dwarf-shrub (on southern slopes). Stands of *Alnus viridis* ssp. *suaveolens* seem to have partly descended into what had been montane.

3.9.3.3 The Number of Taxa

The Corsican flora comprises 2978 vascular taxa (Gamisans and Jeanmonod 1993) of which 80 are cultivated, 375 are naturalised, adventitious or subspontaneous and 2523 are indigenous (2090 species, 264 subspecies, 87 varieties, 2 forms or cultivars and 80 hybrids). The most species-rich families are Asteraceae and Poaceae (ca. 300), Fabaceae (ca. 250), Apiaceae, Brassicaceae and Caryophyllaceae (ca. 100), followed by Cyperaceae, Liliaceae, Orchidaceae, Lamiaceae, Rosaceae, and Scrophulariaceae (ca. 70). Endemic species make up about 12 % of the indigenous flora (Gamisans et al. 1986; Gamisans and Marzocchi 1996). Some species are rare and some have disappeared recently (Gamisans and Jeanmonod 1995). There are about 600 species of Bryophyta and 1200 species of lichens.

In the alpine zone (≥2100 m on N slopes and ≥2300 m in S exposure – 2710 m), 131 native vascular plant species or subspecies have been found (no introduced species). Eleven of them exclusively occur in this zone and a further 23 are at their optimum distribution here. The other 97 are also present at lower elevations, especially in the upper treeline ecotone. Angiospermae are represented by 32 families (87 genera), Pteridophyta by four (nine genera) and Gymnospermae by one (Table 3.9.1). Some taxa are notably absent, such as the Fabaceae, Gentianaceae and Primulaceae. Of the 131 alpine taxa, 57 are endemic (43.5 %), of which 34 are strict Corsican endemics.

Table 3.9.1. Taxonomic richness of the flora in the alpine zone in Corsica

Taxonomic unit	Number of taxa	Taxonomic unit	Number of taxa
Angiospermae	116	Crassulaceae	4
Poaceae	16	Lamiaceae	4
Asteraceae	14	Ranunculaceae	4
Caryophyllaceae	10	Juncaceae	3
Cyperaceae	8	Orchidaceae	3
Brassicaceae	7	Saxifragaceae	3
Liliaceae	5	Other Angiospermae	10
Apiaceae	5	Gymnospermae	1
Rosaceae	5	Pteridophyta	16
Scrophulariaceae	5		

Of the various life forms (sensu Raunkiaer 1934), hemicryptophytes are by far the most abundant (71.8%), followed by chamaephytes (18.3%) and geophytes (7.6%). Phanerophytes (1.5%, represented by 2 nanerophytes/chamaephytes in the lower ecotones) and therophytes (0.8%) are practically absent.

3.9.4 Vegetation Altitude Zones

The upper limit of montane forest is at 1600 m in N exposure and 1800 m on S slopes (Fig. 23 in Gamisans 1999) and mostly coincides with the upper limit of trees. Only *Abies alba* reaches locally 1900 m in the treeline ecotone (1600–2200 m in N exposure), where generally *Alnus viridis* ssp. *suaveolens* scrub dominates the landscape. This is in contrast with the Alps or Pyrenees, where the upper limit of treeline ecotone trees marks the upper limit of the treeline ecotone. In the cryoro-Mediterranean belt (1800–2300 m in S exposure) the treeless landscape is characterised by spiny dwarf-shrub. Palynological and charcoal evidence suggests that human activity slightly lowered the forest limit during the past millennia.

In the alpine zone, north- and south-facing slopes are notably different (Gamisans 1978, 1999). On northern slopes, alpine gives way to the treeline ecotone, where *Alnus viridis* ssp. *suaveolens* communities are interlaced with alpine grasslands (Table 3.9.2).

On the southern slopes and the ridges, insolation is much higher than on the northern ones and here, the alpine zone is replaced by a cryoro-Mediterranean level. The ecotone is a mixture of xerophytic dwarf-shrubs (*Genista salzmannii* var. *lobelioides*, *Astragalus gennargenteus*, *Berberis aetnensis*) and alpine grassland (Table 3.9.2). The lichen or bryophyte assemblages, that are frequent on rocks and boulders in the alpine zone, are poorly documented. In rock crevasses, two associations of flowering plants and ferns occur. Scree is frequent with two distinct communities on north- and south-facing slopes. Grasslands cover moderate areas only (*Geum montanum–Phleum parviceps* community) or occur in snowbeds (*Omalotheca supina–Sibbaldia procumbens*). The spring and mire vegetation is of high conservation value (*Saxifraga stellaris* ssp. *alpigena–Ranunculus marschlinsii*, *Bellis bernardii–Bellium nivale*; Table 3.9.2).

The vegetation of the alpine zone in Corsica is scant and species-poor compared with the Alps or Pyrenees because of the much poorer flora of Corsica. The sharp contrast between the sun-exposed and the shady sides of the mountains from both ecological (insolation, duration of snow cover) and floristic points of view (larger numbers of endemic and Mediterranean plants on the southern slopes) distinguishes two types of alpine with mainly medio-European on north-facing slopes and a greater oro-Mediterranean influence on south-facing slopes. This is similar to that found on the high mountains of

Table 3.9.2. The altitudinal distribution of vegetation types and their species richness in the Corsican high mountains. *N* on north-facing slopes; *S* on south-facing slopes. Data from Gamisans (1976, 1977a–c, 1978)

Vegetation type	Elevation range (m)	Some characteristic species[a]	Number of species (endemics %)
Grasslands (N)	2100–2700	*Geum montanum, Myosotis corsicana*, Luzula spicata* ssp. *italica**, *Plantago sarda**, *Sagina pilifera**, *Cerastium soleirolii**	46 (40)
Grasslands (S)	2200–2650	*Acinos corsicus**, *Leucanthemopsis alpina* ssp. *tomentosa**, *Draba loiseleurii**, *Bellardiochloa variegata, Armeria multiceps**	50 (52)
Snowbed grasslands (N)	2100–2600	*Phleum parviceps**, *Omalotheca supina, Sibbaldia procumbens*	41 (39)
Rock crevices (N and S)	2100–2600	*Asplenium viride, Armeria leucocephala**, *Draba dubia, Erigeron paolii**, *Festuca sardoa**, *Saxifraga pedemontana* spp. *cervicornis**	41 (39)
Scree (S)	2200–2600	*Galium cometerhizon**, *Stachys corsica**	19 (47)
Scree (N)	2100–2600	*Oxyria digyna, Doronicum grandiflorum*	49 (28)
Springs (N)	2000–2500	*Saxifraga stellaris, Ranunculus marschlinsii**	25 (24)
Mires (pozzines)	1600–2400	*Bellis bernardii**, *Bellium nivale**, *Juncus requienii**, *Pinguicula corsica**	36 (33)
Cryoro-Mediterranean Dwarf-shrub (S)	1800–2300	*Juniperus communis* ssp. *nana, Berberis aetnensis**, *Genista salzmannii* ssp. *lobelioides**	51 (44)
Alnus scrub	1600–2200	*Alnus viridis* ssp. *suaveolens**	46 (29)
Tall-herb formations (N)	1600–2300	*Adenostyles briquetii**, *Athyrium distentifolium*	40 (28)

[a] *indicates endemic species; number of species in the last column refers to total numbers recorded; the Mediterranean flora element ranges between 20–36 % in the various vegetation types found in northerly exposure and between 38–54 % in southerly exposure.

N Greece (e.g. Pindos and Olympus). The chionophilous grasslands (snowbeds) of northerly exposure, such as the *Omalotheca supina–Sibbaldia procumbens* communities, are absent from other more southerly Mediterranean mountains. This also holds true for the *Alopecurus gerardi–Omalotheca hoppeana* of the Olympus (Quézel 1967), which is also in an intermediate biogeographical position between central European and Mediterranean. Mountains south of ca. 41°N, especially those on limestone, are much drier and therefore represent true Mediterranean-type high mountains. The alpine vegetation of Corsica may be considered as intermediate between those of eurosiberian and Mediterranean high mountains.

References

Bocquet G, Widler B, Kiefer H (1978) The Messinian model. A new outlook for the floristics and systematics of the Mediterranean area. Candollea 33:269–287

Braun-Blanquet J (1926) Histoire du peuplement de la Corse: les Phanérogames. Bull Soc Sci Hist Nat Corse 45 ("1925") 473/476:237–245

Contandriopoulos J (1962) Recherches sur la flore endémique de la Corse et sur ses origines. Ann Fac Sci Marseille 32:1–351

Contandriopoulos J (1981) Endémisme et origine de la flore de la Corse: mise au point des connaissances actuelles. Boll Soc Sarda Sci Nat 20:187–230

Favarger C (1975) Cytotaxonomie et histoire de la flore orophile des Alpes et de quelques autres massifs montagneux d'Europe. Lejeunia Nov Ser 77:1–45

Gamisans J (1976) La végétation des montagnes corses, I. Phytocoenologia 3:425–498

Gamisans J (1977a) La végétation des montagnes corses, II. Phytocoenologia 4:35–131

Gamisans J (1977b) La végétation des montagnes corses, III. Phytocoenologia 4:133–179

Gamisans J (1977c) La végétation des montagnes corses, IV. Phytocoenologia 4:317–376

Gamisans J (1978) La végétation des montagnes corses, V. Phytocoenologia 4:377–432

Gamisans J (1981) La montagne corse: une montagne subméditerranéenne marquée par l'endémisme. Anales Jard Bot Madrid 37:315–319

Gamisans J (1996). Les altes muntanyes de Còrsega. Entorns naturals i impacte humà. In: Rafa M, Cervera J (eds) Les altes muntanyes de la Mediterrània, protegir Natura protegir Cultura. Congrés internacional, Mountain Wilderness de Catalunya, Barcelona (1995), pp 25–32

Gamisans J (1999) La végétation de la Corse, 2nd edn. Edisud, Aix-en-Provence

Gamisans J, Aboucaya A, Antoine C, Olivier L (1986). Quelques données numériques et chorologiques sur la flore vasculaire de la Corse. Candollea 40:571–582

Gamisans J, Jeanmonod D (1993) Catalogue des plantes vasculaires de la Corse, 2nd edn. Conservatoire et Jardin botaniques, Genève

Gamisans J, Jeanmonod D (1995) La flore de Corse: bilan des connaissances, intérêt patrimonial et état de conservation. Actes du Colloque Connaissance et Conservation de la flore des îles de la Méditerranée. Ecol Medit 21:135–148

Gamisans J, Marzocchi JF (1996) La flore endémique de la Corse. Edisud, Aix-en-Provence

Gauthier A (2000) Corse des sommets. Albiana, Ajaccio

Quézel P (1967) La végétation des hauts sommets du Pinde et de l'Olympe de Thessalie. Vegetatio 14:127–228

Raunkiaer C (1934) The life forms of plants and statistical plant geography. Oxford University Press, Oxford

Reille M, Gamisans J, Beaulieu JL de, Andrieu V (1999) The Holocene at Lac de Creno (Corsica, France), a key-site for the whole island of Corsica. New Phytol 141:291–307

3.10 The High Mountain Vegetation of the Balkan Peninsula

A. STRID, A. ANDONOSKI and V. ANDONOVSKI

3.10.1 Geography

The Balkan Peninsula lies south of the Rivers Drava (46.5°N) and Danube (44°N) and is surrounded by the Adriatic (14°E) and Ionian Seas in the west, the Mediterranean Sea in the south and the Aegean, Marmara and Black Seas in the east (28°E). It has a total land area of 532,000 km^2, of which an estimated 23,260 km^2 (ca. 4.5%) are at and above the treeline ecotone. The main mountain ranges are the Dinaric Alps (Outer and Inner Dinarids, and Pelagonides and Hellenides), the Rila-Rodopi massif, the Stara Planina Mountains, and the East Balkan Uplands (Demek et al. 1984).

The young fold mountains of the Dinarids run from north-west (46.5°N, 13.5°E; Karniski, Karavanki, Kamniski and Julian Alps with Triglav, at 2864 m being the highest peak) to south-east (42.5°N, 19.5°E; Prokletije Mountains, Mount Jezerces 2694 m). About 21% of the Outer and Inner Dinarids lie above the treeline. Mesozoic limestones and dolomite are the main parent rock types; the Outer Dinarids have the largest contiguous area of karst in Europe (Demek et al. 1984). The Pelagonides run from Shar Planina (2748 m) in the north through Korab, its highest peak (2764 m), to Tajget (2409 m) in the south with ca. 24% lying above the treeline. The Shara group of mountains is composed of blue schists, phyllyte and metamorphic schists overlain by large masses of Mesozoic limestone of Triassic age. The Hellenids form the backbone of mainland Greece, running north-north-west to south-south-east. South of the Peloponnese they disappear below sea level and take a sharp turn east, reappearing in the south Aegean island arc, conspicuously as the three large massifs of Crete; further east it continues in the Cilician and Taurus mountain ranges of southern Anatolia. In north central Greece, two shorter, interrupted mountain ranges run more or less parallel to the Hellenids, and in the northeast are differently oriented ranges with a different geological history. The highest mountain is Olympus (2917 m); there are numerous massifs and peaks above 2000 m throughout mainland Greece and Crete.

In the mountains of Crete, most of the Cyclades, southern Attica, southern Evvia (Euboea), and almost all of north central and north-east Greece, there are large areas of old crystalline rocks (e.g. gneiss, amphibolite, marble). The rest of Greece, including Pindos, most of Sterea Ellas, the Ionian Islands, Peloponnese and part of Crete, consists of younger rocks (Triassic to Tertiary). Olympus and Ossa stand out in the area of older crystalline rocks, being composed partly of old metamorphic limestones (marble) and partly of younger (Mesozoic) limestones. A large area of ophiolite (serpentine) occurs at high and moderate altitudes in northern Pindos, with scattered, smaller outcrops in north central Greece (Vourinos), Sterea Ellas, and the island of Evvia; the main ophiolitic area in north-west Greece continues intermittently through Albania to Bosnia. With respect to the mountain flora and vegetation the most important distinction is between areas of limestone (which are generally dry and rocky), granite and schist (which are wetter and have different, often meadow-like plant communities), and serpentine (which supports a special vegetation and flora rich in local and regional endemics).

The Rhodopean system in the south-eastern part of the peninsula is the oldest Balkan mass which began to emerge in the Palaeozoic. It is composed of various crystalline schists, gneiss, mica schist and phyllite. Its highest summit is Rila (2923 m) with several others above 2100 m; about 52 % of the whole system lies above the treeline.

The Balkan system runs from the Black Sea 530 km westwards. The west Balkans or Stara Planina reach 2168 m of altitude; the highest peak in the central Balkans is the Yamrukchal (2380 m) with the eastern Balkans peaking at around 1500 m. About 3 % of the Balkan system lies above the treeline. The main chain of the middle Balkan mountain system is composed of crystalline schists with a core of old igneous rocks, over which younger sedimentary rocks lie sporadically.

3.10.2 Climate

A temperate-continental element from the north and a Mediterranean one from the south shape the climate of the Balkan Peninsula. Locally, relief and altitude are major factors (Table 3.10.1). In the central part of the Balkan Peninsula, the treeline ecotone is estimated to be between 1700 and 2200 m (Filipovski et al. 1996).

The monthly mean precipitation sum during the period between June and September, when mean air temperature is above zero, is ca. 330 mm at Moussala, Rila, Bulgaria, 2925 m (Roussakova 2000). Summer rainfall (July through September) on the peak of Agios Antonios (2850 m) on Mount Olympus ranged between 86 and 542 mm year^{-1} for the period 1966–1973 (Strid 1980).

Table 3.10.1. Climatic data from the treeline ecotone (Popova Sapka, 1750 m a.s.l., Shara Mountains) and alpine zones (Solunska Glava, 2540 m a.s.l., Jakupica Mountains; Moussala, 2925 m a.s.l., Rhodopi-Rila Mts.). Data for Popova Sapka and Solunska from Filipovski et al. (1996); for Moussala from Roussakova (2000)

	Popova Sapka 1750 m	Solunska Glava 2540 m	Moussala 2925 m
Mean annual temperature (°C)	3.5	0.4	−3.1
Annual mean temperature amplitude (°C)	17.4	16.3	16.7
Mean annual maximum (°C)	8.6	2.7	5.5
Absolute maximum (°C)	30.7	30.1	n.d.
Absolute minimum (°C)	−23.7	−27.9	n.d.
Number of frosty days	149	222	n.d.
Growing season (days with t >10 °C)	105	105	n.d.
Annual precipitation (mm)	1001	791	1000–1300

n.d., no data

The highest precipitation in Greece is received on the west-facing slopes of the Pindos (>1800 mm year^{-1}), and precipitation generally decreases in a south-easterly direction. Above 2000 m on Mt. Olympus, there is more or less continuous snow cover, except on steep, wind-blown slopes, from the end of October to mid-May. Even the high mountains of Crete retain patches of snow until early summer.

3.10.3 Soils

The most common soil types in the alpine and subalpine regions of the Balkan Mountains are (1) lithosols (widespread in the mountains of Dinaric and Shara-Pindos systems); (2) regosols on igneous rocks, crystalline schists and crystalline dolomites; (3) rendzina on flysch and dolomites; (4) ranker (humus-silicate soil) on siliceous rocks; calcomelanosol (limestone-dolomite black soil) on limestone and dolomite; (5) podsol on acid igneous rocks; and (6) brown earth on pure limestone and dolomite (Filipovski et al. 1996).

3.10.4 The Flora

The development of the flora of the Balkan Peninsula is closely related to past geological and climatic processes (Horvat 1962). The great heterogeneity and richness of the Balkan flora including the presence of a large number of relicts and endemics may be explained by the lack of a full ice cover at any time, which was periodically characteristic of most of northern and central Europe.

Data are not readily available for the countries of the former Yugoslavia (but see Micevski 1985–1995 for Macedonia), and Albania is poorly known. Greece is the most species rich of the countries of the Balkan Peninsula. A conservative estimate of the number of native and naturalised species is around 5700 (the Flora Hellenica database lists 5862 species belonging to 1079 genera and 163 families, but these figures include some that are not fully naturalised and probably a few are synonymous taxa; Strid and Kit Tan 1997). The Mediterranean zone between sea level and 800 m comprises 5272 species, the montane zone between 800 and 1800 m 3964, and the treeline ecotone and alpine zone above 1800 m 1737 species. The numbers higher up are 934 above 2000 m, 493 above 2200 m, and 246 above 2400 m. The most species-rich mountain flora is found in northern Greece (Epirus, Macedonia and western Thrace) where several phytogeographical elements meet and there are relatively large land areas above 1000 m (Strid 1986; Strid and Kit Tan 1991).

In Bulgaria, there are between 3550 and 3750 vascular plants within 872 genera of 130 families; in the Rila Mountains, there are ca. 400 vascular plants species in the treeline ecotone and above, with about 100 of them being endemic (Bulgarian, Balkanic, or Carpatho-Balkanic). In the alpine proper, i.e. above 2500 m, the upper limit of *Pinus mugo*, there are ca. 100 flowering plants (Roussakova 2000).

3.10.4.1 Endemism and Speciation

The total number of species endemic to Greece is ca. 740 or 13 % of the flora, a high figure by European standards. Crete is particularly well known for its high incidence of local endemism, even in the lowland flora (see survey in Jahn and Schönfelder 1995). The incidence of endemism increases with altitude and in a southerly direction. The Mountain Flora of Greece treats ca. 1600 species and subspecies, ca. 405 of which are endemic to Greece or a smaller area. Among mountain plants the combined frequencies of Greek endemics, single-area endemics and single mountain endemics are 42.0 % in Crete, 26.4 % in the Peloponnese, 21.6 % in Sterea Ellas, and between 10.0 and 6.4 % in the more northerly regions. The latter figures tend to underestimate the incidence of narrow endemism in the north, since species occurring across the borders to Albania, F.Y.R. Macedonia or Bulgaria are classified as

Balkan endemics (not Greek endemics or single area endemics) even if their total distribution area may be quite small. Even with this factor taken into consideration, however, it is obvious that the incidence of endemism increases in a southerly direction, reflecting the age and isolation of oreo-Mediterranean floras.

Edaphic endemism is most pronounced in the serpentine areas in north-west Greece. A large group of perhaps 100 local and regional species occur exclusively or predominantly in the range of serpentine mountains extending intermittently from north-west Greece through Albania to Kosovo and Bosnia. They include species such as *Alyssum smolikanum, Bornmuellera baldaccii, B. tymphaea, Campanula hawkinsiana, Centaurea ptarmicifolia, Crepis guioliana, Euphorbia glabriflora, Fritillaria epirotica, Halacsya sendtneri, Matricaria tempskyana, Minuartia baldaccii, Poa ophiolitica, Saponaria intermedia, Silene schwarzenbergeri* and *Viola dukadjinica*. Some of those currently classified as Greek endemics are likely to be 'lost' as the flora of Albania becomes better known.

Preliminary data indicate that the high incidence of local and regional taxa in the Greek mountains results partly from the survival of ancient, regressive species (paleo-endemics) but mainly from recent speciation through geographical isolation (neo-endemism). The frequency of polyploidy is low among the endemics. A survey of known chromosome numbers in Greek mountain plants (Strid 1993: 431) indicated 10% polyploids among the single mountain endemics and 15% among the single area endemics, but 40–48% among widespread (central European, Euro-Siberian or cosmopolitan) species.

The frequency of arctic-alpine or boreal taxa decreases rapidly as one moves south on the Balkan Peninsula. Large, non-calcareous massifs such as Rila in Bulgaria and Shar Planina on the borders of Kosovo and Macedonia are home to many such taxa not extending to Greece. Some have their southernmost occurrences on schistose or granitic mountains in northern Greece, usually on Voras and/or Varnous, occasionally extending to Vermion, Gramos or Smolikas. Among these are *Achillea clusiana, Carex lasiocarpa, Juncus trifidus, Senecio subalpinus, Stellaria uliginosa, Trifolium badium, Veronica bellidioides* and *Viola palustris*. A few calciphilous species have similar distributions, e.g. *Dryas octopetala*, which extends to Mt. Falakron in north-eastern Greece.

3.10.5 Vegetation

The treeline tree species in the Dinarids and the Balkan and Rhodope Mts. are *Acer heldreichii, Fagus orientalis, F. moesica, Pinus peuce,* and *P. heldreichii*. *Pinus mugo* scrub occurs in the Dinarids (former Yugoslavia) between 1550 and 1750 m in Slovenia in the north and between 1500 and 2200 m in Macedonia in the south, and in the Balkan, Pirin and Rila in Bulgaria between 1900 and 2500 m (Horvat et al. 1974). *Alnus viridis* scrub occurs in the above areas, often as secondary vegetation. Tall-herb vegetation, anthropogenic tall-herb vegetation, and dwarf-shrub heath are the other major vegetation types in the treeline ecotone (Table 3.10.2).

The treeline in northern Greece is mostly formed by *Pinus sylvestris, P. heldreichii* or occasionally *Fagus sylvatica*, in central and southern Greece by *Pinus nigra, Abies borisii-regis* and *A. cephalonica*, and in Crete by *Cupressus sempervirens*. The natural treeline lies at ca. 2300 m on Mount Olympus, elsewhere it ranges between 1800 and 2000 m. The treeline is often artificially lowered to between 1500 and 1900 m as a result of cutting, burning and grazing. In Greece the treeline ecotone vegetation is generally poorly developed, and on most mountains there is no krummholz zone. Large areas above the forest line, especially on southern slopes, are dominated by xerophytic thorny cushion vegetation (Table 3.10.2; Horvat et al. 1974).

The alpine zone ranges from 2000 m (northern Dinarids) and 2200–2500 m (Balkan and the Rila and Pirin Mts.) upwards to the highest peaks, comprising a suite of azonal scree and rock vegetation, calcareous and acid grasslands and sedge heaths, snowbeds, soligenous mires and spring vegetation (Table 3.10.2).

Above the treeline, in Greece, the most important habitats are (1) grasslands associated with late snow lie (>210 days) and irrigation from meltwater in shallow depressions or on more or less flat ground with accumulated fine-grained soil, (2) tussock grasslands on moderately sloping ground, often with some solifluction, (3) stabilised or moving screes, and (4) cliffs and rocks (Quézel 1967). The more mesic *Nardus* or *Bellardiochloa variegata* type grasslands, and to a lesser extent the drier *Festuca-* or *Sesleria*-dominated type, are more important on the more mesic mountains of granite or micaceous schist, especially in northern Greece, whereas screes and rocky habitats are prominent on limestone and serpentine (Horvat et al. 1974).

In the past, a great deal of the alpine pine scrub (*Pinus mugo*) in the Rila and Pirin mountains and nearly all over the Balkan has been destroyed in order to enlarge alpine pastures and to produce charcoal. Today, these areas are covered by communities dominated by various combinations of *Juniperus communis* ssp. *nana, Chamecytisus eriocarpus, Vaccinium myrtillus, V. vitis-idaea, Bruckenthalia spiculifolia, Dryas octopetala* dwarf-shrub heaths, and *Nardus stricta, Agrostis capillaris, Festuca valida, Bellardiochloa variegata, Festuca penzesii,* and *Sesleria korabensis* grasslands. Lately, *Pinus mugo* has been recovering by regeneration from seed in some locations. This process is very important for increasing the biomass and plant cover in the treeline ecotone with respect to ecosystem stabilisation.

Table 3.10.2. Comparative table of the vegetation of the treeline ecotone and the alpine zone of the Alpine vs. Mediterranean high mountains of the Balkan Peninsula. Figures in parentheses are number of communities reported in Horvat et al. (1974). Sources: Quézel (1967), Horvat et al. (1974), Roussakova (2000)

Dinarids, Shar-Korab, Rila-Pirin, Stara Planina	Altitude (m)	Hellenides	Altitude (m)
Snowbeds (11) Creeping willows: *Salix retusa*, *S. reticulata*; dwarf herbs and rosettes: *Saxifraga glabella* (Ca), *Omalotheca* spp., *Soldanella alpina*		Snowbed grasslands (3) Grasses: *Alopecurus gerardii*; dwarf herbs: *Omalotheca* spp.	(2100)–2400
Grasslands Calcareous, primary (22) Grasses: *Sesleria* spp., *Festuca pungens*; sedges: *Carex ferruginea*; herbs: *Edraianthus graminifolius*, *Oxytropis urumovii*	1900–2150	Grasslands Calcareous, summer-dry (5) Grasses: *Sesleria* spp.; thorny cushions: *Acantholimon echinus*	1700–2200
Acid grasslands and sedge heaths (25) Grasses: *Nardus stricta*, *Sesleria comosa*; sedges and rushes: *Carex curvula*, *Juncus trifidus*; herbs: *Jasione orbiculata*	2000–2500	Acid grasslands associated with late snow lie and seep water irrigation (7) Grasses: *Nardus stricta*, *Bellardiochloa variegata*; herbs: *Trifolium parnassi*	1900–2400
Calcareous, secondary grasslands (10) Grasses: *Festuca varia*, *F. xanthina*; herbs: *Onobrychis montana*	1500–1800		
Rock vegetation (24) Forbs and rosettes: *Asplenium* spp., *Silene* spp., *Ramondia nathaliae*, (Ca), *Potentilla haynaldiana* (Si); creeping dwarf shrubs: *Micromeria croatica*	1500–2400	Rock vegetation (14) Forbs: *Silene auriculata*, *Galium* spp., *Ramondia nathaliae*	1500–2000–(2400)
Scree vegetation (13) Forbs and rosettes: *Cardamine glauca*, *Rumex scutatus*, *Saxifraga pedemontana* ssp. *cymosa*; dwarf shrubs: *Drypis spinosa* (Ca)	1800–2200	Scree vegetation (10) Forbs and rosettes: *Silene multicaulis*, *Campanula hawkinsiana*, *C. caesia*	1600–2000–(2400–)

Table 3.10.2. (Continued)

Dinarids, Shar-Korab, Rila-Pirin, Stara Planina	Altitude (m)	Hellenides	Altitude (m)
Springs (3) Rosettes: *Saxifraga aizoides*, *S. stellaris*; mosses: *Cratoneuron* spp. (Ca), *Philonotis* spp. (Si)		Springs (1?)	
Soligenous mires (5) Sedges and rushes: *Carex* spp., *Trichophorum cespitosum*, *Eriophorum vaginatum* (Si)		Soligenous mires (2) Sedges and rushes: *Blysmus compressus*; herbs: *Leontodon hispidus*, *Pinguicula hirtifolia*	
Tall-herb communities (16) *Adenostyles alliariae*, *Cicerbita alpina*, *Veratrum album*, *Rumex alpinus*		Tall-herb communities (1) *Cirsium tymphaeum*, *Veratrum album*	
Dwarf-shrub heath (6) Ericaceous dwarf shrubs: *Bruckenthalia spiculifolia*, *Rhododendron hirsutum*, *Vaccinium* spp.; *Empetrum nigrum*, other dwarf shrubs: *Juniperus communis* spp. *nana*	1700–2500	Thorny cushion vegetation (7) Thorny cushions: *Astragalus* spp., *Centaurea affinis*; Dwarf shrubs: *Marrubium velutinum*; geophytes: *Crocus* spp, *Tulipa* spp, *Scilla* spp.	1700–
Treeline ecotone scrub (3) Prostrate pines and alders: *Pinus mugo*, *Alnus viridis*	1700–1900	Thorny cushion vegetation (2) Thorny cushions: *Astragalus* spp. (sect. *Tragacantha*)	1500–1700

References

Demek J, Gams I, Vaptsarov I (1984) Balkan peninsula. In: Embleton C (ed) Geomorphology of Europe. Macmillan, London, pp 374–386

Filipovski G, Rizovski R, Ristevski P (1996) The characteristics of the climate-vegetation-soil zones (regions) in the Republic of Macedonia. Macedonian Academy of Sciences and Arts, Skopje

Horvat I (1962) Die Vegetation Sudosteuropas in klimatischen und bodenkundlichen Zusammenshang. Mitt Oesterr Geogr Gess 107, I/II Wien

Horvat I, Glavač V, Ellenberg H (1974) Vegetation Südosteuropas. Gustav Fischer Verlag, Stuttgart

Jahn R, Schönfelder P (1995) Exkursionsflora von Kreta. Eugen Ulmer, Stuttgart

Micevski I (ed) (1985–1995) The flora of Macedonia. Macedonian Academy of Sciences and Arts, Skopje

Quézel P (1967) La végétation des hauts sommets du Pinde et de L'Olympe de Thessalie. Vegetetio 14:127–228

Roussakova V (2000) Végétation alpine et sous alpine supérieure de la Montagne de la Rila (Bulgarie). Braun Blanquetia 25:1–132

Strid A (1980) Wild flowers of Mount Olympus. Goulandris Natural History Museum, Kifissia

Strid A (ed) (1986) Mountain flora of Greece vol 1. Cambridge University Press, Cambridge

Strid A (1993) Phytogeographical aspects of the Greek mountain flora. Fragm Florist Geobot Suppl 2:411–433

Strid A, Kit Tan (eds) (1991) Mountain flora of Greece, vol 2. Edinburgh University Press, Edinburgh

Strid A, Kit Tan (eds) (1997) Flora Hellenica, vol 1. Koeltz Scientific Books, Königstein

Strid A, Kit Tan (eds) (2002) Flora Hellenica, vol 2. Ganter Verlag, Lichtenstein

II Plant and Vegetation Diversity

4 Overview: Patterns in Diversity

U. Molau

4.1 Scale and Diversity

The alpine landscape is, simplistically, characterised by its diversity of vegetation, landforms, parent rock types, soils, and water bodies. This diversity distinguishes it from the topographically far less varied, and hence less diverse, tundra region (Chapin and Körner 1995). At a Europe-wide scale, glacial history is an important factor for alpine plant diversity patterns, as exemplified by the high degree of endemism in non-glaciated refugia, particularly in southern and central Europe (Williams et al. 1999). Doubtless, postglacial climatic fluctuations contributed also, however, at local to regional scales and ecological factors alone may explain contemporary species diversity patterns (e.g. Birks 1996). Local (α and β diversities) and landscape-scale patterns (γ diversity) are a function of regional species pools and depend on local biotic and abiotic conditions (Chapin and Körner 1995). Altitude patterns in species richness, for example, show a general decrease in species richness with site-specific variation (Lomolino 2001).

There is substantial variation in species composition, abundance, scale and pattern. Naturally, the scale and pattern of variation observed depends on the organisms sampled. Work on soil microbes may require a sampling unit of 1 cm^3 or less to achieve an adequate representation of genotypes and species. Numerous mosses and lichens coexist within a fraction of a square metre (Körner 1995; Molau and Alatalo 1998), whereas vascular plants are observed at a few m^2 to several 100 m^2 or higher scales. However, perceived patterns differ considerably depending on the different extent of areas and sampling unit sizes used in studies (e.g. Wiens 1989). The scale dependence of observations necessarily limits the inferences that can be made and the applicability of the results. Using different sampling unit sizes and extents of area to investigate if correlations between the environment and vegetation are similar over a range of scales is a way to overcome this limitation (e.g. Allen and Hoekstra 1991), an approach adopted in Chapters 6 and 11.

An observed pattern may result from either spatially determined growth of species (e.g. Körner 1999), especially of those with restricted dispersal, or may be caused by one or several environmental (abiotic and biotic) factors (Legendre and Legendre 1998). A classification of alpine habitats is provided, e.g. by Devilliers and Devilliers-Terschuren (1996) and Grabherr et al. (2000). In alpine environments, glacial and peri-glacial landforms can largely determine the ecological conditions of habitats through their influence on microclimate. Exposed ridges are home to wind-clipped prostrate heath; bare rock, with its fissures and crevices, provides a habitat for small-stature specialists (e.g. Reisigl and Keller 1990; Chaps. 3.1–3.10). The vegetation of screes, which range from relatively fine-grained unstable screes to boulder fields, reflects the stability and microclimate of the habitats; small growing species occur on fine scree and tall herbs take advantage of the shelter afforded by boulders. Glacial cirques or corries and glacial troughs have a range of grasslands and snowbed types (e.g. Körner 1999). The absolute and relative cover of these habitats determines overall species richness of a landscape, often the focus of a search for a single, altitude-determined pattern (e.g. Chaps. 7 and 8). Environmental variables that dominate the picture during part of the year may still linger on in various ways through the rest of the season. Winter snow distribution for instance will govern not only the subsequent summer's flowering phenology of plants, but also the territorial borders of some passerine birds. Small grazing animals, such as voles and lemmings, respond to cues set by the vegetation, and by and large operate within a 100-m radius. Their predators, however, range at the landscape scale. In an alpine landscape, a raven pair may have a territory of several km^2 in one catchment, whereas a single highly mobile top predator, such as the wolverine, can cover 4–5 adjoining catchments with an area of 100 km^2.

Clearly, it is important to establish a relationship between diversity patterns across spatial scales. Until fairly recently, however, biodiversity has mainly been viewed at two contrasting scales, at the local scale from less than a metre square to a hectare (based on field mapping and data sampling within an ecosystem), and the regional scale (based on literature surveys and remote sensing, notably satellite images). The 'mesoscale' has long been the missing link, but with the emergence of GIS techniques our abilities to handle data at the landscape level have greatly improved. Now we are better placed to make mesoscale comparisons, mostly at the landscape/catchment scale across major organism groups for any kind of biodiversity assessment, be it taxonomical, genetic, or ecological.

4.2 Temporal Diversity

Scale in time is equally important (Fig. 4.1). Most alpine plant species are perennial. In fact, they are often extremely long-lived; a standard-size cushion of, e.g., *Diapensia lapponica* or *Silene acaulis*, some 10–20 cm across, may be hundreds of years old (Molau 1997) and genets of clonal species such as the dominating sedges, rushes, and grasses in alpine grasslands may persist even much longer, maybe several thousands of years (Steinger et al. 1996). Thus, a single summer with bad weather resulting in a zero seed set is of little importance, viewed in terms of the lifetime performance of such organisms. Some mire and tussock tundra plants (e.g. *Eriophorum vaginatum*, cottongrass) display synchronous mass flowering in certain years, with the drivers of this cyclicity seemingly controlled by climatic triggers. The observed flowering density is a function of temperature summed over the summers 3 or 4 years in the past, which control the initiation of new daughter tillers. Extreme low densities in cottongrass flowering is mostly the effect of a 'hangover' due to resource depletion from preceding mass flowering events. The astonishing synchrony of flowering over entire regions is probably harmonised by unusu-

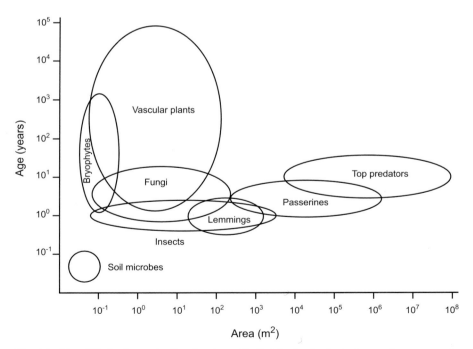

Fig. 4.1. Simplified schematic distribution of individuals of selected organism groups in alpine landscapes in time and space. Note the tendency of plants to group along the time axis whereas animals group along the space axis

ally cold or warm summers, or both (Molau and Shaver 1997). Herbivores undergo population oscillations as well, the best known being the cyclicity of the lemmings (Chap. 20). The triggers of their oscillations are, however, still rather enigmatic. The important difference in the temporal scale lies in the difference in strategy between the long-lived perennial plants and their much more short-lived grazers. An alpine plant, with its clonal being, may undergo several masting events during its lifetime. The animal grazers, such as the lemmings, are short-lived animals, and a population cycle including low and high densities encompasses many consecutive generations of animals. Nevertheless, large-scale synchrony in population numbers or flowering is most prominent in the alpine zone and the arctic tundra and can be compared only with massive performance of desert plants after major rainfalls, having been dormant previously for years or decades in the seed bank. A cold year can trigger synchrony much more effectively when the relative distance to the temperature threshold for life (i.e. 0 °C) is small.

4.3 Diversity at Landscape and Continent Scales

Chapters 5–11, which comprise this section, all touch on landscape level issues in one or several ways. A general feature linking all the papers is the importance of spatial patterns and/or altitudinal differentiation. Chapters 5 and 6 address issues at a European scale, Chapters 7–9 deal with the patterns within selected mountain ranges, and the final two (Chaps. 10 and 11) belong to the case study categories, focusing on a single alpine landscape. All of these papers are based on the spatial patterns of plant species and communities, even though animal diversity patterns and time scales are considered in Chapter 11.

At the continental scale, Chapter 5 makes use of the enormous database of the eleven volumes of *Atlas Florae Europaeae* published to date. Their estimate for total species richness in the European alpine vascular plant flora amounts to ca. 2500 species and subspecies, which comprises about one-fifth of the total European species pool. Since the alpine zone makes up only about 3 % of the European land area, the alpine areas house the highest species richness within the continent. Among all European mountain ranges, the Alps and the Pyrenees are the most species rich, and they also share the highest proportion of species in a correlation matrix. Local endemism is surprisingly high, and 10 % of the taxa are known to occur only within a single 50 × 50 km grid cell. The number of endemics is highest in the Alps and the Pyrenees, whereas the mountains of northern Europe have a much poorer flora. The authors demonstrate the usefulness of approaching species richness at this large scale in order to consider diversity hotspots and nature conservation in an international context.

In a related study, Chapter 6 examines a set of published and unpublished analyses of plant communities and species richness from all major mountain ranges in Europe. Habitats with little, medium, and long-lasting snow protection both on calcareous and siliceous soil substrata are compared. Across all mountain ranges, habitats with medium snow protection have a higher species richness than in exposed sites and snow-beds. Vascular plant richness shows no correlation with the size of mountain ranges, but is higher on calcareous than siliceous soils in the mountain ranges of northern Europe. The authors also demonstrate a major trend of vascular plant species richness declining with increasing latitude, well in accordance with patterns for many other cases (Rosenzweig 1995; Chown and Gaston 2000). Interestingly, the opposite trend is exhibited by bryophytes and macrolichens, indicating a lack of positive covariance between groups of plants (Gaston 2000).

Chapters 7 and 8 explore the spatial variation in plant diversity along altitude gradients in the southern Scandes and the westerns Alps, respectively. In Chapter 7, plant distributions are plotted in relation to two climatic variables, temperature (as a function of altitude) and continentality. Alpine vascular plant species richness reaches a maximum 100–200 m above the treeline in the southern Scandes, and, as pointed out in Chapter 6, species richness is higher on calcareous soils than on siliceous soils at similar altitudes. The species richness decline with altitude is not linear, but instead appears to follow a stepped curve (see also Grabherr et al. 1995). Some of the findings of Chapter 7 from Norway are reflected also in the Alps (Chap. 8). Here, the authors report a consistent decline in vascular species richness with increasing altitudes along all sampled transects, but with the bryophytes exhibiting different and less consistent patterns. The plants are divided into life forms, and the elevation patterns in richness are discussed in relation to various hypotheses. One conclusion in the paper is that vascular plant richness is more directly controlled by temperature than is bryophyte diversity. A surprising phenomenon is that this decline in species richness with altitude correlates with the reduction in available land area in such a way that the resultant species number per unit land area remains fairly constant over a wide range, despite increasingly adverse climatic conditions (Körner 2000; Körner and Spehn 2002).

A somewhat different approach is reported in Chapter 9. The standardised sampling design of the Global Observation Research Initiative in Alpine Environments, GLORIA is applied to address the question whether the relative contribution of endemic plant species should increase with altitude in two contrasting alpine areas (the Hochschwab region in the Austrian Alps and the Sierra Nevada in Spain). The richness of endemics at high altitudes in Sierra Nevada is well documented, and was obvious in this study as well, but in the Hochschwab region there was no such differentiation between endemics and the others. Climate warming is anticipated to be hazardous to the cryophilic species at the highest altitudes in the Alps where no upward distribution shift is feasible. In the Sierra Nevada, on the other hand, drought in combination with an increasingly early spring snowmelt is considered to be the most important threat to the region's endemics.

The section ends with two GIS-based case studies in alpine landscape ecology, one from the Austrian Alps (Chap. 10) and one from the Scandes in northern Swedish Lapland (Chap. 11). Chapter 10 evaluates the applicability of remotely sensed data and analytical methods and their accuracy in vegetation mapping, based on a case study made in the Austrian Alps. The studied landscape in the Hochschwab region encompasses several adjacent catchments (sources for Vienna's drinking water) and amounts to an area of 60 km^2. By using aerial photographs, image analysis, and topographical variables, some 20 plant communities were mapped. The results are discussed in terms of evaluation of the methods and their limitations. The lowest accuracy was found in the mapping of azonal plant communities, communities affected by grazing, and fine-scale mosaics of alpine communities brought about by rugged topography and a consequent mosaic of microclimates. An important message is that automated image processing, compared with traditional field mapping, not only provides a vegetation map, but also offers the possibility of further GIS-based analysis. Importantly, it allows the handling of errors, a key issue in spatial analysis, but largely neglected in cartography.

Many of the general aspects of biodiversity at the landscape level in the alpine zone are found in the case study from the Latnjajaure catchment in the northern Scandes (Chap. 11). A hierarchial GIS approach is adopted in this single-catchment study. Parts of the landscape were staked and mapped in the field, and serve for ground truthing for the analysis of remotely sensed images of the entire catchment. A study of passerine bird territory sizes and spatial distribution (the two most abundant species: wheatear, *Oenanthe oenanthe* and meadow pipit, *Anthus pratensis*) superimposed on the vegetation map served as an example of how plant community diversity can be compared with components of animal diversity. The passerine bird territories embraced several plant communities, with no obvious plant species-specific determinants of the territorial boundaries (see also Currie et al. 2000; Chap. 19). A correlation between territory distribution and snowmelt phenology proved more revealing, however (Molau, Lindblad and Dänhardt, submitted), partly linked to the availability of key prey. Thus, the Latnjajaure study combines considerations of plant and animal diversity and abiotic factors, in an approach which is of general interest in the light of an increasing demand for appropriate methods for biodiversity assessment. The relative structural simplicity of alpine landscapes makes these well suited for studies which assess biodiversity across organism groups and across scales in space and time.

Acknowledgements. The original draft was much improved by G. Grabherr, Ch. Körner, L. Nagy, D.B.A. Thompson and R. Virtanen.

References

Allen TFH, Hoekstra WH (1991) Role of heterogeneity in scaling of ecological systems under analysis. In: Kolasa J, Pickett STA (eds) Ecological heterogeneity. Springer, Berlin Heidelberg New York, pp 47–68

Birks HJB (1996) Statistical approaches to interpreting diversity patterns in the Norwegian mountain flora. Ecography 19:332–340

Chapin FS III, Körner C (eds) (1995) Arctic and alpine biodiversity: patterns, causes and ecosystem consequences. Springer, Berlin Heidelberg New York

Chown SL, Gaston KJ (2000) Areas, cradles and museums: the latitudinal gradient in species richness. TREE 15:311–315

Currie D, Thompson DBA, Burke T (2000) Patterns of territory settlement and consequences for breeding success in the northern wheatear *Oenanthe oenanthe*. IBIS 142:389–398

Devilliers P, Devilliers-Terschuren J (1996) A classification of Palaearctic habitats. Nature and environments. Council of Europe, Strasbourg

Gaston KJ (2000) Global patterns in biodiversity. Nature 405:220–227

Grabherr G, Gottfried M, Gruber A, Pauli H (1995) Patterns and current changes in alpine plant diversity. In: Chapin FS III, Körner C (eds) Arctic and alpine biodiversity: patterns, causes and ecosystem consequences. Springer, Berlin Heidelberg New York, pp 167–181

Grabherr G, Gottfried M, Pauli H (2000) Hochgebirge als 'hot spots' der Biodiversität – dargestellt am Beispiel der Phytodiversität. Ber Reinh Tüxen Ges 12:101–112 (with English summary)

Körner C (1995) Alpine plant diversity: a global survey and functional interpretations. In: Chapin FS III, Körner C (eds) Arctic and alpine biodiversity: patterns, causes and ecosystem consequences. Springer, Berlin Heidelberg New York, pp 45–62

Körner C (1999) Alpine plant life. Springer, Berlin Heidelberg New York

Körner C (2000) Why are there global gradients in species richness? Mountains might hold the answer. TREE 15:513

Körner C, Spehn E (2002) Mountain biodiversity a global assessment. Parthenon, Boca Raton

Legendre P, Legendre L (1998) Numerical ecology, 2nd edn. Developments in environmental modelling 20. Elsevier, Amsterdam

Lomolino MV (2001) Elevation gradients of species-density: historical and prospective views. Global Ecol Biogeogr 10:3–13

Molau U (1997) Age-related growth and reproduction in *Diapensia lapponica*, an arctic-alpine cushion plant. Nordic J Bot 17:225–234

Molau U, Alatalo JM (1998) Responses of subarctic-alpine plant communities to simulated environmental change: biodiversity of bryophytes, lichens, and vascular plants. Ambio 27:322–329

Molau U, Shaver GR (1997) Controls on seed production and seed germinability in *Eriophorum vaginatum*. Global Change Biol 3 (Suppl 1):80–88

Reisigl H, Keller R (1990) Fiori e ambienti delle Alpi. Arti Grafiche Saturnia, Trento

Rosenzweig ML (1995) Species diversity in space and time. Cambridge University Press, Cambridge

Steinger T, Körner C, Schmid B (1996) Long-term persistence in a changing climate: DNA analysis suggests very old ages of clones of alpine *Carex curvula*. Oecologia 105:94–99

Wiens JA (1989) Spatial scaling in ecology. Funct Ecol 3:385–397

Williams P, Humphries C, Araujo M (1999) Mapping Europe's biodiversity. Making a start with species: combining atlas data for plants and vertebrates. European Centre for Nature Conservation and the Natural History Museum, London

5 Taxonomic Diversity of Vascular Plants in the European Alpine Areas

H. VÄRE, R. LAMPINEN, C. HUMPHRIES and P. WILLIAMS

5.1 Introduction

Using a relatively broad species concept, *Flora Europaea* describes about 11,500 native species (Tutin et al. 1964, 1968–1980). The richest areas are around the Alps, Pyrenees and the Balkan Peninsula, where altitudinal zonation results in diverse habitats with many different species or subspecies within small areas (Ozenda 1983; Lahti and Lampinen 1999; Williams et al. 2000). Although there have been previous attempts to understand the relationships and hence the history of the mountain systems in Europe, a comprehensive analysis by comparing mountain ranges across Europe is difficult. The distribution data are compiled in various local floras that use administrative or political boundaries rather than natural biogeographical divisions. Moreover, workers at different times, even in similar areas, have used different species concepts, thus making comparisons difficult. As a new initiative, we examine here the richness and distribution of European alpine species using data from the first 11 volumes of *Atlas Florae Europaeae* (AFE). AFE (Jalas et al. 1996) provides individual maps for more than 3000 vascular plant taxa from Lycopodiaceae to Brassicaceae, covering about 20% of all taxa treated in the five volumes of *Flora Europaea*. What makes this particular analysis possible is that these data have been digitised recently, enabling the use of the necessary diversity measures and area-selection analyses of alpine taxa using the WORLDMAP software (Humphries et al. 1999; Lahti and Lampinen 1999; Williams et al. 2000).

Several factors make comparisons of alpine taxa difficult. Different definitions of the term alpine, and, in turn, the extent of the alpine zone, causes problems, especially when data are compared between major biogeographical regions, such as the Alps and northern Scandinavia. Closer to the Arctic Ocean, it is not always apparent whether an area represents an arctic or an alpine ecosystem. For the purpose of this work we define alpine as referring to areas above the altitudinal treeline, although this varies considerably in different parts of Europe and is a questionable definition for Mediterranean high mountains. Scattered trees may occur far above the treeline on mountain slopes and rocky slopes may be treeless at lower altitudes. However, by using this convention, overlap between alpine and arctic areas as defined by Talbot et al. (1999) is reduced. The alpine habitats represent about 3 % of the surface area of Europe (cf. Chaps. 1 and 6).

5.2 Methods

5.2.1 The Selection of Taxa

Our analyses on vascular plant richness distribution patterns used digitised distribution data for over 630,000 records imported into WORLDMAP 4.1 for Windows (Humphries et al. 1999), on the ca. 50 × 50 km grid cells used by the AFE. Analyses included those taxa from the AFE (Vols. 1–11) which grow predominantly above the altitudinal tree limit, although many alpine taxa occurring in upper montane woods were also included. This selection included families which are common in the alpine zone such as Brassicaceae and Caryophyllaceae, but excluded those from later volumes of *Flora Europaea* such as Asteraceae, Cyperaceae and Poaceae. Several taxa found in alpine habitats were also excluded because their distribution ranges also extend way below the treeline. Local floras were consulted in order to check data on altitudinal limits (Pawlowski 1956; Villar 1980; Strid 1986; Landolt 1992; Engelskjøn and Skifte 1995). Areas north of latitudinal tree limits, i.e. the coasts of the Arctic Ocean and Iceland, were excluded from the selection process. Several alpine species also grow in these regions, representing the arctic-alpine element. The Caucasus is not included in AFE.

5.2.1.1 Taxon Richness

The simplest and most popular measure of biodiversity is taxon richness (number of taxa per grid cell), which in this study is simply a count of species and subspecies per unit area. Relative richness is provided for all of the included 501 alpine taxa (Fig. 5.1), separately for alpine Brassicaceae, *Draba* spp., *Minuartia* spp. and *Salix* spp. (Figs. 5.2–5.5). The taxa of the Alps, the mountains of Grecce, the Pyrenees, the Scandes and of the Tatra Mountains are mapped separately also (Figs. 5.6–5.10), and may include local species additional to the 501 taxa in Fig. 5.1. In regional comparisons the selection of species is based on compatible data sets. The colour scale (red to blue) uses

Taxonomic Diversity of Vascular Plants in the European Alpine Areas

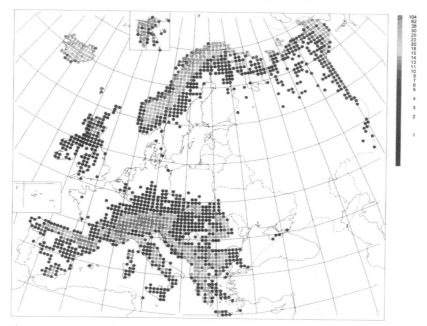

Fig. 5.1. Taxon richness for 501 predominantly alpine species and subspecies mapped in the first 11 volumes of *Atlas Florae Europaeae* (Jalas and Suominen 1972–1994; Jalas et al. 1996). Each cell is approximately 50 × 50 km. *Red cells* indicate high richness, *blue cells* low richness

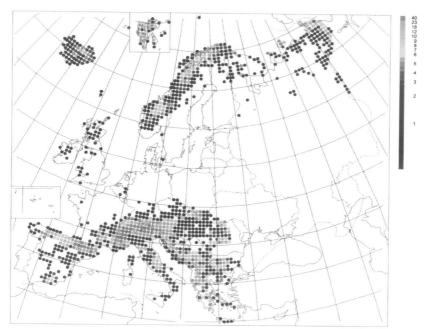

Fig. 5.2. Taxon richness for 125 alpine Brassicaceae. For explanations, see Fig. 5.1

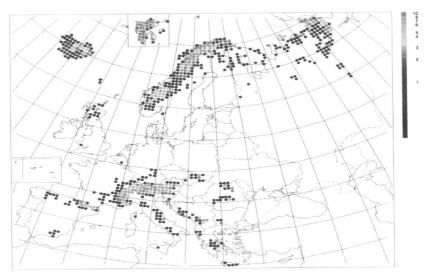

Fig. 5.3. Taxon richness for 22 alpine *Minuartia* spp. For explanations, see Fig. 5.1

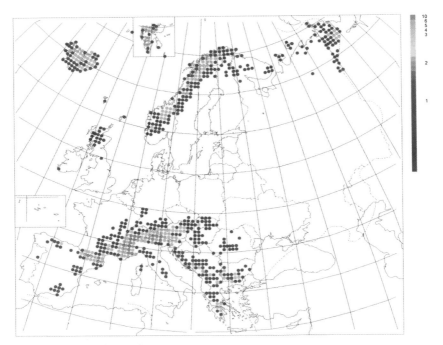

Fig. 5.4. Taxon richness for 22 alpine *Salix* spp. For explanations, see Fig. 5.1

Taxonomic Diversity of Vascular Plants in the European Alpine Areas

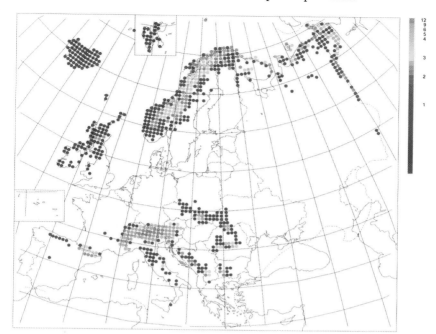

Fig. 5.5. Taxon richness for 27 alpine *Draba* spp. For explanations, see Fig. 5.1

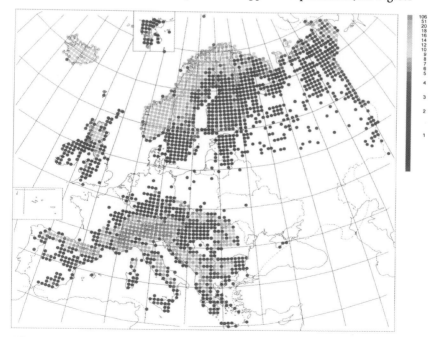

Fig. 5.6. Taxon richness in Europe for 191 alpine vascular plant species and subspecies of the Alps mapped in the first 11 volumes of *Atlas Florae Europaeae* (Jalas and Suominen 1972–1994; Jalas et al. 1996). Each cell is approximately 50 × 50 km. *Red cells* indicate high richness, *blue cells* low richness

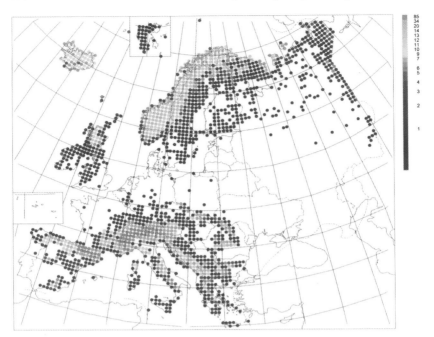

Fig. 5.7. Taxon richness for 118 alpine species and subspecies in the Pyrenees. For explanations, see Fig. 5.6

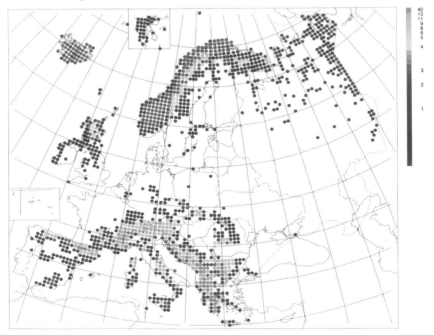

Fig. 5.8. Taxon richness for 89 alpine species and subspecies in the Greek mountains. For explanations, see Fig. 5.6

Taxonomic Diversity of Vascular Plants in the European Alpine Areas

Fig. 5.9. Taxon richness for 54 alpine species and subspecies in the Scandes. For explanations, see Fig. 5.6

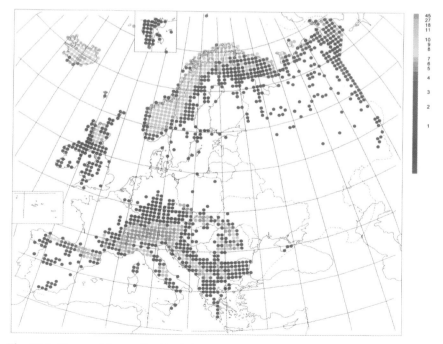

Fig. 5.10. Taxon richness for 53 alpine species and subspecies in the Tatra Mountains. For explanations, see Fig. 5.6

Table 5.1. **A** Total number of alpine taxa from *Atlas Florae Europaea* vols. 1–11 (AFE) in the studied mountain ranges of Europe with (estimates of total numbers of alpine taxa per mountain range – based on the assumption that about 20 % of all European taxa have been mapped in AFE); **B** the maximum number of taxa per AFE grid cell with (the number of AFE taxa in the richest cell expressed as the percentage of the total AFE taxa in the mountain range)

Area	A	B
Alps	190 (950)	106 (56 %)
Pyrenees	115 (575)	85 (74 %)
Mountains of Greece	90 (540)	40 (44 %)
Scandes	50 (250)	36 (72 %)
Tatra Mountains	53 (265)	45 (85 %)

Table 5.2. Jaccard's similarity index values between the studied mountain ranges:

Area[a]	P	G	S	T
A	25	6	5	20
P		8	6	17
G			3	4
S				11

[a] *A* Alps; *P* Pyrenees; *G* Mountains of Greece; *T* Tatra Mountains; *S* Scandes

red for high taxon richness and blue for low. The number of taxa in these mountain ranges is given in Table 5.1 and pair-wise similarities between these ranges (Table 5.2) were calculated using Jaccard's similarity index (JS).

$JS = a/(a+b+c) \times 100$,

where *a* is the number of species common to two areas, *b* occurs in area 1 only and *c* in area 2 only. Note that the Jaccard index is scaled between 0 (when no species are common to a pair of areas) to 100 (when the two areas are identical in species composition).

5.2.1.2 Range-Size Rarity

Species with relatively restricted ranges (e.g. rare and endemic taxa) do not necessarily coincide with areas that have the greatest overall taxon richness. Two approaches are presented following Humphries et al. (1999). The first is

Taxonomic Diversity of Vascular Plants in the European Alpine Areas

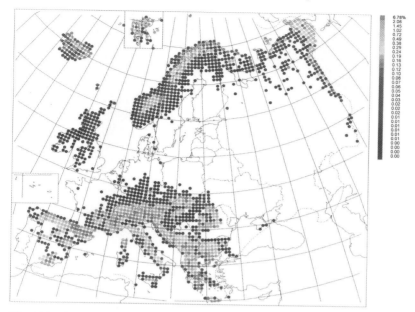

Fig. 5.11. Range-size rarity in the alpine zone in Europe using the *Atlas Florae Europaeae* (AFE) data set. The scores for each taxon are calculated as 1/number of cells occupied by each taxon. Scores are summed for all the taxa in each cell and assigned an equal frequency colour code. *Red cells* indicate high scores, *blue cells* low scores

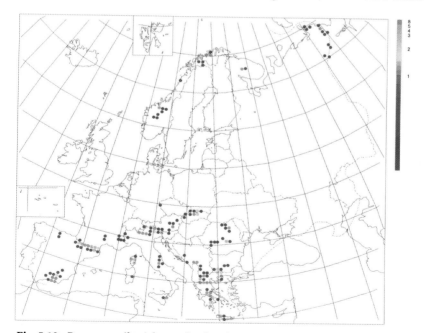

Fig. 5.12. Rare-quartile richness for the alpine AFE data set. The most restricted 25 % of taxa are included, within which the richness has been counted. Red cells show high richness, blue cells indicate low richness

to measure rarity as a continuous function of range size. The scores are calculated as a fraction, 1/number of cells occupied by each taxon. The range-size rarity score is given by the sum of the scores of the taxa present within a 50 × 50 km grid cell. It is most strongly influenced by the most restricted species, and has the advantage that all species contribute to the outcome (Fig. 5.11). The second approach is a discontinuous measure, mapping richness in just those taxa at the more restricted distribution ranges, (such as the lower quartile or the 25 % most restricted European plants in this study; Fig. 5.12). This is a simulation of the kind of method (e.g. ICBP 1992) which suggests that conservation of the areas with the range-restricted taxa would also cover most of the widespread taxa. The threshold, or cut off point, is arbitrary.

5.2.2 Area Selection

5.2.2.1 Near-Minimum Area Set to Represent All

A minimum set contains irreplaceable and flexible areas. Irreplaceable areas with the most narrowly distributed taxa are essential to any minimum set (Humphries et al. 1999). Flexible areas are those which may or may not be used in a minimum set but which contribute to the goal of representing every taxon at least once. A near-minimum set contains areas to represent a taxon more than once, up to a pre-set limit. There are numerous ways of determining the near-minimum set. The method here is a progressive rarity algorithm based on Margules et al. (1988) but differs by taking account of the concept of irreplaceable and flexible areas (see Pressey et al. 1993). The near-minimum area set algorithm is similar to the greedy algorithm (Kirkpatrick 1983; Vane-Wright et al. 1991) but first selects all areas with taxa that are equally or more restricted than the representation goal. For a pre-set goal of representation (with a minimum of each species occurring at least once), the algorithm selects all areas that have unique records for endemic species: these areas are irreplaceable to achieving the goal. The algorithm then reiterates the search and follows a simple set of rules, to select areas richest in the rarest taxa. This begins by selecting grid cells with the greatest complementary richness in just the rarest taxa (ignoring less rare taxa). If there are tied areas, it proceeds by selecting those grid cells with the greatest complementary richness in the next rarest taxa, and so on. If there are persistent ties, it then selects areas among persistent ties with the lowest grid cell number. This is an arbitrary rule used rather than random choice among ties in order to ensure repeatability in tests. (Other criteria, such as proximity to previously selected cells or number of records in surrounding cells, can be added by modifying the algorithm.) The final step repeats the first three as necessary until the representation goal is achieved. The minimum set satis-

Taxonomic Diversity of Vascular Plants in the European Alpine Areas 143

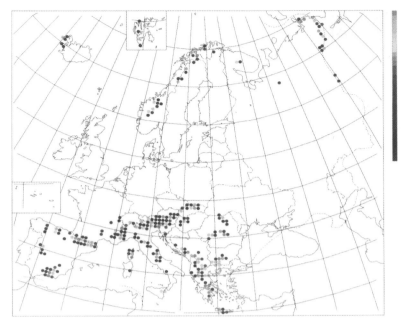

Fig. 5.13. Near-minimum set for 501 predominantly alpine species and subspecies in the AFE data set. Map areas by order displays the area sequence colour-coded from *red* for first choices to *dark blue* for the last ones

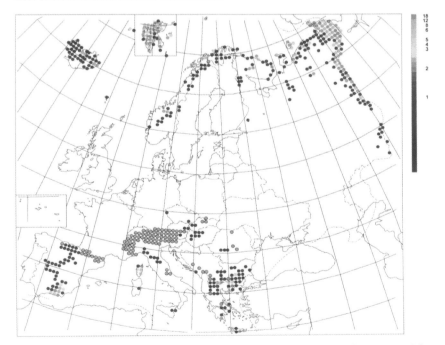

Fig. 5.14. Top 100 hotspot alpine areas (*grey cells*) of Europe based on taxon richness of the AFE cells. The richness of the remaining cells consists of those taxa not present in the top 100

fies the goal of representing every taxon at least once but a representation goal can be set to any number of the cells occupied. In this example, the goal was set for five representations (Fig. 5.13), which ensured the capture of every cell for taxa that occurred in less than five cells. Within WORLDMAP, analyses can provide heuristic solutions for both the minimum set (obtaining representation of all taxa in a near-minimum area set) and for maximum coverage (obtaining representation of as many taxa as possible within a given number of areas).

5.2.2.2 Hotspot Areas

Hotspot analysis was used to find areas exceptionally rich in taxa. For this purpose 100 grid cells richest in taxa are presented in grey in Fig. 5.14. The number of taxa not present in the top 100 are shown by the colour scale (red to blue), with red for high taxon richness and blue for low.

5.3 Results and Discussion

The alpine AFE 1–11 data set contained 19,550 grid cell records (3 % of the potential maximum cover of 630,000). The median number of grid cells for all taxa was 13 and the average 40. In total 48 species or subspecies occurred in only a single cell (about 10 % of all species and subspecies), while *Diphasiastrum alpinum* had the widest distribution with 622 grid cell records out of 4419.

5.3.1 Taxon Richness

A total of 501 taxa were included in the analysis. Thus, if the present digitised AFE represents 20 % of all plants, it might be estimated that the approximate number of European alpine vascular plant species and subspecies is around 2500. As all taxa included in the analyses are European natives, the result indicates that 20 % of the European natives are predominantly alpine. The highest score was 104 taxa for one 50 × 50-km grid cell in the Maritime Alps.

The highest taxon richness occurs in the Alps and Pyrenees followed by the mountains of the Balkans (the Dinaric Alps, Rhodope Massif, and the Pindos Mountains). High richness is found in the Cantabrian Mountains, the Carpathians (incl. the Tatra Mountains), Central Apennines, the Scandes, Mt. Khibiny and the North Ural Mountains. Less species-rich alpine areas include the Sierra Nevada, Scottish Highlands, Central Massif and Corsica. The distribution of alpine Brassicaceae (Fig. 5.2), *Minuartia* spp. (Fig. 5.3) and *Salix* spp.

(Fig. 5.4), which is likely to represent most vascular plant genera that are rich in alpine taxa, shows a similar pattern. At the genus level, the diversity of *Draba* spp. (Fig. 5.5) is highest on the North Ural Mountains, a feature of the Arctic connection.

5.3.2 Regional Mountain Range Richness

Within the European mountains, the Alps (Fig. 5.6) and Pyrenees (Fig. 5.7) are floristically the richest (Table 5.1), and show the highest similarity between any two ranges (Table 5.2). Although not analysed separately, the similarities between the Cantabrian Mountains and Pyrenees are also high (Fig. 5.7). Our estimate of the total number of alpine taxa in the Alps (950) is very close to the estimate of about 1000 obtained by Ozenda (1983).

Table 5.1 shows that the number of rare species or narrow endemics is highest in the Alps and in the mountains of Greece. The floristic similarities of the latter are highest with the Maritime Alps and the Pyrenees (Fig. 5.8) and lower with others (Table 5.2), e.g., Dinaric Alps and Southern Carpathian Mountains (Fig. 5.8). Polunin (1980) discussed historical reasons for this pattern. It is worth noting that the Dinaric Alps and the South Carpathian Mountains are also floristically rather rich (Fig. 5.1).

The Scandes (Fig. 5.9) and the Tatra Mountains (Fig. 5.10) are considerably poorer than the Alps or the Pyrenees (Fig. 5.1, Table 5.1). The Tatra Mountains have an intermediate geographical position in Europe and therefore their floristic similarities are relatively high with the Alps, the Pyrenees, and with the Scandes (Fig. 5.10, Table 5.2). The Scandes and the Scottish Highlands are especially closely related (with a Jaccard's similarity index score of about 45), because a substantial proportion of the British arctic-alpine flora (79 in total) grows in the Scandes (Preston and Hill 1997). Similarities between the Scandes and those of central or southern Europe are low (Table 5.2). Figure 5.9 shows a close connection between the European arctic and the Scandinavian alpine flora, and the bicentric distribution pattern of vascular plants in the Scandinavian mountains (e.g. Gjærevoll 1956), which have a separate distribution area in the south and the north along the mountain chain.

5.3.3 Area Selections

The range-size rarity analysis showed the areas with concentrations (shown in red) of alpine AFE taxa with narrow distribution area (Fig. 5.11). These centres are the Sierra Nevada Mountains, Pyrenees, Maritime Alps, Eastern Alps, Central Apennines, Sicily, Crete and the Balkan Mountains. Conversely, taxa present in the Scottish Highlands, for example, have a broad distribution (shown in blue). Similar results were obtained using all AFE taxa (Humphries et al. 1999).

Rare-quartile richness (Fig. 5.12) and near-minimum area sets (Fig. 5.13) further emphasise those areas with narrowly distributed AFE taxa, usually endemics. In the latter case, 265 cells were required to meet the requirement for representing species at least five times. Hotspot analysis (Fig. 5.14) shows the 100 cells richest in AFE taxa (grey cells). In addition, those cells with taxa not present in these top 100 cells are shown. Most hotspots of high species richness are concentrated in the Alps (67) and the Pyrenees (12), which together account for 80% of the cells (Figs. 5.1 and 5.14). The third richest area is the Balkan Mountains. The only area chosen in the Scandes was ranked at 98th place. One important point to note is that if the selection of priority areas for conservation was made by selecting sites of highest taxon richness alone, then most sites would be replicating each other in terms of taxon composition, ignoring other unprotected taxa absent in the chosen hotspots. By contrast, the first spots selected in the near-minimum-set analysis (red spots in Fig. 5.13) are not concentrated in the Alps, Pyrenees and Balkans but are widely dispersed additionally in the Sierra Nevada, Corsica, the Balkan Mountains, one in the Tatra and another in the Ural Mountains. For conservation, this means that, to represent a maximum diversity of taxa, a strategy such as a near-minimum set is required to ensure complementarity among areas. At an European scale the Sierra Nevada and the mountains of Greece are not very rich (Figs. 5.1 and 5.14) but possess endemics that make certain grid cells irreplaceable and thus ranked high priority for conservation.

5.4 Conclusions

1. The digitised AFE taxa represent about 20% of all native plants in Europe at present. If the proportion of alpine taxa is similar for the other 80% then the approximate number of European alpine vascular plant species and subspecies is around 2500. This figure makes 20% of the European native plants predominantly alpine, which, when the corresponding surface area is considered (3%), makes the alpine zone the richest in vascular plants in Europe.
2. The Alps and the Pyrenees are the richest in vascular plant species among the European mountains, followed by the Balkan Mountains, Cantabrian Mountains, Carpathians, Central Apennines and the Scandes. Although the floristic similarity is considerable, e.g., between the Alps and Pyrenees, each of the major mountain ranges has its peculiar, unique character.
3. About 10% of the species and subspecies occur in a single 50 × 50-km grid cell, indicating a high proportion of narrow range endemics. The number

of endemics and narrow-range taxa are highest in the Alps and Pyrenees. Other areas outside these include the Balkan Mountains, Crete and Sierra Nevada and, e.g., the Central Massif, Corsica and Central Apennines are invaluable also. The number of endemics and narrow-range taxa are low in the other mountain ranges of Europe.

4. The Carpathians, a crescent from south to north, is correspondingly floristically intermediate in Europe. The mountains of northern Europe are floristically poorer compared with central or southern Europe and their species composition is markedly arctic-alpine.
5. The 50 × 50-km grid cell system used by AFE and incorporated within WORLDMAP helps to identify the areas of particular importance for detailed conservation programmes. The selection of priority areas for conservation cannot be determined by using taxon richness alone, as hotspot selections result in considerable redundancy and replication. Methods that target sites of complementary floristic composition are more valuable. For example, in Europe, in addition to highly species-rich regions such as the Alps and the Pyrenees, sites of lower diversity with unique and restricted range species should be protected.

Acknowledgements. We thank an anonymous reviewer for clarifying comments on the manuscript.

References

Engelskjøn T, Skifte O (1995) The vascular plants of Troms, North Norway. Revised distribution maps and altitude limits after Benum: the flora of Troms fylke. Tromura, Naturvitenskap nr. 80. Universitet i Tromsø, Institutt for museumvirksomhet, Tromsø

Gjærevoll O (1956) The plant communities of the Scandinavian alpine snowbeds. K norske Vidensk Selsk Skr 1956(1):1–405

Humphries C, Araújo M, Williams P, Lampinen R, Lahti T, Uotila P (1999) Plant diversity in Europe: Atlas Florae Europaeae and WORLDMAP. Acta Bot Fennica 162:11–21

ICBP (1992) Putting diversity on the map: priority areas for global conservation. International Council for Bird Preservation, Cambridge

Jalas J, Suominen J (eds) (1972–1994) Atlas Florae Europaeae, vols 1–10. The Committee for Mapping the Flora of Europe and Societas Biologica Fennica Vanamo, Helsinki

Jalas J, Suominen J, Lampinen R (eds) (1996) Atlas Florae Europaeae, vol 11. The Committee for Mapping the Flora of Europe and Societas Biologica Fennica Vanamo, Helsinki

Kirkpatrick JB (1983) An iterative method for establishing priorities for the selection of nature reserves: an example from Tasmania. Biol Conserv 25:127–134

Lahti T, Lampinen R (1999) From dot maps to bitmaps: Atlas Florae Europaeae goes digital. Acta Bot Fennica 162:5–9

Landolt E (1992) Unsere Alpenflora. Gustav Fischer Verlag, Stuttgart Jena

Margules CR, Nicholls AO, Pressey RL (1988) Selecting networks of reserves to maximise biological diversity. Biol Conserv 21:79–109

Ozenda P (1983) The vegetation of the Alps. European Committee for the Evaluation of Nature and Natural Resources, Strasbourg

Pawłowski B (1956) Flora Tatr Tom 1. Panstwowe Wydawnictwo Naukowe, Warszawa

Polunin O (1980) Flowers of Greece and Balkan. A field guide. Oxford University Press, Hong Kong

Pressey RL, Humphries CJ, Margules CR, Vane-Wright RI, Williams PH (1993) Beyond opportunism: key principles for systematic reserve selection. TREE 8:124–128

Preston CD, Hill MO (1997) The geographical relationships of British and Irish vascular plants. Bot J Linn Soc 124:1–120

Strid A (1986) Mountain flora of Greece, vol 1. Cambridge University Press, Cambridge

Talbot SS, Yurtsev BA, Murray DF, Argus GW, Bay C, Elvebakk A (1999) Atlas of rare endemic vascular plants of the Arctic. Conservation of Arctic Flora and Fauna (CAFF). Technical Report no 3. US Fish and Wildlife Service, Anchorage

Tutin TG, Heywood VH, Burges NA, Valentine DW, Walters SM, Webb DA (eds) (1964) Flora Europaea, vol 1. Cambridge University Press, Cambridge

Tutin TG, Heywood VH, Burges NA, Moore DM, Valentine DW, Walters SM, Webb DA (eds) (1968–1980) Flora Europaea, vols 2–5. Cambridge University Press, Cambridge

Vane-Wright RI, Humphries CJ, Williams PH (1991) What to protect? – Systematics and the agony of choice. Biol Conserv 55:235–254

Villar L (1980) Catálogo florístico del Pirineo Occidental español, vol 11. Publicaciones del Centro Pirenaico de Biologia Experimental, Jaca

Williams PH, Humphries CJ, Araújo MB, Lahti T, Lampinen R, Uotila P, Vane-Wright RI (2000) Important plant areas of Europe; exploring the consequences of selection criteria. In: Synge H, Akeroyd J (eds) Planta Europaea. Proceedings of the 2nd European Conference on the Conservation of Wild Plants, 9–14 June 1998, Uppsala, Sweden. Cigam Group, Newcastle-upon-Tyne

6 Patterns in the Plant Species Richness of European High Mountain Vegetation

R. VIRTANEN, T. DIRNBÖCK, S. DULLINGER, G. GRABHERR, H. PAULI, M. STAUDINGER and L. VILLAR

6.1 Introduction

There are 14 main mountain regions in Europe ranging from the Mediterranean to the Arctic between 35–80°N (Chap. 1). The high mountains across Europe differ in extent and altitude, glaciation history, geology, and ecological conditions (Chap. 1), and their plant species composition varies considerably (Chaps. 3.1–3.10).

In most high mountains, there are three types of habitats according to winter snow cover: exposed wind-blasted, snow-protected and late lying snow affected sites (snowbeds). Each of these habitats sustains comparatively similar plant community types across Europe (Grabherr 1995; Körner 1999; Grabherr et al. 2000). There is a large botanical literature describing these plant community types which provides a previously unutilised opportunity to examine geographical and ecological patterns in the species richness of alpine vegetation. It is also possible to address how species richness in local communities is related to the regional (mountain or mountain range) species richness (see Cornell and Lawton 1992; Zobel 1997).

The interest in comparing variation in species richness in the alpine communities of Europe has a big history (e.g. Szafer 1924; Eurola 1974; Dullinger et al. 2000). However, no earlier attempts to document the variation in species richness in mountain communities of the whole of Europe exist. We investigated three habitat types (exposed sites, snow protected sites and snowbeds) that occur in most European mountain areas. We analysed these data in relation to: (a) the size of a particular mountain area, (b) latitude, (c) soil substratum, and (d) snow cover.

The relationship between local and regional species richness was also examined. The analyses were based on data from about 100 published and unpublished vegetation studies where the numbers of species could be estimated (Table 6.1).

Table 6.1. Study sites, locations and sources of materials used in the analyses. Empty cells in[dicate] that data were not available

No.	Mountain chain	Mountain area	Geographic co-ordinates (°N and E)				Vegetation type on calcareous soil	
			Latitude	Longitude	Lower limit (m a.s.l.)	Peaks (m a.s.l.)	Wind-swept	Snow-prote[cted]
1	Alps	Hautes Alpes	45.00	7.00	2200	4103		Festuca viol[acea] - Trifolium th[alii]
2	Alps	Maritime Alps	44.20	7.00	2200	3297		Sesleria cae[rulea] – Helicotrich[on] sedenense
3	Alps	Vanoise Alps	45.20	6.40	2200	3852	Kobresia myosuroides–Carex rosae	Helianthem[um] oelandicum alpestre – Sesleria al[bicans]
4	Alps	Valais	46.10	7.20	2000	4634	Kobresia myosuroides	Astragalus f[rigidus] – Sesleria ca[erulea]
5	Alps	Montafon	47.10	9.30	2000	2965	Kobresia myosuroides	Sesleria caer[ulea] – Carex sempervirens
6	Alps	Rätische Alps	46.30	9.40	2000	3417	Kobresia myosuroides	Nardus stric[ta] – Trifolium
7	Alps	Davos	46.50	9.40	2300	2636		Mesic grass[land]
8	Alps	Ötztaler, Schrankogel	47.00	11.10	2200	3497	Oxytropis halleri – Kobresia myosuroides	
9	Alps	Dolomites	46.30	12.00	2000	2847	Carex firma	Sesleria cae[rulea] – Carex sempervirens
10	Alps	SE Alps	46.10	12.30	1800	2250	Dryas octopetala	Carex ornith[opoda] – Sesleria ca[erulea]
11	Alps	Hohe Tauern, Grossglockner	47.10	12.40	2100	3797	Kobresia myosuroides	
12	Alps	Niedere Tauern	47.20	13.40	2000	2863	Kobresia myosuroides	Sesleria caer[ulea] – Carex sem[pervirens]
13	Alps	Karawanken	46.30	14.20	2000	2558	Carex firma	Sesleria caer[ulea] –Carex semp[ervirens]
14	Alps	Austrian Kalkalpen (W)	47.30	15.00	1500	2365	Carex firma	Sesleria caer[ulea] – Carex sem[pervirens]
15	Alps	Austrian Kalkalpen (E)	47.30	15.40	1500	2277	Carex firma	Sesleria caer[ulea] Salix retusa
16	Apennines	Central	42.30	13.20	2000	2914	Leontopodium alpinum –Kobresia myosuroides	Carex laevis – Sesleria ca[erulea]

	Vegetation types on siliceous soil			References
owbed	Wind-swept	Snow-protected	Snowbed	
ix retusa reticulata		Centaurea-Festuca paniculata	Salix herbacea	Dalmas (1972)
		Festuca halleri		Guinochet (1938)
		Carex curvula – Tanacetum alpinum	Salix herbacea– Sibbaldia procumbens	Gensac (1977, 1979)
	Sempervivum arachnoideum – Pulsatilla halleri		Carex lachenalii – Salix herbacea	Richard (1975, 1985); Richard and Geissler (1979)
ix retusa reticulata	Loiseleuria procumbens – Carex curvula	Carex curvula	Salix herbacea	Grabherr (1985)
		Carex curvula	Salix herbacea	Braun-Blanquet (1948b, 1949, 1969, 1975); Galland (1979)
			Salix herbacea	Gigon (1971); Vetterli (1982)
	Loiseleuria procumbens – Vaccinium uliginosum	Nardus stricta – Festuca paniculata	Salix herbacea	Dullinger (1999)
				Pignatti and Pignatti (1983); Feoli-Ciapella and Poldini (1993); Poldini and Martini (1993)
x retusa reticulata				Feoli-Ciapella and Poldini (1993); Poldini and Martini (1993)
		Carex curvula	Salix herbacea	Braun-Blanquet (1931); Zollitsch (1968)
		Carex curvula		Pauli (1993)
				Aichinger (1933)
x retusa mogyne discolor				
				Greimler (1997)
x retusa				
				Dirnböck et al. (1999)
eticulata rex sempervirens				
lotheca eana intago a		Vaccinium myrtillus		Lüdi (1943); Feoli-Chiapella and Feoli (1977); Pedrotti (1982)

Table 6.1. (*Continued*)

No.	Mountain chain	Mountain area	Geographic co-ordinates (°N and E)				Vegetation type on calcareous soil	
			Latitude	Longitude	Lower limit (m a.s.l.)	Peaks (m a.s.l.)	Wind-swept	Snow-protected
17	Apennines	Apuanes	44.00	10.20	1800	1945	*Empetrum nigrum – Vaccinium uliginosum*	*Sesleria caerule – Carex semper*
18	Carpathians	Low Tatra	48.55	19.30	1700	2045	*Carex firma*	*Sesleria caerul*
19	Carpathians	High Tatra incl. Biela Tatra	49.15	20.00	2000	2663	*Carex firma*	*Festuca versico*
20	Carpathians	Eastern (Rumanian)	47.35	25.00	2000	2305		*Festuca carpat*
21	Carpathians	Southern; Retezat	45.20	22.30	2000	2509		
22	Carpathians	Southern; Bucegi	45.30	25.30	2000	2507	*Kobresia myosuroides*	*Sesleria caerul – Carex sempe*
23	Carpathians	Southern; Fagaras	45.30	24.30	2000	2544		
24	Corsica		42.30	8.50	2300	2710		*Festuca sardo – Phyteuma se*
25	Dinarids		43.30	18.00	2000	2396	*Kobresia myosuroides – Edraianthus* sp.	*Edraianthus* s –*Dryas octope*
26	Peloponnisos Mts	Mt. Killini	37.50	22.10	2000	2376	*Aster parnassi – Globularia stygia*	*Alopecurus ge – Omalotheca hoppeana*
27	Píndos Mts		40.00	21.30	2200	2900	*Sesleria coerulans – Thymus cherlerioides*	*Alopecurus ge – Omalotheca hoppeana*
28	Pyrenees	Pyrenees	42.00	0.00	2200	3367	*Kobresia myosuroides*	
29	Rhodope	Rila-Pirin Planina Mts.	41.50	23.30	2000	2925		*Onobrychis n – Sesleria ten*
30	Shara Mts		41.50	20.50	2000	2704	*Carex firma*	*Carex sempe –Sesleria ten*
31	Polar Ural	Rai-Iz Mts.	66.00	65.00	400	1100		
32	Finnish Lapland	Mt. Ounastunturi	68.20	23.40	500	821	*Dryas octopetala*	*Bistorta vivip – Thalictrum*

	Vegetation types on siliceous soil			References
bed	Wind-swept	Snow-protected	Snowbed	
				Barbero and Bono (1973)
		Juncus trifidus	Salix herbacea	Klika (1932); Leps et al. (1985)
reticulata		Juncus trifidus – Oreochloa disticha	Salix herbacea	Szafer et al. (1927); Krajina (1933); Palowski (1935)
retusa	Empetrum nigrum – Vaccinium	Carex curvula		Pawlowski and Walas (1949); Resmerita (1979a,b, 1981); Coldea (1991)
	Loiseleuria procumbens	Carex curvula	Luzula alpinopilosa	Borza (1934); Resmerita (1970, 1979a,b); Hodisan and Boscaiu (1986); Coldea (1991)
reticulata	Carex curvula	Carex curvula	Salix herbacea	Domin (1933); Puscaru et al. (1956); Beldie (1967); Resmerita (1976); Coldea (1991)
			Salix herbacea	Huml et al. (1979); Resmerita (1979a); Puscaru-Soroceanu et al. (1981); Coldea (1991)
		Huperzia selago – Carex ornithopoda	Omalotheca supina – Sibbaldia procumbens	Gamisans (1977)
retusa		Carex curvula	Ranunculus crenatus	Horvat (1930); Lakusic (1970)
				Quezel (1964); Dimopoulos and Georgiadis (1994)
				Quezel (1967)
retusa		Carex curvula	Anthelia juratzkana – Salix herbacea	Braun-Blanquet (1948a); Negre (1969, 1970); Gruber (1978); Montserrat (1986); Ballesteros and Canalis (1991); Rivas-Martinez et al. (1991); Carrillo and Ninot (1992); Carreras et al. (1993); Vigo (1996)
retusa		Sesleria comosa	Omalotheca supina	Horvat et al. (1937, 1974); Mucina et al. (1990)
		Sesleria comosa	Salix herbacea	Horvat et al. (1974)
ampsia folia				Igoshina (1966)
	Empetrum nigrum – Flavocetraria nivalis	Phyllodoce caerulea – Stereocaulon sp.	Salix herbacea	Kalliola (1939)

Table 6.1. (*Continued*)

No.	Mountain chain	Mountain area	Geographic co-ordinates (°N and E)				Vegetation type on calcareous soil	
			Latitude	Longitude	Lower limit (m a.s.l.)	Peaks (m a.s.l.)	Wind-swept	Snow-protected
33	Scandes	Raisduottar	69.00	21.00	700	1324	*Dryas octopetala – Carex rupestris*	*Cassiope tetrag – Dryas octope*
34	Scandes	Mt. Peltsa	69.00	20.20	700	1518	*Dryas octopetala – Rhytidium rugosum*	*Dryas octopeta*
35	Scandes	Torne Lapland	68.30	19.00	700	1765	*Dryas octopetala*	*Dryas octopeta – Cassiope tetr*
36	Scandes	Sylene Mts.	63.00	12.10	900	1762	*Dryas octopetala*	*Potentilla cra – Bistorta vivi*
37	Scandes	Sikilsdalen	61.30	9.10	1000	1843	*Carex rupestris – Encalypta rhaptocarpa*	
38	Scandes	Rondane	61.50	9.40	1000	2183		
39	Scandes	Dovre Mts.	62.20	9.20	1100	2286	*Kobresia myosuroides*	*Potentilla cra – Bistorta viv*
40	Scotland		57.00	−5.00	750	1343	*Dryas octopetala*	*Dryas octope*
41	Sierra Nevada		37.00	−3.00	2900	3478		
42	Spitsbergen		78.20	16.00	10	901	*Dryas octopetala – Carex rupestris*	*Cassiope tetr*
43	Sudetes Mts	High Sudetes, Krkonose Mts.	50.45	15.45	1500	1603		

	Vegetation types on siliceous soil			References
...bed	Wind-swept	Snow-protected	Snowbed	
...olaris ...e acaulis	Empetrum nigrum – Cassiope tetragona – Alectoria ochroleuca	Empetrum nigrum – Cassiope tetragona	Salix herbacea	Virtanen and Eurola (1997)
		Vaccinium myrtillus	Salix herbacea	Nordhagen (1955); Hedberg et al. (1952); R. Virtanen (unpubl.)
		Deschampsia flexuosa – Anthoxanthum odoratum	Salix herbacea	Bringer (1961a,b); Gjaerevoll (1956)
reticulata	Loiseleuria procumbens – Flavocetraria nivalis – Alectoria ochroleuca	Deschampsia flexuosa – Anthoxanthum odoratum	Salix herbacea	Nordhagen (1928)
	Loiseleuria procumbens – Vaccinium uliginosum – Alectoria ochroleuca	Juncus trifidus	Salix herbacea	Nordhagen (1943)
	Arctostaphylos alpina – Flavocetraria nivalis	Deschampsia flexuosa – Anthoxanthum odoratum	Salix herbacea	Dahl (1957)
olaris		Deschampsia flexuosa – Anthoxanthum odoratum	Salix herbacea – Kiaeria starkei	Resvoll-Holmsen (1920); Nordhagen (1928); Gjaerevoll (1956)
	Arctostaphylos alpina – Calluna vulgaris	Juncus trifidus – Festuca ovina	Gymnomitrion concinnatum – Salix herbacea	McVean and Ratcliffe (1962); McVean (1964)
	Festuca clementei – Erigeron frigidus	Vaccinium uliginosum – Ranunculus acetosellifolius		Quezel (1953); El Aallali et al. (1998)
olaris	Salix polaris – Ochrolechia sp.	Salix polaris – Cetraria islandica	Salix polaris – Sanionia uncinata	Eurola (1968); Elven and Elvebakk (1996); Virtanen et al. (1997)
		Cetraria islandica – Festuca airoides	Carex bigelowii	Zlatnik (1925); Jenik (1961)

6.2 The Data

Published or otherwise available vegetation tables were used to assemble a database on wind-exposed heaths and dry grasslands, snow-protected sites including mesic grasslands, and snowbed sites on both siliceous (such as granite or gneiss) and calcareous (such as limestone or dolomite) substrata (Table 6.1). The data give an uneven coverage for the European alpine zone and is heterogeneous because the original geobotanical investigations used a variety of sampling methods. Quadrat samples (relevés) of the original studies were the main source of data. It was assumed that the listing of species occurrences in the sample quadrats was sufficiently comparable among investigators, and therefore these data provided an adequate account of the numbers of vascular plants present. To reduce the effect of quadrat area, cumulative species curves were calculated for each set of original quadrat data and species numbers for 10 m^2 and 100 m^2 were estimated by interpolation. Unfortunately, in many cases, data were insufficient to have both 10- and 100-m^2 values for all communities. In some cases, the original quadrat data were not available or the original study did not report the exact sizes of the sampling areas and the cumulative species curves could not be calculated as described above. For the Hautes Alpes (Dalmas 1972), the Vanoise Alps (Gensac 1977, 1979), most of the Balkan Mts. (Horvat et al. 1974) and the Polar Urals (Igoshina 1966), species richness had to be estimated from summary tables and the sample numbers given.

The estimation of the regional species pools for bryophytes, macrolichens and vascular plants assessed their richness in the alpine zone of a given mountain area (regional species pool). These estimates were obtained at first by counting all species listed in vegetation tables representing the alpine vegetation types within the altitude range given in Table 6.1. For some lower mountains (e.g. the Sudetes), records from treeline ecotone communities near summit areas were included. Secondly, records from other published literature sources (listed in Table 6.1) were added. In compiling the regional species pool, plants confined to azonal habitats such as vertical rock walls, mires and watercourses were excluded. This method of compiling the regional species pool gives fairly similar results to that based on ecological attributes (Dupré 2000). We assume that for vascular plants these data are reliable and give only little biased estimates of species richness at the regional scale. It is not the case, however, for bryophytes and macrolichens. They have been overlooked in a number of studies and their numbers may be underestimated. The size of the alpine zone for the various mountain ranges was obtained from a digital database (GLOBE: Global Land One-kilometre Base Elevation) by using geographic information systems (D. Moser and M. Staudinger, Univ. of Vienna).

Patterns in species richness in relation to latitude, mountain range area (alpine zone), soil type, snow cover, and glaciation history (glaciated or not during the last glacial maximum) were analysed by using generalised additive models (GAM, Hastie and Tibshirani 1990). At first, GAM models with one or more explanatory variable(s) and their interactions were constructed. The significance of explanatory variables was tested by deletion. A term was dropped from the model if the deletion did not cause a significant change of deviance ($p<0.05$). The species richness data showed usually over-dispersion and therefore F-tests were used in model simplification. The fitted trend lines obtained from the Poisson family GAMs show the relationship between species richness and latitude. In fitting algorithms, smoothing splines with three degrees of freedom were used. Although GAMs were regarded as a sufficient way to analyse the essential patterns in the data, some generalised linear models (Poisson regression, S-PLUS 2000) were also run to test trends in subsets of the data. It should be noted that the species richness values were probably not fully independent observations because of spatial autocorrelations and the statistical results may show a tendency to an increased type I error rate.

6.3 Geographical and Ecological Trends in Plant Species Richness

Vascular plants and cryptogams showed opposite latitudinal trends. The number of vascular plants declined from south to north (Fig. 6.1, deletion test for change of deviance: F=5.4, d.f. 4.1, 37.9). Meanwhile, the number of bryophytes increased towards the north (F=18.8, d.f. 3.94, 38.06), as did the number of macrolichens (Fig. 6.1, F=18.13, d.f. 3.9, 38.1). For all species groups combined, no latitudinal trend occurred (F=1.2, d.f. 4.1, 37.9).

There was no significant relationship between the number of plants in a mountain (range) and the size of that mountain range (Fig. 6.2).

Species richness on calcareous soil was higher than on siliceous soil (F=59.3, d.f. 1.0, 239.0). As indicated by a significant [soil type×latitude] interaction (F=9.8, d.f. 1.0, 236.9), species richness in relation to soil type was not constant along the latitudinal gradient. Fitted lines in Fig. 6.3 suggested that on calcareous soils, species richness remained relatively constant from south to north while on siliceous soils there was a declining trend. The curved trend on calcareous soils showed that species richness was low on oro-Mediterranean high mountains, peaked in the Alps and then declined towards the north (curvilinear trend supported by the deletion test of term [latitude]^2 in a second-order Poisson regression; F=11.3, d.f. 1, 129).

Species richness differed in communities in relation to snow cover (F=5.86, d.f. 2.0, 237.9); the snow-protected communities were the most species rich followed by exposed and snowbed communities. There was no significant interaction between latitude and richness (snow × latitude, F=1.16, d.f. 2.0, 235.9); however, the fitted splines (Fig. 6.4) and deletion test ([latitude]^2 in a

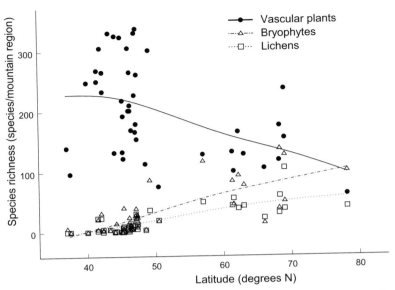

Fig. 6.1. Latitudinal trends for vascular plant, bryophyte and lichen species richness in European mountain areas. The fitted trend lines were derived from generalised additive models with spline smoothing

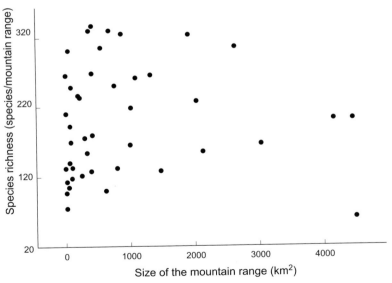

Fig. 6.2. Relationship between the size of mountain areas and species richness of European mountains

Patterns in the Plant Species Richness of European High Mountain Vegetation

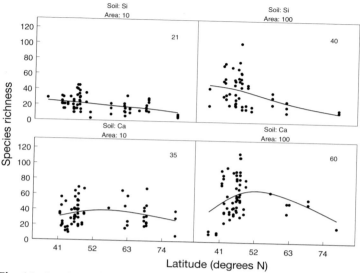

Fig. 6.3. Species richness of European alpine areas estimated for 10 m² and 100 m² on calcareous (Ca) and on siliceous soils (Si) in relation to latitude. The mean species richness for each plot is given in the top right hand corner. The *fitted lines* are based on a generalised additive model with spline smoothing

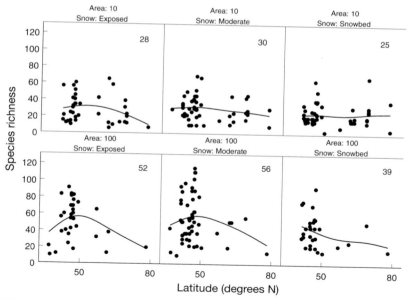

Fig. 6.4. Species richness of European alpine areas estimated for 10 m² and 100 m² on exposed, snow-protected (moderate snow cover) and snowbed sites in relation to latitude. The mean species richness for each plot is given in the top right hand corner. The *fitted lines* are based on a generalised additive model with spline smoothing. Note that the latitudinal trends (*fitted lines*) were not statistically significant

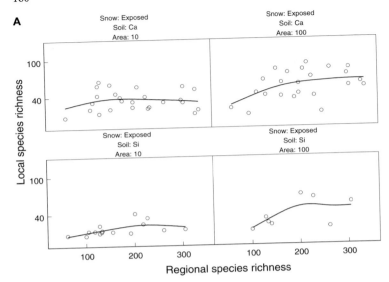

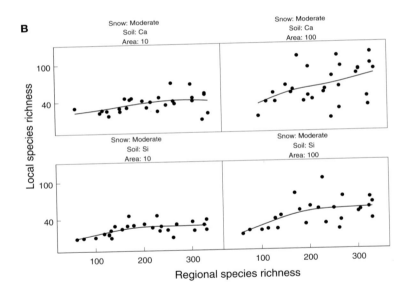

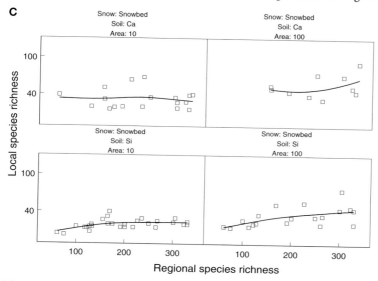

Fig. 6.5a–c. Relationship between local (species richness estimates for 10 m² and 100 m²) and regional species richness (species richness of the mountain area) in European alpine plant communities on different soil substrata and in relation to snow cover. The codes are as in Figs. 6.3 and 6.4. The *fitted lines* are based on a generalised additive model with spline smoothing

second-order Poisson regression; F=8.89, d.f. 1, 69) suggested a curved relationship between species richness and latitude in wind-blasted communities. Exposed communities had a unimodal pattern with a peak at latitudes around 50°N (Fig. 6.4). No latitudinal trends were obvious for the communities on mesic snow-protected sites or for snowbed communities.

The relationship between local and regional species richness for the two soil types and three types of communities is presented in Fig. 6.5. As an ordinary regression statistics has not been regarded as a valid way to assess the relationship (e.g. Cresswell et al. 1995; Herben 2000), it was examined only by fitting smoothing splines to show the pattern in data. On siliceous soils, in most cases, the curves tended to level off at higher values of regional species richness in the three habitat types. For exposed communities on calcareous soils, the curves were asymptotic, while, particularly in the 100-m² local richness data, a nearly linear positive trend was seen. The curves for snowbed communities on calcareous soil showed no obvious trends and these plots suffered from a paucity of data points.

6.4 Discussion

6.4.1 Latitudinal Trends

The observed latitudinal trend for vascular plants of the European alpine areas conforms to the pattern reported from the Nearctic area (Billings 1973). The latitudinal variation in vascular plant species richness can largely be explained by low rates of extinction in the southern part of Europe during the Pleistocene, when many parts of the mountains of southern Europe (including some parts of the Alps) were unglaciated and alpine species survived only in periglacial refugia. Meanwhile, on mountains of northern Europe, glacial survival was very unlikely and the colonisation of plants must have taken place after deglaciation (Comes and Kadreit 1998; Stehlik 2000; see also Whittaker 1965; Grubb 1987; Tilman and Pacala 1993). Contrary to that for vascular plants, opposite latitudinal trends were detected for bryophytes and macrolichens. Although this pattern may partially reflect incomplete sampling in data from southern Europe, it agrees with available evidence that the abundance and taxonomic richness of cryptogams is lower in the alpine areas of southern and many central European mountains than in those of northern Europe (e.g. Ozenda 1950; Ochsner 1954; Söyrinki 1954; Reccia and Villa 1996; Virtanen and Eurola 1997; Janovicova et al. 1999). The contrasting trends of vascular plants and cryptogams and reasons for this pattern have not been much discussed earlier in the context of high altitude mountain vegetation. One possible explanation is that glacial extinctions and subsequent recolonisation were dissimilar in different groups of organisms. Bryophytes and lichens, producing spores, more effectively colonised northern mountains after the glacial retreat than vascular plants. On the other hand, it is possible that cryptogams and vascular plants react differently to environmental conditions such as habitat availability, moisture and temperature conditions, and seasonality of growth and light regimes which all show latitudinal differentiation (Eurola 1974; Oksanen and Virtanen 1995; Väisänen 1998; Hedenäs 1999; Körner 1999, 2000; Chap. 29).

6.4.2 The Size of Mountain Areas

There was a poor fit between species richness and the size of a mountain area. The probable reason for this is that factors other than area (and factors correlating with it) influence species richness more. In particular, variable ecological conditions and glaciation history across Europe may have influenced the detection of a relationship between species richness and area. Detecting a relationship between species richness and area would require similar ecological conditions among the studied systems and a state of equilibrium between extinctions and immigration (MacArthur and Wilson 1967). The poor fit in species richness to area relationship implies that area has only limited predictive power for plant species richness in European high mountains. Studies elsewhere showed it having explanatory power (Riebesell 1982; White et al. 1984; White and Miller 1988).

6.4.3 Soil Substratum

In the present analyses, plant communities on calcareous soils had higher vascular plant richness than those on siliceous substrates, agreeing with earlier studies in different European mountain areas (Wahlenberg 1814; Blytt 1876; Domin 1928; Nordhagen 1943; Gjærevoll 1956; Wohlgemuth 1993; Körner 1995; Birks 1996; Coldea and Cristea 1998). One possible explanation for the higher richness on calcareous soil is provided by the reservoir or species pool hypothesis according to which higher local species richness results from a larger regional species pool adapted to a particular soil type (Grime 1979; Grubb 1987; Zobel 1997). According to this scenario, high mountain plants adapted to high concentrations of calcium in the soil have presumably diversified in cold steppe-like environments (Chapin and Körner 1995). These species and their descendants form the core element of plants in European alpine environments while there are fewer taxa adapted to siliceous soils. Recent phylogenetic analyses do support this scenario. For example, among alpine Gentians, calcicolous taxa are more numerous (*G. occidentalis* Jak., *G. angustifolia* Vill., *G. ligustica* R. Vilm. & Chop., *G. clusii* Perr. & Song. and *G. dinarica* G. Beck) and calcifuge taxa, often evolved from calcicole ancestors, (e.g. *Gentiana acaulis* L. and *G. alpina* Vill.) are fewer (Gielly and Taberlet 1996). A corresponding substratum-driven adaptive radiation may have played a role in the evolution of the taxonomic diversity of alpine Saxifrages (Conti et al. 1999).

There was evidence on latitudinal differentiation in vascular plant richness on different soil substrata. Deviating from the general trend across Europe, species richness on calcareous and siliceous soil was about equal on the mountains of southernmost Europe. We thus consider that plant richness on southernmost mountains is not so markedly high on calcareous soil substrata as in the northern mountains. Possibly, drought stress during growing seasons on mountains near the Mediterranean may reduce plant diversity and cancel species richness differences between calcareous and siliceous soils (Gigon 1971; Grime and Curtis 1976; Grime 1979). This contrasts with the situation in most other mountains of Europe. For example, on the Alps, vascular plant species richness on calcareous soils seems to be markedly higher than on siliceous soils. Here, calcareous rock deposits commonly occur, and the relatively high species richness on calcareous substratum may reflect the pool effect (see above). Also, in northern Europe (the Scandes and Scotland), communities on calcareous substrata are markedly rich in vascular plants. It is noteworthy and somewhat paradoxical that these communities only locally dominate, being confined to relatively small areas of mountain terrain where calcareous substrata maintain soil pH around neutral. According to Blytt (1876), today's rich flora on calcareous soil substrata on some Norwegian mountains reflects a past when a flora rich in calciphilous plants of a '*Dryas* vegetation' had a wide distribution after the glacial retreat. As noted later also by Gjærevoll (1963) and Godwin (1975), the area suitable for these plants has then contracted owing to leaching of soils and the high dominance of acidophilous plants during the post-glacial. Only on the most calcareous substrata, such floristic impoverishment has not taken place and a rich flora has been preserved. This line of reasoning would explain why vegetation on calcareous substrata, which cover a proportionally small area, has more species than can be expected on the basis of latitudinal trends and habitat area (i.e. are supersaturated, sensu Wilcox 1978). We put forward this kind of considerations to recall that, while ecological conditions have played and continue to play a marked role in determining patterns in species richness, historical factors also have to be considered.

6.4.4 Snow Cover

It is well established that snow gradient on mountains is of prime importance for plant survival and functioning of alpine plants (Körner 1999), and these data show that high altitude communities across European mountains differing in exposure and snow cover differ in their species richness, too. Overall, it seems that sites with moderate snow cover sustain the highest species richness in alpine areas, while exposed and snowbed communities have fewer species. This trend is compatible with other studies showing that species density is at its highest at intermediate positions along environmental gradients (Grime 1979; Virtanen et al. 2002). It should be noted that there were relatively small differences in richness among the communities studied here. There may be several reasons: the data combines several study areas differing in species richness, and the late snowbeds, which are very poor in species (e.g. Gjærevoll 1956), were excluded. In addition, the communities of exposed sites tend to have more sheltered microsites allowing less wind-hardy species to coexist. These analyses were restricted to vascular plants, but it is likely there are marked differences in other plant groups and lichens (see e.g. Bültmann and Daniëls 2001), and even functionally differentiated species groups of vascular plants show distinct patterns.

6.4.5 Local and Regional Species Richness

It has been of considerable interest to examine the relationship between local and regional species richness (Lawton 1999). Alpine systems (representing an insular model system; Körner 2000) obviously contribute interesting insight into this question: there are similar community types found in different mountain areas that differ in regional species pools. Indeed the data from the European mountains showed different types of relationship between local and regional richness of alpine vascular plants: non-saturating (Type I) and asymptotically saturating (type II; Cornell and Lawton 1992). An asymptotic relationship was detected for community types on siliceous substrata while in snow-protected communities on calcareous soils a non-saturated pattern was found. Such deviating patterns suggest that the processes structuring alpine communities may differ depending on habitat conditions. The underlying processes for different patterns remain contentious and only some hypotheses can be offered. The finding from moderately snow-protected communities on calcareous substrata suggests that there is a positive correlation between local and regional richness. The simplest message is that local communities of this vegetation type are richer in species when the overall regional scale species richness is higher. Following Cornell and Lawton (1992), this finding also implies that for this type of vegetation there does not seem to be strong abiotic or biotic limitations to species co-existence. However, this type of relationship between local and regional richness probably cannot be regarded as the norm in European alpine vegetation as the trend from most other types of vegetation suggest a asymptotic or saturating relationship. This latter relationship suggests that there is an upper limit for local species richness that is independent of the regional species pools. At present, we can offer two likely explanations. Firstly, local species richness can be limited by a high dominance of certain plants that can be found in communities with mat-forming graminoids or dwarf shrubs. In such communities, competitive exclusion (spatial dominance through strong vegetative growth) may be relatively important to reduce species richness. Secondly, on exposed sites with no or little protective snow cover, or in sites with very late-lying snow, environmental harshness may filter species not adapted to cope with such conditions (Körner 1995). There are still rather few analyses so far to compare these results in other systems (Cornell 1999). Recently, Safford et al. (2001) found positive correlation between local and regional species richness, but they did not analyse different types of communities separately. Nevertheless, these still rather few studies on the relationship between local and regional richness indicate that alpine systems can provide useful insight into diverse ecological research problems.

6.5 Conclusions

This attempt to analyse overall patterns in plant species richness of European mountains is by no means definitive, as it was limited to a few types of alpine vegetation, and there are even some methodological points to be considered.

The use of historical data for some mountain areas (samples collected before the 1970s), which were collected largely to describe vegetation units and used subjective sampling, makes it questionable if the use of inferential statistics was fully justified (e.g. Palmer and White 1994, see also Chytry 2001). The placement and spatial arrangement much determined cumulative species numbers, i.e. species richness. It is also likely that the extent of area sampled and distance between sampling stations varied according to individual workers and local conditions. It is likely that there were differences in the recording of species and this may have introduced some bias (Gaston 1998). However, these data are an invaluable source of information to examine the most obvious patterns and construct testable hypotheses for more rigorous analyses. In brief, the study can be summarised as follows:

1. Multi-scale patterns in plant species richness of European alpine areas were analysed. The analyses were based on species richness values for exposed, snow-protected and snowbed communities and estimated sizes of species pools of the mountain areas.
2. The number of vascular plants declines towards north, whereas bryophytes and macrolichens increase. We detected no clear relationship between vascular plant richness and the size of mountain area.
3. Vascular plant richness is greater on calcareous than siliceous soil, particularly in the mountains of northern Europe. Moderately snow-protected communities were the most species rich followed by exposed and snowbed communities.
4. The relationship between local and regional species richness shows patterns differing among community types. Most communities on siliceous substrates show a saturating relationship (type II), snow-protected communities on calcareous soils conform to the proportional sampling model (type I), predicting a positive correlation between local and regional species richness.

Acknowledgements. V. Andonovski, D. Goméz, K. Laakso, A. Marinas, L. Nagy and F. Pedrotti helped with the data collection. Didi Moser helped with the GIS analyses and graphics. We thank L. Nagy for comments on earlier drafts and M. O. Hill for some critical remarks on the manuscript. The work was supported through grants from the Finnish Research Council of Environment and Natural Resources and the European Science Foundation (ALPNET network).

References

Aichinger E (1933) Vegetationskunde der Karawanken. Pflanzensoziologie 2:1–329
Ballesteros E, Canalís V (1991) La vegetatió uluminal dels massisos de besiberris i de mulleres (Pirineus Centrals Catalans). Bull Inst Catalana Hist Nat 59(8):95–106
Barbero M, Bono G (1973) La végétation orophile des Alpes apuanes. Vegetatio 27:1–48
Beldie A (1967) Flora si vegetatia Muntilor Bucegi. Ed Acad, Bucuresti

Billings WD (1973) Arctic and alpine vegetations: similarities, differences, and susceptibility to disturbance. Bioscience 23:697-704

Birks HJB (1996) Statistical approaches to interpreting diversity patterns in the Norwegian mountain flora. Ecography 19:332-340

Blytt A (1876) Forsög til en theori om invandringen af Norges flora. Nytt Mag Naturvidensk 21:279-362

Borza A (1934) Studii fitosociologice în muntii Retetzatului. Étud Physociol Monts du Rétézat Cluj 14:1-84

Braun-Blanquet G (1931) Recherches phytogéographiques sur le Massif du Gross Glockner (Hohe Tauern). Stn Int Géobot Medit Alpine Montpellier 13:1-65

Braun-Blanquet J (1948a) La Végétation alpine des Pyrénées Orientales. Monografía de la Estación de Estudios Pirenaicos, Barcelona

Braun-Blanquet J (1948b) Übersicht der Pflanzengesellschaften Rätiens. Vegetatio 1:29-41, 285-316

Braun-Blanquet J (1949) Übersicht der Pflanzengesellschaften Rätiens (IV). Vegetatio 2:20-37

Braun-Blanquet J (1969) Die Pflanzengesellschaften der rätischen Alpen im Rahmen ihrer Gesamtverbreitung. I Teil. Bischofsberger and Co, Chur

Braun-Blanquet J (1975) Fragmenta Phytosociologica Raetica. I. Jahrb Natf Ges Graubünden 96:42-71

Bringer K-G (1961a) Den lågalpina Dryas-hedens differentiering och standortsekologi inom Torneträskområdet. I. Svensk Bot Tidskr 55:349-375

Bringer K-G (1961b) Den lågalpina Dryas-hedens differentiering och standortsekologi inom Torneträskområdet. II. Svensk Bot Tidskr 55:551-584

Bültmann H, Daniëls FJA (2001) Lichen richness-biomass relationship in terricolous lichen vegetation on non-calcareous substrates. Phytocoenologia 31:537-570

Carreras J, Carrillo E, Masalles R, Ninot JM, Vigo J (1993) El poblament vegetal de les valls de Barravés i de Castanesa. I. Flora i vegetació. Acta Bot Barcinonensia 42:1-392

Carrillo E, Ninot JM (1992) Flora y vegetació de les valls d´Espot i de Boí. Institut d´Estudis Catalans, Barcelona

Chapin FS III, Körner C (eds) (1995) Arctic and alpine biodiversity, patterns, causes and ecosystem consequences. Springer, Berlin Heidelberg New York

Chytry M (2001) Phytosociological data give biased estimates of species richness. J Veg Sci 12:439-444

Coldea G (1991) Prodrome des associations végétales des Carpates du sud-est (Carpates roumaines). Doc Phytosociol NS 13:317-539

Coldea G, Cristea V (1998) Floristic and community diversity of sub-alpine and alpine grasslands and grazed dwarf-shrub heaths in the Romanian Carpathians. Pirineos 151-152:73-82

Comes HP, Kadereit JW (1998) The effect of Quaternary climatic changes on plant distribution and evolution. Trends Plant Sci 3:432-438

Conti E, Soltis DE, Hardig TM, Schneider J (1999) Phylogenetic relationships of the silver Saxifrages (Saxifraga, Sect. Ligulatae Haworth): implications for the evolution of substrate specificity, life histories, and biogeography. Mol Phylog Evol 13:536-555

Cornell HV (1999) Unsaturation and regional influences on species richness in ecological communities: a review of the evidence. Écoscience 6:303-315

Cornell HV, Lawton JH (1992) Species interactions, local and regional processes, and limits to the richness of ecological communities: a theoretical perspective. J Anim Ecol 61:1-12

Cresswell JE, Vidal-Martinez VM, Crichton NJ (1995) Investigation of saturation in the species richness of communities: some comments on methodology. Oikos 72:301-304

Dahl E (1957) Mountain vegetation in south Norway and its relation to the environment. Skr Nor Vidensk Akad Oslo I. Mat Naturvidensk kl 1956(3):1-374

Dalmas J-P (1972) Etudes phytosociologique et ecologique de l'etage alpin des Alpes Sud-Occidentales francaises et plus particulierement de la region de Vars-Ecreins (H.A.). Thèse, Laboratoire de Phytosociologie et Cartographie Vegetale, Universite de Provence, Centre Saint Charles

Dimopoulos PD, Georgiadis T (1994) Ecological evaluation of the above the timberline communities of Mount Killini (NE Peloponnisos – Greece), in the perspective of nature conservation. Ann Bot 52:151–165

Dirnböck T, Dullinger S, Gottfried M, Grabherr G (1999) Die Vegetation des Hochschwab – Subalpine und Alpine Stufe. Mitt Naturwiss Ver Steiermark 129:111–251

Domin K (1928) The relations of the Tatra mountain vegetation to the edaphic factors of the habitat. Acta Bot Bohemica 6/7:133–163

Domin (1933) Die Vegetationsverhältnisse des Bucegi in den rumänischen Südkarpathen. Veröff Geobot Inst Rübel 11:96–144

Dullinger S (1999) Vegetation des Schrankogel, Stubaier Alpen. Diplomarbeit, Univ Wien

Dullinger S, Dirnböck T, Grabherr G (2000) Reconsidering endemism in the north-eastern limestone Alps. Acta Bot Croat 59:55–82

Dupré C (2000) How to determine a regional species pool: a study in two Swedish regions. Oikos 89:128–136

El Aallali A, Nieto LJM, Raya FP, Molero Mesa J (1998) Estudio de la vegetación forestal en la vertine sur de Sierra Nevada (Alpujarra Alta granina). Itin Geobot 11:3 87–402

Elven R, Elvebakk A (1996) A catalogue of Svalbard plants, fungi, algae and cyanobacteria, part 1. Vascular plants. Norsk Polarinst Skr 198:9–55

Eurola S (1968) Über die Fjeldheidevegetation in den Gebieten von Isfjorden und Hornsund in Westspitzbergen. Aquilo Ser Bot 7:1–56

Eurola S (1974) Plant ecology of Northern Kiölen: arctic or alpine ? Aquilo Ser Bot 13:10–22

Feoli-Chiapella L, Feoli E (1977) A numerical phytosociological study of the summits of the Majella massive (Italy). Vegetatio 34:21–39

Feoli-Chiapella L, Poldini L (1993) Prati e Pascoli del Friuli (NE Italia) su substrati basici. Stud Geobot 13:3–140

Galland P (1979) Note sur le Caricetum firmae du Parc National Suisse. Doc Phytosoc NS 4:279–287

Gamisans J (1977) La végétation des montagnes corses, II. Phytocoenologia 4:35–131

Gaston KJ (1998) Species richness: measure and measurement. In: Gaston KJ (ed) Biodiversity – a biology of numbers and difference. Blackwell, Oxford, pp 77–113

Gensac P (1977) Les groupements végétaux a Carex curvula All. dans le massif de la Vanoise. Trav Sci Parc Nat Vanoise 8:67–94

Gensac P (1979) Les pelouses supraforestières du massif de la Vanoise. Trav Sci Parc Nat Vanoise 10:111–242

Gielly L, Taberlet P (1996) A phylogeny of the European gentians inferred from chloroplast *trn*L (UAA) intron sequences. Bot J Linn Soc 120:57–75

Gigon A (1971) Vergleich alpiner Rasen auf Silikat- und auf Karbonatboden; Konkurrenz- und Stickstofformenversuche sowie standortkundliche Untersuchungen im Nardetum und im Seslerietum bei Davos. Veröff Geobot Inst Rübel Zürich 48:1–159

Gjærevoll O (1956) The plant communities of the Scandinavian alpine snowbeds. Kongel Nor Vidensk Selsk Skr 1956(1):1–405

Gjærevoll O (1963) Survival of plants on nunataks in Norway during the Pleistocene glaciation. In: Löve À, Löve D (eds) North Atlantic biota and their history. Pergamon Press, Oxford, pp 261–283

Godwin H (1975) The history of the British flora, 2nd edn. Cambridge University Press, Cambridge

Grabherr G (1985) Numerische Klassification und Ordination in der alpinen Vegetationsökologie als Beitrag zur Verknüpfung moderner "Computermethoden" mit der pflanzensoziologischen Tradition. Tuexenia NS 5:181–190

Grabherr G (1995) Alpine vegetation in a global perspective. In: Box EO (ed) Vegetation science in forestry. Kluwer, The Hague, pp 441–451

Grabherr G, Gottfried M, Pauli H (2000) Hochgebirge als 'hot spots' der Biodiversität – dargestellt am Beispiel der Phytodiversität. Ber Reinh Tüxen Ges 12:101–113

Greimler J (1997) Pflanzengesellschaften und Vegetationsstruktur in den südlichen Gesäusebergen (nordöstliche Kalkalpen, Steiermark). Mitt Landesmus Joann Graz 25/26:1–238

Grime JP (1979) Plant strategies and vegetation processes. Wiley, Chichester

Grime JP, Curtis AV (1976) The interaction of drought and mineral nutrient stress in calcareous grassland. J Ecol 64:975–988

Grubb PJ (1987) Global trends in species-richness in terrestrial vegetation: a view from the northern hemisphere. In: Giller JHR, Giller PS (eds) Organisation of communities: past and present. Blackwell, Oxford, pp 99–118

Gruber M (1978) La végétation des Pyrénées ariégeoises et catalanes occidentales. Thèse, Marseille

Guinochet M (1938) Etudes sur la végétation del'étage alpin dans le bassin supérieur de la Tinée (Alpes Maritimes). Thèse, Grenoble

Haapasaari M (1988) The oligotrophic heath vegetation of northern Fennoscandia and its zonation. Acta Bot Fenn 135:1–219

Hastie TJ, Tibshirani RJ (1990) Generalized additive models. Chapman and Hall, London

Hedberg O, Mårtensson O, Rudberg S (1952) Botanical investigations in the Pältsa region of northermost Sweden. Bot Not Suppl 3(2):1–209

Hedenäs L (1999) Altitudinal distributions in relation to latitude; with examples among wetland mosses in the Amblystegiaceae. Bryobrothera 5:99–115

Herben T (2000) Correlation between richness per unit area and the species pool cannot be used to demonstrate the species pool effect. J Veg Sci 11:123–126

Hodisan I, Boscaiu M (1986) Floristic diversity indices of some plant associations in the Retezat National Park. Stud Univ Babes Bolyai B 31:14–18

Horvat I (1930) Vegetationsstudien in den kroatischen Alpen. I. Die alpinen Rasengesellschaften. Bull Int Acad Yougosl Sci Art Cl Math Nat 24:1–96 (in Croatian)

Horvat I, Pawlowski B, Walas J (1937) Phytosociologische Studien über die Hochgebirgsvegetation der Rila Planina in Bulgarien. Bull Acad Polon Krakow 8:159–189

Horvat I, Glavac V, Ellenberg H (1974) Vegetation Südosteuropas. Gustav Fisher, Stuttgart

Huml O, Lepš J, Prach K, Rejmánek M (1979) Zur Kenntnis der Quellfluren, alpinen Hochstaudenfluren und Gebüsche des Fagaras-Gebirges in den Südkarpaten. Preslia 51:35–45

Igoshina KN (1966) The peculiarities of flora and vegetation on ultrabasic rocks of the Polar Urals (Rai-Iz as an example). Bot Zhurhn 51:322–337 (in Russian)

Janovicova K, Kubinska A, Soltes R (1999) Bryophytes of the Cervené vrchy Mts and the Tichá dolina valley (the Západne Tatry Mts, Slovakia) – threat and apophytic tendencies in local bryophyte flora. Biol Bratislava 54:369–378

Jeník J (1961) Alpinská vegetace Krkonoč, Králichého Snezníku a Hrubého Jeseníku. Naklad. CSAV, Praha

Jovanovic B, Lakusic R, Rizovski R, Trinajstic S, Zupancic M (1986) Prodromus Phytocoenosum Jugoslaviae. Map 1: 200 000. Scientific Council of Vegetation Map of Yugoslavia, Bribir-Ilok

Kalliola R (1939) Pflanzensoziologische Untersuchungen in der alpinen Stufe Finnisch-Lapplands. Ann Bot Soc Zool Bot Fenn Vanamo 13:1–328

Klika J (1932) Der Seslerion coeruleae-Verband in den Westkarpathen. Beih Bot Centralbl 49:133–175

Körner C (1995) Alpine plant diversity: a global survey and functional interpretations. In: Chapin FS III, Körner C (eds) Arctic and alpine biodiversity, patterns, causes and ecosystem consequences. Springer, Berlin Heidelberg New York, pp 45–62

Körner C (1999) Alpine plant life. Springer, Berlin Heidelberg New York

Körner C (2000) Why are there global gradients in species richness? Mountains might hold the answer. Trends Ecol Evol 15:513–514

Krajina V (1933) Die Pflanzengesellschaften des Mlynica-Tales in den Vysoké Tatry (Hohe Tatra). I. Beih Bot Centralbl 50:774–957

Lakusic R (1970) Die Vegetation der südöstlichen Dinariden. Vegetatio 21:321–373

Lawton JH (1999) Are there general laws in ecology? Oikos 84:177–192

Lepš J, Prach K, Slavíková J (1985) Vegetation analysis along the elevation gradient in the Nízké Tatry Mountains (Central Slovakia). Preslia 57:299–312

Lüdi W (1943) Über Rasengesellschaften und alpine Zwergstrauchheide in den Gebirgen des Apennin. Ber Geobot Forsch Inst Rübel Zürich 1942:23–68

MacArthur RH, Wilson EO (1967) The theory of island biogeography. Princeton University Press, Princeton

McCullagh P, Nelder JA (1989) Generalized linear models, 2nd edn. Chapman and Hall, London

McVean DN (1964) Dwarf shrub heaths. In: Burnett JH (ed) The vegetation of Scotland. Oliver Boyd, Edinburgh, pp 481–498

McVean DN, Ratcliffe DA (1962) Plant communities of the Scottish highlands. Monographs of the Nature Conservancy No 1. HMSO, London

Montserrat G (1986) Flora y Vegetación del Macizo de Cotiella y la Sierra de Chía. Tesis, Universidad de Barcelona

Mucina L, Valachovic K, Jarolímek I, Seffer J, Kubinská A, Pisút I (1990) The vegetation of rock fissures, screes, and snow-beds in the Pirin Planina mountains (Bulgaria). Stud Geobot 10:15–58

Negre R (1969) Le Gentiano-Caricetum curvulae dans la région luchonaise (Pyrénées centrales). Vegetatio 18:167–202

Negre R (1970) La végétation du Bassin de l'One (Pyrénées centrales). Portug Acta Biol B 11 (1–2):51–166

Nordhagen R (1928) Die Vegetation und Flora des Sylenegebietes. I. Die Vegetation. Skr Nor Vidensk Akad Oslo. I. Mat Naturvidensk kl 1927(1):1–612

Nordhagen R (1943) Sikilsdalen og Norges fjellbeiter. En plantesosiologisk monografi. Bergens Mus Skr 22:1–607

Nordhagen R (1955) Kobresieto-Dryadion in northern Scandinavia. Sv Bot Tidskr 49:63–87

Ochsner F (1954) Die Bedeutung der Moose in alpinen Pflanzengesellschaften. Vegetatio 5–6:279–291

Oksanen L, Virtanen R (1995) Topographic, altitudinal and regional patterns in continental and suboceanic heath vegetation of northern Fennoscandia. Acta Bot Fenn 153:1–80

Ozenda P (1950) Matériaux pour la flore lichénelogique des Alpes-Maritimes. Bull Soc Bot Fr 97:29–50

Palmer MW, White PS (1994) Scale dependence and the species-area relationship. Am Nat 144:717–740

Pauli H (1993) Untersuchungen zur phytosoziologischen und ökologischen Stellung von Festuca pseudodura in den Niederen Tauern. Diplomarbeit, Univ Wien

Pawlowski B (1935) Über die Klimaxassociation in der alpinen Stufe der Tatra. Bull Int Acad Pol Sci Lett Ser B 1935:115–146

Pawlowski B, Walas J (1949) Les associations des plantes vasculaires des Monts de Czywczyn. Bull Int Acad Pol Sci Lett Ser B 1948:117–180

Pedrotti F (1982) La vegetation des monts de la Laga. In: Pedrotti F (ed) Guide – Itinéraire de l'Excurcion Internationale de Phytosociologie en Italie centrale (2–11 juillet 1982). Univ Studi, Camerino, pp 571–579

Pignatti E, Pignatti S (1983) La vegetazione delle vette di Feltre al di Sopra del limite Degli Alberi. Stud Geobot 3:7–47

Poldini L, Martini F (1993) La vegetatzione delle vallete nivali su calcare, dei conoidi e delle alluvioni Nel Friuli (NE Italia). Stud Geobot 13:141–214

Puscaru D, Puscaru-Soroceanu E, Pauca A, Serbanescu I, Beldie A, Stefureac T, Cernescu N, Saghin F, Cretu V, Lupan L, Tascenco V (1956) Panusile alpine din Muntii Bucegi. Ed Acad, Bucuresti

Puscaru-Soroceanu E, Csürös S, Puscaru R, Popova-Cucu A (1981) Die Vegetation der Wiesen und Weiden des Fagaras-Gebirges in den Südkarpaten. Phytocoenologia 9:257–309

Quezel P (1953) Contribution à l'etude phytosociologique et géobotanique de la Sierra Nevada. Mem Soc Brot (Coimbra) 9:5–77

Quezel P (1964) Végétation des hautes montagnes de la Grèce Méridionale. Vegetatio 12:289–385

Quezel P (1967) La végétation des hauts sommets du Pinde et de l'Olympe de Thessalie. Vegetatio 14:127–228

Recchia F, Villa S (1996) A first contribution to the lichen flora of Abruzzi (C. Italy). Flora Medit 6:5–9

Resmerita I (1970) Flora, vegetatia si potentiaul productiv pe Masivul Vládeasa. Ed Acad, Bucuresti

Resmerita I (1976) La classe des Salicetea herbacea Br. Bl. 1947 des Carpates Roumaines. Doc Phytosoc 15–18:123–135

Resmerita I (1979a) La vegetation chionophile des Carpathes Roumaines. Doc Phytosoc NS 4:871–881

Resmerita I (1979b) Flora rezervatiei naturale 'Pietrosul Mare'. Stud Univ Babes-Bolyai B 24:8–14

Resmerita I (1981) Vegetatia rezervatiei naturale "Pietrosul Mare" (II). Stud Univ Babes-Bolyai B 26:3–11

Resvoll-Holmsen H (1920) Om fjeldvegetationen i det østenfjeldke Norge. Arch Math Naturvid 37:1–266

Richard J-L (1975) Premiere approche de la vegetation de l'etage alpin du val d'Anniers (Alpes Valaisannes, Suisse). Doc Phytosoc 9–14:223–236

Richard J-L (1985) Pelouses xérophiles alpines des environs de Zermatt (Valais, Suisse). Bot Helv 95:194–211

Richard J-L, Geissler P (1979) A la découverte de la végétation des bords de cours d'eau d'étage alpin du Valais (Suisse). Phytocoenologia 6:183–201

Richardson DM, Cowling RM, Lamont BB, van Hensbergen HJ (1995) Coexistence of Banksia species in southwestern Australia: the role of regional and local processes. J Veg Sci 6:329–242

Riebesell JF (1982) Arctic-alpine plants in mountaintops: agreement with island biogeography theory. Am Nat 119:657–674

Rivas-Martínez S, Báscones JC, Díaz TE, Fernández-González F, Loidi J (1991) Vegetación del Pirineo occidental y Navarra. Itin Geobot 5:5–455

Safford HD, Rejmánek M, Hadac E (2001) Species pools and the 'hump-back' model of plant species diversity: an empirical analysis at a relevant spatial scale. Oikos 95:282–290

Scharfetter R (1909) Über die Artenarmut der ostalpinen Ausläufer der Zentralalpen. Österr Bot Z 1909:1–7

Söyrinki N (1954) Vermehrungsökologische Studien in der Pflanzenwelt der bayerischen Alpen. Ann Bot Soc Zool Bot Fenn Vanamo 27:1–232

S-PLUS (2000) Guide to statistics, vol 1. Data Analysis Products Division, MathSoft, Seattle
Stehlik I (2000) Nunataks and peripheral refugia for alpine plants during Quaternary glaciation in the middle part of the Alps. Bot Helv 110:25-30
Strid A (1995) The Greek mountain flora, with special reference to the central European element. Bocconea 5:99-112
Szafer W (1924) Zur soziologischen Auffassung der Schneetälchenassoziationen. Veröff Geobot Inst Rübel Zürich 1924:300-310
Szafer W, Kulczynski S, Pawlowski B, Stecki K, Sokolowski M (1927) Die Pflanzenassoziationen des Tatra-Gebirges. – III-V. Teil. Die Pflanzenassoziationen des Chocholowska-Tales. Bull Int Acad Pol Sci Lett Ser B Suppl 2 1926:1-144
Tilman D, Pacala SW (1993) The maintenance of species richness in plant communities. In: Ricklefs RE, Schluter D (eds) Species diversity in ecological communities: historical and geographical perspectives. University of Chicago Press, Chicago, pp 13-25
Väisänen R (1998) Current research trends in mountain biodiversity in NW Europe. Pirineos 151-152:131-156
Vetterli L (1982) Alpine Rasengesellschaften auf Silikatgestein bei Davos. Veröff Geobot Inst Rübel Zürich 76:1-92
Vigo J (1996) El poblament vegetal de la Vall de Ribes. Institut Cartogràfic de Catalunya, Barcelona
Virtanen R, Eurola S (1997) Middle oroarctic vegetation in Finland and middle-northern arctic vegetation on Svalbard. Acta Phytogeogr Suec 82:1-60
Virtanen R, Lundberg PA, Moen J, Oksanen L (1997) Topographic and altitudinal patterns in plant communities of European arctic islands. Polar Biol 17:95-113
Virtanen R, Dirnböck T, Dullinger S, Pauli H, Staudinger M, Grabherr G (2002) Multi-scale patterns in the plant species richness of European high mountain vegetation. In: Körner C, Spehn E (eds) Alpine biodiversity. Parthenon, Boca Raton, pp 91-101
Wahlenberg G (1814) Flora Carpatorum principalium. van den Hoeck, Göttingen
White PS, Miller RI (1988) Topographic models of vascular plant richness in the southern Appalachian high peaks. J Ecol 76:192-199
White PS, Miller RI, Ramseur GS (1984) The species-area relationship of the southern Appalachian high peaks: vascular plant richness and rare plant distributions. Castanea 49:47-61
Whittaker RH (1965) Dominance and diversity in land plant communities. Science 147:250-260
Wilcox BA (1978) Supersaturated island faunas: a species-age relationship for lizards on post-Pleistocene land-bridge islands. Science 199:996-998
Wohlgemuth T (1993) Der Verbreitungsatlas der Farn- und Blütenpflanzen der Schweiz (Welten und Sutter 1982) auf EDV. Die Artenzahlen und ihre Abhängigkeit von verschiedenen Faktoren. Bot Helv 103:55-71
Zlatník A (1925) Les associations de la végétation des Krkonoče et le pH. Extrait des Mémoires de la Société des Sciences de Bohême. Classe des Sci 1925:1-67
Zobel M (1997) The relative role of species pools in determining plant species richness: an alternative explanation of species coexistence? Trends Ecol Evol 12:266-269
Zollitsch B (1968) Soziologische und ökologische Untersuchungen auf Kalkschiefern in hochalpinen Gebieten. I. Die Steinschuttgesellschaften der Alpen mit besonderer Berücksichtigung der Gesellschaften auf Kalkschiefer in den mittleren und östlichen Zentralalpen. Ber Bayer Bot Ges 40:67-100

7 Altitude Ranges and Spatial Patterns of Alpine Plants in Northern Europe

J. I. Holten

7.1 Introduction

Plant species richness, composition, or turnover along an altitude gradient may be continuous (Odland and Birks 1999), or discontinuous (Grabherr et al. 1995). Discontinuity arises when species are distributed over a narrow altitude range, e.g. in the Alps (Grabherr et al. 1995). Where a gradient in macroclimate exists such as along a coast–inland gradient in central Norway, the lower and upper limits of plant distribution (e.g. Jørgensen 1933, 1937; Kilander 1955; Gjærevoll 1990) are different and continuous along the gradient (Holten 1986).

Various phytogeographical (Bøcher 1938; Dahl 1951, 1998; Gjærevoll 1959; Berg 1963; Danielsen 1971; Holten 1986; Grabherr 1995; Birks 1996), physiological (Crawford and Palin 1981; Crawford 1997; Crawford and Wolfe 1998; Gauslaa 1984, 1985) and other hypotheses (Moe 1995; Holten 1998a) have been presented about the distribution limitation of alpine plants at the local and regional scales, without any satisfactory explanation.

The main aims of the project have been to map regional altitude ranges of vascular plants along two complex environmental gradients in western central Scandes, i.e. (1) along the altitude gradient, and (2) along the coast–inland gradient. Both gradients are complex thermic-hygric gradients. The resulting patterns of distributions are related to the main patterns of macroclimates (see study area below), thus following a correlative approach (see Sect. 7.3).

A grouping of alpine vascular plants based on their altitude ranges is made below.

7.1.1 Study Area

The study area extended 135 km inland from the west coast of Norway (centre of study area 63°N 9°E) on the western macroslope of the southern Fennoscandian mountain range (southern Scandes), between Kristiansund on the coast and the main watershed at Dovrefjell. The topography is heterogeneous, with relatively low mountains (maximum altitude about 1000 m) in the coastal and middle fjord districts (0–50 km from the coast) and typical alpine relief in the inner fjord and western valley districts (50–90 km from the coast, with a maximum altitude of 1900 m). Further inland (90–135 km from the coast, with a maximum altitude of 2286 m), the relief is gentler. The bedrock is acidic gneiss in the west and calcareous and schistose in the eastern third of the transect. Both the temperature and precipitation showed a steep gradient from the coast inland (Fig. 7.1). The January mean temperature is +1.5 °C on the coast as opposed to –12 °C at the watershed of Dovrefjell. The annual precipitation is about 2000 mm on coastal mountains and is <400 mm in the Drivdalen–Dovrefjell area.

7.1.2 Methods

The case study has involved mapping of approximately 500 vascular plant species along the coast–inland transect, a transect having a main direction NW–SE. Thirty mountains were selected in central Norway in 1992 to represent a complex coast–inland gradient in macro- and local climate on each mountain. The collection of data was standardised by using vertical belts, 100 m wide, and distance zones from the coast (Fig. 7.1). Each distance zone

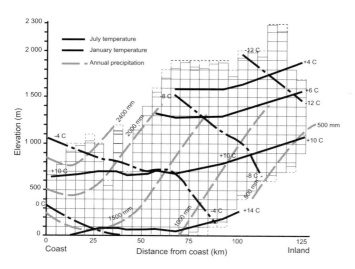

Fig. 7.1. Vertical projections of mean annual precipitation, July mean temperature and January mean temperature into the coast–inland transect in central Norway. Precipitation figures based on Førland (unpubl.)

Altitude Ranges and Spatial Patterns of Alpine Plants in Northern Europe

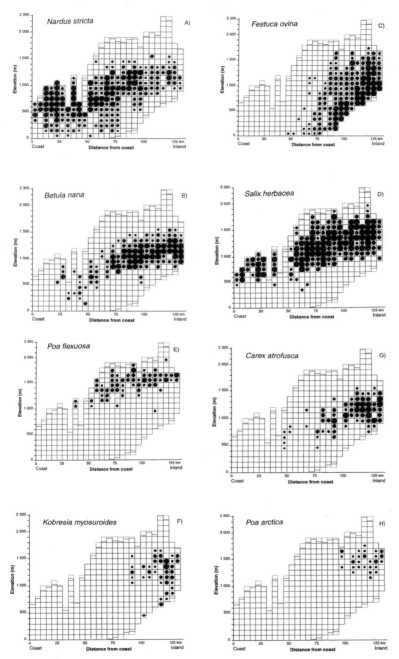

Fig. 7.2A–H. Altitude ranges of alpine species in a coast–inland transect in central Norway. **A** Boreal-alpine group: *Nardus stricta*; **B** boreal-alpine group: *Betula nana*; **C** boreal-alpine group: *Festuca ovina*; **D** widespread group: *Salix herbacea*; **E** widespread group: *Poa flexuosa*; **F** centric group: *Kobresia myosuroides*; **G** centric group: *Carex atrofusca*; **H** centric group: *Poa arctica*. The size of the filled circles is proportional to the abundance (rare, scattered, common to dominant) of each species within cells

had a horizontal breadth of 5 km; each of the 414 cells in the figures (Fig. 7.2A-H) therefore represents an area of 5 × 0.1 km. Floristic lists were collected for each 100 m vertical belt on the 'warmer' slopes of the mountains, i.e. slopes facing from eastern, through southern, to western aspects of the selected mountains. The abundance of the species was scored in each 100-m belt: 4 – dominant, 3 – common to dominant, 2 – scattered, 1 – rare ('dominant' and 'common to dominant' are put together in the maps below). The time used for each list was 30–45 min, largely depending on the local topography. In one day usually 8–10 lists were produced, from the lowest possible (e.g. valley floor, seaside) to the highest possible altitude on a selected mountain (the summit area).

7.2 Altitude Ranges and Patterns

The pattern of the altitude range of a plant species can be characterised by the altitude variation in its abundance; the total vertical range; and its upper and lower limits. Except for the upper and lower limits, the information from this case study is regarded as being representative for the whole Scandes range. The groups below are classified according to these attributes.

7.2.1 Altitude Ranges of Alpine Plant Species

From the total material of distribution data (ca. 500 species of vascular plants), a special case study is carried out below on taxa defined as alpine (ca. 200 species) in the collected material.

7.2.1.1 Boreal-Alpine (Montane) Group

This group has a very wide distribution in northern Europe, both horizontally and vertically (Figs. 6.2 and 6.3). Most species dominantly occur in the north boreal and low alpine vegetation zones in the Scandes, with extensions into the mid-alpine and the lower part of the high-alpine zones. Those species which are most wide-ranging vertically are also very wide-ranging ecologically, so-called euryoicous species (e.g. *Polygonum viviparum* with a range of 1700 m and *Solidago virgaurea* with 1600 m). Humidiphilous and more continental subgroups of the boreal-alpine (montane) group may also be distinguished. The humidiphilous subgroup has a frequency optimum in the humid, coastal uplands and mountains. The most typical humidiphilous species in the snow-rich coastal uplands are *Nardus stricta* (dominant, see Fig. 7.2A), *Blechnum spicant*, *Cornus suecica* and *Thelypteris limbosperma*.

The humidiphilous subgroup has its main occurrence below the birch treeline in the coastal and fjord region.

Another subgroup of boreal-alpine species is found on the steep, mesic to wet inland valley slopes. These slopes are more eutrophic than the coastal slopes with their gneissic bedrock, and are characterised by stands of the tall herb *Aconitum septentrionale*, frequently associated with *Viola biflora*, *Myosotis decumbens*, *Salix lapponum*, *Salix glauca* and *Salix lanata*. Some eutrophic mire species occur in the same vegetation complex, such as *Saxifraga aizoides*, *Carex capillaris*, *Salix myrsinites*, *Salix arbuscula*, *Epilobium davuricum*, *Juncus triglumis* and the rare *Carex capitata*. A smaller number of species grow in drier, low alpine sites inland (e.g. *Betula nana*, Fig. 7.2B, and *Arctostaphylos uva-ursi*).

The geographically and altitudinally most restricted species, having an altitude range of 300–600 m, are found inland. They are stenoicous species with a narrow environmental niche. Examples include *Botrychium boreale* (range 600 m), *Vahlodea atropurpurea* (range 500 m) and *Carex capitata* (range 300 m). Some xerophilous species form a drier subelement of the boreal-alpine group in the easternmost dry valleys. Except for *Festuca ovina* (Fig. 7.2C), the xerophilous members are not very prominent, because most of them are close to the north-western margin of their European range. A xerothermic boreal-alpine meadow and rock community with *Juniperus communis*, *Cotoneaster scandinavicus*, *Poa glauca* and *Veronica fruticans* occurs on steep south-facing slopes in the north boreal and low-alpine zone (Nordhagen 1943).

7.2.1.2 Widespread (Ubiquitous) Group

The species belonging to this group (Fig. 7.3) typically display a uniform distribution along the whole Scandes range. They are among the most dominant species in the mountain landscape in both the southern and northern Scandes, especially westwards of the main watershed, and mainly on hard, acidic rock. The group has its main occurrence above the birch treeline. Most species have their absolute lower limits in the north boreal zone. They mostly have a very wide altitude range; that of *Salix herbacea*, for example, is as much as 1600 m (Fig. 7.2D). Most species show a variable vertical range along the coast–inland gradient, having their maximum vertical range 50–80 km from the coast (inner fjord districts). From this area, the vertical range decreases both westwards and eastwards. On the extreme coastal mountains (0–10 km from the coast), the lower limits of some mountain heath species rise slightly towards the coast. The widespread group has a subelement with an optimum in the mid- and high-alpine zones, with *Ranunculus glacialis*, *Poa flexuosa* and *Luzula arcuata* being the most prominent. These species have a narrow vertical range with *Carex rufina* at 400 m being the narrowest. *Poa flexuosa*

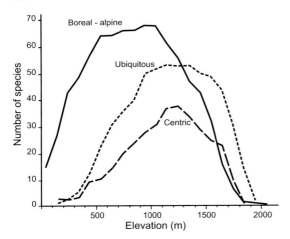

Fig. 7.3. Altitude range of the boreal-alpine (montane), widespread (ubiquitous) and centric groups in the south Scandes

(Fig. 7.2E) has a vertical range of 100–200 m at both its coastal and inland limits. *Trisetum spicatum* shows a different pattern, having a broad vertical range near its western limit in the inner fjord districts. Some species of the widespread group grow at the upper limit of vascular plant life in this part of the southern Scandes, which is at about 2000 m at Snøhetta (summit 2286 m). Due to the general drop of altitude limits towards the coast, the altitude range of most species on a single mountain is much less and very often about half of the total range. *Poa flexuosa* has locally on most mountains an altitude range of about 500 m. However, its total altitude range from coast to inland is about 1000 m (see Fig. 7.2E).

7.2.1.3 Centric Group

The term 'centric' has its origin in the fact that some 80 alpine plant species in the Scandes have characteristic distribution patterns in two separate areas, a larger northern area (65–70°N), and a smaller southern area (61–63°N). Those occurring in both areas are called bicentric, and those occurring in only the northern or southern areas are called northern or southern unicentric. Påhlson (1998) has listed 42, 29 and 11 species in the bicentric, northern unicentric and southern unicentric groups, respectively.

The centric group species are heterogeneous with regard to their lower limits, but homogeneous with respect to their western and upper limits. The western limits coincide with the western boundary of calcareous bedrock and the eastern boundary of gneiss in the transect. The centric group has the same optimum altitude as the widespread group (Fig. 7.3), at about 1250 m, which is in the upper part of the low-alpine zone. A characteristic distributional feature of the centric group is the generally narrower altitude ranges of the species than for the widespread group. Centric species have lower upper limits than the two former groups. The latter is mainly due to the lower summit altitudes of calcareous mountains.

In the wooded zones, centric species are found in exposed rocky habitats, on calcareous screes and calcareous mires. The most typical centric species are found in the mid-alpine zone, at 1100–1600 m. They are often quite rare and of relict nature. They also have a narrow (100–500 m) altitude and phytosociological range (see point 3, Sect. 7.4; Holten 1998a,b).

Some centric species grow in drier conditions and show an affinity with the central Asian steppic mountains, where *Kobresia* communities have their centre of diversity (see Grabherr 1995). Examples are *Kobresia* (=*Elyna*) *myosuroides* (Fig. 7.2F), *Potentilla nivea*, *Gentianella tenella*, *Elymus alskanus* and *Oxytropis lapponica*.

7.2.2 Alpine Vascular Plant Species Richness vs. Altitude

Total vascular plant species richness (Fig. 7.4) and alpine vascular plant species richness (Fig. 7.5) vs. altitude are shown separately. For the former, eight mountains were selected to compare the pattern of altitude variation in the total species richness along the coast–inland gradient (Fig. 7.4). Species richness on the coastal mountains was more or less constant in the altitude interval of 0–800 m. Between 800 and about 1300 m, there was a steep decrease, followed by a less steep decrease levelling off at 12–15 species

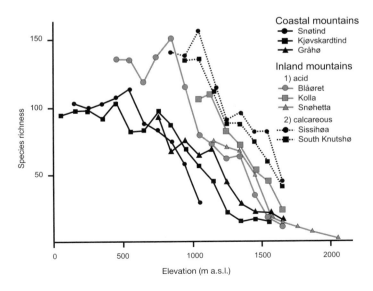

Fig. 7.4. Altitudinal changes in total vascular plant species richness on eight mountains along a coast–inland transect

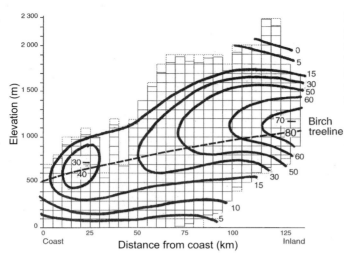

Fig. 7.5. Vertical projection of species richness of alpine vascular plant species along a coast–inland transect

(between 1300 and 1700 m) because few summits reach higher than 1700 m. On the inland mountains, there was little change up to about 1000 m, above which the decrease in species richness was very steep. In any given 100-m altitude band, the species richness is generally 20–30 species higher where the bedrock is calcareous than where it consists of hard, acidic rock types. In the inland part of the transect, a typical summit of 1650 m therefore has 40–45 species where the bedrock is calcareous and schistose and only 10–20 species where it is acidic (Fig. 7.4). A two-dimensional vertical projection of the species richness of alpine vascular plants in the coast–inland transect showed that, from a maximum of 70–80 alpine species in the Drivdalen-Knutshø mountains (115–130 km from the coast), the richness decreased in all directions (Fig. 7.5). Along the altitude gradient, maximum species richness was found about 100–200 m above the birch treeline. The local maximum of species richness on steep coastal mountains was half of that on the inland mountains.

7.3 What Limits the Altitude Ranges and Patterns in the Alpine Zone?

There are two main environmental gradients, one temeperature gradient from the lowlands to the mountan tops, and one humidity gradient from the coastal mountains to the drier inland valley floors and dry slopes. The latter gradient is illustrated by the change from the coastal mountains with the humid, oligotrophic *Nardus–Carex bigelowii* and *Phyllodoce-Vaccinium* type communities to the drier *Veronica-Poa glauca* type communities (Nordhagen 1943) on the drier interior valley slopes, often dominated by *Festuca ovina*. Common or dominant occurrence of dry *Kobresia* heaths indicates drier conditions in the low-alpine zone of the eastern part of the transect (see Grabherr 1995). In the mid- and high-alpine zones the distribution and depth of snow is the dominant factor for vegetation pattern as opposed to moisture availability. Cold winters (see Fig. 7.1) combined with thin snow cover cause deep penetration of alpine permafrost in the easternmost part of the transect. The *Cassiope–Salix herbacea* communities (Gjærevoll 1956) and the dominant occurrence of *Ranunculus glacialis* have their optima in snow-rich coastal mountains.

7.3.1 Limiting Factors

7.3.1.1 High Summer Temperatures and Drought

Dahl (1951) discussed the importance of high summer temperatures for the lower limits of alpine plants in the Scandes. He proposed that alpine plants had different sensitivity to high temperature, which resulted in their different lower limits. To support his hypothesis, Dahl (1951) demonstrated that 149 alpine species in Scandinavia show a distribution that could be delineated by the mean annual maximum temperature. For example, the rare, mid-alpine species *Campanula uniflora* in the southern Scandes correlates well with the 22 °C isotherm (Dahl 1986, 1998). Gauslaa (1984, 1985) modified Dahl's (1951) hypothesis and concluded that many alpine species seem to be restricted by moisture shortage in the lowlands, and others (mostly cushion plants) by high temperatures. By calculating energy budgets, Gauslaa (1984, 1985) concluded that there was a close connection between water availability and the danger of over-heating.

7.3.1.2 Winter Temperatures

Many of the centric plants have a western limit that shows some correlation with winter isotherms. The correlation with mean annual precipitation isohyets is much weaker (Fig. 7.1). Most centric species have limiting January isotherms from −8 to −10 °C. Gauslaa (1985) pointed out that nearly all bi- and unicentric species were 'south-west coast avoiding' (sensu Dahl 1951). However, the western limitation of alpine species by mild winters is still a vague hypothesis that needs substantiating experimentally. According to Crawford and Palin (1981), northern (arctic-alpine) species have higher respiration rates than thermophilous ones and higher carbohydrate loss during winter (Crawford and Wolfe 1998). A retreat to colder environments is a consequence of this. Mild periods may trigger premature shoot growth and subsequent dieback of non-hardy shoots (Crawford 1997). It may be speculated that most alpine plants in the Scandes, and especially centric species, favour more seasonal climates with high thermic continentality, with more non-seasonal (thermic oceanic) coastal climates being less favourable (see also Moe 1995).

7.3.1.3 Alpine Permafrost

Alpine permafrost results from a thin snow cover combined with low winter temperatures. Such physical conditions occur in the Drivdalen–Dovrefjell area. According to Ødegaard et al. (1996), the lower limit of discontinuous permafrost on flat, exposed sites is 1450–1500 m on Snøhetta, Dovrefjell. Sporadic permafrost, for example in the form of palsas, has been described on Dovrefjell from an elevation interval of 1000–1400 m (Sollid and Sørbel 1998). On Snøhetta, a sharp decrease in the species richness of vascular plants has been found between 1500 and 1550 m (Fig. 7.4), accompanied by a change in life forms abundance from dwarf shrubs to graminoids (Jonasson 1986; Holten 1998b). The altitude band of 1500–1600 m was highlighted by Odland and Birks (1999) as one of the major biotic boundaries along a vertical gradient of species richness in Aurlandsdalen, south-western Scandes. The lower limit of discontinuous permafrost on Dovrefjell is about 1400 m. Only one species, *Sagina caespitosa*, has its absolute lower limit at this elevation. However, many other centric species occur in the altitude band of 1150–1650 m. The species with the most strongly arctic character in the centric group are most frequent at around 1300–1600 m, the altitude belt most strongly affected by active cryoturbation.

7.4 Conclusions

1. Macroclimate controls the altitude ranges of alpine plants at a regional scale; on a local scale, microhabitat conditions often overrule macroclimate effects. The species richness of alpine vascular plants in the Scandes varies with distance from the coast, with altitude and bedrock. It decreases in all directions from a maximum of about 100–200 m above the birch treeline. In a given 100 m altitude band, the species richness is usually 20–30 species higher on calcareous bedrock than on acidic bedrock.
2. The lower limits of alpine vascular plants are determined by a combination of competitive exclusion, drought and tolerance of high summer temperatures. Cryoturbation and solifluction are probably important factors determining the lower limits of some 'cold-soil' species in the centric group.

3. Oceanic climate, especially the mild winter temperature, is generally unfavourable for alpine plants in the boreal and temperate zones, and probably explains the very low alpine species richness on coastal mountains in the southern Scandes.
4. The altitudinal decrease of total vascular plant species richness in the southern Scandes is not steady but shows a plateau between 0–900 m; followed by a steep and more or less linear decrease from 900–1500 m and by an exponential decrease from 1500 to the upper limit at 2000 m in the Dovrefjell area.

References

Berg RY (1963) Disjunctions in the Norwegian alpine flora and theories proposed for their explanation. Blyttia 21:133–177

Birks HH (1993) Is the hypothesis of survival on glacial nunataks necessary to explain the present-day distributions of Norwegian mountain plants? Phytocoenologia 23:399–426

Birks HH (1996) Statistical approaches to interpreting diversity patterns in the Norwegian mountain flora. Ecography 19:332–340

Bøcher T (1938) Biological distribution types in the flora of Greenland. Meddelelser Grønland 106:1–339

Bøcher T (1951) Distributions of plants in the circumpolar area in relation to ecological and historical factors. J Ecol 39:376–395

Crawford RMM (1997) Oceanity and the ecological disadvantages of warm winters. Bot J Scotl 49:205–221

Crawford RMM, Palin MA (1981) Root respiration and temperature limits to the north south distribution of four perennial maritime species. Flora 171:338–354

Crawford RMM, Wolfe DW (1998) Temperature: cellular to whole plant and ecosystem responses. In: Mooney HA, Seaman J, Luo Y (eds) Stress effects on future terrestrial carbon fluxes. Academic Press, New York

Dahl E (1951) On the relation between the summer temperature and the distribution of alpine vascular plants in Fennoscandia. Oikos 3:22–52

Dahl E (1986) Zonation in arctic and alpine tundra and fellfield ecobiomes. In: Polunin N (ed) Ecosystem theory and application. Environmental Monographs and Symposia, Wiley, New York, pp 35–62

Dahl E (1998) The phytogeography of northern Europe (British Isles, Fennoscandia and adjacent areas). Cambridge University Press, Cambridge

Danielsen A (1971) Scandinavia's mountain vascular flora in the light of late quaternary history of vegetation. Blyttia 29:183–209

Gauslaa Y (1984) Heat resistance and energy budget in different Scandinavian plants. Holarctic Ecol 7:1–78

Gauslaa Y (1985) Climatic limitations on the distribution of alpine plants. A historical review. Blyttia 43:75–86

Gjærevoll O (1956) The snow-bed vegetation of the Scandinavian alpine snow-beds. K norske Vidensk Selsk Skr 1:1–405

Gjærevoll O (1959) Overvintringsteoriens stilling i dag. K norske Vidensk Selsk Forh 32:36–71

Gjærevoll O (ed) (1990) Maps of distribution of Norwegian vascular plants. II. Alpine plants. Tapir Publishers, Trondheim

Grabherr G (1995) Alpine vegetation in a global perspective. In: Box EO, Peet RK, Masuzawa T, Yamada I, Fujiwara K, Maycock PF (eds) Vegetation science in forestry. Kluwer, Dordrecht, pp 441–451

Grabherr G, Gottfried M, Gruber A, Pauli H (1995) Patterns and current changes in alpine plant diversity. In: Chapin FS, Koerner C (eds) Arctic and alpine biodiversity. Springer, Berlin Heidelberg New York, pp 167–181

Holten JI (1986) Autecological and phytogeographical investigations along a coast-inland transect in central Norway. PhD Thesis, University of Trondheim, Trondheim

Holten JI (1998a) Vertical distribution patterns of vascular plants in the Fennoscandian mountain range. Ecologie 29:129–138

Holten JI (1998b) Vascular plant species richness in relation to altitudinal and slope gradients in mountain landscapes. In: Beniston M, Innes JL (eds) The impacts of climate variability on forests. Springer, Berlin Heidelberg New York, pp 231–239

Jonasson S (1986) Influence of frost heaving on soil chemistry and on the distribution of plant growth forms. Geogr Ann 68A:185–195

Jørgensen R (1933) Karplantenes høidegrenser i Jotunheimen. Nytt Mag Naturvidensk 72:1–130

Jørgensen R (1937) Die Hohengrenzen der Gefasspflanzen in Troms fylke. K norske Vidensk Selsk Skr 1936 8:1–106

Kilander S (1955) Upper limits of vascular plants on mountains in southwestern Jämtland and adjacent parts of Härjedalen (Sweden) and Norway. Acta Phytogeogr Suec 35:1–108

Moe B (1995) Studies of the alpine flora along an east-west gradient in central Norway. Nord J Bot 15:77–89

Nordhagen R (1943) Sikkilsdalen og Norges fjellbeiter – en plantesosiologisk monografi. Bergens Mus Skrifter 22:1–607

Odland A, Birks HH (1999) The altitudinal gradient of vascular plant species richness in Aurland, western Norway. Ecography 22:548–566

Ødegaard RS, Hoelzle M, Vedel Johansen K, Sollid JL (1996) Permafrost mapping and prospecting in southern Norway. Norsk Geogr Tidsskr 50:41–53

Påhlson L (1998) Vegetationstyper i Norden. TemaNord 1998 510:1–706

Sollid JL, Sørbel L (1998) Palsa bogs as a climate indicator – examples from Dovrefjell, southern Norway. Ambio 27:287–291

8 Vascular Plant and Bryophyte Diversity along Elevation Gradients in the Alps

J.-P. Theurillat, A. Schlüssel, P. Geissler, A. Guisan, C. Velluti and L. Wiget

8.1 Introduction

A decrease in plant species richness has been observed with increasing altitude for high mountains in Europe (Ozenda 1985, 1997; Grabherr et al. 1995; Körner 2000), and different hypotheses put forward to explain them include space limitation, ecological and historical factors. This paper presents the synthesis of studies on the elevation patterns in the distribution of vascular plants and bryophytes from three transects spanning from the treeline ecotone to the subnival zone in the Valais, Western Alps, Switzerland. Our prediction was that the generality of a temperature-related decline in species richness does not hold for poikilohydric plants such as mosses and liverworts. Furthermore, we expected species with a narrow elevation range (specialists) to become more abundant at high elevations.

8.1.1 Study Areas

Three transects were chosen at two sites in the Valaisian Alps from the treeline ecotone to the upper limit of the alpine vegetation (=subnival zone). For a detailed description and preliminary results, see Geissler and Velluti (1997), Theurillat and Schlüssel (1997), Theurillat et al. (1997, 1999) and Geissler (1999). Two opposite transects (south vs. north) on granite bedrock were located in the Val d'Arpette, at the eastern limit of the Mont Blanc Massif (7°04'18"E; 46°02'54"N). Arpette South is along a steep rocky, south to southeast oriented slope running from 1720 to 2815 m a.s.l. Arpette North has a gentler slope and runs from 1800 to 2700 m a.s.l. It is dominated by colluviums and boulders that prevent at many places the development of a closed vegetation at the treeline ecotone; in the high alpine, extensive permafrost is likely to account for the open vegetation. The third transect is located at Belalp in the Aletsch region (7°58'18"E; 46°22'78"N) on a mainly gneiss bedrock. It runs along an easterly slope from 1940 to 2855 m a.s.l. and has a vegetation of mires, snowbeds and mesophilous alpine meadows. Belalp and Arpette differ from each other climatically, with Belalp being slightly more continental than Arpette.

8.1.2 Methodology

8.1.2.1 The Transects

The transects were wedge-shaped with their base in the treeline ecotone (about 1 km wide at Arpette and 2.5 km at Belalp) and the thin edge in the high alpine zone. Each of the three transects were divided into 100-m elevation sections along the contours. The transects were bounded by conspicuous geomorphological features, such as gullies or ridges. Owing to the differences in the steepness of the slopes, the width of the transects at base, the irregular shape of the mountain sides, and the bordering geomorphological features, the projected surface areas of the 100-m altitude sections varied within and between the transects. From 2000 to 2599 m, the mean projected surface of the sections was 0.11 km^2 at Arpette South, 0.16 km^2 at Arpette North and 0.46 km^2 at Belalp. Above 2600 m, the sections were between 0.01–0.14 km^2 at Belalp and ranged from <0.01 to 0.03 km^2 at Arpette South.

8.1.2.2 Analyses

Each 100-m altitude section was surveyed over several years to record all the species of vascular plants and bryophytes. Vascular plants were identified by using Aeschimann and Burdet (1989) and bryophytes by mainly using Frey et al. (1995) or other specialised literature for particular groups (e.g. *Grimmia*). The elevation patterns of vascular plants were analysed for (1) taxonomic diversity (species, genera, families), (2) life forms, (3) elevational distribution, (4) vegetation similarity, and (5) cumulative species richness. Raunkier's main life forms for vascular plants followed Aeschimann and Burdet (1989) and bryophytes were allocated to four growth forms (turfs, mats, cushions and wefts). Vascular plants were allocated into four elevation-based categories (alpine, A; treeline ecotone-alpine, S-A; colline-alpine, C-A; and colline-montane-treeline ecotone species, C-M-S) after Aeschimann and Burdet (1989). Sørensen's similarity index was calculated for the 2000–2099 m section and each 100 m section above in each transect (Whittaker 1972). Cumulative species richness was calculated by adding new species in each 100 m altitude section to the baseline numbers (2000–2099 m) to compare Arpette South, Arpette North, and Belalp for the altitude vs. regional pool (gamma-diversity) relationship.

For spatial gradients such as the ones described above, spatial autocorrelation may influence the results as well as the different size of the successive sampling plots (i.e. 100-m sections). However, after omitting the three uppermost sections, i.e. from 2600 m upwards at Arpette South and Belalp, no correlation was found between elevation and decrease of surface area whilst, at Arpette North, there was no correlation over the entire transect (Kendall's rank correlation tau, not shown).

8.2 Patterns of Distribution

8.2.1 Elevation and Species Richness

Vascular plant species richness decreased with increasing elevation consistently in each transect (Fig. 8.1). Not so that of the bryophytes, whose decrease in species richness was not consistent and differed between the transects. When considering the overall richness for all three transects, the most remarkable result is the much higher vascular plant, as well as bryophyte, species richness at Belalp than at either of the Arpette sites (e.g. 330 vs. 256 for the vascular plants, and 207 vs. 147 for the bryophytes for the Belalp vs. South Arpette). This difference cannot be interpreted by the different sizes of the 100-m sections alone (about four-fold larger in Belalp), as no correlation between species richness and section size was found along the elevation gradient. The numbers for the uppermost sections (>2600 m) were similar for all three transects (Fig. 8.1).

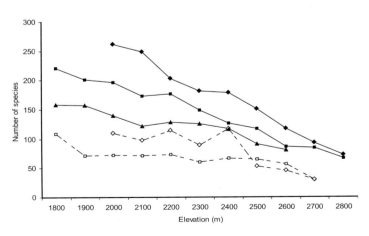

Fig. 8.1. Vascular plant (*filled symbols*) and bryophytes (*open symbols*) species richness with elevation in 100 m altitude sections in the transects at Belalp (*diamonds*), Arpette South (*squares*) and Arpette North (*triangles*)

8.2.2 Life Forms, Growth Forms and Elevation

The patterns of life forms were similar among the three transects. The frequency of hemicryptophytes was around 70 % in every 100-m altitude section in the three transects, similar to the frequency found for the whole nival zone in the Swiss Alps (Jenny-Lips 1948). Chamaephytes consistently increased with elevation. Their frequency was 22–30 % in the upper sections, close to the 24.5 % found for the nival zone in the Swiss Alps (Jenny-Lips 1948). The majority of bryophytes formed turfs. They had a constant frequency (60 %), as well as wefts (7 %), in each 100-m altitude section.

8.2.3 Distribution of Elevational Species Groups

In general, with increasing elevation, there was a decrease in the number of species that occupy a wide range of altitude zones (C-M-S: colline-montane treeline ecotone; C-A: colline up to alpine; S-A: treeline ecotone-alpine). The decrease in the number and proportion of C-M-S was linear, and was similar in each transect (Fig. 8.2a). Proportionally, the S-A-species – although also occurring in the (upper) treeline ecotone – showed a maximum in the upper

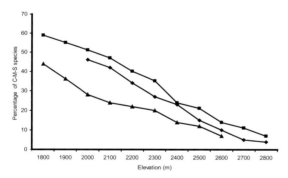

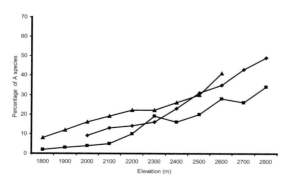

Fig. 8.2. The percentage contribution of a wide ranging colline-montane treeline ecotone (C-M-S) and b alpine vascular plant species to total species richness with elevation in 100-m altitude sections in the three transects at Belalp (2000–2855 m a.s.l.), Arpette South (1800–2815 m a.s.l.) and Arpette North (1800–2699 m a.s.l.). *Symbols* are identical to those used in Fig. 8.1

alpine zone. Alpine species (A-species = specialists) became more prominent with increasing elevation (Fig. 8.2b). The number of species in each of the four elevation groups appeared specific to each transect.

8.2.4 Similarity and Elevation

There was a consistent, linear decrease, in a similar manner in all three transects, in the similarity coefficient between the 2000–2099 m section and the subsequent sections, and no distinct change was observed when moving from the treeline ecotone to the alpine zone in any of the three transects (Fig. 8.3). For bryophytes, on the contrary, the pattern of the similarity coefficient between the 2000–2099 m section and the higher sections showed a plateau (Fig. 8.3), particularly at Arpette, similar to that for species numbers (Fig. 8.1), implying a less consistent turnover along the altitude gradient.

8.2.5 Cumulative Species Richness and Elevation

The cumulative species richness pattern was similar for both vascular plants and bryophytes in the three transects with an asymptotic increase with elevation tending to level off at the low nival zone (Fig. 8.4). This pattern corresponds to a saturation sampling (sensu Cornell 1993; Lawton 1999) with pool exhaustion. The process of pool exhaustion for vascular plants is very similar in all three transects, suggesting that it is independent of the regional pool and its peculiarities. On average, the difference between the curves corre-

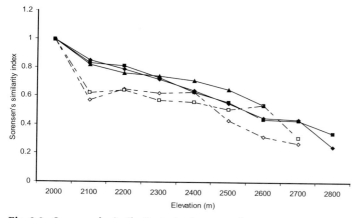

Fig. 8.3. Sørensen's similarity index between the 2000–2099 m section and each individual 100-m altitude section for vascular plants (*filled symbols*) and bryophytes (*open symbols*) in the transects at Belalp (*diamonds*), Arpette South (*squares*) and Arpette North (*triangles*)

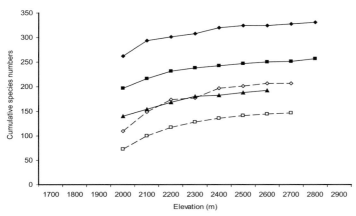

Fig. 8.4. Cumulative species richness with elevation in the 100-m altitude sections for vascular plant (*filled symbols*) and bryophyte species (*open symbols*) in the transects at Belalp (*diamonds*), Arpette South (*squares*) and Arpette North (*triangles*)

sponds to the difference in the number of species between the two transects, especially for vascular plants. Hence, the difference in the species number between Arpette South and Arpette North between 2000 and 2099 m a.s.l. is 60, and the mean difference between the cumulative numbers of the two curves at each successive 100 m altitude from 2100 m a.s.l. upward is 60. Between Arpette South and Belalp the difference in species numbers is 74 and the mean difference between the two curves for $n-1$ is 74.5.

The case for bryophytes is similar to that for vascular plants corresponding to a saturation sampling pattern (Fig. 8.4). However, pool exhaustion for bryophytes is different for the two transects. Nonetheless, a better convergence between the two curves occurs from the low alpine (2400 m) upward where the exhaustion process is similar in the two transects and, therefore, is independent of the species pool, as for vascular plants.

8.3 Discussion

8.3.1 Theoretical Background

Many hypotheses have been put forward to explain species richness and its variation (e.g. Whittaker 1972; Palmer 1995; Huston 1999; Schläpfer and Schmid 1999; Waide et al. 1999; O'Brien et al. 2000). Space availability, species interactions, productivity, and habitat heterogeneity have often been considered as driving forces. Competitive exclusion in particular (Palmer 1995) could be invoked to explain the decrease in species richness with increasing elevation. However, only two of Palmer's (1995) conditions for competitive exclusion fit, namely the limitation of growth by a single factor (temperature in alpine environments), and the survival of rarer species and their opportunity to compete. The relative increase of alpine species, which may outcompete species from lower down, may be taken as proof for the latter. Other conditions would even predict the opposite such as shorter instead of longer time for exclusion, the temporally changing environment (as opposed to a constant one), the high spatial variation, and the relatively high immigration rates, here driven by the strong winds. Thus Huston's productivity hypothesis (Huston 1999) might fit better as biomass production is higher at lower elevation.

8.3.2 The Temperature–Physiography Hypothesis

Obviously, colder temperatures determine the pool of the potential species which are able to withstand the increasingly severe conditions in the alpine zone, independent of the regional peculiarities. This is in agreement with the general temperature hypothesis posed by Rahbek 1995 (see also Whittaker 1999; Gaston 2000), as similar trends can also be found along the latitudinal gradient (O'Brien et al. 2000). But how to explain in a unified way the elevational distribution patterns of vascular plants and the deviating one of bryophytes? We suggest that microrelief (i.e. 1–100 m; Leser 1977) affects the potential species pool as it determines habitat-type richness. In general, habitat richness decreases as the terrain becomes steeper with increasing elevation (Theurillat et al. 1998; Theurillat and Guisan 2001). A steeper relief also accelerates all gravitational processes, which enhance habitat convergence. In parallel the reduction of fluvial processes with increasing winter length, and an increase of periglacial processes at higher elevation, homogenise the landscape, and, therefore, diminish habitat-type richness. In addition, the effect of vegetation itself on habitat heterogeneity is lower at high elevations. This combination of temperature-related effects on the species pool as well as on geomorphological processes, added to a changing physiography, i.e. a steeper landscape at high elevations, led us to that what we call the temperature–physiography hypothesis.

The distribution pattern of bryophytes along the elevation gradient deviates from this hypothesis because at lower elevations their distribution depend on the availability of special microhabitats (e.g. trunks, humid, shaded rocks, rivulets). However, in the highest elevations, microhabitats follow more and more the processes of the temperature–physiography hypothesis. Therefore, vascular plant richness may be much more affected by temperature itself than bryophyte richness.

8.4 Conclusions

1. This paper presents the synthesis of studies on the elevation patterns in the distribution of vascular plants and bryophytes from three transects extending from the treeline ecotone to the subnival zone in the Valais, Western Alps, Switzerland.
2. Vascular plant species richness decreased consistently with increasing elevation in each transect as opposed to bryophytes where the decrease occurred in a less consistent manner and differed between the transects. There was an indication of constant species turnover along the altitude gradient for vascular plants, whilst bryophytes did not follow such a pattern.
3. The distribution of vascular plants along elevation gradients is primarily governed by temperature-related processes alongside a gradual change in the physical environment. Bryophytes are less coupled to such a direct temperature gradient and their distribution is much modified by the presence of microhabitats.

Acknowledgements. The present study has been carried out mainly within the 'Ecocline' project and the integrated project CLEAR of the Priority Programme Environment supported by the Swiss National Science Foundation (project FNRS 5001-35343 to P. Geissler; projects 5001-35341 and 5001-44604 to J.-P. Theurillat). We thank G. Grabherr and L. Nagy who improved the manuscript, two anonymous referees for helpful comments, and Julie Warrillow for revising the English.

References

Aeschimann D, Burdet HM (1989) Flore de la Suisse. Le Griffon, Neuchâtel
Cornell HV (1993) Unsaturated patterns in species assemblages: the role of regional processes in setting local species richness. In: Ricklefs RE, Schluter D (eds) Species diversity in ecological communities historical and geographical perspectives. University of Chicago Press, Chicago, pp 243–252
Frey W, Frahm J-P, Fischer E, Lobin W (1995) Kleine Kryptogamenflora. IV. Die Moos- und Farnpflanzen Europas. Fischer, Stuttgart
Gaston KJ (2000) Global patterns in biodiversity. Nature 405:220–227
Geissler P (1999) Altitudinal distribution of bryophytes. ESF Alpnet News 1:20–22
Geissler P, Velluti C (1997) L'écocline subalpin-alpin: approche par les bryophytes. Bull Murith Soc Valais Sci Nat 114:171–177
Grabherr G, Gottfried M, Gruber A, Pauli H (1995) Patterns and current changes in alpine plant diversity. In: Chapin A, Körner C (eds) Arctic and alpine biodiversity: patterns, causes and ecosystem consequences. Ecological studies, vol 113. Springer, Berlin Heidelberg New York, pp 167–181
Huston MA (1999) Local processes and regional patterns: appropriate scales for understanding variation in the diversity of plants and animals. Oikos 86:393–401

Jenny-Lips H (1948) Vegetation der Schweizer Alpen. Büchergilde Gutenberg, Zürich
Körner C (2000) Why are there global gradients in species richness? Mountains might hold the answer. TREE 15:513
Lawton JH (1999) Are there general laws in ecology? Oikos 84:177–192
Leser H (1977) Feld- und Labormethoden der Geomorphologie. Gruyter, Berlin
O'Brien EM, Field R, Whittaker RJ (2000) Climatic gradients in woody plant (tree and shrub) diversity: water-energy dynamics, residual variation, and topography. Oikos 89:588–600
Ozenda P (1985) La végétation de la chaîne alpine dans l'espace montagnard européen. Masson, Paris
Ozenda P (1997) Aspects biogéographiques de la végétation des hautes chaines. Biogeographica 73:145–179
Palmer MW (1995) Variation in species richness: towards a unification of hypotheses. Folia Geobot Phytotaxon 29:511–530
Rahbek C (1995) The elevational gradient of species richness: a uniform pattern? Ecography 18:200–205
Schläpfer F, Schmid B (1999) Ecosystem effects of biodiversity: a classification of hypotheses and exploration of empirical results. Ecol Appl 9:893–912
Theurillat J-P, Guisan A (2001) Impact of climate change vegetation in the European Alps: a review. Climatic Change 50:77–109
Theurillat J-P, Schlüssel A (1997) L'écocline subalpin-alpin: diversité et phénologie des plantes vasculaires. Bull Murith Soc Valais Sci Nat 114:163–169
Theurillat J-P, Felber F, Geissler P, Guisan A, Gobat J-M (1997) Le projet "Ecocline" et le programme prioritaire "Environnement". Bull Murith Soc Valais Sci Nat 114:151–162
Theurillat J-P, Felber F, Geissler P, Gobat J-M, Fierz M, Fischlin A, Küpfer P, Schlüssel A, Velluti C, Zhao G-F (1998) Sensitivity of plant and soil ecosystems of the Alps to climate change. In: Cebon P, Dahinden U, Davies HC, Imboden D, Jäger CC (eds) Views from the Alps Regional perspectives on climate change. MIT Press, London, pp 225–308
Theurillat J-P, Schlüssel A, Wiget L, Guisan A (1999) Elevational floristic gradient of vascular plants at the subalpine-alpine ecocline in the Valais (Switzerland). ESF Alpnet News 1:19–20
Waide RB, Willig MR, Steiner CF, Mittelbach G, Gough L, Dodson SI, Juday GP, Parmenter R (1999) The relationship between productivity and species richness. Annu Rev Ecol Syst 30:257–300
Whittaker RH (1972) Evolution and measurement of species diversity. Taxon 21:213–251
Whittaker RJ (1999) Scaling, energetics and diversity. Nature 401:865–866

9 Assessing the Long-Term Dynamics of Endemic Plants at Summit Habitats

H. PAULI, M. GOTTFRIED, T. DIRNBÖCK, S. DULLINGER and G. GRABHERR

9.1 Introduction

Evidence from high summits in the Alps, that mountain plants have migrated upwards (Gottfried et al. 1994; Grabherr et al. 1994, 2001; Pauli et al. 1996), prompted the initiation of a Global Observation Network (GLORIA, see www.gloria.ac.at) to study climate change induced effects on alpine biodiversity (Grabherr et al. 2000; Pauli et al. 2001). Mountain tops or summits form comparable environmental units, where habitats of every exposure (N, E, S, and W) are present within a small area and are little affected by shading from neighbouring land features. Mountain summits often have a high habitat diversity. They are of particular interest for detecting any upward migration of species. The summits are prominent landmarks that can be readily relocated for re-investigations and the highest summit points can be characterised by an average climate at any given altitude.

This paper draws on the results of a GLORIA pilot study made in the Hochschwab region, north-eastern Limestone Alps, Austria, in 1998, and in the Sierra Nevada, Spain, in 1999. Our focus was the distribution of endemic species in relation to elevation. It is well documented that vascular plant species richness generally decreases with altitude; however, 74% of the endemic species of the north-eastern Limestone Alps are orophytes (Pawłowski 1970) and 80% of them are in the Sierra Nevada (Molero Mesa et al. 1996). Using a standardised sampling design, we examined at mountain summits of different altitudes (Pauli et al. 2001) whether the relative contribution of endemic plant species increased with altitude in the two climatically contrasting regions. The main concern of this study was the potential impacts of forecast climate change on rare and locally distributed high mountain species. We discuss implications for the survival of high mountain endemics of an upward migration of common widely distributed species

9.2 Study Areas

The Sierra Nevada of southern Spain extends over about 80 km in an E-W direction and reaches its highest point at 3479 m a.s.l. (Fig. 9.1). Its climate is Mediterranean with winter rainfall (cf. zonobiome IV, Walter 1985). There is a pronounced summer drought at all altitudes, and, at altitudes above 2500 m, precipitation falls almost exclusively as snow (Molero Mesa 1998). The four summits studied were in the western, higher part of the range on siliceous bedrock. The summits ranged from 2778 m in the upper oro-Mediterranean zone (sensu Molero Mesa and Pérez-Raya 1987) to 3327 m in the upper cryo-oro-Mediterranean zone (Fig. 9.1).

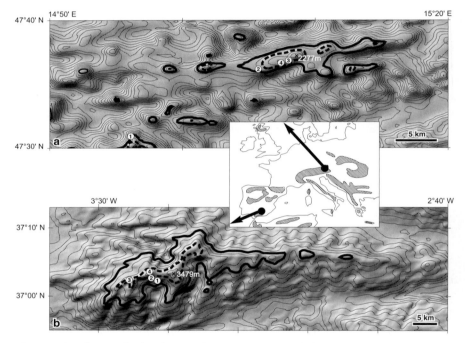

Fig. 9.1a, b. The Hochschwab, NE-Alps, Austria (a) and the Sierra Nevada, Spain (b), each with four study summits. **a** *1* Rössel (1855 m); *2* Zinken NW-summit (1910 m); *3* G'hacktkogel (2214 m); *4* Zagelkogel NW-summit (2255 m). Contour lines at 100-m intervals; *bold continuous line* 1800-m isoline indicating the estimated average altitude of the montane/alpine transition zone; *broken line* 2000-m isoline. **b** *1* Pulpitito (2778 m); *2* Cúpula (2968 m); *3* Pico del Tosal Cartujo (3150 m); *4* Cerro de los Machos (3327 m). The first two names were chosen arbitrarily as the summits are unnamed in the national 1:50,000 map. Contour lines at 200-m intervals; *bold continuous line* 2300-m isoline indicating the supra-/oro-Mediterranen transition zone; *broken line* 2900-m isoline showing the estimated upper transition area between oro- and cryoro-Mediterranean belts

Assessing the Long-Term Dynamics of Endemic Plants at Summit Habitats   197

The Hochschwab region is located in the north-eastern Limestone Alps. The range runs 35 km in an E-W direction and reaches its highest point at 2277 m a.s.l. (Fig. 9.1). It has a temperate climate with no dry season (zonobiome VI, Walter 1985). The four summits studied ranged from 1855 m in the upper treeline ecotone to 2255 m in the alpine zone (Fig. 9.1).

9.3 Methods

Within each region, summits of the same bedrock type were chosen to minimise variation arising from substratum. Extreme landforms such as plateaux and steep rocky or scree summits were excluded to ensure the applicability of a standardised sampling regime. The summit habitats were largely natural where human land use was likely to have had minimal interference.

On each summit area, four 3 × 3-m permanent plots, subdivided into 1 × 1-m quadrats, were established. The plots were positioned at each point of the compass (N, S, E and W), with their lower boundaries at the 5-m isoline below the summit (Fig. 9.2). In some cases, the positions deviated from the exact compass direction to avoid steep inhospitable unvegetated areas. The four corner 1 × 1-m quadrats in each 3 × 3-m plot were recorded, giving a total of 16 1-m² quadrats on each summit. In each quadrat, the top cover (see Greig-

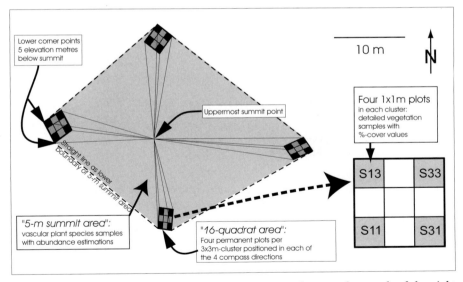

Fig. 9.2. Sampling design for the multi-summit approach as used on each of the eight summits (top view, example from Zagelkogel). Four quadrats of 1 × 1-m in each of the four 3 × 3-m plots together form the 16-quadrat area

Smith 1983) of each surface type (vascular plants, solid rock, scree, bare ground, and cryptogams occurring on soil separately from vascular plants) and percentage cover of vascular plant species were visually estimated. The data from the sixteen 1 × 1-m quadrats were averaged for each summit (16-quadrat area) and used to compare the summits for plant species cover against altitude and for percentage of endemism.

A second set of samples consisting of an estimated cover of species in the whole area above the 5-m isoline at each summit (5-m summit area, see Fig. 9.2) was used to compare altitude gradients of total and endemic species richness.

Flora Europaea (Tutin et al. 1964, 1968–1980), *Flora Iberica* (Castroviejo et al. 1986–1998), Molero Mesa and Pérez-Raya (1987), Molero Mesa et al. (1996), Meusel et al. (1965–1992) and Adler et al. (1994) were used to define the distribution ranges of species. Nomenclature of species follows *Flora Europaea* (Tutin et al. 1964, 1968–1980).

9.4 Results

9.4.1 Vegetation Cover and Vascular Plant Species Cover

Vegetation top cover showed a gradual decrease with altitude in both regions (Table 9.1). Whereas in the Sierra Nevada almost no cryptogams were recorded from soil, in the NE-Alps soil cryptogams at the treeline had a cover >20 % (however, their cover decreased to <1 % at the highest peak). There was a lower vegetation cover (including the complete absence of bryophytes) in the Sierra Nevada than in the NE-Alps. This is likely to have been the result of a regular pronounced summer drought at the Sierra Nevada summits and their higher altitudes.

Percent cover values of all species indicated a pronounced decrease of vegetation with altitude (Table 9.1); considering endemic species only, no such gradient could be observed.

In the Sierra Nevada, the percentage of the total species cover composed of Sierra Nevada endemics increased from 33 % (36 % when Baetic endemics were included) at the lowest summit to 88 % (95 %) at the second highest peak at 3150 m. At the highest peak, they accounted for 6 %. The very low cover values for endemics on the highest peak resulted from the permanent quadrats containing only 5 of all 11 species (10 of which were Sierra Nevada endemics) present in the 5-m summit area (cf. Fig. 9.3). The only non-endemic species, *Hormathophylla spinosa*, accounted for 94 % of the total cover in the quadrats.

In the NE-Alps, endemics contributed far less to the cover; the highest contribution was on the second lowest summit in the upper treeline ecotone owing to the high cover of *Festuca versicolor* ssp. *brachystachys*.

Table 9.1. Top cover of vegetation and cover of vascular plant species on the summits of the Sierra Nevada, Spain, and the NE-Alps, Austria. Values refer to the 16-quadrat area at each summit. For the cryptogams, values include the top cover of lichens and bryophytes on soil not covered by vascular plants as well as the top cover of those below the vascular plants (the latter was estimated separately in the field). Endemics (s.l.) include those of the Baetic Cordillera and of the Eastern Alps; endemics sensu stricto are those of the Sierra Nevada and of the NE-Alps

Summit	Altitude (m)	Top cover		Cover of vascular plant species				
		Vascular plants	Cryptograms on soil	All species	Endemics s.l.	% of all spp.	Endemics s. str.	% of all spp.
Sierra Nevada								
Cerro de los Machos	3327	2.4	0	2.4	0.2	6	0.2	6
Pico del Tosal Cartujo	3150	11.1	0	11.7	11.2	95	10.3	88
Cúpula	2968	14.3	0	15.3	8.0	52	5.1	33
Pulpitito	2778	18.0	0.004	18.4	6.6	36	6.1	33
N-E Alps								
Zagelkogel NW	2255	46.6	0.2	50.5	1.2	2	0.2	0.4
G'hacktkogel	2214	57.8	2.8	60.5	3.0	5	1.1	2
Zinken NW	1910	79.7	5.5	92.7	14.8	16	14.5	16
Rössel	1855	89.8	23.1	113.2[a]	3.8	3	0.2	0.1

[a] Value exceeds 100 % because plants partly overlap.

9.4.2 Species Richness and Endemic Species

In both regions, vascular plant species richness decreased with altitude (Fig. 9.3) except for the second lowest peak in the Sierra Nevada, which was similar to the lowest one. This was probably due to the ecotonal position of the second peak (2968 m), where a number of cryoro-Mediterranean species occurred together with oro-Mediterranean species.

The percentage of endemic species increased with altitude and their number increased up to the second highest summit in both regions (Fig. 9.3). However, the two mountain ranges were rather different concerning the importance of endemic species for high mountain biodiversity. In the Sierra Nevada, the percentage of endemics showed a steady rise from the lowest to the highest summit (28, 45, 64, and 91 %; Fig. 9.3). Almost all Sierra Nevada endemics found on the summits were restricted to the cryoro- and oro-Mediterranean zones, where they made up the majority of species (60 % of all species restricted to these belts, Fig. 9.4). None of the five species that were documented to be largely limited to the cryoro-Mediterranean zone occurs outside the Sierra Nevada. In contrast, 19 of the 21 species that have an alti-

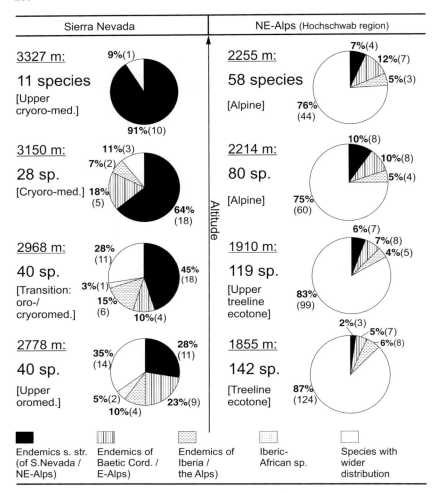

Fig. 9.3. Chorological spectra of vascular plants of the four summits (5-m summit area) in the Sierra Nevada and in the NE Alps (*values in parentheses* are numbers of species). For the NE Alps, endemics as well as subendemics are indicated in the three categories NE Alps, E Alps, and Alps. The total species number of all Sierra Nevada summits was 67 with 43% S. Nevada, 15% Baetic, 12% Iberic, 3% Iberic-African, 27% wider distribution; the corresponding values for the 206 species of the NE Alps were 5% NE Alps, 7% E Alps, 5% Alps, 83% wider distribution

tude amplitude from the meso-/supra- up to the oro-/cryoro-Mediterranean zones are species with a wider distribution.

In the NE-Alps, the percentage and number of endemics increased with altitude up to the second highest peak, after which the percent value decreased slightly (Fig. 9.3). The endemics made up much lower percentages of all species than those in the Sierra Nevada.

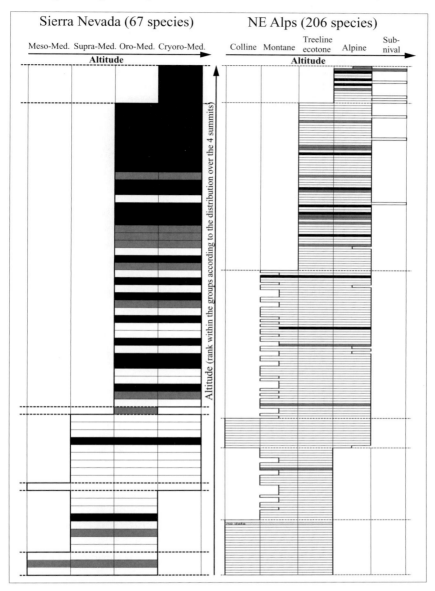

Fig. 9.4. Rank of vascular plant species on the four summits of Sierra Nevada and the NE-Alps according to their altitudinal distribution. Each band represents one species. First rank according to the vertical species distribution along altitudinal vegetation belts (for Sierra Nevada: after Molero Mesa et al. 1996; Molero Mesa 1998; for NE-Alps: after Adler et al. 1994). Second rank within the groups (divided by *dashed lines*) according to the species distribution over the four summits. *Black bands* Endemics of Sierra Nevada/ of NE Alps, respectively; *grey* endemics of the Baetic Cordillera/of the E Alps; *white* species with wider distribution. In the endemic categories NE Alps and E Alps, species with outposts in close neighbouring mountain areas were included. The endemic taxa in rank order: Sierra Nevada endemics: *Poa minor* ssp. *nevadensis, Erigeron* cf. *frigidus, Festuca*

Fig. 9.4. (*Continued*) *clementei, Trisetum glaciale, Iberis embergeri, Saxifraga nevadensis, Jasione amethystina, Hormathophylla purpurea, Arenaria tetraquetra* ssp. *amabilis, Viola crassiuscula, Chaenorrhinum glareosum, Reseda complicata, Lepidium stylatum, Plantago nivalis, Dactylis juncinella, Carduus carlinoides* ssp. *hispanicus, Leucanthemopsis pectinata, Herniaria boissieri, Artemisia granatensis, Linaria aeruginea* ssp. *nevadensis, Trisetum antonii-josephii, Thymus serpylloides, Draba hispanica* ssp. *laderoi, Pimpinella procumbens, Festuca pseudoeskia, Campanula willkommii, Erysimum nevadense, Sempervivum minutum, Genista versicolor.* Endemics of the NE-Alps: *Draba stellata, Draba sauteri, Soldanella austriaca, Festuca versicolor* ssp. *brachystachys, Dianthus alpinus, Campanula pulla, Thlaspi alpinum* s.str., *Leucanthemum atratum* s.str., *Primula clusiana, Alchemilla anisiaca*

The large majority of endemic species are restricted to the upper treeline ecotone and the alpine zone (Fig. 9.4). Of the 15 species being almost restricted to the alpine zone, six (40%) are considered as endemic or subendemic to the NE-Alps and the E-Alps. Of the 68 species occurring in the upper treeline ecotone and the alpine zone, 14 (21%) are endemic and, of the 123 species with wider altitudinal amplitudes (down to the montane or colline zone), only five species (4%) are endemic or subendemic.

9.5 Discussion

The average contribution of endemics to the flora on the four Sierra Nevada summits (43%; Fig. 9.3) was higher than the 31% reported by Molero Mesa et al. (1996) for the cryoro-Mediterranean belt and nearly three times higher than the percentage of all orophilous species represented by endemics (15%). The high values for the summits suggest that the relative frequency of endemic species is higher in summit habitats than in other habitat types on the same elevation level. The comparison of the chorological spectra of the four summits of the Hochschwab region and their average spectrum (Fig. 9.3) with those of all the species recorded from 1800 m upwards within the entire Hochschwab range (Dirnböck et al. 1999) indicated no differences. This suggests that in the north-eastern Limestone Alps, endemism at summits does not exceed the average at the same altitude in a non-summit position.

9.5.1 Orographic Isolation and the Cryophilic Nature of Endemics

The outstanding richness of endemic species in high elevation habitats in the Sierra Nevada has long been recognised (e.g. Quézel 1953; Rivas Goday and Mayor López 1966). For the Alps, Pawłowski (1970) reported that the majority of endemic species were orophytes. Körner (1999) argued that endemic species are usually less important in the summit floras and reach their greatest abundance in the lower alpine zone. This holds true when the whole of the Alps range is considered, as most of the endemics occur in the Pre-Alps, the summits of which reach moderate heights compared with the central Alps. Only few endemics, usually widely distributed over the entire range, are found in the subnival and nival zone of the central Alps. However, we found that in both the Sierra Nevada and the NE Alps, the proportion of endemics in the flora increased from the treeline ecotone to the highest (or at least to the second highest) summit. The majority of endemic species found on the summits is restricted to the upper elevation levels. In contrast, most of the widerspread species, occurring at the same summits, have wider altitudinal ranges or extend to lower elevation levels (Fig. 9.4).

Of the non-geographical-distance related factors, low temperature is considered to be the main factor to contribute to orographic isolation, at least from the treeline upwards. Other factors specific to mountain environments that cause habitat separation, such as frequent disturbance in scree habitats, or reduced space and accessibility in rock fissures, are thought to be of secondary importance. Dullinger et al. (2000) showed that many alpine endemics of the NE Alps are not restricted to distinct plant communities, but are distributed over a variety of vegetation types such as snowbeds, grasslands, and rock and scree communities. In the Sierra Nevada, endemics were found to be of fundamental importance for the zonal vegetation of the upper elevation levels (compare Molero Mesa 1998), where they accounted for up to about 90 % of vegetation cover (Table 9.1).

The difference in endemic percentages between the two regions was consistent with the general north-south gradient of mountain endemism across Europe and over to the Atlas Mountains reported by Favarger (1972). This gradient appears to have been caused mainly by the more pronounced current and historic geographic isolation of the southern mountains. Small islands of cold environments in the south remained more isolated during the Pleistocene than the mountains further north, which maintained connections and offered numerous migration pathways. In the Alps, patterns of glaciation during the Pleistocene provided barriers that shaped the distribution centres of endemism on the outer and lower mountain ranges which remained largely free of ice (Merxmüller 1952).

Notwithstanding the importance of historic determinants, we consider the temperature component of orographic isolation as being the most important factor favouring endemism in high mountain areas today. Most of the endemic species recorded from the two regions are cryophilic plants restricted to low-temperature environments. These species are restricted to cold environments as a result of either physiological requirements or weak competitive abilities.

9.5.2 Climate Warming and the Vulnerability of Cryophilic Endemic Species

A warming of the climate may pose threats to the survival of cryophilic species occurring at the highest altitudes where no upward shift in their distribution is feasible. Although there is no evidence of recent extinction of European mountain plants related to climatic change, model predictions lead to the assumption that serious biodiversity losses may occur in the future. Projections based on Global Circulation Models indicate a diurnal temperature increase by 2100 of 2.5–4.5 °C for northerly latitudes of Europe and 1.5–4.5 °C for southern Europe. Warming has been particularly pronounced during the end of the twentieth century, being most apparent in a belt from Spain through central Europe into Russia (Beniston and Tol 1998). At some high elevation locations in the Alps, increases of >1 °C were recorded during the 1980s (Auer and Böhm 1994). In the Sierra Nevada, the 1990s were particularly dry (Molero Mesa 1998) and the first half of 1999 was the driest in a century, causing a very early melting of snow (J. Molero Mesa, pers. comm.).

Environments that are currently temperature limited may become available for colonisation by invading species in the future. Also, plants distributed from low to high elevations may expand in their upper distribution areas. Both processes may rapidly increase competition pressure experienced by slow-growing high mountain plants (Grabherr et al. 1995). From the Alps, there is evidence that an upward migration is underway (Grabherr et al. 1994). No such data are available for the Sierra Nevada. However, in 1999, some widespread species were observed at much higher altitudes than their documented maximum: *Arabidopsis thaliana* 3140 vs. 2400 m, *Hornungia petraea* 2770 vs. 2200 m and *Sedum amplexicaule* 2965 vs. 2065 m (M. Gottfried and H. Pauli, pers. observ.; Castroviejo et al. 1986–1998).

High mountain plants experience high temperature maxima even in current climates (Moser et al. 1978; Körner 1999). Thus, an enhanced heat stress alone might be of minor importance. However, an earlier snowmelt, caused by higher temperatures, may prolong the drought period to an extent that would reduce the vitality of plants.

In the Sierra Nevada, climate warming is likely to be coupled with more rapid snowmelt. Thus, high mountain plants, which derive most of their water from snowmelt, may suffer from a too early onset of the dry summer season. Arguably, drought would also affect potential invaders; however, many species from lower altitudes are adapted to arid conditions.

In addition, migration corridors for invaders are generally less interrupted in the Sierra Nevada than are in most parts of the Alps, because local geomorphology is characterised by gentle and more or less uninterrupted slopes all the way up to the ridges. The open vegetation, which occurs even at low altitudes in the Sierra Nevada, may facilitate the upward migration of invaders. In contrast, in the Alps, late successional closed montane forest and alpine grassland vegetation have been shown to change very slowly (Körner 1999) and therefore may act as a barrier to invasion.

In the Alps, endemic species currently restricted to the tops of mountains are considered to be under threat from extinction (cf. Grabherr et al. 1995; Theurillat 1995). For example, *Draba sauteri*, being disjunctly distributed over some small alpine areas in the NE Alps, only rarely occurs below 2000 m (see also Dullinger et al. 2000). In addition, endemics that preferentially occur in snowbeds, such as *Soldanella austriaca* or *Campanula pulla*, may face an increased competition pressure when snow melts earlier.

This outline of potential climate-induced threats for cryophilic mountain endemics is still rather general and of a hypothetical character. On the other hand, the observed altitudinal distribution patterns of endemic species clearly demands intensive research focusing on high mountain biodiversity. This will not only be a matter of conservation biology at a regional scale. The unique distribution of high mountains over all major biomes provides a huge potential for tracing climate-change-induced impacts on biodiversity along all fundamental climatic gradients – altitude as well as latitude and longitude (Körner 1999; Grabherr et al. 2000).

9.6 Conclusions

1. The proportion of endemic vascular plants in the flora increases along an elevation gradient within the alpine life zone of the NE-Alps and the Sierra Nevada. This was documented by data from standardised sampling of mountain summits at different elevations. The contribution of endemics to the summit floras was much higher in the Sierra Nevada, with 28 % on the lowest and 91 % on the highest summit, compared with the NE-Alps, with 2 % on the lowest and a maximum of 10 % on the second highest summit.
2. The endemic species have narrower altitudinal distribution ranges than most of the widespread species recorded at the summits. The temperature component of orographic isolation is considered to be a crucial factor favouring mountain endemism today. Most of the recorded endemic species are cryophilic plants restricted to low-temperature environments.
3. Predicted warming may seriously threaten the survival of cryophilic plants. An enhanced heat stress alone might be of minor importance, compared to a prolonged growing season caused by an earlier snowmelt. In the Sierra Nevada, where precipitation during summer is almost absent, this may be coupled with a more pronounced drought period, reducing the vitality of plants.
4. In addition to the direct impacts of increased temperature on the cryophilic species, their habitats are likely to be opened up for competitive

invaders from lower altitudes. This would further threaten the survival of the cryophilic high mountain species.
5. Migration routes for invading species appear to be generally less interrupted in the Sierra Nevada than in the Alps, owing to a more gentle geomorphology and to the absence of closed montane forest and alpine grassland vegetation, which may act as barriers against invasion from the lowlands.

Acknowledgements. The current study was financed by the Austrian Academy of Sciences (IGBP 17/98-99 GCTE) and the Austrian Federal Ministry of Science and Transport (GZ 30.733/1-III/A/4/99). We are grateful to Juan Montes and his colleagues (Junta de Andalucía, Granada) for logistic support in Sierra Nevada National Park, and to Dr. Joaquín Molero Mesa and coworkers (University of Granada) for identifying critical plant taxa and for their kind help in organising and carrying out fieldwork in the Sierra Nevada.

References

Adler W, Oswald K, Fischer R (1994) Exkursionsflora von Österreich. Verlag Eugen Ulmer, Stuttgart

Auer I, Böhm R (1994) Combined temperature-precipitation variations in Austria during the instrumental period. Theor Appl Climat 49:161-174

Beniston M, Tol RSJ (1998) Europe. In: Watson RT, Zinyowera MC, Moss RH, Dokken DJ (eds) The regional impacts of climate change - an assessment of vulnerability. Special Report of IPCC Working Group II. Cambridge University Press, Cambridge, pp 149-185

Castroviejo S, Laínz M, López Gonzáles G, et al (eds) (1986-1998) Flora Iberica, vols I-VI and VIII. Real Jardín Botánico, CSIC, Madrid

Dirnböck T, Dullinger S, Gottfried M, Grabherr G (1999) Die Vegetation des Hochschwab (Steiermark) - alpine und subalpine Stufe. Mitt Naturwiss Ver Steiermark 129:111-251

Dullinger S, Dirnböck T, Grabherr G (2000) Reconsidering endemism in the northeastern Limestone Alps. Act Bot Croat 59:55-82

Favarger C (1972) Endemism in the montane floras of Europe. In: Valentine DH (ed) Taxonomy and evolution. Academic Press, London, pp 191-204

Gottfried M, Pauli H, Grabherr G (1994) Die Alpen im 'Treibhaus': Nachweise für das erwärmungsbedingte Höhersteigen der alpinen und nivalen Vegetation. Jahrb Ver Schutz Bergwelt München 59:13-27

Grabherr G, Gottfried M, Pauli H (1994) Climate effects on mountain plants. Nature 369:448

Grabherr G, Gottfried M, Gruber A, Pauli H (1995) Patterns and current changes in alpine plant diversity. In: Chapin FS, Körner C (eds) Arctic and alpine biodiversity: patterns, causes and ecosystem consequences, Springer, Berlin Heidelberg New York, pp 167-181

Grabherr G, Gottfried M, Pauli H (2000) GLORIA: a global observation research initiative in alpine environments. Mountain Res Dev 20:190-191

Grabherr G, Gottfried M, Pauli H (2001) Long-term monitoring of mountain peaks in the Alps. In: Burga C, Kratochwil A (eds) Biomonitoring. Tasks for vegetation science. Kluwer, Dordrecht, pp 153–177

Greig-Smith P (1983) Quantitative plant ecology, 3rd edn. Blackwell, Oxford

Körner C (1999) Alpine plant life – functional plant ecology of high mountain ecosystems. Springer, Berlin Heidelberg New York

Merxmüller H (1952) Untersuchungen zur Sippengliederung und Arealbildung in den Alpen, Teil I. Jahrb Ver Schutze Alpenpflanzen Tiere 17:96–133.

Meusel H, Jäger E, Rauschert S, Weinert E (1965–1992) Vergleichende Chorologie der zentraleuropäischen Flora, 3 Bd. Gustav Fischer Verlag, Jena

Molero Mesa J (1998) La vegetación de alta montaña de Sierra Nevada. Trabajo Original de Investigación, Universidad de Granada

Molero Mesa J, Pérez-Raya F (1987) La flora de la Sierra Nevada – Avance sobre el catálogo florístico nevadense. Universidad de Granada, Diputacion Provincial de Granada

Molero Mesa J, Pérez Raya F, González-Tejero MR (1996) Catalogo y análisis florístico de la flora orófila de Sierra Nevada. In: Chacón Montero J, Rosúa Campos JL (eds) 1a Conferencia Internacional Sierra Nevada – Conservación y Desarrollo Sostenible, vol II: Suelos; Biodiversidad de Flora y Vegetatión. Conservación y Restauración, Granada, pp 271–290

Moser W, Brzoska W, Zachhuber K, Larcher W (1978) Ergebnisse des IBP-Projekts 'Hoher Nebelkogel 3184 m'. Sitzungsber Oesterr Akad Wiss Wien Math Naturwiss Kl Abt I 186:387–419

Pauli H, Gottfried M, Grabherr G (1996) Effects of climate change on mountain ecosystems – upward shifting of alpine plants. World Resource Rev 8:382–390

Pauli H, Gottfried M, Reiter K, Grabherr G (2001) High mountain summits as sensitive indicators of climate change effects on vegetation patterns: the 'Multi Summit-Approach' of GLORIA (Global Observation Research Initiative in Alpine Environments). In: Visconti G, Beniston M, Iannorelli E, Barba D (eds) Global change and protected areas. Kluwer, Dordrecht, pp 45–51

Pawłowsky B (1970) Remarques sur l'endémisme dans la flore des Alpes et des Carpates. Vegetatio 21:181–243

Quézel P (1953) Contribution a l'étude phytosociologique et geobotanique de la Sierra Nevada. Mem Soc Brot 9:5–77

Rivas Goday S, Mayor López M (1966) Aspectos de la vegetación y flora oróphila del Reino de Granada. Anal Real Acad Farm 31:345–400

Theurillat J-P (1995) Climate change and the alpine flora: some perspectives. In: Guisan A, Holten JI, Spichiger R, Tessier L (eds) Potential ecological impacts of climate change in the Alps and Fennoscandian mountains. Conserv Jard Bot, Genève, pp 121–127

Tutin TG, Heywood VH, Burges NA, Valentine DW, Walters SM, Webb DA (eds) (1964) Flora Europaea, vol 1. Cambridge University Press, Cambridge

Tutin TG, Heywood VH, Burges NA, Moore DM, Valentine DW, Walters SM, Webb DA (eds) (1968–1980) Flora Europaea, vols 2–5. Cambridge University Press, Cambridge

Walter H (1985) Vegetation of the earth and ecological systems of the geo-biosphere, 3rd edn. Springer, Berlin Heidelberg New York

10 Mapping Alpine Vegetation

T. Dirnböck, S. Dullinger, M. Gottfried, C. Ginzler and G. Grabherr

10.1 Introduction

Vegetation mapping has played an important role in studying the diversity of alpine vegetation (e.g. Coldea and Cristea 1998; Ozenda 1985). Mapping techniques were traditionally based on the manual delineation of units on aerial photographs and the field identification of corresponding vegetation types. However, the rapid development of software to handle geo-referenced data and the increased availability of remotely sensed data revolutionised the methods for vegetation mapping in the 1980s and 1990s. Remote sensing and image analysis not only provide a tool for rapidly mapping large areas, but they also implement objective criteria, whereas manual aerial photographic interpretation ultimately remains a subjective operation (Treitz and Howarth 1993; Goodchild 1994; Franklin 1995).

Various types of remotely sensed data are currently in use and the procedures that are applied to derive vegetation maps vary from case to case depending on the objectives of the surveys, the available data and the map scale. In this paper, we use as an example the mapping of the vegetation in an alpine landscape using aerial photographs, image analysis and topographic variables. The results are described and their evaluation and limitations explained with respect to the methods used, and the suitability of the data to study physiographic and vegetation diversity. The assumption underlying the work was that specific ecological habitat conditions would predetermine the vegetation types to be mapped (e.g. Franklin 1995; Austin and Meyers 1996; Dirnböck et al. 2002). By finding the range of habitat characteristics, in combination with the specific spectral characteristics of vegetation types, we mapped all of the important plant communities in the study area: five open pioneer communities, 10 different grassland types (including natural alpine swards, mires and pastures), tall herb, dwarf-shrub heath and krummholz vegetation (Table 10.1). The study area contains some of the catchments supplying Vienna with drinking water. The maintenance of high water quality in

Table 10.1. Mapped treeline ecotone and alpine plant communities of the Hochschwab mountain range. User's accuracy and producer's accuracy are shown for each plant community. Overall accuracy and Cohen's kappa of the final vegetation map are shown in the last two rows. The column n indicates the number of reference points (field samples) used for accuracy assessment

Name	Vegetation structure and important plant species	User's accuracy	Producer's accuracy	n
Vegetation on rocks	Graminoids, small perennial herbs and hummocks in fissures and on small terraces (e.g. *Trisetum distichophyllum, Draba stellata, Potentilla clusiana, Minuartia cherlerioides*); vegetation cover 5–30 %	69.5	62.9	105
Calcareous scree vegetation	Specifically adapted graminoids and herbs (e.g. *Thlaspi rotundifolia, Rumex scutatus*); vegetation cover 5–30(80) %	46.9	60.0	32
Snowbed vegetation	Creeping *Salix* spp. or minute herbs (e.g. *Arabis coerulea, Achillea clusiana, Achillea atrata, Campanula pulla*); vegetation cover 10–95 %	77.5	46.3	40
Carex firma grassland, open	Pioneer grassland dominated by the sedge *Carex firma*, additional plant species are *Saxifraga caesia* or *Crepis terglouensis*; vegetation cover 40–70 %	58.0	78.4	50
Carex firma grassland, closed	Closed grassland of *Carex firma* with cushion-plants such as *Silene acaulis* and prostate dwarf-shrub heath (e.g. *Dryas octopetala*); vegetation cover 70–100 %	80.0	83.3	50
Sesleria albicans–Carex sempervirens grassland, open	Pioneer grassland dominated by *Carex sempervirens* and *Sesleria albicans*; vegetation cover 40–60 %	44.4	47.1	18
Sesleria albicans–Carex sempervirens grassland, closed	Steep micro-terraced grassland dominated by *Carex sempervirens* and *Sesleria albicans*; vegetation cover 60–100 %	39.5	75.0	38
Carex ferruginea grassland	Moist grassland dominated by *Carex ferruginea* and some herbs (e.g. *Achillea anisiaca, Astrantia major*); vegetation cover 80–100 %	80.0	66.7	5

Name	Vegetation structure and important plant species	User's accuracy	Producer's accuracy	n
Festuca pumila–Agrostis alpina grassland	Alpine grassland dominated by two graminoids (*Festuca pumila*, *Agrostis alpina*); vegetation cover 85–100 %	50.0	42.9	6
Oreochloa disticha grassland	Acid grassland dominated by *Oreochloa disticha* and *Valeriana celtica*; vegetation cover 75–95 %	100.0	50.0	3
Acid Sesleria albicans grassland	Alpine grassland dominated by *Sesleria albicans* and *Festuca pumila*; vegetation cover 95–100 %	–	–	0
Festuca rubra pasture grassland	Pasture dominated by *Festuca rubra*, *Leontodon hispidus* and *Crepis aurea*; vegetation cover 80–100 %	70.0	77.8	10
Nardus stricta grassland	Acid grassland dominated by *Nardus stricta*; vegetation cover 90–100 %	61.5	80.0	13
Deschampsia cespitosa grassland	Acid semi-natural grassland dominated by *Deschampsia cespitosa*; vegetation cover 80–100 %	87.5	63.6	8
Hydrophilus tall herb communities	Various, mostly semi-natural assemblages with, e.g., *Rumex alpinus*, *Aconitum napellus*, *Senecio subalpinus*; vegetation cover 100 %	–	–	0
Mires	Moist grasslands dominated by, e.g., *Carex rostrata*, *Carex nigra*, *Deschampsia cespitosa*; vegetation cover 80–100 %	100.0	50.0	3
Rhododendron hirsutum dwarf heath	Dwarf heath dominated by *Rhododendron hirsutum*, *Erica herbacea*, *Calamagrostis varia*; vegetation cover 80–100 %	–	–	0
Krummholz	2-m tall scrub dominated by *Pinus mugo*; vegetation cover 90–100 %	78.3	81.0	120
Overall accuracy		67.9		
Cohen's kappa		0.63		

the long term is of high importance and an intensive ecological research programme is being carried out. The specific use of the vegetation data set was in the spatial prediction of the hydrological characteristics of the plant communities (e.g. water-holding capacity and evapotranspiration) to evaluate the sensitivity of the karst area to environmental impacts (Dirnböck and Grabherr 2000). High-resolution maps of plant communities (phytosociologically defined) were thought to best facilitate this application.

10.2 Study Site

The study area was in the Hochschwab mountain range (2277 m), NE Calcareous Alps, Austria (47°34' to 47°38'N and 15°00' to 15°18'E, Fig. 10.1). The massive plateau with its rugged relief in the alpine zone is in stark geomorphological contrast with the steep and needle-like Northern Calcareous Alps (Grabherr 1997). Grazing and pasture improvement over many years, as well as intensive forest utilisation, has shaped today's vegetation patterns, especially at the treeline. The main vegetation types are calcareous alpine grasslands, alpine snowbed communities, scree and rock vegetation, pastures, mires and *Pinus mugo* krummholz (Table 10.1). For a detailed description of the plant communities, see Dirnböck et al. (1999).

10.3 Methods

The vegetation map was derived from spatial data sets that included 40 false-colour infrared (CIR) orthophotographs with a 25-cm horizontal resolution, a digital elevation model (DEM) with a horizontal resolution of 50 m, and 455 vegetation samples taken from the field. The technical and modelling procedures of deriving the vegetation map are not presented in detail, but the sequence of processing is summarised in Fig. 10.2.

TWINSPAN, a divisive cluster algorithm (Hill 1979), was applied to classify the vegetation samples (Table 10.1). Vegetation samples were assigned to plant community types following Grabherr and Mucina (1993). In addition to spectral band values of the CIR orthophotographs (visible green, 500–600 nm; visible red, 600–700 nm; and near infrared, 750–1000 nm), spectral texture measures (second-order variance) and image band transformations (normalised difference vegetation index) were used. Topographical variables were derived spatially from the DEM using ARC/Info-GRID surface functions. These variables represent surrogates for ecological features (e.g. steepness, wind exposure, soil accumulation or erosion) at various scales (macro-, meso- and microrelief); for details, see Gottfried et al. (1998).

Mapping Alpine Vegetation

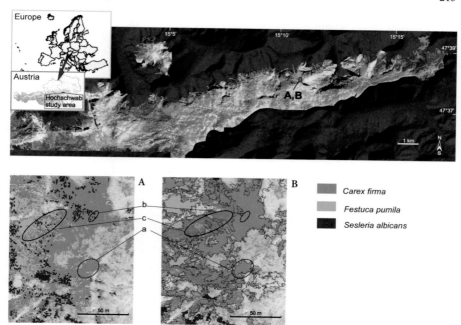

Fig. 10.1A, B. The Hochschwab mountain range in the NE Calcareous Alps, Austria. False-colour infrared aerial orthophotograph of the total study area overlain on a DEM shaded from NE. **A** The result of per pixel modelling of an enlarged area of three important alpine plant communities; the different grasslands appear either as separate homogeneous polygons or in mosaics due to fine-scale relief contrasts. **B** The final map after post-processing using image segmentation with the same plant communities as examples (*black lines* are polygons derived from image segmentation); *a* homogeneous areas remain as map units comprising just one plant community; *b* isolated predictions are eliminated during post-processing; *c* mosaics comprise two or three plant communities. Zonal *Carex firma* grassland dominates; *Festuca pumila–Agrostis alpina* grassland occurs on patches of acid soils frequently constituting mosaics with the former plant community; *Sesleria albicans–Carex sempervirens* grassland appear either on south-facing slopes (lower part of A and B) or scattered in rugged areas within a matrix of *Carex firma* grassland, again identified as mosaics since patch size fall below the minimum mapping unit size on the final map

We used canonical correspondence analysis (CCA) to analyse correlation between species composition and habitat qualities of the vegetation samples (ter Braak 1986; Fig. 10.2). [This technique was successfully applied previously by Gottfried et al. (1998, 1999) to study the fine-scale distribution of plant species at a site in the alpine-nival zone.] After deriving spectrally homogenous polygons by image segmentation of the CIR orthophotographs, the polygons were matched up by the output from the CCA (Fig. 10.1). All polygons were clustered according to the plant communities found and then assigned to the final mapping units. This last step was to generate a polygon vegetation

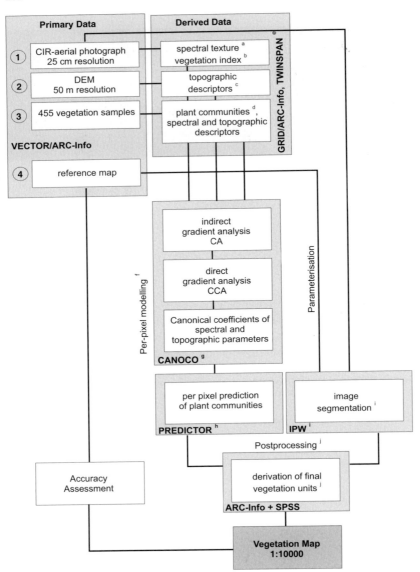

Fig. 10.2. Flow diagram showing primary data (*1–4*), derived data, processing steps (*white boxes*) and software applied (*grey boxes*) to produce the vegetation map. *a* Second-order variance of each band in a moving window of 3 × 3, 5 × 5, 7 × 7 and 15 × 15 pixels; *b* NDVI (normalised difference vegetation index; Wiegand et al. 1991); *c* according to Gottfried et al. (1998); *d* see Table 10.1; *e* Hill (1979); *f* for prediction procedure, see Gottfried et al. (1998); *g* ter Braak and Smilauer (1998); *h* hybrid code of Visual Basic 6.0 and Visual C++ 6.0 using ARC/Info-GRID Ascii files as input and output format; *i* region growing multiple pass segmentation algorithm developed by Woodcock and Harward (1992) implemented in IPW (Image Processing Workbench); *j* the segments were subsequently clustered (Ginzler 1997) using prediction results as variables, agglomerative hierarchical cluster algorithm and k-means clustering (SPSS 7.5 for Windows)

map, to reduce scattered errors from the CCA, and to identify areas with mosaic vegetation (Fig. 10.1).

Following the method described by Congalton and Green (1999), an error matrix was used to assess the accuracy of the derived map using field verification data. Approximately 700 reference points were used in the derivation of the error matrix. The percentage of the total units of a reference vegetation type assigned to its true type (user's accuracy), the percentage of a reference vegetation type assigned to other than its true type (producer's accuracy), and overall accuracy were calculated in addition to Cohen's kappa coefficient (Cohen 1960; Table 10.1).

10.4 Results and Discussion

10.4.1 The Map

Patchy vegetation typically occurs in high alpine regions largely owing to the rugged relief, which causes contrasts in the microclimate and soils (e.g. Grabherr 1997; Körner 1999). To separate a large number of vegetation units at a low minimum mapping unit size is thus necessary to capture major vegetation properties (species composition, community and species pattern as well as diversity). Consequently, a total of 18 vegetation units were mapped (see Table 10.1, Fig. 10.1), with a minimum mapping unit size of 80 m^2, resulting in a total number of about 20,000 vegetation polygons. The dominant plant communities were *Pinus mugo* scrub, *Carex firma* grassland, and *Sesleria albicans–Carex sempervirens* grassland (Table 10.1, Fig. 10.1). In addition to this typical series of calcareous plant communities, there were fragmented stands of acid (*Festuca pumila–Agrostis alpina*, *Oreochloa*, *Nardus* and *Deschampsia* types), and semi-natural grasslands (e.g. *Festuca rubra* type). In total, mosaics of six mapping units covered 23% of the total study area. Such systematic fine-scaled patterns of distinct plant communities (e.g. *Carex firma* grassland and *Sesleria albicans–Carex sempervirens* grassland, scree vegetation and snowbed vegetation) are typical for the region (Fig. 10.1).

10.4.2 Accuracy Assessment

In terms of total spatial accuracy, the distribution of dominant plant communities was predicted with an accuracy of almost 70% and with a kappa value of about 0.65 (Table 10.1). The accuracy achieved was comparable to other studies, despite a higher spatial resolution and a larger number of vegetation units mapped (Treitz and Howarth 1993; Franklin and Woodcock 1997; Congalton and Green 1999). Overall, producer's accuracy, on average, was slightly lower than user's accuracy; consequently, the predictive model had better results for the inclusion of the correct classes than on their omission. The highest accuracy obtained was shown for the krummholz vegetation and the closed *Carex firma* swards (Table 10.1). Both may be called zonal vegetation types; the former at the treeline, the latter in the alpine zone, and both represent the predominant landscape matrix in their respective areas (see Grabherr 1997; Fig. 10.1). Prediction errors of vegetation such as alpine snowbeds, initial pioneer swards and pastures propagated through all of the processing steps until the production of the final vegetation map. All these vegetation types are found in extreme, so-called azonal, environments. It is likely that the lack of more visually distinctive characteristics of some of these vegetation types, as well as their high spatial variability and small size, may have been below the resolution of the minimum mapping unit and restricted the accuracy with which they were mapped (see also Fig. 10.1).

To evaluate the results presented with those derived from a traditional approach, an analysis of the accuracy of each approach would have been required. However, the relevant independent vegetation data were not available. Traditional vegetation maps are usually characterised by uncertain data quality comprising errors originated from geographic inaccuracies, fuzzy boundaries, scale limitations, inaccessible areas and stochastic failures during the field recording and image-capturing phase (Goodchild 1994). In contrast, the technique described intrinsically operates with errors as a tool to parameterise and improve the outcome. Consequently, we know the statistical error of each vegetation unit as well as the entire map by cross-validation against independent field points. This knowledge does not necessarily mean that processed maps contain less error but it provides a tool to deal with error propagation in further GIS analysis.

10.4.3 Remotely Sensed Maps and Biodiversity Research

The applicability of vegetation maps for biodiversity research is often limited by their inconsistent nomenclatures, classification schemes and methodology of derivation. The use of remotely sensed data alone relies on the classification of overstorey species (Treitz et al. 1992). However, the inclusion of environmental response models may help distinguish spectrally similar plant communities (Franklin 1995; Dirnböck et al. 2002). Environmental response models have proved especially suitable in alpine ecosystems where human influence on vegetation is weak and steep environmental gradients cause a close relationship between plants, plant communities and their environment (Grabherr 1997; Körner 1999). Some recent studies on plant species (Gottfried et al. 1998; Guisan et al. 1998) and plant community distribution (del Barrio et al. 1997; Ostendorf and Reynold 1998; Zimmermann and Kienast 1999) exemplified the feasibility of environmental response models to study spatial patterns of arctic and alpine vegetation. Although models for plant communities achieved better results (Zimmermann and Kienast 1999), the focus on community-based approaches to study biodiversity has been criticised (Austin 1999). In the light of possible impacts of environmental change on biodiversity, modelling approaches that handle communities as entities with intrinsic features are questionable. The method chosen for the study presented in this paper allows for both community and species predictions, since each sample comprises the information of its distinct species collection (Gottfried et al. 1998).

The method used in this study suggests that it has a valuable role in future work on biodiversity research. Firstly, the use of automated processing can accommodate an analysis of the potential errors in the production of vegetation maps, which improves the basis on which they may be used in further applications, which is rarely possible with manually derived vegetation maps. Secondly, the study showed that spatially explicit environmental response models are able to improve the accuracy of vegetation maps. Biodiversity research, and especially the impact of environmental change on the diversity of the biota, needs flexible survey tools. Environmental response models capable of predicting species as well as community distributions could provide valuable information on potential future changes of the vegetation in alpine ecosystems (e. g. Dirnböck et al. 2003; Gottfried et al. 1999).

10.5 Conclusions

1. Today's vegetation mapping is largely based on the use of remote sensing, image analysis and modern GIS techniques. Effective and fast computer equipment, high-resolution image data and robust statistical methods have been developed. Furthermore, the combination of image processing and environmental response models provide a basic tool for spatially explicit vegetation analysis applicable far beyond sole mapping purposes. It is especially the flexibility of the presented method that allows for both community and species predictions that qualifies it for theoretical as well as applied biodiversity research application.
2. Map accuracy is biased towards zonal plant communities characterising the landscape matrix. Azonal plant communities, which are found in extreme environments, and those affected by grazing, show the lowest accuracy. Moreover, fine-scaled patterns of distinct alpine plant communities, largely due to the rugged relief which causes contrasts in the microclimate and soils, often fall below the resolution obtained by the procedure. This bias may be problematic when using the maps to evaluate vegetation diversity.

3. However, automated processing, compared with traditional field mapping, has not only the advantages that it speeds up the mapping process and its results are objective and reproducible, but also that it is accompanied by an intrinsically consistent and constructive handling of errors. Cross-validation against field points provides a tool that can deal with error propagation in further analysis of the GIS data.

Acknowledgements. We are grateful to Laszlo Nagy, Christian Körner and an anonymous reviewer for valuable comments on the manuscript. We thank the Karst Research Group, and in particular Gerhard Kuschnig, for support. The study was funded by the Department for Water Supply, Vienna. Travelling costs were covered by the European Science Foundation (ALPNET network).

References

Austin MP (1999) The potential contribution of vegetation ecology to biodiversity research. Ecography 22:465–484
Austin MP, Meyers JA (1996) Current approaches to modelling the environmental niche of eucalyptus: implication for management of forest biodiversity. For Ecol Manage 85:95–106
Cohen J (1960) A coefficient of agreement for nominal scales. Educ Psych Measurement 20:37–46
Coldea G, Cristea V (1998) Floristic and community diversity of sub-alpine and alpine grasslands and grazed dwarf-shrub heaths in the Romanian Carpathians. Pirineios 151–152:73–82
Congalton RG, Green K (1999) Assessing the accuracy of remotely sensed data – principles and practices. Lewis Publishers, Boca Raton
del Barrio G, Alvera B, Puigdefabregas J, Diez G (1997) Response of high mountain landscape to topographic variables: central Pyrenees. Landscape Ecol 12:95–115
Dirnböck T, Grabherr G (2000) GIS assessment of vegetation and hydrological change in a high mountain catchment of the northern Limestone Alps. Mount Res Dev 20:172–179
Dirnböck T, Dullinger S, Gottfried M, Grabherr G (1999) Die Vegetation des Hochschwab (Steiermark) – Alpine und Subalpine Stufe. Mitt Naturwiss Ver Steiermark 129:111–251
Dirnböck T, Hobbs RJ, Lambeck RJ, Caccetta PA (2002) Vegetation distribution in relation to topographically driven processes in south-western Australia. J Appl Veg Sci 5:147–158
Dirnböck T, Dullinger S, Grabherr G (2003) A reginal impact assessment of climate and land use change on alpine vegetation. J Biogeography (in press)
Franklin J (1995) Predictive vegetation mapping: geographic modelling of biospatial patterns in relation to environmental gradients. Progr Phys Geogr 19:447–499
Franklin J, Woodcock CE (1997) Multiscale vegetation data for the mountains of southern California – spatial and categorical resolution. In: Quattrochi DA, Goodchild MF (eds) Scale in remote sensing and GIS. CRC Press, Boca Raton, pp 141–171
Ginzler C (1997) Analyse digitaler Orthophotos im Rahmen der Erfolgskontrolle Moorbiotopschutz Schweiz. Salzb Geogr Mat 26:441–442

Goodchild MF (1994) Integrating GIS and remote sensing for vegetation analysis and modelling: methodological issues. J Veg Sci 5:615–626

Gottfried M, Pauli H, Grabherr G (1998) Prediction of vegetation patterns at the limits of plant life: a new view of the alpine-nival ecotone. Arc Alp Res 30:207–221

Gottfried M, Pauli H, Reiter K, Grabherr G (1999) A fine-scaled predictive model for changes in species distribution patterns of high mountain plants induced by climate warming. Div Distr 5:241–251

Grabherr G (1997) Polar and alpine tundra. In: Wielgolaski FE (ed) Ecosystems of the world, vol 3. Elsevier, Amsterdam, pp 97–121

Grabherr G, Mucina L (1993) Die Pflanzengesellschaften Österreichs, Teil 2. Fischer, Jena

Guisan A, Theurillat J-P, Kienast F (1998) Predicting the potential distribution of plant species in an alpine environment. J Veg Sci 9:65–74

Hill MO (1979) TWINSPAN – a Fortran program for two-way-indicator-species-analysis. Cornell University Press, New York

Körner C (1999) Alpine plant life: functional plant ecology of high mountain ecosystems. Springer, Berlin Heidelberg New York

Ostendorf B, Reynolds FJ (1998) A model of arctic tundra vegetation derived from topographic gradients. Landscape Ecol 13:187–201

Ozenda P (1985) La végétation de la chaîne alpine dans l'espace montagnard européen. Masson, Paris

ter Braak CFJ, Smilauer P (1998) CANOCO reference manual and user's guide to Conoco for Windows. Centre for Biometry, Wageningen

Treitz P, Howarth P (1993) Remote sensing for forest ecosystem characterisation: a review. Northern Forestry Programme, Canada, Ontario

Treitz PM, Howarth PJ, Suffling RC (1992) Application of detailed ground information to vegetation mapping with high spatial resolution digital imagery. Remote Sensing Environ 42:65–82

Wiegand CL, Richardson AJ, Escobar DE, Gerbermann AH (1991) Vegetation indices in crop assessment. Remote Sensing Environ 35:105–119

Woodcock C, Harward VJ (1992) Nested-hierarchical scene models and image segmentation. Int J Remote Sensing 13:3167–3187

Zimmermann NE, Kienast F (1999) Predictive mapping of alpine grasslands in Switzerland: species versus community approach. J Veg Sci 10:469–482

11 A GIS Assessment of Alpine Biodiversity at a Range of Scales

U. Molau, J. Kling, K. Lindblad, R. Björk, J. Dänhardt and A. Liess

11.1 Introduction

Biodiversity is a complex concept that covers diversity in a continuum of organisational levels from genetic diversity within populations to functional diversity at a global scale. The scale problem is central, and organism groups reach their maximum diversity (taxonomic richness) per unit area at very different geographical scales. Cryptogams and soil invertebrates may have high genetic and taxonomic diversity at a patch size of <1 m^2, whereas other organism groups need to be assessed on a much larger spatial scale (grain size and extent, sensu Wiens 1989). Positive correlation between organism groups with regard to diversity (species richness) has rarely been investigated. The few existing data sets from alpine environments show a lack of correlation between the vascular plants of the herb layer and soil surface dwelling cryptogams at the 1 m^2 scale for species richness and diversity indices (Molau and Alatalo 1998). In addition to spatial aspects, longevity of individual organisms and dormancy (e.g. of seeds and meristems) highlight the importance of considering temporal scale in assessments of biodiversity.

For many organism groups, such as birds and predatory mammals, diversity (taxonomic richness) assessments cannot be made at the grain size at which botanists usually work. Furthermore, at higher organisational levels of biodiversity (e.g. ecosystem diversity), meso-scale approaches are required. Most work in the twentieth century was made at two widely differing scales: small-scale (small grain size and small to large spatial extent) in most botanical studies and large-scale (large grain size and large spatial extent) ecosystem diversity assessments based on remote sensing. This wide gap between scales is being bridged by the increasing use of meso-scale techniques developed by landscape ecologists (Johnston 1998). The recent development in geographical information systems (GIS) has facilitated this process.

Here, we use as an example a recent analysis of diversity patterns in space and time from the Latnjajaure catchment (68°22'N 18°29'E), a subarctic alpine glacial valley. The GIS-based study, 'Tundra Landscape Dynamics' (TLD), was initiated in 1998 to (1) create a dynamic model of the vegetation and ecosystems in the entire Latnjajaure catchment (ca. 12 km^2), and (2) elucidate the pattern of taxonomic and ecosystem diversity at various spatial and temporal scales.

11.2 GIS in Ecology

Although GIS has traditionally been developed by geographers, it became a useful tool for ecologists during the 1990s. Integrating GIS and remote sensing for vegetation analysis and modelling is a new challenge for ecologists (Goodchild 1994). GIS is well suited for analysing geographical relationships between biological entities or how the environment influences their functioning. For example, where does a certain plant community A exist and in what relation to plant community B; how do environmental factors x, y, z influence the distribution of plant community A; how will the distribution of plant community A change if environmental factor x is altered?

In landscape ecology, GIS is a fundamental tool especially for manipulating models and real data and transferring information from implicit to explicit analysis (Johnston 1998). With the use of GIS, different spatial scales which are of interest to ecology can be linked: α-diversity is the species richness of a particular community or sample, β-diversity the degree of differentiation between communities or stands, and γ-diversity – the combined diversity of a number of stands, or the sum of β-diversities over a region. The potential distribution of plant species and communities over a larger area can be predicted using a sample set with known geographical coordinates (Brzeziecki et al. 1995; Gottfried et al. 1998; Guisan et al. 1998; Nilsen et al. 1999; Zimmerman and Kienast 1999).

11.3 The Latnjajaure Catchment

The Latnjajaure catchment, situated in the Scandes of northern Swedish Lapland (Fig. 11.1), encompasses mid- and high-alpine ecosystems in a U-shaped glacial valley with the surrounding mountains reaching 1500 m a.s.l. The main large lake, Latnjajaure, ca. 1 km^2, is at 986 m a.s.l. The Latnjajaure field station (LFS), originally a limnological station established in 1965, was transformed into a terrestrial ecology field station in 1990 for use in the International Tundra Experiment (ITEX; see Molau and Mølgaard 1996; Molau 2001). There are near-complete data sets for climatic variables for 1990–2000. In addition, there are a variety of long-term data sets on the responses of arctic

A GIS Assessment of Alpine Biodiversity at a Range of Scales

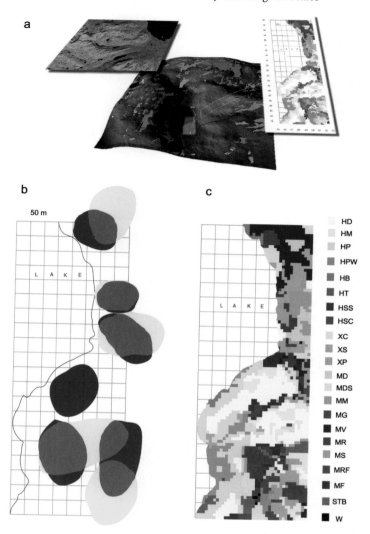

Fig. 11.1a–c. Geographical information sources at successive scales used in the Tundra Landscape Dynamics project. **a** Satellite image of the Abisko area with Latnjajaure and Lake Torneträsk (*upper left*); areas above the treeline are *red* and *light blue*. In the satellite image of the Abisko area, Latnjajaure is located just NW of the image centre. The *red line* in the 3D image in the middle delimits the Latnjajaure catchment with area of the main grid in *red*. At the more fine-grade scale, a main grid in the valley along the east side of Lake Latnjajaure, a 400 × 1000 m area, is used for mapping of plant communities (c) and two passerine bird species' territories in June 1999 (b). **b** Territories of the wheatear (*Oenanthe oenanthe*) and the meadow pipit (*Anthus pratensis*) are given as *deep blue* and *lemon yellow*, respectively (overlap areas in *green*). **c** In the vegetation map (10 × 10 m pixel size) grasslands are *green* (the driest kinds are those with the *lightest colour*; the wettest ones are *bluish*) and heath communities vary from *light brown* (tussock tundra) to *yellowish white* (dry heath); wet patterned ground is *bright red*, and late-melting snowbeds are *maroon* or *cerise*

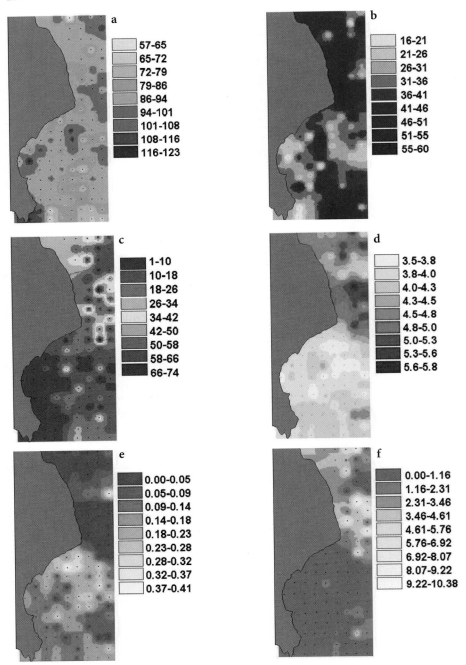

Fig. 11.2a–f. Spatial models of the length of snow-free period and soil properties in the main grid at Latnjajaure in 1999. Sampling points are indicated by *black dots*. An inverse distance weighted interpolator (Ormsby and Alvi 1999) was used to estimate values for the different variables between sample points. **a** Length of snow-free period (days), varying from very long in the windblown ridges (*dark green*) to very short in the snowbeds

and alpine plants to observed and experimentally induced climate variations (e.g. Henry and Molau 1997; Arft et al. 1999; Molau 2001).

The catchment is an alpine glacial valley totalling ca. 12 km^2. Of that area, about 1 km^2 is made up of lakes and 2 km^2 of permanent snowfields or ice patches. Of the remaining area, half is located in the mid-alpine zone, the other in the high alpine, and the altitude ranges from 900–1500 m a.s.l. (the treeline is at ca. 700 m a.s.l.).

11.4 Biodiversity Assessment by GIS

The initial concept was based on the method described by Walker et al. (1993) as hierarchical geographic information systems (HGIS), a nested set of GIS databases at several spatial scales. Sampling over scales at Latnjajaure included mapping the species within 1-m^2 field plots, and at 50-m interval sample points in a grid totalling 0.4 km^2. A vegetation map of the grid area with a 10 × 10 m pixel size (Fig. 11.1) was developed by Ließ (1998) based on an overlapping series of colour photographs taken from helicopter at 200 m above ground, an IR aerial photograph (scale ca. 1:10,000), and ground truthing plant community classification at all grid points. Additional grids in other parts of the valley were implemented in 1999 and 2000 (not presented here). In the final step of TLD our aim is to merge our multi-scale data sets with satellite images in order to analyse the landscape over a larger region (see Fig. 11.1). This paper is a first overview of TLD and its aims, supplemented with some results from the initial stage.

11.4.1 Field Sampling

Classification of plant communities at the grid points was carried out in the field by one of us (U. M.) in 1998 according to Påhlsson (1994), based on nutrient availability, soil moisture, and snow protection criteria. A more detailed survey of the plant community structure is now being undertaken by K. L., using standard ITEX point frames (see Molau and Mølgaard 1996). Snow depth, and later soil moisture, were recorded at all grid points at 15-day intervals from 25 May each year, and final snowmelt at the grid points was checked

---

(*yellow*). **b** Soil water content (%) from dry (*light mauve*) to wet (*deep lavender*). **c** Loss on ignition (%) from almost pure mineral soils; dark green to peat; brown. **d** Soil pH from very acidic (*pale pink*) to almost neutral (*deep purple*). **e** Exchangeable aluminium (mg Al/g soil), with the lowest Al contents in the mesic meadows (*purple*). **f** Exchangeable calcium (mg Ca/g soil), with the lowest Ca contents in the heath communities (*brownish ochre*)

every other day. Snow depth was recorded with a 3-m avalanche probe and soil moisture with a Theta probe (Delta-T Devices). Soil samples were collected by R. B. in August 1999 at all grid points except boulder fields and unmelted snowbeds and analysed for pH, exchangeable Al, Ca, Mg, and K, base saturation, and organic matter content (Björk 2000). A survey of territory size of all breeding birds was made by J. D. in the main grid in June of 1999 and 2000. Grazing experiments simulating the impacts of the main herbivores in the area, reindeer (*Rangifer tarandus* L.) and lemming (*Lemmus lemmus* L.), by selective plant clipping were initiated in 1999 in rich meadow communities. Weather observations at LFS follow the ITEX standards (Molau and Mølgaard 1996) with a manual station throughout the summer and an automatic station operating all year round. (For climate records and further information about LFS, projects, publications, and natural history of the area, see http://www.systbot.gu.se/ research/latnja/latnja.html.)

11.4.2 GIS Database and Modelling

The information from each grid point was stored in an ArcView GIS database (ESRI 1996); for spatial modelling, Spatial Analyst was used (ESRI 1996). In spatial modelling, values for continuous variables between any two sample points were estimated by interpolation (inverse distance weighted interpolator; Ormsby and Alvi 1999), resulting in variable maps (Fig. 11.2). The geophysical part of the GIS database encompassed detailed information on soil properties and processes, hydrology, and permafrost for all co-ordinates of the grid which are connected to vegetation and species diversity.

11.5 Results and Conclusions

The results from the Latnjajaure catchment study indicate little coupling between diversity of various organism groups, at least at spatial scales pertaining to bird species and plant communities (Fig. 11.2). The same holds true for cryptogams and vascular plants, as previously shown by Molau and Alatalo (1998). Therefore, a major focus for future action within the project will be inventorying other groups of organisms and assessing edaphic variables and productivity measures at various scales. The overall purpose is to gather as much knowledge as possible about the ecosystems of the catchment and their functional interactions. Such information is in most cases still rather patchy, even in well-studied areas.

11.5.1 Plants

Snow-cover duration is a main determinant of plant communities and productivity in alpine environments and, because of wind re-distribution of the winter snow-pack, spatial and temporal variation is best studied at the landscape level. Owing to the hilly relief of the valley bottom, winter redistribution of snow by wind results in a mosaic of ridges with little or no snow cover and deep snowbeds. A survey in late August 1999 revealed no correlation between snow cover and soil moisture, i.e. snowbeds varied from dry to very wet and small-scale variations in humidity did not appear to limit plant production.

Seed flux, including seed rain and the seed bank (total and germinable), was studied along an altitude transect in 1996–1998 using equal numbers of random samples at every 100 m of altitude. Seed production varied greatly among years (Molau and Larsson 2000; Molau 2001). The species diversity in the seed rain and seed bank was highly correlated with the plant community structure within a radius of 10 m at each site. There were also many extra-zonal recoveries of seeds 200–400 m above the present distributional limit of the species. Scattered saplings of mountain birch, previously unobserved, were found in the low- and the mid-alpine zones, probably as a result of the warm summers of the 1990s. Similar observations have recently been reported from the southern Scandes (Kullman 2000). This indicates that the treeline ecotone will progress upwards in abrupt events of invasion where entire alpine zones change in species composition more or less simultaneously rather than as a slow continuous process.

11.5.2 Animals

The median territory size for birds in 1999 was 18,000 m^2 for the wheatear (*Oenanthe oenanthe* L.) and 30,000 m^2 for the meadow pipit (*Anthus pratensis* L.) as opposed to a median patch size of 600 m^2 for plant communities (Fig. 11.2). Each bird territory included typically mosaics of three to five plant communities. The territories of the main predators of the catchment, the raven (*Corvus corax* L.) and the wolverine (*Gulo gulo* L.), were estimated to be 10 and 50 km^2, respectively. The species number per unit area (α-diversity) in the catchment was 50–100 km^{-2} for higher plants and 5–10 km^{-2} for birds. When considering the diversity variation in time, it is important to note that, whereas plant communities remain almost constant in their spatial distribution among years, bird territories may vary dramatically in number and extent even between subsequent years (Molau and Dänhardt, unpubl. data; Chap. 19). Mammal populations may fluctuate considerably over time such as the more or less cyclic population fluctuations in lemmings and voles. Populations of hares appeared to vary dramatically between years at our study site as a function of winter snow conditions and predator pressure (Molau, unpubl.). Insect populations in plant communities of mountain tundra are also known to vary manifold in numbers and extent among years, as observed for seed predators of alpine herbs (Molau et al. 1989). Thus, a momentary investigation may provide a proper data set for the assessment of plant species richness and community diversity, but be of little value when dealing with animal communities.

Despite its early stage of development, the TLD research program has already shown the importance of the meso-scale approach. In particular, when correlating soil chemistry and plant productivity measures with animal performance, the landscape level is optimal. An additional benefit arising from a landscape ecology approach with a close collaboration among botanists, zoologists, and physical geographers is a better understanding of how assemblages of diverse organisms form.

Acknowledgements. We thank the Abisko Scientific Research Station and its staff for help and hospitality. Special thanks are due to Per Folkesson for devoted field assistance at LFS in 1999 and 2000. The TLD project is funded by the Swedish Natural Science Research Council to U. M.; NFR, grant no. B-AA/BU 08424. Fieldwork for the GIS development was supported by the Royal Society of Arts and Sciences, Göteborg (to K. L.), and the Royal Swedish Academy of Sciences (Abisko scholarship to K. L.).

References

Arft AM, Walker MD, Gurevitch J, Alatalo JM, Bret-Harte MS, Dale M, Diemer M, Gugerli F, Henry GHR, Jones MH, Hollister RD, Jónsdóttir IS, Laine K, Lévesque E, Marion GM, Molau U, Mølgaard P, Nordenhäll U, Raszivin V (1999) Responses of tundra plants to experimental warming: meta-analysis of the international tundra experiment. Ecol Monogr 69:491–511

Björk R (2000) Soil properties and plant community types at Latnjajaure. B.Sc. thesis, Gothenburg University, Gothenburg

Brzeziecki B, Kienast F, Wildi O (1995) Modelling potential distribution of climate change on the spatial distribution of zonal forest communities in Switzerland. J Veg Sci 6:257–268

ESRI (1996) ArcView Version 3.2. Spatial Analyst: Environmental Systems Research Institute, Inc., Redlands

Goodchild MF (1994) Integrating GIS and remote sensing for vegetation analysis and modeling: methodological issues. J Veg Sci 5:615–626

Gottfried M, Pauli H and Grabherr G (1998) Prediction of vegetation patterns at the limits of plant life: a new view of alpine-nival ecotone. Arct Alp Res 30:207–221

Guisan A, Theurillat J-P, Kienast F (1998) Predicting the potential distribution of plant species in an alpine environment. J Veg Sci 9:65–74

Henry GHR, Molau U (1997) Tundra plants and climate change: the International Tundra Experiment (ITEX). Global Change Biol 3:1–9

Johnston CA (1998) Methods in ecology. Geographic information systems in ecology. Blackwell, Oxford

Kullman L (2000) Tree-limit rise and recent warming: a geographical case study from the Swedish Scandes. Norsk Geogr Tidsskr 54:49–59

Ließ A (1998) A vegetation map of the Latnjajaure area: mapping, analysis, and recommendations for optimal sampling. B.Sc. thesis, Botanical Institute, Göteborg University, Göteborg

Molau U (2001) Tundra plant responses to experimental and natural temperature changes. Mem Natl Inst Polar Res 54:445–466

Molau U, Alatalo JM (1998) Responses of subarctic-alpine plant communities to simulated environmental change: biodiversity of bryophytes, lichens, and vascular plants. Ambio 27:322–329

Molau U, Larsson E-L (2000) Seed rain and seed bank along an alpine altitudinal gradient in Swedish Lapland. Can J Bot 78:728–747

Molau U, Mølgaard P (eds) (1996) ITEX manual, 2nd edn. Danish Polar Center, Copenhagen

Molau U, Eriksen B, Teilmann Knudsen J (1989) Predispersal seed predation in *Bartsia alpina*. Oecologia 81:181–185

Nilsen L, Brossard T, Joly D (1999) Mapping plant communities in a local Arctic landscape applying a scanned infrared aerial photograph in a geographical information system. Int J Remote Sensing 20:463–480

Ormsby T, Alvi J (1999) Extending Arc View GIS. ESRI Press, Redlands, CA

Påhlsson L (ed) (1994) Vegetationstyper i Norden. Tema Nord 665. Nordic Council of Ministers, Copenhagen [in Swedish]

Walker DA, Halfpenny JC, Walker MD, Wessman CA (1993) Long-term studies of snow-vegetation interactions. BioSci 43:287–301

Wiens JA (1989) Spatial scaling in ecology. Funct Ecol 3:385–397

Zimmermann NE, Kienast F (1999) Predictive mapping of alpine grasslands in Switzerland: species versus community approach. J Veg Sci 10:469–482

III Invertebrates

12 Overview: Invertebrate Diversity in Europe's Alpine Regions

P. Brandmayr, R. Pizzolotto and S. Scalercio

Our knowledge of alpine invertebrate taxa is rather patchy. Some insect groups, such as the Lepidoptera, Coleoptera and Arenaea, are far better documented (Chaps. 13–17) than are, for example, soil-dwelling organisms (e.g. Broll 1998). The need for a coordinated approach to providing a comprehensive inventory of the biological richness of alpine areas in Europe has been highlighted in several recent works (e.g. Nagy 1998; Nagy et al. 1998).

12.1 Macrolepidoptera

The best-known invertebrates are, without doubt, the Macrolepidoptera; however, comparisons are mostly possible involving butterflies only. Chapters 13 and 16 show how powerfully the highly mobile Lepidoptera may be used both in biogeographical surveys of European mountain chains and in habitat evaluation for conservation.

Chapter 13 provides a broad geographical insight into macrolepidopteran species assemblages of the European mountain chains. The distinction between elements of the humid-alpine (zono-)biomes and the xeromontane faunas of more or less eremial origin reflects the classical dualism found also in the alpine vegetation belts. Some of the phytophagous insect groups show similar faunal origins (e.g. Orthoptera and Lepidoptera). However, it is regrettable that data of comparable consistency are unavailable for non-phytophagous invertebrate groups. Ground beetles (Carabidae) and perhaps spiders (Aranaea) could be the next taxa to be analysed in relation to the existence of a Mediterranean-xeromontane faunal element more or less restricted to the thorny scrub vegetation that extends from the Maghreb to the western Asiatic countries. In Italy, for example, an invertebrate survey of the thorny cushion vegetation (Pignatti 1979; Pignatti et al. 1980) of the Apennines and of Mount Etna would be of particular interest.

Of the 275 Italian native butterfly species, 119 (43 %) occur in and above the treeline ecotone in the South Eastern Alps and Apennines (Chap. 16). The high mobility of the adult butterflies probably contributes to such a high percentage having been recorded from the alpine zone. At the regional scale, there is a decrease in the proportion of the alpine taxa from the Alps to the Apennines. This is presumably caused by the reduction of available habitats for stenotopic high mountain taxa. The observed guilds or species assemblages each contain 10–15 species, are strongly dependent on vegetation type, and are influenced by several ecological factors and human pressures. For example, heavy grazing can cause a strong decrease in the species richness and density of individuals. Further studies are required to clarify the effective maintenance of these populations (Chap. 16) in the Italian Alps and Apennines.

Chapter 14 shows that in the Urals, also, the alpine diurnal butterfly fauna decreases from the north to the south in species numbers (from 49 in the Polar Ural to five in the south of northern Ural), whereas soil arthropod abundance and biomass increase along the same gradient. Few tundra butterflies and soil arthropods with a circumpolar distribution occur in the alpine zone of the middle or southern parts of the Urals.

12.2 Carabids

Available data for ground beetles in alpine areas are not so exhaustive as for Lepidoptera. The reasons for this include the laborious nature of assemblage studies and the lack of a universally used standardised methodology. Many studies on arthropods use absolute density estimates by, e.g., sieving or using eclectors, instead of pitfall-traps. From a biogeographical point of view the available data rarely allow meaningful comparisons between countries or mountain ranges because, in most cases, these are no more than species lists without reference to habitat type or elevation. An exception is the case of Fennoscandia (Lindroth 1949). The comparison of the altitude distribution of the Carabidae of Fennoscandia (Lindroth 1949) and Italy was made by assigning each of the 1280 species recorded from the Italian Peninsula in the topographic catalogue of Magistretti (1965, 1968) to a broad vegetation type corresponding to altitude zones (Fig. 12.1). For each vegetation type, the number of endemic species is reported (endemic to Italy in the strict sense). Of the 354 species listed by Lindroth (1949) for the Scandes, 316 are present in the beech and oak forests of Italy. The numbers are fewer in the coniferous forests of Scandinavia (more or less corresponding to the upper montane conifer forests of Italy, including treeline ecotone scrub where present); in the lower alpine belt there are no more than 70 species; and there are few in the upper parts of the alpine zone. The alpine zone of the Italian Alps and Apennines is inhabited by at least 180 ground beetle species with 109 (60%) of them being endemic. The conifer forests are less species rich (168 ground beetle species)

Overview: Invertebrate Diversity in Europe's Alpine Regions 235

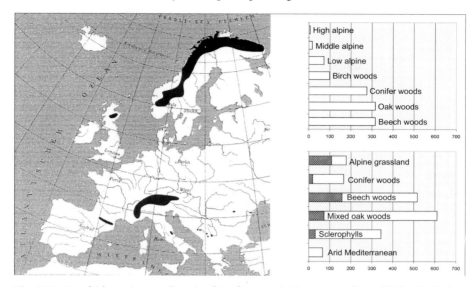

Fig. 12.1. Carabid species numbers in altitude zones in Fennoscandia and Italy. For Italy, the numbers of strictly Italian endemic taxa (from Vigna Taglianti 1993) are indicated by *cross-hatched part of the bars*. The map of Europe is redrawn from Holdhaus (1954), and shows the classic arctic alpine distribution of *Zygaena exulans*

with only 19 (11%) endemics. The highest number of endemic species is found in the beech forests (159 out of 518 species, 31%) because the beech forests occupy mostly glacial refugial areas in the mountains. The low incidence of endemics in the upper montane conifer forests at the lower edge of the treeline ecotone is interesting. This habitat type became widespread in the Alps only at the end of the Ice Age and its wet peaty soils were scarcely suitable for beech-forest-dwelling petrophilic endemics (which are adapted to well-drained shallow or skeletal soils; Brandmayr 1991) and would have been too dark and cool for stenotopic alpine beetles.

12.3 A Comparison of the Diversity Patterns of Arachnids, Carabids and Lepidoptera

Spiders and other Arachnids (Chap. 15) belong to the best-known invertebrate groups in the Alps, especially in the Eastern Alps of Austria, where they have been studied for over a century. Chapter 15 is an overview with an important ecological synthesis. In many other countries of Europe arachnid studies have been less detailed (e.g. in Italy the ecological description of spider communities is scarce, because of the faunal richness and the scarcity of taxon specialists). The altitudinal and diversity patterns of arachnids seem to

Table 12.1. Species richness of carabids, butterflies and spiders in some regions of Europe. Figures are for regions including lowland areas, the alpine zone and individual habitats

Taxon (region)	Total number of species	Total number of species in the alpine zone with (% of total)	Number of species in single habitats with (% of total)
Butterflies (Italy)	275 (Chap. 16)	119 (43%)	10–15 (3–5%)
Carabid beetles (Italy)	1280 (Magistretti 1965, 1968)	180 (14%)	6–15 (0.5–1.2%)
Carabids (Fennoscandia)	354 (Lindroth 1949)	72 (20%) lower alpine 16 (4.6%) middle alpine	
Spiders (Eastern Alps)	683 (Thaler 1998)	90 (ca. 13%)	10–25 (1.4–3.6%)

be similar to that of carabids, with 30% of spider species endemic to the Alpine system (Chap. 15).

A comparison of species diversity is given for butterflies, carabids and spiders at a range of scales (Table 12.1). In Italy, butterflies are the most species-rich invertebrate group in the alpine zone, but their proportion of endemics is low in comparison with soil-dwelling carabids (60%). In the Scandes, there are no endemic carabids. About 20% of the species are present in the lower alpine zone and less than 0.05% in the middle alpine zone (the latter value is similar to that found in the less diverse habitats of the alpine zone in Italy). In comparison with other European alpine regions, the Scandes are clearly at an early stage of faunal colonisation.

In order to improve our understanding of the nature of alpine invertebrate diversity we need more comparative studies of habitat affinities and diversity estimates at the habitat and landscape scales within geographical regions. Such information would highlight geographical differences and could be used for monitoring purposes [such as proposed for Arthropods and phytophagous Chrysomelids (leaf beetles) in Chap. 14].

12.4 Conclusions

The alpine zone is clearly not the most species-rich zone for invertebrates. The vegetation below the timberline normally has a higher species richness (with the possible exception of some mobile taxa such as Dipterans and butterflies). However, alpine invertebrate communities are interesting and important for conservation not simply for their species numbers, but also for their adaptations to the habitat conditions. The alpine biome harbours the final specialisation for many taxa, and in some ways is an evolutionary 'blind alley'. In the event of a rapid large-scale climate change the survival of these invertebrate specialists is threatened.

References

Brandmayr P (1991) The reduction of metathoracic alae and of dispersal power of carabid beetles along the evolutionary pathway into the mountains. In: Lanzavecchia G, Valvassori R (eds) Form and function in zoology. U.Z.I. Selected Symp and Monographs 5. Mucchi Editore, Modena, pp 363–378
Broll G (1998) Diversity of soil organisms in Alpine and Arctic soils in Europe. Review and research needs. Pirineos 151/152:43–72
Holdhaus K (1954) Die Spuren der Eiszeit in der Tierwelt Europas. Abh Zool Bot Ges Wien 18:1–493
Lindroth CH (1949) Die Fennoskandischen Carabidae. Kungl Vetensk Vitterh Samh Handl SB4, 3 Allgemeiner Teil. Bröderna Lagerström Boktryckare, Stockholm
Magistretti M (1965) Coleoptera, cicindelidae, carabidae. Catalogo topografico. Fauna D'Italia, 8. Calderini, Bologna
Magistretti M (1968) Catalogo topografico dei Coleoptera cicindelidae e Carabidae d'Italia. I supplemento. Mem Soc Ent Ital 47:177–217
Nagy J (1998) European mountain biodiversity: a synthetic overview. Pirineos 151/152:7–41
Nagy L, Thompson DBA, Martinez-Rica JP (1998) Trends in European mountain biodiversity: an introduction. Pirineos 151/152:3–5
Pignatti S (1979) I piani di vegetazione in Italia. Giorn Bot It 113:411–428
Pignatti E, Pignatti S, Nimis P, Avanzini A (1980) La vegetazione degli arbusti spinosi emisferici: contributo all'interpretazione della fasce di vegetazione delle alte montagne dell'Italia mediterranea. Quad Prog Finalizzati CNR AQ/1/71:1–130
Thaler K (1998) Die Spinnen von Nordtirol (Arachnida, Araneae): Faunistische Synopsis. Veröff Mus Ferdinandeum (Innsbruck) 78:37–58
Vigna Taglianti A (1993) Coleoptera Archostemata, Adephaga 1 (Carabidae). In: Minelli A, Ruffo S, La Posta S (eds) Checklist delle specie della fauna italiana, 44. Calderini Publ., Bologna

13 The Geographical Distribution of High Mountain Macrolepidoptera in Europe

Z. Varga

13.1 Introduction

This paper covers the traditional Macrolepidoptera. Several important families and genera, rich in endemic Microlepidoptera in the Alps (e.g. Huemer and Tarmann 1993; Huemer 1998), are not considered because of the lack of comparable data sets on their distribution in most south European high mountains. An overview of the alpine and xeromontane types is given including their biogeography, speciation and endemism. Biogeographical connections among European orobiomes are discussed on the basis of the distribution of the alpine and arctic-alpine Macrolepidoptera. The taxonomy and nomenclature follow Huemer and Tarmann (1993), with the exception of some of the Lycaenidae and Noctuidae.

13.2 Faunal Types in the European Orobiomes

The definition of faunal types is often discussed in the biogeographical, and especially, in the lepidopterological literature (see Table 13.1 for a list of relevant references). Faunal types have been often differentiated based on the shape and extension of their distribution ranges (e.g. the Palaearctic and Holarctic faunal elements in numerous faunistical publications). A traditional interpretation of faunal elements is given, e.g., by Caradja (1933, 1934) and Rebel (1932). More dynamic attempts have been made based on 'centres of dispersal' by de Lattin (1957, 1967) and Varga (1977), or on 'faunal types' by de Jong (1972). Quantitative methods of biogeography using large databases have rarely been used. Using multivariate methods, Dennis (1993) and Dennis et al. (1991) differentiated several types of species from endemic to 'extent' (i.e. widespread) and identified some faunal structures for butterflies. Balletto

Table 13.1. Selection of relevant faunistic publications on European high mountain Lepidoptera

Mountain ranges by geographical location	Publications by country
Central and western Europe: Alps, Pyrenees, Cantabrian Mts., Apennines, Mountains of the Iberian peninsula	Austria: Thurner (1948); Mack (1985); Reichl (1992); Huemer and Tarmann (1993)[a] Switzerland: Rappaz (1979); Schweitzerischer Bund f. Naturschutz (1987); Cupedo (1996) Spain: Fernandez-Rubio (1991); Vives Moreno (1991) France: Lhomme (1923–1963); Dufay (1961); Leraut (1980, 1996) Italy: Mariani (1940–1941); Verity (1943–1953); Wolfsberger (1966b, 1971); Scheuringer (1972); Balletto (1983)
Eastern-central Europe: Carpathians	Poland: Niessiolowsky (1929, 1936); Niessiolowsky and Woytusiak (1937) Slovakia: Hruby (1964) Romania: Diószeghy (1929–1930, 1936); Goltz (1935–1936); Popescu-Gorj (1952, 1962, 1963, 1964); König (1959, 1975, 1982); Alexinschi (1960); Alexinschi and König (1963); Rákosy (1992a,b, 1995, 1997, 1998)
Scandinavian Mts.	Scandinavia: Nordström (1955); Nordström et al. (1961, 1969); Henricksen and Kreutzer (1982) Finland: Kaisila (1947, 1962) Norway: Opheim (1958, 1972) Sweden: Nordström and Wahlgren (1935–1941)
Northern Balkan Peninsula	States of the former Yugoslavia: Rebel (1903, 1904, 1911); Daniel (1964); Thurner (1964); Pinker (1968); Sijarić (1971, 1980); Sijarić et al. (1984); Jaksić (1988); Schaider and Jaksić (1989) Bulgaria: Drenowsky (1925; 1928 vertical distribution); Buresch and Tuleschkow (1929–1943 horizontal distribution); Varga and Slivov (1977); Abadjiev (1992, 1993, 1995); Beshkow (1995, 1996) Albania: Rebel and Zerny (1931); Popescu-Gorj (1971)
Southern Balkan Peninsula, Aegean islands	Greece: Thurner (1964); Coutsis (1969, 1972); Koutsaftikis (1974); Hacker (1989); Coutsis and Ghavalas (1991); Koutroubas (1994); Coutsis et al. (1997); Pamperis (1997);

Table 13.1. (*Continued*)

Mountain ranges by geographical location	Publications by country
Crimea, Asia Minor, Caucasus, Transcaucasia	Crim0ea: Nekrutenko (1985) Caucasus: Woytusiak and Niessolowsky (1947); Alberti (1969); Nekrutenko (1990) Transcaucasia: Asarjan et al. (1970) Asia Minor: Hesselbarth et al. (1996)
General works treating Lepidopteran groups in Europe and the West Palaearctic	General: Karsholt and Razovski (1995) Zygaenidae: Naumann et al. (1984) Bombyces and Sphinges: Freina and Witt (1987) *Erebia*: Warren (1936) Noctuidae: Fibiger and Hacker (1990); Svendsen and Fibiger (1992); Fibiger (1993a,b, 1997); Ronkay and Ronkay (1994, 1995); Ronkay et al. (2001); Hacker et al. (2002)

[a] With complete references on Lepidoptera of the Austrian Alps.

(1995) analysed the species diversity (747 species) and areas of endemism (sensu Harold and Mooi 1994) of West Palaearctic butterflies.

Quantitative analyses based on rough-scaled 'surface' maps, published in popular books (e.g. Higgins and Riley 1975; Tolman and Lewington 1997, used by Hengeveld 1990; Dennis et al. 1991; Dennis 1993), reflect only some general trends or gradients in the distribution of the European butterflies. For a fuller analysis more grid or point maps are needed (see e.g. Naumann et al. 1984; Jakšić 1988; Reichl 1992; Svendsen and Fibiger 1992), based on taxonomical revisions of suprageneric units (families or subfamilies, genus groups, e.g. of Papilionoidea, Geometridae, Noctuidae).

This paper deals with the oreal fauna, i.e. the set of species that is attached to orobiomes (sensu Walter and Breckle 1991: 15-27). I follow the distinction of the vertical belts of the biota proposed by Walter and Straka (1970, pp. 340-341) from general horizontal zonation. Consequently, the oreal fauna is considered here as a major biogeographical unit correlated with orographically determined non-arboreal ecosystems (Varga 1996).

Alpine vs. xeromontane (eremoreal) faunal types can be distinguished in the European orobiomes (Varga 1976, 1996). The alpine type, which is characteristic of humid high mountains, is closely connected with the Quaternary glaciations, resulting in some long-distance area translocations and disjunctions, and, as a consequence, a great number of arctic-alpine species. On the other hand, there are also numerous stenotopic species that survived the last glaciations in unglaciated parts of the south European high mountains (e.g. Pyrenees, SW and SE Alps, Balkans).

The xeromontane type prevails in the summer-arid Mediterranean high mountains and the arid high mountains of central Asia. Its members have a more continuous evolutionary history, less influenced by the Quaternary glaciations. They often have a high potential for speciation in taxonomic groups that are adapted to cold and arid conditions (e.g. many Orthoptera, 'cutworm' Noctuidae, rodents: Microtinae, Dipodidae, Zapodidae) and have close connections with the eremic zonobiome. The distribution of the two faunal types overlap in, for example, the central (Wallis, Upper Engadin, Upper Inn valley) and SW Alps. There are also areas of overlap in the Balkan high mountains (e.g. Dinarids, Pindos, Pirin and Rhodope Mts.).

13.3 The Alpine Faunal Type

Glacial history resulted in a rapid differentiation within the alpine type. Arctic-alpine species generally have a Eurasian distribution (Eurasian and Palearctic Asian, for an extensive tabulated list of species, see http://www.stir.ac.uk/departments/naturalsciences/DBMS/nagy/alpine_biodiversity/) and have been shown to have widely dispersed in the periglacial zone during the last glacial phases (mostly Riss and/or Würm). The alpine part of the distribution of these species is restricted either to the Eurasiatic high mountains having typical alpine and sub-nival elevations or to the high mountain systems of Palaearctic Asia. Alpine, as opposed to arctic-alpine, taxa occur in all or parts of Europe where they are often endemic (e.g. Pyrenees, Alps, and the highest massifs of the Balkan Peninsula) and in the Altai and Sayan Mountains of the Eastern Palaearctic. Alpine taxa survived the last glacial periods mostly in refugia in the less glaciated parts of the Alps, Pyrenees and Carpathians, at nunataks, and in the high mountains of the Mediterranean peninsula (Varga 1976, 1989a, 1995a, 1996).

Numerous widely distributed alpine species are absent from and few alpine endemics occur in the N Calcareous Alps or the N Carpathians. A possible explanation is the relatively small and discontinuous alpine zone of these ranges, combined with unfavourable climatic conditions. Numerous species groups which are mostly distributed in the alpine-type S European high mountains are often represented by vicariant species pairs or disjunct subspecies in SW and SE Europe. There are high levels of faunal similarities between parts of the Alps and Pyrenees and within the Balkan ranges (see Sect. 13.6). The above are also the areas with the highest number of alpine and arctic-alpine butterfly and moth species.

The faunal diversity of isolated mountain ranges in Europe depends, in some cases, on the distance from the (assumed) source of colonisation. For example, owing to their proximity to the SE Alps, the NW Dinarids, although relatively low and small in extent, have a rich fauna of terricolous insects with a low mobility such as small, flightless Carabidae and Curculionidae (Buresch and Arndt 1926; Holdhaus and Lindroth 1939; Maran 1946; Holdhaus 1954). In other cases, conditions favourable for long-term survival appear to be more important. For example, the remote but extensive high massifs of the Balkan Peninsula are richer in alpine and arctic-alpine Orthoptera, butterflies (e.g. species of *Erebia* Dalman) and moths (e.g. *Pygmaena fusca* Thunberg, *Sciadia tenebraria* Esper, species of *Glacies* Milliére [= *Psodos* auct.], *Elophos* Boisduval and *Charissa* Curtis [= *Gnophos* auct. partim]) than the Dinarids or the Carpathians (Ander 1949; Warnecke 1959; Varga 1975, 1995a). Sites that are advantageous for a species' survival may facilitate differentiation into allopatric subspecies or autochthonous neo-endemic species, often vicariants to related ones isolated in some other remote massifs (Mařan 1946; Varga 1976, 1989a, 1996).

13.4 Speciation and Endemism in the Alpine Faunal Type

The distribution and speciation/subspeciation patterns of the alpine and arctic-alpine species are diverse. Populations of some allopatric subspecies belonging to the genera *Glacies* Milliére and *Erebia* Dalman in the N Alps show close connections to those in Scandinavia, while the populations of the S Alps and Balkans are more differentiated (Warren 1936; Lorkovic 1952, 1953, 1957; Popescu-Gorj 1952, 1962; Povolnỳ and Moucha 1956, 1958; Wolfsberger 1966a; Varga 1975, 1995a). A close relationship between the populations in the SW Alps and those in the E Pyrenees and between populations in the SE Alps and those in the Dinaric Mountains has been reported. Meanwhile, the populations of the E and S Carpathians seem to be closely related to those of the Stara Planina, Rila, and Pirin Mts. (Varga 1975).

Several types of endemism have been described in the European alpine fauna (Table 13.2). Some Balkan endemics have vicarious sibling species in the Alps (e.g. *Erebia rhodopensis* Nicholl in the Balkans and *E. aethiopella* Hoffmannsegg

Table 13.2. Some main types of the endemic Lepidoptera in European high mountains

Mountain range	Species
Pyrenees, Cantabrian Mts.[1], Sierra Nevada[2]	*Erebia gorgone* Boisduval, *E. lefebvrei* Boisduval[1], *E. palarica* Chapman[1], *E. hispania* Butler[2], *E. sthennyo* Graslin, *Entephria caerulata* Guenée
SW Alps (and E Pyrenees)	*Erebia scipio* Boisduval, *E. aethiopella* Hoffmannsegg, *E. neoridas* Boisduval
Western and southern central Alps	*Mellicta varia* Meyer-Dür, *Erebia flavofasciata* Heyne, *E. christi* Rätzer, *E. mnestra* Hübner, *Colostygia kitschelti* Rebel, *Glacies wehrlii* Vorbrodt
Eastern central Alps	*Mellicta asteria* Freyer, *Erebia claudina* Borkhausen, *E. nivalis* Lorković and de Lesse, *Elophos zierbitzensis* Pieszczek, *Glacies alticolaria* Mann, *G. chalybaeus* Zerny, *G. burmanni* Tarmann
SE Alps	*Erebia calcaria* Lorković, *Glacies spitzi* Rebel, *G. baldensis* Wolfsberger
Eastern central Alps, SE Alps and N Carpathians	*Colostygia austriacaria* Herrich-Schaeffer, *Glacies noricana* Wagner, *G. bentelii* Rätzer (also in S Carpathians)
Balkan high mountains (and Southern Carpathians)	*Colias balcanica* Rebel, *Erebia orientalis* Elwes, *E. rhodopensis* Nicholl, *E. melas* Herbst, *Coenonympha rhodopensis* Elwes
SW[1] or SE[2] Alps and Balkans (disjunct areas)	*Boloria graeca* Staudinger[1], *Erebia ottomana* Herrich-Schaeffer[1], *Aplocera simpliciata* Treitschke[2], *Xestia ochreago* Hübner[2]
Caucasus, Transcaucasia	*Colias caucasica* Staudinger, *Boloria caucasica* Staudinger, *E. melancholica* Herrich-Schaeffer, *E. hewittsoni* Lederer, *Apamea rjaboviana* Mikkola and Varga

Superscripts in each row with multiple mountain ranges are used to specify a species' locality. For example in row 1, [1] denotes that *Erebia lefebvrei* is endemic to the Cantabrian Mts.

in the SW Alps). Others form disjunct subspecies (e.g. *Boloria graeca graeca* Staudinger: Mts. of Greece, *B. graeca balcanica* Rebel: Balkans, *B. graeca tendensis* Higgins: SW Alps; *Erebia ottomana ottomana* Herrich-Schaeffer: Ulu Dagh, *E. ottomana balcanica* Rebel: Balkans, *E. ottomana benacensis* Dannehl: Mt. Baldo, *E. ottomana tardenota* Praviel: Massif Central of France). The distribution of the western sibling taxa is mostly restricted to the 'massifs de refuge' of the S-SW Alps. The disjunctions and vicariant patterns indicate inter- or preglacial connections and the position of some possible glacial refugia.

Large numbers of restricted (nunatak) refugia must have existed in the heavily glaciated central Alps; some stationary Microlepidoptera, which often have flightless females (e.g. *Dahlica* Enderlein, *Siederia* Meier, *Taleporia* Hübner spp.) or rather strict food-plant specialisation (e.g. *Kessleria* Nowicki spp. on *Saxifraga* plants) attest to this. In these genera there are numerous strictly localised endemic species and parthenogenetic races (Huemer and Tarmann 1993; Huemer 1998). The survival of some endemic Geometridae in the central and southern Alps have probably also been connected with ice-free refugia of high altitudes (e.g. *Glacies wehrlii* Vorbrodt, *G. perlinii* Turati, *G. spitzi* Rebel, *G. baldensis* Wolfsberger, *Elophos zirbitzensis* Pieszczek).

13.5 Biogeographical Characterisation of the Xeromontane Faunal Type

They form two distinct groups, Mediterranean and continental central Asian (Fig. 1 in Varga 1995b), which are separated by the limit of the equinoctial type of precipitation (Agakhanjanz 1981; Agakhanjanz and Breckle 1995). Core areas of endemic species of the Mediterranean type are dispersed in the Mediterranean high mountains from the Atlas and Sierra Nevada to Daghestan, Transcaucasia, W-and N-Iran and Transcaspia (Kopet-dagh). The xerophytic oro-Mediterranean formations (sensu Ozenda 1988; *Astragalo-Acantholimetalia* in Horvat 1962; Horvat et al. 1974) of the mountains of S Greece are especially rich in xeromontane species. Some Mediterranean xeromontane species also occur in the SW Alps (e.g. *Euxoa hastifera* Donzel, *Dichagyris [Yigoga] celsicola* Bellier, *Chersotis fimbriola* Esper, *Ch. larixia* Guenée, *Ch. elegans* Eversmann, *Ch. anatolica* Draudt, *Hadena clara* Staudinger, *Heterophysa dumetorum* Geyer). This is probably the consequence of the refugium character of this region.

Outside the Mediterranean, the distribution of xeromontane species lacks a geographical pattern in Europe and is influenced by local edaphic conditions. They are likely to have established during the extreme continental late-glacial and early postglacial phases (Varga 1996). Little is known about how the xeromontane oro-Mediterranean vegetation and fauna will be affected by the current climate change. Observations near to the southernmost distribution of alpine vegetation/fauna and northernmost boundary of the oro-Mediterranean one (e.g. in the Pindos or Orvilos Mts. in Greece) are likely to be useful to study changes in distribution patterns.

13.6 Biogeographical Connections and Similarities of the European High Mountains

A multivariate classification (for details, see http://www.stir.ac.uk/departments/naturalsciences/DBMS/nagy/alpine_biodiversity/) of 30 European high mountain areas, including subregions of the Alps, Carpathians, Balkans the Caucasus and Transcaucasia, for the presence of Papilionoidea only and for all Macrolepidoptera showed that:

- the different parts of the Alps are usually similar to each other and different from the other European mountains. Most closely related are the SW Alps, S Alps, W Central Alps and the E Central Alps (Macrolepidoptera, many endemic alpine species), and the W and E Central Alps and N Calcareous Alps (butterflies, many arctic-alpine species, relatively few endemics);
- the similarity between the Central and E Pyrenees is moderate, probably because of their different geological and climatic conditions. The faunal connections of the Cantabrian Mountains, the Sierra Nevada and the Massif Central are uncertain owing to the relatively low numbers of alpine species in these mountains;
- the Carpathians also form a compact group with the highest similarity between the southern part of the E Carpathians and S Carpathians, which are the highest and have the richest fauna in the whole of the range;
- the high mountains of the North Balkans have a European alpine character with numerous widely distributed European alpine and arctic-alpine species and with some Balkanic oreal species. Closest connections have been found between the different parts of the Dinarids on one hand, and between the North Albanian Alps and the Šar-Korab massif on the other. The Rila-Pirin massif seems to be mostly connected to the latter because of the numerous common alpine and arctic-alpine species that occur in the Balkans only at the highest elevations;
- the Stara Planina shows some similarities with the more southern mountains of the Balkan peninsula through the lack of numerous alpine and arctic-alpine species, typical for the highest altitudes. This seems to be a typical insular phenomenon that has appeared at the periphery of distribution of the alpine species;
- the high mountains of the South Balkans, mostly the mountains of Epirus/Pindos and Peloponnessos, display some similarity with the high mountains of Asia Minor. However, this connection may not be very close,

because of the relatively low numbers of alpine species. The alpine Lepidoptera of the North Caucasus and Transcaucasia are few in number and mostly different from the European mountains.

The above interpretations are subject to some uncertainties arising from using presence-absence data and the rather uneven species numbers in the different high mountains, ranging from 2-119. The similarities obtained for mountains with low species numbers are obviously highly uncertain. It should also be emphasised that the above comparison was made on regional orobiomes, i.e. areas of regions with alpine vegetation, as opposed to Dennis et al. (1991) who separated subregions comprising both lowlands and mountains. As a consequence, all widespread ('extent', sensu Dennis et al. 1991) species and, generally, all species connected with the zonobiomes (sensu Walter and Straka 1970; Walter and Breckle 1991) were excluded from the analyses.

13.7 Geographical and Phylogenetic Patterns in the European High Mountain Macrolepidoptera

Geographical patterns of core areas and their significance in the survival, speciation and dispersal of biota can be confirmed by phylogenetic (cladistic) analysis (e.g. Rosen 1978; Nelson and Platnick 1980; Oosterbroek 1980; Cracraft 1982, 1983; Cracraft and Prum 1988; Oosterbroek and Arntzen 1992; Enghoff 1993, 1995; Choi 1997, 2000).

The geographical and phylogenetic patterns of alpine faunal type are highly diverse. There are numerous spot-like endemics, especially in the groups of restricted dispersal capacities (see Huemer 1998). Spot-like endemics also occur, however, in some genera of butterflies (e.g. *Erebia*) and moths (especially *Glacies, Charissa, Elophos*). On the other hand, species that were widely dispersed in the periglacial belt during the last glacial phases could acquire a transcontinental Eurasiatic or even Holarctic distribution, which has been recently dissected due to the postglacial re-invasion of arboreal zonobiomes, i.e. woodlands.

Numerous spot-like endemics occur in, e.g., the Alps, Carpathians, Balkans in genera with probably European-alpine origin (e.g. *Glacies*, see Povolny and Moucha 1955, 1958), or in which at least a European-alpine centre of diversification-speciation can be accepted [e.g. *Erebia*, with 13 species, endemic in the Alps, the geometrid genera *Entephria* Hübner (Aubert 1959), *Charissa, Elophos*; the Arctiid genus *Setina* Schrank].

On the other hand, some genera are underrepresented in the European high mountains compared with the East Palaearctic (and North America). Examples include some Arctiid (*Acerbia* Sotavalta, *Borearctia* Dubatolov, *Holarctia* Ferguson, *Pararctia* Sotavalta) and Noctuid (e.g. *Estimata* Kozhantshikow, subgenera of *Xestia* Hübner: *Anomogyna* Staudinger, *Pachnobia* Guenée, *Schoeyenia* Aurivillius; see Lafontaine et al. 1983, 1987) moth genera or the butterfly genera *Parnassius* Latreille, *Colias* Fabricius, *Oeneis* Hübner. In some Arctiidae and Noctuidae a northern (e.g. Beringian) origin is likely (Lafontaine and Wood 1988; Mikkola et al. 1991), whilst, in others, a xeromontane origin is more probable (see below).

A major part of the xeromontane fauna appears to be stationary with numerous relict-like species, especially in some core areas of Central and Inner Asia. The restricted areas of allopatric sister-species completely fulfil the criteria of the areas of endemism (sensu Harold and Moor 1994; Balletto 1995). Thus, these taxa can be regarded as monophyletic species groups, suitable for phylogenetic-biogeographical studies. Studies of this nature on the Noctuidae have been started recently (e.g. Lafontaine 1981, 1987, 1998; Mikkola et al. 1987; Varga 1989b, 1996, 1998; Varga et al. 1990; Ronkay and Ronkay 1995; Mikkola 1998; Ronkay and Varga 1999).

It has been proposed that the ancient groups of the Mediterranean xeromontane fauna were derived from two major sources. Partly, they originated from the primary bifurcation of the continental xeromontane faunal complex (Fig. 1 in Varga 1995b) and partly by the adaptation of diverse Mediterranean xerophilous arboreal groups to the arid conditions during the late Tertiary. This hypothesis is strongly supported by the macro-taxonomical duality of the Mediterranean xeromontane Noctuidae. Those genera which belong to the Noctuinae and have cutworm-type larvae (e.g. *Euxoa* Hübner, *Agrotis* Ochsenheimer, *Dichagyris* Lederer, *Chersotis* Boisduval, *Rhyacia* Hübner, *Standfussiana* Boursin) probably originated in the continental orobiomes. Their Mediterranean representatives often belong to different, divergently derived phyletic lines within these polytypic genera. Their Mediterranean-Anatolian taxa also display western and central Asiatic connections, and only a marginal speciation in the Mediterranean ranges.

Other genera were probably connected to xerophilous scrub formations (*Eugnorisma* Boursin, *Metagnorisma* Varga and Ronkay, *Auchmis* Hübner, *Lophoterges* Hampson, *Ostheldera* Nye, see: Ronkay and Varga 1993) or thorny cushion-scrub communities (*Xenophysa* Boursin, *Copiphana* Hampson) originally. Their Mediterranean and Anatolian taxa also display western and central asiatic connections, and only a marginal speciation in the Mediterranean ranges. Other Mediterranean xeromontane genera show an essentially autochtonous evolution which had been influenced by the upper Tertiary dry conditions in the Mediterranean basin. In such genera the Ponto-Mediterranean (incl. parts of Anatolia), the Atlanto-Mediterranean and Maghreb areas usually display a high level of species diversity, and also have some pairs or groups of vicariant species. Examples of such genera are *Calophasia* Stephens, *Ompalophana* Hampson, *Copiphana* Hampson, *Metopoceras* Guenée (subfamily *Cuculliinae*) and also some *Hadeninae* s.l., *Xylenini*-genera (*Leucochlaena* Hampson, *Aporophila* Guenée, *Antitype* Hübner, *Ammoco-*

nia Lederer, subgenera of *Mniotype* Franclemont, *Polymixis* Hübner, *Agrochola* Hübner, *Conistra* Hübner). This biogeographical group is equivalent to the Palaeomediterranean xeromontane faunal type of the ornithologists (Stegmann 1938) and to the ancient Mediterranean faunal type of the Russian biogeographical school (Kryzhanovsky 1965).

13.8 Conclusions

1. The European oreal fauna can be subdivided into two major types: alpine (*s. l.*) and xeromontane. The alpine type is characteristic of humid high mountains, and is closely connected with the Quaternary glaciations, resulting in a great number of arctic-alpine species. On the other hand, there are also numerous stenotopic species that survived the last glaciations in unglaciated parts of the south European high mountains.
2. Several types of endemism have been described in the European alpine fauna. Some Balkan endemics have vicarious sibling species in the Alps, others form disjunct subspecies. The distribution of the western sibling taxa is mostly restricted to the 'massifs de refuge' of the S-SW Alps. The disjunctions and vicariant patterns indicate inter- or preglacial connections and the position of some possible glacial refugia.
3. The xeromontane type prevails in the summer-arid Mediterranean high mountains and the arid high mountains of central Asia, less influenced by the Quaternary glaciations. The distribution of the two major faunal types overlap in, e.g., the Central and SW Alps and in some Balkan high mountains.
4. The xeromontane type forms two distinct groups, Mediterranean and continental central Asian, which are separated by the limit of the equinoctial type of precipitation. Core areas of endemic species of the Mediterranean xeromontane type are dispersed from the Atlas and Sierra Nevada to Daghestan, Transcaucasia, W-and N-Iran and Transcaspia (Kopet-dagh). Some Mediterranean xeromontane species also occur in the SW Alps which probably is the consequence of the refugium character of this region.
5. High values of faunal similarity have been found between some parts of the Alps and some Balkan high mountains because of the numerous common alpine and arctic-alpine species. On the other hand, the high mountains of the South Balkans display some similarity with the high mountains of Asia Minor.
6. Some genera of butterflies and moths (e.g. *Erebia, Entephria, Glacies, Charissa, Elophos, Setina*) are rich in European endemic species with probably European alpine origin, or at least with a European alpine centre of diversification-speciation. On the other hand, some genera are underrepresented in the European high mountains compared with the East Palaearc-

tic and North America, e.g. some Arctiid and Noctuid moth genera or the butterfly genera *Parnassius, Colias, Oeneis,* etc. In some Arctiidae and Noctuidae a Beringian origin is likely, whilst in others a xeromontane origin is more probable.

7. It has been proposed that the ancient groups of the Mediterranean xeromontane fauna were derived from two major sources. Partly, they originated from the primary bifurcation of the continental xeromontane faunal complex and partly by the adaptation of diverse Mediterranean xerophilous arboreal groups to the arid conditions during the late Tertiary. This hypothesis is strongly supported by the macro-taxonomical duality of the Mediterranean xeromontane Noctuidae.

Acknowledgements. This paper is dedicated to the late Dr. h.c. Karl Burmann and to Mr. Josef Wolfsberger, outstanding investigators of the Alpine Lepidoptera. I thank C. Chemini, L. Nagy, L. Ronkay and R. Vaisänen and two referees, whose comments largely improved the manuscript. I thank L. Nagy for linguistic revision of the text, and I.A. Rácz and P. Kozma for help with preparing the tables.

References

Abadjiev S (1992) Butterflies of Bulgaria I. Veren, Sofia
Abadjiev S (1993) Butterflies of Bulgaria II. Veren, Sofia
Abadjiev S (1995) Butterflies of Bulgaria III. Veren, Sofia
Agakhanjantz OE (1981) The arid high-mountains of the URSS. Mysl, Moscow (in Russian)
Agakhanjantz OE, Breckle S-W (1995) Origin and evolution of the mountain flora in Middle Asia and neighbouring mountain regions. In: Chapin FS, Körner C (eds) Arctic and alpine biodiversity. Springer, Berlin Heidelberg New York, pp 63–80
Alberti B (1969) Neue oder bemerkenswerte Lepidopteren-Formen aus dem Großen Kaukasus. Dtsch Ent Z NF 16:189–203
Alexinschi A (1960) Contributiuni la cunoaşterea faunei Macrolepidopterelor din Masivul Rodna, cu consideraţii sistematice, ecologice şi zoogeografice. Anal stiint Univ I Cuza Iaşi 6:729–754
Alexinschi A, König F (1963) Contributiuni la cunoaşterea faunei de Lepidoptere din Munţii -Lotru şi Parâng. Comun Zool 2:137–149
Ander K (1949) Die boreo-alpinen Orthopteren Europas. Opusc Entomol Lund 14:89–104
Asarjan GC, Gevorkjan MR, Miljanowski ES (1970) Materialen zur Zusammensetzung, Biologie und Ökologie der Eulen (Lepidoptera, Noctuidae) der Rasdan-Region, Armen. SSR. Arb Inst Pflanzenschutz Eriwan 1:5–44
Aubert JF (1959): Les Géométrides paléarctiques du genre Entephria Hb. Z Wiener Ent Ges 44:178–202
Balletto E (1983) Le comunitá di Lepidotteri ropaloceri come strumento per la clasificazione e l'analisi della qualitá degli alti pascoli montani. Atti XII Congr Nazion Ital Entomol Roma 1980:285–293

Balletto E (1995) Endemism, areas of endemism, biodiversity and butterfly conservation in the Euro-Mediterranean area. Boll Mus Reg Sci Nat Torino 13:445–491

Beshkov S (1995) Contribution to the knowledge of the Bulgarian lepidoptera fauna (Lepidoptera, Macrolepidoptera). Phegea 23:201–218

Beshkov S (1996) A new subspecies of Erebia cassioides (Reiner and Hohenwarth, 1792) from Bulgaria: *Erebia cassioides kinoshitai* ssp.n. (Lepidoptera, Nymphalidae: Satyrinae). Phegea 24:109–124

Buresch I, Arndt W (1926) Die Glazialrelicte stellenden Tierarten Bulgariens und Mazedoniens. Z Morphol Ökol Tiere 5:381–405

Buresch I, Tuleschkow K (1929–43) Die horizontale Verbreitung der Schmetterlinge in Bulgarien. Teile I–V. Mitt Naturwiss Inst Sofia 2:145–250, 3:145–248, 5:211–288, 8:289–347, 9:349–422, 16:79–188

Caradja A (1933) Gedanken über Herkunft und Evolution der europäischen Lepidopteren. Ent Rundsch 50:213–217, 236–240, 245–248

Caradja A (1934) Herkunft und Evolution der palaearktischen Lepidopterenfauna. Int Ent Z 28:217–224, 233–236, 261–264, 287–292, 361–366, 381–385

Choi S-W (1997) A phylogenetic study of the Cidariini from the Holarctic and the Indo-Australian areas (Lepidoptera, Geometridae, Larentiinae). Syst Entomol 22:287–312

Choi S-W (2000) Cladistic biogeography of the moth tribe Cidariini (Lepidoptera, Geometridae) in the Holarctic and Indo-Chinese regions. Biol J Linnean Soc 71:529–547

Coutsis J (1969) List of Grecian butterflies. Entomologist 102:264–268

Coutsis J (1972) List of Grecian butterflies. Additional records. Entomol Rec 84:145–151, 165–167

Coutsis J, Ghavalas N (1991) *Agriades pyrenaicus* (Boisduval, 1940) from N. Greece and notes on *Apatura metis* (Freyer, 1829) from NE. Greece (Lepidoptera, Lycaenidae, Nymphalidae). Phegea 19:133–138

Coutsis J, Dils J, Ghavalas N, van der Poorten D (1997) A new *Erebia* species for the Greek fauna (Lepidoptera, Nymphalidae, Satyrinae). Phegea 25:169–172

Cracraft J (1982) Geographic differentiation, cladistic and vicariance biogeography: reconstructing the tempo and mode of evolution. Am Zool 22:453–471

Cracraft J (1983) Cladistic analysis and vicariance biogeography. Am Sci 71:273–281

Cracraft J, Prum RO (1988) Patterns and processes of diversification: speciation and historical congruence in some Neotropical birds. Evolution 42:603–620

Cupedo F (1996) Die morphologische Gliederung des *Erebia melampus*-Komplexes, nebst Beschreibung zweier neuen Unterarten: *Erebia melampus semisudetica* ssp.n. und *Erebia sudetica belladonnae* ssp.n. (Lepidoptera: Satyridae). Nota Lepid 18:95–-125

Daniel F (1964) Die Lepidopterenfauna Jugoslavisch Mazedoniens. III. *Bombyces* and *Sphinges*. Prirodonaučen Muzej Skopje, Skopje

Dennis RLH (1993) Butterflies and climate change. Manchester University Press, Manchester

Dennis RLH, Williams WR, Shreeve TG (1991) A multivariate approach to the determination of faunal structures among European butterfly species (Lep.: Rhopalocera). Zool J Linn Soc 101:1–49

Diószeghy L (1929–30) Die Lepidopterenfauna des Retyezat-Gebirges I. Verh Mitt Siebenbürg Ver Naturwiss Herrmannstadt 79–80:188–289

Diószeghy L (1936) Die Lepidopterenfauna des Retyezat-Gebirges II. Verh Mitt Siebenbürg Ver Naturwiss Herrmannstadt 86:84–85

Drenovsky AK (1925) Die vertikale Verteilung der Lepidopteren in den Hochgebirgen Bulgariens. Dtsch Ent Z 1925:29–125

Drenovsky AK (1928) Die Lepidopterenfauna auf den Hochgebirgen Bulgariens. Sammelwerk Bulgar Akad Wissensch Sofia 23:1–120

Dufay C (1961) Lepidopteres I. Macrolepidopteres. Faune terrestre et l'eau douce des Pyrenées Orientales.

Enghoff H (1993) Phylogenetic biogeography of a Holarctic group: the julidan millipedes. Cladistic subordinatedness as an indicator of dispersal. J Biogeogr 20:525–536

Enghoff H (1995) Historical biogeography of the Holarctic: area relationships, ancestral areas and dispersal of non-marine animals. Cladistics 11:223–263

Fernandez-Rubio F (1991) Guia de mariposas diurnas de la peninsula Iberica, Baleares, Canarias, Azores y Madeira, vols I-II. Piramide, Madrid

Fibiger M (1993a) Noctuinae I. In: Noctuidae Europeae, vol 1. Entomological Press, Sorø

Fibiger M (1993b) Noctuinae II. In: Noctuidae Europeae, vol 2. Entomological Press, Sorø

Fibiger M (1997) Noctuinae III. In: Fibiger M (ed) Noctuidae Europeae, vol 3. Entomological Press, Sorø

Fibiger M, Hacker H (1990) Systematic list of the Noctuidae of Europe. Esperiana (Schwanfeld) 2:1–109

Freina JJ de, Witt TJ (1987) Die Bombyces und Sphinges der West-Palaearktis (Insecta, Lepidoptera). vol I, Ed. Forschung und Wissenschaft, München

Freina JJ de, Witt TJ (1990) Die *Bombyces* und *Sphinges* der West-Palaearktis (Insecta, Lepidoptera), vol II. Ed. Forschung und Wissenschaft, München

Goltz H (1935–36) Die Erebien Siebenbürgens. Verh Mitt Siebenbürg Ver Naturwiss Herrmannstadt 86:1–30

Gómez-Bustillo MR, Arroyo Varela M (1981) Catalogo sistematico de los lepidopteros Ibericos. INIA, Madrid

Hacker H (1989) Die Noctuidae Griechenlands. Mit einer Übersicht über die Fauna der Balkanraumes (Lepidoptera, Noctuidae). Herbipoliana (Würzburg) 2. U Eitschberger, Marktleuthen

Hacker H (1990) Die Noctuidae Vorderasiens (Lepidoptera). Systematische Liste mit einer Übersicht über die Verbreitung unter besonderer Berücksichtigung der Fauna der Türkei. Neue entomol Nachr 27:1–766

Hacker H, Ronkay L, Hreblay M (2002) Hadeninae I. In: Fibiger M (ed) Noctuidae Europeae, vol 4. Entomological Press, Sorø

Harold AS, Mooi RD (1994) Areas of endemism: definition and recognition criteria. Syst Biol 43:261–266

Hengeveld R (1990) Dynamic biogeography. Cambridge University Press, Cambridge

Henricksen HJ, Kreutzer IB (1982) The butterflies of Scandinavia in nature. Skandinavisk Bogforlag, Odense

Hesselbarth G, Oorschot H van, Wagener S (1996) Die Tagfalter der Türkei unter Berücksichtigung der angrenzenden Länder, vols I-III. Selbstverl S Wagener, Bocholt

Higgins LG, Riley ND (1975) A field guide to the butterflies of Britain and Europe, 3rd edn. Collins, London

Holdhaus K (1954) Die Spuren der Eiszeit in der Tierwelt Europas. Abh Zool Bot Ges Wien 18:1–493

Holdhaus K, Lindroth C (1939) Die europäischen Koleopteren mit boreo-alpiner Verbreitung. Ann Nat Hist Mus Wien 50:123–293

Horvat I (1962) Die Vegetation Südosteuropas im klimatischen und bodenkundlichen Zusammenhang. Mitt Österreich Geogr Ges 1:136–160

Horvat I, Glavać V, Ellenberg H (1974) Die Vegetation Südosteuropas. G Fischer, Stuttgart

Hruby K (1964) Prodromus Lepidopter Slovenska. Slovakian Academy of Sciences, Bratislava

Huemer P (1998) Endemische Schmetterlinge der Alpen – ein Überblick (Lepidoptera). Stapfia (Linz) 55:229–256.

Huemer P, Tarmann G (1993) Die Schmetterlinge Österreichs. Veröff Mus Ferdinandeum 73 Beil 5:1-224
Illies J (1967) Limnofauna Europaea. G Fischer, Stuttgart
Jaksic P (1988) Provisional distribution maps of the butterflies of Yugoslavia (Lepidoptera, Rhopalocera). Jugosl Entomol Drustvo Zagreb, Posebno izd. I.
Jong de R (1972) Systematics and geographic history of the genus *Pyrgus* in the Palaearctic region (Lepidoptera, Hesperiidae). Tijdschr Entomol 115:1-121
Kaisila J (1947) Die Makrolepidopterenfauna des Aunusgebietes. Soc Entomol Fennica, Helsinki
Kaisila J (1962) Immigration und Expansion der Lepidopteren in Finnland in den Jahren 1869-1960. Acta Entomol Fennica 18:1-452
Karsholt O, Razovski J (1996) The Lepidoptera of Europe. A distributional checklist. Apollo Books, Stenstrup
König F (1959) Răspânderea orizontală şi verticală a lepidopterelor din Retezat, Godeanu-Tarcu şi Pietrii-Petreanu. Stud Cerc Acad RPR (Timişoara) 6:126-139
König F (1975) Catalogul colectiei de Lepidoptere a Muzeului Banatului. Muzeului Banatului, Timişoara
König F (1982) Montane, subalpine, alpine und boreo-alpine Schmetterlingsarten aus den rumänischen Karpathen. Stud Comun (Reghin) 2:229-236
Koutroubas AG (1994) Erebia rhodopensis (Nicholl, 1900), espèce nouvelle pour la Grèce (Lep.: Nymphalidae, Satyrinae). Phegea 22:9-13
Koutsaftikis A (1974) Systematics, ecology and zoogeography of the Rhopalocera (Lepidoptera) in Greece (gr.). Museum Goulandris, Athens
Kryzhanovsky O (1965) Composition and origin of the terrestrial fauna of central Asia. Nauka, Moscow (in Russian)
Lafontaine JD (1981) Classification and phylogeny of the *Euxoa detersa*-group (Lepidoptera: Noctuidae). Quaest Entomol 17:1-120
Lafontaine JD (1987) Noctuoidea, Noctuidae (part Noctuinae I). In: Dominick RB (ed) The moths of America north of Mexico, 27.2. The Wedge Entomological Research Foundation, Washington
Lafontaine JD (1998) Noctuoidea, Noctuidae (part Noctuinae II). In: Dominick RB (ed) The moths of America north of Mexico, 27.3. The Wedge Entomological Research Foundation, Washington
Lafontaine JD, Wood DM (1988) A zoogeographic analysis of the Noctuidae (Lepidoptera) of Beringia and some inferences about past Beringian habitats. Mem Ent Soc Canada 144:109-123
Lafontaine JD, Mikkola K, Kononenko VS (1983) A revision of the genus *Xestia* subgen. Schoeyenia Aurivillius (Lepidoptera, Noctuidae) with descriptions of four new species and a new subspecies. Ent Scand 14:337-369
Lafontaine JD, Mikkola K, Kononenko VS (1987) A revision of the genus *Xestia* subgen. *Pachnobia* Guenée (Lepidoptera, Noctuidae) with descriptions of two new subspecies. Ent Scand 18:305-331
Lattin de G (1957) Die Ausbreitungszentren der holarktischen Landtierwelt. Verh Dtsch Zool Ges Hamburg (1956):380-410
Lattin de G (1967) Grundriß der Zoogeographie. G Fischer, Jena
Leraut P (1980, 1996) Liste systématique et synonymique des Lepidoptères de France, Belgique et Corse. Ed Soc Ent de France, Paris
Lhomme L (1923-1963) Catalogue des Lépidoptères de France, Belgique et Corse. Suppl Alexanor Bull Soc Ent Fr:1-334
Lorković Z (1952) Beiträge zum Studium der Semispezies. Spezifität von Erebia styrius Godt. and styx Frr. (Satyridae). Z Lepid (Krefeld) 2:159-176

Lorković Z (1953) Spezifische, semispezifische und rassische Differenzierung bei Erebia tyndarus Esp. (I-II.) Trav Inst Biol Exp Acad Youg 11–12:163–192, 193–224

Lorković Z (1957) Die Speziationsstufen in der Erebia tyndarus-Gruppe. Biol Glasnik (Zagreb) 10:61-104

Mack W (1985) Lepidoptera II., Rhopalocera, Hesperiidae, Bombyces, Sphinges, Noctuidae, Geometridae. In: Franz H (ed) Die Nordost-Alpen im Spiegel ihrer Landtierwelt. Wagner, Innsbruck

Maran J (1946) Le rôle importante de la variation géographique des insectes pour les questions zoogéographiques et évolutives. Sborn ent Odd nar Mus v Praze 23:23–128

Mariani M (1940–41) Fauna Lepidopterorum Italiae. Giorn Sci Nat Econ Palermo 42:81–235

Mikkola K (1998) Revision of the genus *Xylomoia* Staudinger (Lepidoptera: Noctuidae), with descriptions of two new species. Syst Ent 23:173–186

Mikkola K, Lafontaine JD, Grotenfelt P (1987) A revision of the holarctic *Chersotis andereggi* complex (Lep.: Noctuidae). Nota Lepid 10:140–157

Mikkola K, Lafontaine JD, Kononenko VS (1991) Zoogeography of the Holarctic species of Noctuidae (Lepidoptera): importance of Beringian refuge. Entomol Fenn 2:158–173

Naumann CM, Feist R, Richter G, Weber V (1984) Verbreitungsatlas der Gattung *Zygaena* Fabricius, 1775 (Lepidoptera: Zygaenidae). Thes Zool (Braunschweig) 5:1–47

Nekrutenko YP (1985) Bulavousye chesuekrylye Kryma (The butterflies of Krymaea). Naukova Dumka, Kiev

Nekrutenko YP (1990) Dnyevnye babochki Kavkaza (The butterflies of the Kaukasus). Naukova Dumka, Kiev

Nelson G, Platnick NI (1980) A vicariance approach to historical biogeography. BioScience 30:339–343

Niessolowsky W (1929) Motyle wiekse Tatr polskich. Pol Akad Umetn Prace Monog Fizjogr (Krakow) 5:1–88

Niessolowsky W (1936) Pieris napi L. subsp. bryoniae Ochs. Unter besonderer Berücksichtigung der Karpathen-Formen. Ann Mus Zool Pol 11:213–236

Niessolowsky W, Woytusiak RJ (1937) Über die Verbreitung der geographischen Formen von *Erebia manto* Esp. in den Karpaten. Bull Acad Pol Sci Ser B 11:111–126

Nordström F (1955) De fennoskandiska dagfjärilarnas utbredning. Lepidoptera Diurna. CWK Gleerup, Lund

Nordström F, Wahlgren E (1935–41) Svenska fjärilar. Stockholm

Nordström F, Opheim M, Sotavalta O (1961) De fennoskandiska svärmarnas och spinnarnas utbredning (Sphinges, Bombycomorpha). CWK Gleerup, Lund

Nordström F, Kaaber S, Opheim M, Sotavalta O (1969) De fennoskandiska och danska nattflynas utbredning (Noctuidae). CWK Gleerup, Lund

Oosterbroek P (1980) The western Palaearctic species of *Nephrosoma* Meigen, 1803 (Diptera, Tipulidae). Part 5. Phylogeny and biogeography. Beaufortia 29 (358):311–394

Oosterbroek P, Arntzen JW (1992) Area-cladograms of circum-Mediterranean taxa in relation to Mediterranean palaeogeography. J Biogeogr 19:3–20

Opheim M (1958) Catalogue of the Lepidoptera of Norway. Part I. Rhopalocera, Grypocera, Sphinges and Bombyces. Norsk lepidopterologisk selskap, Zoologisk Museum, Oslo

Opheim M (1972) Catalogue of the Lepidoptera of Norway. Part III. Geometrae, Arctiina, Zygaenina, Psychina and Jugatae. Norsk lepidopterologisk selskap, Zoologisk Museum, Oslo

Ozenda P (1988) Die Vegetation der Alpen im europäschen Gebirgsraum. G Fischer, Stuttgart

Pamperis L (1997) The butterflies of Greece. Bastas and Plessas, Athens

Pinker R (1968) Die Lepidopterenfauna Jugoslavisch Mazedoniens. III. Geometridae Prirodonauč. Muzej Skopje posebno izd 4:1–21

Popescu-Gorj A (1952) Revizuirea speciilor genului *Erebia* Dalm. din Carpaţii Româneşti (grupa epiphron). Anal Acad RPR Ser Geol Geogr Biol 4:161–175

Popescu-Gorj A (1962) Revision des espèces du genre Erebia Dalm. des Carpathes de la Roumanie (groupes pluto et tyndarus). Trav Hist Nat Mus Grigore Antipa 3:205–223

Popescu-Gorj A (1963) Genul Erebia Dalm. (Lepidoptera, Satyridae) şi răspânderea sa verticală în masivul Bucegi. Acad RPR Ocrotirea Nat 7:53–62

Popescu-Gorj A (1964) Catalogue de la collection de Lépidopteres "Prof. A. Ostrogovich" du Muséum d'Histoire Naturelle 'Grigore Antipa'. Muséum d'Histoire Naturelle 'Grigore Antipa', Bucharest

Popescu-Gorj A (1971) Ergebnisse der Albanien-Expedition 1961 des "Deutschen Entomologischen Institutes" 82. Beitr.: Lepidoptera, Satyridae I (genus *Erebia* Dalman). Beitr Ent 21 (3/6):509–516

Povolny D, Moucha J (1956) On the high mountain Geometridae of the genus Psodos Treitschke, 1828 (Lepidoptera, Geometridae). Acta Ent Mus Nat Pragae 30:140–179

Povolny D, Moucha J (1958) Kritischer Nachtrag zur Kenntnis der Taxonomie und Zoogeographie der Gattung Psolos Tr. (Lepidoptera, Geometridae). Acta Ent Mus Nat Pragae 32:181–190

Rákosy L (1992a) Tagfaltergemeinschaften des Retezat-Gebirges (Rhopalocera + Grypocera) (Karpaten, Rumänien). Nota Lepid Suppl 4:118–128

Rákosy L (1992b) Macrolepidoptere din Parcul naţional de Retezat. In: Popovici J (ed) Parcul National Retezat – Studii ecologice (Braşov), pp 254–280

Rákosy L (1995) Die Noctuidae Siebenbürgens. Nachr Ent Ver Apollo Frankfurt am Main Suppl 13:1–109

Rákosy L (ed) (1997) Entomofauna parcurilor nationale Retezat şi Valea Cernei. Soc lepidopt roman, Cluj-Napoca

Rákosy L (1998) Die endemischen Lepidopteren Rumäniens. Stapfia (Linz) 55:257–280

Rappaz R (1979) Les Papillons du Valais. Martigny

Rebel H (1903) Studien über die Lepidopterenfauna der Balkanländer I. Bulgarien und Ostrumelien. Ann Naturhist Hofmus Wien 18:11–346

Rebel H (1904) Studien über die Lepidopterenfauna der Balkanländer II. Bosnien und Herzegovina. Ann Naturhist Hofmus Wien 19:8–377

Rebel H (1911) Die Lepidopterenfauna von Herkulesbad und Orsova. Ann Naturhist Hofmus Wien 25:353–430

Rebel H (1932) Zur Frage der europäischen Faunenelemente. Ann Naturhist Mus Wien 46:49–55

Rebel H, Zerny H (1931) Die Lepidopterenfauna Albaniens. Denkschr Akad Wiss Wien 103:39–161

Reichl ER (1992) Verbreitungsatlas der Tierwelt Österreichs. Bd. 1, Lepidoptera – Diurna. Forschungsinst f Umweltinformatik, Linz

Ronkay G, Ronkay L (1994) Cuculliinae I, Noctuidae Europaeae, vol 6. Entomological Press, Sorø

Ronkay G, Ronkay L (1995) Cuculliinae II, Noctuidae Europaeae, vol 7. Entomological Press, Sorø

Ronkay L, Varga Z (1993) On the taxonomy of the genus *Ostheldera* Nye, 1975 (Lepidoptera, Noctuidae, Cuculliinae). Acta Zool Hung 40:157–170

Ronkay L, Varga Z (1999) Revision of the genus *Eugnorisma* Boursin, 1946, part V. New genera and species of the Eugnorisma genus-group from Pakistan and from China (Lepidoptera, Noctuidae). Acta Zool Hung 45:345–373

Ronkay L, Yela JL, Hreblay M (2001) Hadeninae II. In: Fibiger M (ed) Noctudidae Europeae, vol 5. Entomological Press, Sorø

Rosen D (1978) Vicariant patterns and historical explanations in biogeography. Syst Zool 27: 159–188

Schaider P, Jakšić P (1989) Die Tagfalter von Jugoslavisch-Mazedonien. Selbstverlag Schaider, München

Scheuringer E (1972) Die Macrolepidopteren-Fauna des Schnalstales (Vintschgau, Südtirol). Ed. Museo Tridentino di Scienze Naturali, Trento

Schweitzerischer Bund f Naturschutz (ed) (1987) Tagfalter und ihre Lebensräume. Arten, Gefährdung, Schutz. Schweiz und angrenzende Gebiete. Selbstverlag, Basel

Sijarić R (1971) Faunisticka istrazivanija Rhopalocera (Lepidoptera) na kompleksu hercegovackih visokih planina (Prenj, Cvrstnica i Cabulja). Glasnik Zemaljsk Muzeja (Sarajevo) 10:163–184

Sijarić R (1980) Fauna Lepidoptera Bosne i Hercegovine. Akad Nauka i Umjetnosti Bosne i Hercegovine Sarajevo odj prirodnih i matemat nauka Knjiga 47:83–98

Sijarić R, Lorković Z, Carnelutti J, Jakšić P (1984) Fauna Durmitora I. Rhopalocera (Lepidoptera). Crnogorska Akademija Nauka i Umjetnosti Titograd, Posebno izd 13: 95–184

Stegmann B (1938) Grundzüge der ornithogeographischen Gliederung des paläarktischen Gebietes. Fauna SSSR Nov Ser 19:1–158

Svendsen P, Fibiger M (eds) (1992) The distribution of European Macrolepidoptera. Noctuidae, vol 1, Noctuinae 1. European Faunistical Press, Sorø

Thurner J (1948) Die Schmetterlinge Kärntens und Osttirols. Carinthia 2:1–200

Thurner J (1964) Die Lepidopterenfauna Jugoslavisch Mazedoniens. I. Rhopalocera, Grypocera and Noctuidae. Prirodonauč. Mus Skopje, posebno Izd, pp 1–159

Tolman T, Lewington P (1997) Field guide of butterflies of Britain and Europe. Harper Collins, London

Varga Z (1975): Geographische Isolation und Subspeziation bei den Hochgebirgs-Lepidopteren der Balkanhalbinsel. Acta Entomol Jugosl 11:5–39

Varga Z (1976) Zoogeographische Gliederung der paläarktischen Orealfauna. Verh 6 Int Symp Entomofaun Mitteleuropa (Lunz a S., 1975). Junk, The Hague, pp 263–294

Varga Z (1977) Das Prinzip der areal-analytischen Methode in der Zoogeographie und die Faunenelemente-Einteilung der europäischen Tagschmetterlinge. Acta Biol Debrecina 14:223–285

Varga Z (1989a) The origin and division of the northern-montane disjunct areas in Palaearctic Lepidoptera: their importance in solving zoogeographical and evolutionary problems. Acta Biol Debrecina 21 (1988–89):91–116

Varga Z (1989b) Zweiter Beitrag zur Kenntnis der Gattung *Xenophysa* Boursin 1969 (Lep.: Noctuidae) mit der Beschreibung fünf neuer Arten. Z Arbeitsgem Österr Ent 41:1–18

Varga Z (1995a) Isolates of arctic-alpine Lepidoptera in South-eastern Europe. Proceedings of the 9th Congress of EIS, Helsinki (1993). Finnish Entomological Society, Helsinki, pp 140–151

Varga Z (1995b) Geographical patterns of biodiversity in the Palaearctic and in the Carpathian Basin. Acta Zool Hung 41:71–92

Varga Z (1996) Biogeography and evolution of the oreal Lepidoptera in the Palaearctic. Acta Zool Hung 42:289–330

Varga Z (1998) Sibling species and species groups in the genus *Chersotis* Boisduval 1840 (Lepidoptera, Noctuidae: Noctuinae) with description of two new species. Acta Zool Hung 44:341–372

Varga Z, Slivov AV (1977) Beitrag zur Kenntnis der Lepidopterenfauna der Hochgebirge in Bulgarien. Terrestrial fauna of Bulgaria, Bulg Acad Sci, Sofia, pp 167–190

Varga Z, Ronkay L, Yela JL (1990) Revision of the genus *Eugnorisma* Boursin 1946 (Lep.: Noctuidae), Part II. Taxonomic news, biogeographic and phylogenetic considerations

and descriptions of two new genera: *Ledereragrotis* and *Pseudohermonassa* (Lep.: Noctuidae). Acta Zool Hung 36:331-360
Verity R (1943-53) Farfalle diurne d'Italia, vols I-IV. Marzocco, Firenze
Vives Moreno A (1991) Catalogo sistematico y sinonimico de los lepidopteros de la peninsula Iberica y Baleares. Ministerio di Agricultura, Madrid
Walter H, Breckle S-W (1991) Ökologie der Erde. Ökologische Grundlagen in globaler Sicht. G Fischer, Stuttgart
Walter H, Straka H (1970) Arealkunde. Floristisch-historische Geobotanik. Eugen Ulmer, Stuttgart
Warnecke G (1959) Verzeichnis der boreoalpinen Lepidopteren. Z Wiener Ent Ges 44:17-26
Warren BCS (1936) Monograph of the genus *Erebia*. Natural History Museum, London
Wolfsberger J (1966a) Eine neue Art der Gattung Psodos Tr. vom Monte Baldo in Oberitalien (Lep., Geometridae). Mem Mus Civ Storia Nat 14:449-454
Wolfsberger J (1966b) Die Makrolepidopteren-Fauna des Gardaseegebietes. Ed Mus civ Storia Natur, Verona
Wolfsberger J (1971) Die Makrolepidopteren-Fauna des Monte Baldo in Oberitalien. Ed Mus Civ Storia Natur, Verona
Woytusiak RJ, Niessolowsky W (1947) Lepidoptera of the Central Caucasus, collected during the Polish Alpine Expedition in 1935 with ecological and zoogeographical remarks. I. Macrolepidoptera. Prace Mus Prirodonaucz (Krakow) 6:1-74

14 High Altitude Invertebrate Diversity in the Ural Mountains

Y.E. Mikhailov and V.N. Olschwang

14.1 Introduction

The Urals extend over 2000 km in a north-south direction from ca. 70°N to ca. 51°N with a width of up to 250 km. The five orographic regions, polar, subpolar, northern, central, and southern Urals, cross several zonobiomes from arctic tundra to steppe (Fig. 14.1). The highest peak, Mt. Narodnaya (1896 m), is in the subpolar Urals. The largest areas with alpine vegetation are on plateaux, formed after the Quaternary uplift (Parmuzin 1985). The uplift was least pronounced in the central Urals, where mountains are the lowest and do not reach the treeline, except for the Basegi Mts. The submeridional disposition of the ridges, which form a barrier and largely influence precipitation distribution, results in an asymmetrical landscape structure. This asymmetry is a result of uneven tectonic uplift and is intensified by today's climatic barrier situation, which modifies the structure of latitudinal zonation. The climatic differences between the western and eastern macroslopes result in contrasting vegetation, subalpine meadows and forests on the western macroclope, and grass-moss-lichen heath on the central ridges and plateaux and on the eastern macroslope. Alpine landscapes are found on every peak above 800 m in the north and above 1000–1100 m in the south (Makunina 1974). The western and eastern macroslopes are contrasting climatically because the western macroslope receives Atlantic air masses whilst the eastern one is often influenced by Arctic air (Parmuzin 1985). This climatic difference has been the reason for traditional faunistic comparisons between the two macroslopes.

The altitude effect of the Ural Highlands is equivalent to about 200 km in latitude, i.e. zonobiomes advance southwards by about 200 km from their northernmost limit in the lowlands (Fig. 14.1). This penetration of northerly zonobiomes at high altitudes into more southerly latitudinal zonobiomes and the presence of various transitional zones led Olschwang (1992) to describe the animal communities of the Urals of an extrazonal ecotone nature.

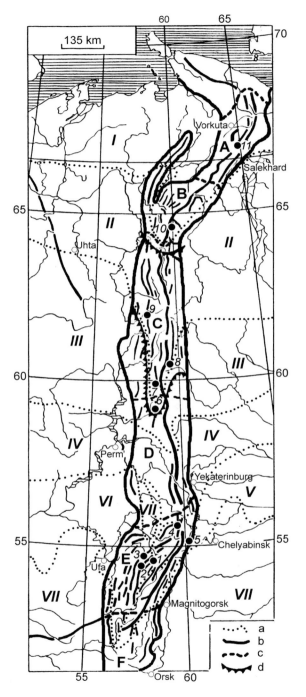

Fig. 14.1. The Urals and its environs (a–d in the right hand lower corner of the map).
a Zonobiomes or latidudinal vegetation zones from the Arctic Ocean to the South Urals (after Gorchakovskij 1975):
I tundra and forest-tundra zone; *II* northern taiga subzone; *III* middle taiga subzone; *IV* southern taiga subzone; *V* pine and birch forests subzone; *VI* broadleaf forest zone; *VII* forest-steppe zone;
b The extent of the Ural Mountains;
c Natural subregions (after Olenev 1965): *A* polar; *B* subpolar; *C* northern; *D* central; *E* southern; *F* southern Ural Plateau;
d Main zoogeographical borders (after Esyunin and Efimik 1994; Olschwang and Gorbunov 1996);
black circles regions which have been the main focus of zoological research in the Urals: *1* Iremel' massif; *2* Zigalga ridge; *3* Nurgush ridge; *4* Taganai massif; *5* Ilmensky Nature Reserve; *6* Basegi Mts.; *7* Konzhakovsky Kamen' massif; *8* Denezhkin Kamen' massif; *9* Pechero-Ilychsky Nature Reserve; *10* Neroika Mt.; *11* Pai-Er and Rai-Iz ridges

14.2 Species Diversity: State of Knowledge

14.2.1 Historical Origins of Zoological Research

Early zoological research in the Urals included the expeditions of Pallas (southern Urals, 1768–1773) Eversmann (southern Urals, 1830s), Zhuravsky (subpolar Urals, 1904–1908), and the Kuznetsov brothers (polar Urals, 1908). It was continued during the Complex Expeditions of the Academy of Sciences of the USSR (northern and southern Urals 1924–1938). E.G. Rodd's expeditions in 1896–1899 (southern Urals) yielded three new endemic species of chrysomelids, *Chrysolina roddi* Jcbs., *C. poretzkyi* Jcbs., *C. kuznetsovi* Jcbs. (Jacobson 1897), and one carabid *Nebria uralensis* Glas. Later, the interest of systematic entomologists in the Urals declined and it was another 100 years before the next endemic species [*Carabus karpinskii* Kryzh. et Matv. (from southern Urals, in 1993) and *Chrysolina hyperboreica* Mikh. (from northern Urals, in 2002)] were described.

Collections between 1970 and 2000 resulted in a large amount of material, which shows a diverse alpine fauna in the Urals. However, no comprehensive review has been made of the animals of the highlands of the Urals, vertebrates, or invertebrates to date. In the monograph of Mani (1968) on the alpine insects of the world, only few species are mentioned for the Urals. We present some summary information (Tables 14.1, 14.2) to illustrate the general nature of animal diversity in the Ural Highlands. The lists are far from complete, particularly for insects; however, they reflect today's knowledge and available data.

14.2.2 A Brief Inventory of the Fauna

Large herbivorous mammals are absent from the alpine zone, except for reindeer (*Rangifer tarandus* L.), which is common in the polar and subpolar Urals (Marvin 1977). However, it was a common species throughout the highlands including the southern Urals until the beginning of the twentieth century (Fridolin 1936; Bolshakov 1977). Predators such as red fox (*Vulpes vulpes* L.), brown bear (*Ursus arctos* L.) and common stoat (*Mustela erminea* L.) are rare but occur throughout the Ural Highlands; wolverine (*Gulo gulo* L.) occurs in the Polar Urals. Rodents, such as the northern redback vole (*Clethrionomys glareolus* Schreber), large-toothed redback vole (*C. rufocanus* Sundevall), mouse-hare (*Ochotona alpina hyperborea* Pall.) and several species of shrews (*Sorex* spp.), play an important role in alpine ecosystems (Bolshakov 1977; Marvin 1977; Bolshakov et al. 1996).

Table 14.1. Species diversity of animals in the alpine zone of the Urals

Animal group	Polar Urals	Subpolar Urals	Northern Urals	Southern Urals
Invertebrata				
Mollusca[a]	(+)	(+)	(+)	(+)
Annelida[a]				
Oligochaeta	1	1	1	3
Arthropoda[a]				
Pauropoda	–	–	–	–
Diplopoda	–	–	–	–
Chilopoda	(+)	(+)	(+)	(+)
Crustacea	–	–	–	–
Arachnida	?	66	168	229
Tardigrada	(+)	(+)	(+)	(+)
Insecta[b]	533	*291*	*285*	611
Ephemeroptera	?	?	?	–
Odonata	–	1	?	2
Plecoptera	–	?	?	1
Blattodea	–	–	–	1
Orthoptera	2	2	7	2
Homoptera	11	5	1	26
Psocoptera	(+)	(+)	–	(+)
Heteroptera	45	2	?	17
Thysanoptera	?	?	?	3
Mecoptera	–	–	–	3
Rhaphidioptera	–	–	–	1
Neuroptera	1	?	?	3
Megaloptera	2	?	?	?
Coleoptera	49	53	260	110
Strepsiptera	–	–	–	1
Trichoptera	?	(+)	(+)	(+)
Lepidoptera	160	47	35	265
Hymenoptera	81	5	14	27
Diptera	186	176	33	148
Vertebrata[c]				
Pisces	4	–	–	–
Amphibia	–	–	–	–
Reptilia	–	–	–	1
Aves	7	6	10	11
Mammalia	11	14	14	17

[a] Fridolin (1936); Voronova (1973); Perel (1979); Esyunin and Efimik (1997); Olschwang and Malozemov (1987); V.N. Olschwang (pers. comm.).
[b] Navas (1914); Becker et al. (1915); Esben-Petersen (1916); Karavaev (1916); Kirichenko (1916); Klapalek (1916); Martynov (1916); Friese (1918); Enslin (1919); Riedel (1919); Becker (1923); Fridolin (1936); Kharitonov (1976); Olschwang (1978); Olschwang and Malozemov (1987); Malozemov (1989, 1992); Nikolaeva (1990); Malozemov et al. (1997); W.J. Bicha (pers. comm.); A.I. Ermakov (unpubl.); Y.E. Mikhailov and V.N. Olschwang (unpubl. collection data).
[c] Fridolin (1936); Schwartz et al. (1951); Portenko (1958); Bolshakov (1977); Marvin (1977); Arkhipova and Yastrebov (1990); Bolshakov et al. (1996).
? no data (not investigated); – no records (area investigated taxon not found); (+) findings not determined to species level; figures in italics indicate obviously incomplete data.

Table 14.2. The species diversity of the Coleoptera and Lepidoptera, the two largest insect orders in the alpine zone of the Urals. Data from Kuznezov (1925); Fridolin (1936); Zaizev (1952); Sedykh (1974); Olschwang (1980); Grosser (1985); Olschwang and Malozemov (1987); Korobejnikov et al. (1990); Olschwang and Bogacheva (1990); Korobejnikov (1991); Putz (1992); Gorbunov and Olschwang (1993, 1997); Kozyrev et al. (1993); Korshunov and Gorbunov (1995); Keskula and Luig (1996); Keskula et al. (1996); Kozyrev (1997); Mikhailov (1997; unpubl. data); Tatarinov and Dolgin (1997); Bogacheva and Olschwang (1998); Ermakov (1998; unpubl. data); Nupponen et al. (2000); Y.E. Mikhailov and V.N. Olschwang (unpubl.)

	Polar Urals	Subpolar Urals	Northern Urals	Southern Urals
Coleoptera	49	53	260	110
Carabidae	18	42	43	31
Dytiscidae	1	–	3	–
Staphylinidae	5	–	59	6
Sylphidae	1	–	2	–
Leiodidae	1	–	7	1
Cantharidae	2	2	7	5
Malachiidae	–	–	–	1
Cleridae	–	–	–	1
Elateridae	1	1	9	3
Buprestidae	–	–	–	2
Byrrhidae	1	1	4	1
Hydrophilidae	1	–	4	–
Heteroceridae	–	–	–	1
Tenebrionidae	–	–	1	1
Coccinellidae	3	1	8	8
Oedemeridae	–	–	1	3
Mordellidae	–	–	2	1
Melandryidae	–	–	1	1
Bostrychidae	–	–	–	1
Anthicidae	–	–	1	2
Lagriidae	–	–	–	1
Cerambycidae	2	1	10	8
Scarabaeidae	–	–	7	1
Chrysomelidae	8	5	30	14
Curculionidae	5	?	26	15
Attelabidae	–	–	1	1
Anthribidae	–	–	–	1
Lepidoptera	160	47	35	265
Micropterygidae	1	–	–	1
Eriocraniidae	1	–	–	–
Hepialidae	1	–	–	–
Nepticulidae	–	–	–	–
Incurvariidae	2	–	–	2
Adelidae	–	–	–	2
Psychidae	–	–	–	2
Tineidae	1	–	–	3
Zygaenidae	–	–	–	1
Gracillariidae	1	–	–	1

Table 14.2. (*Continued*)

	Polar Urals	Subpolar Urals	Northern Urals	Southern Urals
Sesiidae	–	–	–	3
Glyphipterigidae	–	–	–	1
Yponomeutidae	–	–	–	2
Douglasiidae	1	–	–	–
Ethmiidae	–	–	–	1
Scythrididae	–	–	–	1
Plutellidae	5	–	–	–
Oecophoridae	–	–	–	2
Coleophoridae	10	–	–	2
Gelechiidae	7	–	–	7
Tortricidae	21	–	–	21
Pterophoridae	2	–	–	–
Phycitidae	2	–	–	1
Crambidae	2	–	–	3
Hesperiidae	4	2	1	3
Papilionidae	2	1	1	3
Pieridae	8	2	5	7
Lycaenidae	9	–	2	7
Nymphalidae	13	6	6	14
Satyridae	13	5	6	9
Lasiocampidae	–	–	–	3
Cymatophoridae	–	–	–	4
Drepanidae	1	–	–	4
Geometridae	23	19	6	57
Sphingidae	2	–	–	4
Notodontidae	–	–	–	12
Lymantriidae	1	–	–	1
Pantheidae	–	–	–	1
Noctuidae	16	12	8	53
Nolidae	–	–	–	1
Arctiidae	9	1	2	6

–, No record; figures in italics indicate obviously incomplete data.

Lakes in the alpine zone are known only from the polar Urals (Bolshoe Shchuchje lake is the deepest in the Urals with –136 m). Four species of fish are known from this region (Fridolin 1936; Arkhipova and Yastrebov 1990) including the Arctic grayling (*Thymallus arcticus* Pall.) and Chekanowsky's minnow (*Phoxinus czekanowskii* Dyb.).

The spider fauna of the Urals is considered to be one of the best studied in Russia (Esyunin and Efimik 1994), less so for the highlands and virtually unknown for the polar Urals (Esyunin and Efimik 1997).

Of the earthworms (Lumbricidae), *Eisenia nordenskioldi* Eisen inhabits the alpine zone throughout the Urals and, in the Southern Urals, another two species are found, the widespread *Dendrobaena octaedra* Savigny and the endemic *Allobophora* (*Svetlovia*) *diplotetratheca* Perel (Voronova 1973; Perel 1979).

For insects, the situation depends largely on the systematic group (Table 14.2). Near-complete species lists are available for carabid beetles (Korobejnikov et al. 1990; Korobejnikov 1991; Kozyrev 1997; Ermakov 1998; Kozyrev et al. 2000), diurnal butterflies (Gorbunov et al. 1992; Gorbunov and Olschwang 1993, 1997), and for some moths (Ahola et al. 1997). Chrysomelids have been revised by Mikhailov (1997, 2000) based on previous collections and recent expeditions by the one of authors (Y.M.) to the ranges of the southern Urals (Taganai, Iremel', Nurgush, Zigalga Mts.) and northern Urals (Konzhakovsky Kamen') in 1994–2000. Scorpion-flies (Mecoptera) have been investigated and found to date only in meadows at the treeline ecotone of the Taganai and Zigalga ridges (southern Urals); two and three species, respectively (W.J. Bicha, pers. comm.) and one species in the alpine zone of the Denezhkin Kamen' massif, northern Urals (A.I. Ermakov, pers. comm.). However, many other groups typical of alpine ecosystems have been under-investigated or still remain to be studied. Just how much an expert can add to the knowledge of poorly known groups is illustrated by the family Scythrididae (Lepidoptera) for which 14 new species have been described from the Southern Ural mountains in the last few years (Nupponen et al. 2000).

14.3 The Composition and Origin of the High Altitude Fauna

The botanical subdivision of high mountain systems into true alpine systems (e.g. Alps, Pyrenees, Caucasus; Ozenda 1985) and northern mountain systems, bearing tundra-like vegetation (e.g. Scandes and the Urals; Gorchakovskij 1975) is supported by the scarcity of European alpine Chrysomelids in the Urals. For example, the genus *Oreina* Chevr. and two subgenera of the genus *Chrysolina* Motsch. (*Colaphoptera* Motsch. and *Heliostola* Motsch.) are considered to be characteristic elements of alpine ecosystems. One widespread *Oreina* species, *O. luctuosa suffriani* Bontems (=*rugulosa* Suffr.), of the 22 known from the mountains of Europe is distributed from central Europe to the lowlands of the Middle and Southern Urals. *Colaphoptera* is restricted to Europe, largely to the Alps and Carpathians, but occurs in the Caucasus and the South Crimea also; *Heliostola* is known from the Alps, Carpathians and from the Altai Mountains, and both are absent from the Urals.

We follow here the classification proposed for carabids by Korobejnikov et al. (1990) to divide the high altitude fauna of the Urals into (1) widespread, (2) endemic and subendemic, (3) boreo-montane and arctic-alpine, and (4) steppe elements (Table 14.3). These groups reflect the history of the regional fauna.

Table 14.3. The number of species in the main groups of high altitude insect fauna of the Urals (with examples in chosen groups)

Insect group (Ural subregions)	Widespread (Number in subregions)	Endemic and sub-endemic species (Number in subregions)	Boreo-montane and arctic-alpine species (Number in subregions)	Interglacial remnants and tundra-steppe species (Number in subregions)	Source of data
Ground beetles (north/south)	21/20	2/4 *Carabus karpinskii* Kryzh. et Matv., *Nebria uralensis* Glas., *Pterostichus uralensis* Mts., *P. urengaicus* Jur.	12/6 *Carabus loshnikovi* F.-W., *C. odoratus* Motsch., *Curtonotus alpinus* F., *Nebria nivalis* Payk., *Pterostichus brevicornis* Kirby, *P. kaninensis* Pop., *P. kokeili* Popp.	1/0 *Carabus sibiricus* F.-W.	Korobejnikov et al. (1990); Ermakov (1998)
Leaf beetles (polar/north/south)	4/18/11	0/1(2)/1(2) *Chrysolina hyperboreica* Mikh., *C. roddi* Jcbs. Gebl., *C. orotshena* Jcbs.	3/3/2 *Chrysolina relucens* Ros., *C. septentrionalis* Men., *Cryptocephalus krutovskii*	2/0/1 *Cercyonops caraganae* Gebl., *Chrysolina exanthematica gemmifera* Motsch	Olschwang and Malozemov (1987); Mikhailov (1997, 2000); Bogacheva and Olschwang (1998)
Weevils (polar/north/south)	2/9/12	0/0/0	3/4/5 *Donus opanassenkoi* Leg., *D. lepidus* Cap, *Dorytomus winteri* Kor, *Lepyrus arcticus* Pk., *Otiorhynchus dubius* Str, *O. nodosus* Mull., *Prisistus olgae* Korot., *Trichalophus korotyaevi* Zher. et Naz.	–/–	Olshvang and Malozemov (1987); Olschwang and Bogacheva (1990); A.I. Ermakov (unpubl.); Y.E. Mikhailov (pers. comm.)

High Altitude Invertebrate Diversity in the Ural Mountains 267

Insect group (Ural subregions)	Widespread (Number in subregions)	Endemic and sub-endemic species (Number in subregions)	Boreo-montane and arctic-alpine species (Number in subregions)	Interglacial remnants and tundra-steppe species (Number in subregions)	Source of data
Butterflies (polar/south)	30/37	0/0	14/3 *Agriades glandon* (Prun.), *Boloria alaskensis* Holl., *Erebia disa* (Beck.), *E. rossi* Curt., *E. euryale* (Esp.), *Oeneis bore* (Sch.), *O. norna* (Beck.), *O. oene* (Boisd.), *O. semidea* (Say), *Parnassius phoebus* (F.)	0/3 *Coenonympha amaryllis* Stoll, *Synchloe callidice* Hb., *Tryphysa phryne* Pall.	Korobejnikov et al. (1990); Korshunov and Gorbunov (1995); Tatarinov and Dolgin (1999)
Tiger moths (polar/south)	0/8	1/0 *Arctia olschwangi* Dubat	5/1 *Acerbia alpina* (Quens.), *Dodia albertae* Dyar, *Grammia quenseli* (Pk.), *Holoarctia puengeleri* (B.-H.), *Pararctia lapponica* (Thnb.), *P. subnebulosa* (Dyar.)	–/–	Olschwang (1980, pers. comm.); Korobejnikov et al. (1990)
Crane flies (polar/ north/south)	2/2/2	0/0/0	3/2/6 *Tipula dulkeiti* Sav., *T. excisa* Sch., *T. kaisilai* Mannh., *T. subexcisa* Lund, *T. T. trispinosa* Lund, *T. tristriata* Lund.	0/0/1 *Nephrotoma stackelbergi* Sav.	Olschwang (1978); Korobeinikov et al. (1990)

Although the Urals did not become entirely glaciated in the Quaternary, they could not be the centre of origin of an ancient alpine fauna because, in the Cretaceous and the Palaeocene, the Urals had not reached treeline elevations. Today's relief dates from the upper Tertiary and Quaternary (Kryzhanovskij 1969), the low numbers of endemic species (Table 14.3) being proof of the comparatively young age of the fauna. All endemic and subendemic species date back to the first half of the Pleistocene or later (Korobejnikov et al. 1990). The essential factor in the originality of the alpine fauna of the Urals is that the last uplift of the highest ridges practically coincided with glaciation. This affected colonisation less rather than the later survival of some species. The vegetation of the periglacial zone was rather homogenous over large areas, spreading from East Siberia to the Urals (Panova 1990). This made it possible for Angarian (East Siberian) forms to establish in the Urals, and, following isolation, become endemic species of relict character (Korobejnikov et al. 1990).

The interrupted distribution of boreo-montane and arctic-alpine species is a result of the alternation of glacial and postglacial periods. The initial connection of tundra-bearing mountains (cf. mountain tundra of Gorchakovskij 1975 and Ozenda 1985) with zonal Arctic tundra was severed during the last periglacial period and populations in the Urals became isolated (Korobejnikov et al. 1990). Some boreo-montane species with much interrupted distribution may be treated as 'oreothetes' – relic species that survived climate changes, which caused the extinction of their relatives in other regions (Semenov-Tien-Shansky 1937). Examples are the disjunct distribution of weevils, such as *Dorytomus winteri* Kor. (Urals – Transbaikalia – Kamchatka), *Trichalophus korotyaevi* Zher. et Naz. (Urals – Central Yakutia) and *Prisistus olgae* Korot. (Urals – Vrangel Island).

Today's altitudinal vegetation zones in the southern Urals formed in the Boreal period of the Holocene (ca. 8000 years ago). In the montane zone, the periglacial forest-steppe vegetation was replaced by montane forest, forcing the steppe vegetation to the foothills and the tundra-like vegetation to highlands (Panova 1990). This influenced the group of tundra-steppe species (Table 14.3) whose populations in the alpine and steppe, respectively, have been isolated since.

The role of Angarian elements in the formation of the majority of the characteristic groups of alpine fauna in the Urals suggests that the highlands of the Urals are a relict part of the typical Angarian distribution area (Mikhailov 2000). This establishes a connection between West-Palaearctic and East-Palaearctic faunas. The border is considered to be in a meridional direction along the Enisei River. Recently, it has been shown that in fact it crosses the Urals (Olschwang and Gorbunov 1996). The actual line of equal proportion of western and eastern species passes along the southern border of the northern taiga (in Fennoscandia – near the Polar circle, in the Pechora basin – at 65°N). Then, in the northern Urals, this line descends to 59°N and crosses it according to the shift of the boundary between northern and middle taiga zonobiomes (Fig. 14.1).

The proportion of western and eastern species varies two-fold or more from 200–400 km from the described line. For example, around Ekaterinburg or Omsk, the number of western species is three-fold higher than that of the eastern ones, and, in the Bashkirsky Reserve, southern Urals, it is nearly ten-fold higher. In the northern Urals, the number of eastern species is twice as many as the western ones. In the polar Urals and central Altai (Katunskii Mts.) they prevail ten-fold.

14.4 Dynamics of Species Diversity: Long-Term Observations

New observations are common in the Urals; insect species new for the region and large increases in the numbers of formerly rare species are recorded. For example, *Chrysolina* (*Anopachys*) *relucens* Rosenh. had been considered to be an endemic to Tyrol (Warchalowski 1993) until its discovery near the White Sea, East Siberia and the Far East (Bienkowski 1997) and, more recently, in the northern and southern Urals (Mikhailov 2000).

There are indications that some species may have become extinct since the first half of the twentieth century. For example, the rare subendemic leaf-beetle *Chrysolina roddi* Jacobs. is known from few collections only from mountain steppe habitats in the Ilmensky Nature Reserve, southern Urals (1937 and 1958) and the Zhiguli Mountains near the river Volga (1935). All recent attempts to re-find it either in the Zhiguli Mts. (Pavlov 1988), in the Ilmeny Mts., or from the place where it was originally described by Jacobson (1897) have failed.

A constant increase in the number of boreal forest species widely distributed in the forest zone of the Palaearctics has been observed in the Polar Urals since 1976 (V.N. Olshvang, pers. observ.). This phenomenon is poorly investigated in the Urals; however, it has been well documented from Finland since 1869. In addition to a constant increase, peak periods coinciding with warm periods were identified in the 1910s, 1930s and 1960s (Kaisila 1962).

Whilst boreal butterflies were not found in the Polar Urals between 1970–1975, a marked presence was observed during the hot summers of 1976–1977 (Bogacheva and Olschwang 1978). In 1989–1992 a large number of *Aporia crataegi* L. were observed. In the valleys, nymphalid butterflies such as *Araschnia levana* L., *Argynnis paphia* L. and *A. aglaja* L. appeared. There are fluctuations in numbers and, in particularly favourable seasons (or the year after a hot summer), often numerous offspring of migrating species survive after successful hibernation. Such phenomena had been described (Zhuravsky 1909); however, these common boreal species were not reported later from the same sites (Fridolin 1936).

A migration in the opposite direction, i.e. penetration of tundra and alpine species to the south, may also take place on the odd occasion. Ecological pathways for such migrations are alpine vegetation and *Sphagnum* bogs, along which arctic species can migrate nearly to the central Urals. Few tundra butterfly species (family Satyridae) migrate ever. One of the few examples of a southward migration was the rapid and extensive expansion of the pierid butterfly *Synchloe callidice* Hb., a rare species known from the polar Urals. In 1990–1991, it was observed in all parts of the Urals, from the polar to the southern (V.N. Olshvang, pers. observ.).

14.5 Altitudinal Gradients in Species Diversity and Abundance of Soil Fauna

Voronin and Esyunin (1990) reported species richness and composition of carabids in altitudinal zones from forests to alpine vegetation in the Basegi Mts., Central Urals (without full species lists and relative figures). 'Spring' species group dominated in the alpine rather than in lower zones. This group of insects is active only until the snowbeds melt, as they need high humidity.

Average daily prey weight taken by the ant *Formica truncorum* L. was found to be five times lower in the alpine zone than in the subalpine in the Basegi Mts. Prey species numbers were also lower in the alpine zone than in the treeline ecotone. *F. truncorum* is thought to be restricted to the alpine because of competition from *F. lugubris* L. (Batalin and Gridina 1983).

Malozemov (1992) showed for muscid flies a disproportionate increase in the relative dominance of certain species with altitude. The dominant species reached their highest absolute density at the highest altitude. Species richness was not lower in the alpine zone than at lower altitudes because of species replacement.

The species richness of truly alpine diurnal butterflies was found to decrease from 49 species in the polar Urals to seven in the northern part of the northern Urals and five in the southern tip of the Northern Urals. Montane forest species along the same latitudinal gradient showed a relative constancy with 26, 30, and 36 species (Tatarinov and Dolgin 1997). The decrease in alpine butterflies was owing to the fact that the alpine zone is in direct contact with the tundra zonobiome in the northern part of the range (Polar and Subpolar Urals), where the butterfly fauna is enriched with circumpolar species. In addition, only few tundra species reach as far south as the northern Urals (Olschwang and Malozemov 1987).

Everywhere in the alpine zone of the Urals, soil invertebrates are concentrated in the short moss and grass vegetation and in the upper ca. 5-cm soil layer. Their biomass is dominated by earthworms: more than 90% in polar and subpolar Urals (4–7 g m^{-2}), 85–95% in northern Urals (6–30 g m^{-2}), and

Table 14.4. Abundance and biomass of soil Arthropods in the alpine zone of the Urals. Data from Olschwang (1981); Fileva (1983); Ermakov (1997)

	Polar and subpolar	Northern	Southern
Abundance (individuals/m^{-2})	150	72.4–395[a]	1500
Biomass (g/m^{-2})	0.4–0.5	0.5–2.0	Up to 6
Main components	Larvae of crane-flies	Larvae of crane flies, ground beetles and millipedes	Ensign cochineals, larvae of crane-flies

[a] On various ridges.

up to 80% in southern Urals (10–11 g m^{-2}). The rest of the biomass is unevenly distributed among various Arthropods and cocoons of earthworms (Olschwang 1981; Ermakov 1997).

Abundance and biomass of soil Arthropods in the alpine zone of the Urals is shown in Table 14.4. Olschwang (1981) found the same number of dominating species (10–15 species with abundance >1 individual m^{-2}) everywhere in the alpine zone of the Urals. The same dominant species occur with several times higher abundance and biomass in the southern Urals than in more northern regions.

14.6 Herbivore-Vegetation Interactions in the Alpine Zone

Bogacheva (1990) noticed that abundance of sucking insects (Homoptera: aphids and leaf hoppers) decreases along altitudinal gradients in the polar Urals similarly to that in the mountains of Norway (Hagvar 1976). Meanwhile, some species of insects (e.g. some saw-flies, miners, and leaf-beetles) are more abundant in the mountains (e.g. outbreaks of miners in the polar Urals, Bogacheva 1990, and in Norway, Koponen 1981, were recorded only in the mountains). The above species may cause a higher level of damage in the mountains than in neighbouring lowland areas. However, Bogacheva (1990) reported that dwarf-shrubs, which have their upper limit in the alpine zone in the Polar Urals, suffered less damage than at lower altitudes. Conversely, prostrate *Salix* species that do not grow at low altitudes are usually severely damaged by leaf-beetles. Adults and larvae of *Chrysomela lapponica* L. heavily feed on *Salix* species in the northern Urals and are usually very numerous. They usually leave only nerves and, on some plants, all the leaves may be damaged. *Gonioctena arctica* Mannh. (=*affinis* Gyll.), another boreo-montane species, has its relict populations in the Iremel massif, southern Urals, where it is abundant in the alpine zone and feeds on *Salix* species. In the Northern Urals (Konzhakovsky Kamen' massif), a whole assemblage of leaf-feeding insects was found on prostrate *Salix*, including leaf-beetles *Gonioctena pallida* L. and *Phratora polaris* Schnd., large weevil *Lepyrus arcticus* Pk., caterpillar *Cerura vinula* L, and various gall-makers (Y.E. Mikhailov, pers. observ.). The same genera (*Chrysomela, Gonioctena, Phratora*) of leaf-beetles were described from willow scrub in the Arctic (Medvedev 1996).

Most typical arctic and alpine Chrysomelids are wingless, feeding at night and keeping under cover of stones during daytime. Arctic species of leaf-beetles are typically polyphagous, feeding on various herbaceous plants including species of the Asteraceae, or oligophagous, feeding on Ranunculaceae (Medvedev 1996). Host plants have been identified so far for two alpine Chrysomelids in the Urals (Y.E. Mikhailov, pers. obs.). *Chrysolina septentrionalis* Men. feeds on *Lagotis uralensis* Schischk. (Scrophulariaceae) and *Chrysolina cf. subcostata* Gebl. on *Anemonastrum biarmiensis* (Juz.) Holub. (Ranunculaceae). Both plants are alpine endemics of the Urals (Gorchakovskij 1975), and both leaf-beetles may represent endemic subspecies although are of still unclear taxonomic rank (Mikhailov 1997).

14.7 Phytophagous Insects as Potential Bioindicators in Alpine Ecosystems

Plants are usually viewed as the prime indicator organisms for the biological monitoring of climate change in alpine habitats. However, other groups of organisms appear to respond more rapidly and more predictably. The herbivorous insects specifically associated with common and widely distributed boreal plants provide opportunities to develop long-term sampling networks. They appear to be more sensitive biosensors of climate change than their host plants (Hodkinson and Bird 1998). Traditionally, investigations of biodiversity have been limited to the species level. However, biodiversity manifests itself also at the population level and the most common example here is polymorphism, an ecologically selected variability to effectively exploit environmental heterogeneity. For example, various metallic colours and melanistic forms common in alpine populations of leaf-beetles have ecological importance, because they absorb solar radiation more efficiently for warmth and simultaneously decrease the penetration of ultraviolet radiation (Lopatin 1996). Therefore, colour polymorphism may serve well for monitoring populations under climate change. If species track environmental changes, it will be reflected in their population structure and, by using appropriate methods, it will be possible to follow such changes.

14.8 Conclusions

1. The fauna of the Urals is very heterogeneous showing strongest connection with the Altai Mts. but related to the arctic, boreal and alpine regions of northern and central Europe, Siberia and the Far East.
2. The composition of the fauna of the Urals differs from the faunas of adjacent lowlands and it is possible to consider it as an integral biogeographic unit (Baranchikov and Olschwang 1979). The fauna of high altitude mountain ranges appears rich; however, it has yet to be studied in detail.
3. In all vegetation types of the alpine zone of the Urals, the biomass of soil invertebrates is dominated by earthworms. The composition and abundance of the soil invertebrate fauna are similar in the northern, subpolar and polar Urals; however, it is much poorer in species and lower in biomass than in the southern Urals.
4. Two main borders subdivide the Urals into three zoogeographic regions based on the analysis of local insect and spider faunas. The first border is between the boreal (taiga) and hypoarctic zones and coincides with the suggested geobotanical border. The second border crosses the Northern Urals and coincides with the boundary between northern and middle taiga zonobiomes.
5. During the last 70 years the alpine zone in the southern Ural Mts. has reduced by 10–30 % (Chap. 26) and global warming is likely to affect all the invertebrate groups. Endemic and subendemic species together with boreal-montane and arctic-alpine species are likely to become rarer which will endanger some, already rare, species. Meanwhile, widespread species may increase in their abundance and numbers of species, further accelerating faunal changes.
6. Recognising plants as the basic indicator group of organisms for climate change monitoring in alpine ecosystems, we propose phytophagous insects as a group of sensitive indicator organisms. Several species of leaf-beetles are suitable because of their high abundance, comparatively convenient collecting and complex of research methods already elaborated for monitoring of their population dynamics.

Acknowledgements. We are much obliged to our colleague Alexander Ermakov (Institute of Plant and Animal Ecology RAS, Ekaterinburg), who kindly provided unpublished data, based on his own long-term explorations on insect diversity in the northern Urals (Denezhkin Kamen' massif).

References

Ahola M, Kaitila J-P, Nupponen K, Junnilainen J, Olschwang VN, Mikhailov YE (1997) Materialy k poznaniyu babochek Urala. Nauchnye resultaty russko-finskoi lepidopterologitcheskoi expeditsii na Yuzhnom Urale v 1996 [Contributions to Urals's butterflies and moths. Scientific results of the Russian-Finish Lepidopterological expedition to the southern Urals in 1996. Macrolepidoptera]. In: Olschwang VN, Bogacheva IA, Nikolajeva NV (eds) Uspekhi entomologii na Urale. Aerokosmoekologiya, Ekaterinburg, pp 98–104

Arkhipova NP, Yastrebov EV (1990) Kak byli otkryty Ural'skie gory [How the Ural Mountains were discovered]. Sredne-Uralskoje Publishers, Sverdlovsk

Baranchikov YN, Olschwang VN (1979) Zoogeograficheskij analiz fauny bulavousyh cheshuekrylyh Ural'skogo hrebta [A zoo-geographical analysis of the butterfly fauna of the Ural Mountains]. Zool Zh 57: 612–614 (English summary)

Batalin AV, Gridina TI (1983) Khishchnicheskaya deyatelnost' murav'ev roda Formica v raznyh poyasah hrebta Basegi [Predation activity of the ants from the genus *Formica* in various altitude zones of the Basegi Mts.]. In: Nikolaeva NV (ed) Dinamika chislennosti i rol' nasekomyh v biogeocenozah Urala. UB AS USSR, Sverdlovsk, pp 7–8

Becker T (1923) Arktische Ural-Dipteren. Wien Ent Zeit 40:111–115

Becker T, Dziedzicki H, Schabl J, Villeneuve J (1915) Diptera. Scientific results of Kuznezov brothers' expedition to Polar Urals. Zap Imp Akad Nauk Ser 8 6 (Separata):1–19

Bienkowski AO (1997) Some surprising discoveries of *Chrysolina relucens* (Coleoptera, Chrysomelidae) on the White Sea shore, in Siberia and in the Far East. Ent Fen 7:195–199

Bogacheva IA (1990) Vzaimootnosheniya nasekomyh-fitofagov i rastenij v ekosistemah Subarktiki [Relationships of phytophagous insects and plants in ecosystems of the subarctic]. UB AS USSR, Sverdlovsk

Bogacheva IA, Olschwang VN (1978) O proniknovenii nekotoryh yuzhnyh vidov nasekomyh v lesotundru [On the penetration of some southern species of insects to the forest-tundra]. In: Vigorov YL (ed) Fauna, ecologiya i izmenchivost' zhivotnyh. UNC AN SSSR, Sverdlovsk, pp 16–18

Bogacheva IA, Olschwang VN (1998) Listoedy (Coleoptera, Chrysomelidae) Priobskogo Severa [Leaf beetles (Coleoptera, Chrysomelidae) of the northern Ob area]. Entomol Obozr 77:775–786 (English summary)

Bolshakov VN (1977) Zveri Urala [The mammals of the Urals]. Sredne-Uralskoje Publishers, Sverdlovsk

Bolshakov VN, Vasiljev AG, Sharova LP (1996) Fauna i populyatsionnaya ekologiaya zemleroek Urala (Mammalia, Soricidae) [The shrew (Mammalia, Soricidae) fauna and their population ecology in Urals]. Yekaterinburg Publishers, Yekaterinburg

Chernov YI, Medvedev LN, Khruleva OA (1993) Zhuki-listoedy (Coleoptera, Chrysomelidae) v Arktike [Leaf beetles (Coleoptera, Chrysomelidae) in the Arctic]. Zool Zh 72:78–92. (English summary)

Enslin E (1919) Tenthredinidae. Scientific results of Kuznezov brothers' expedition to Polar Urals. Zap Imp Akad Nauk Ser 8 28 (Separata):1–10

Ermakov AI (1997) Kolichestvennyi sostav bespozvonochnyh mohovo-lishainikovogo yarusa v gornyh tundrah Severnogo Urala [Invertebrate composition of moss-lichen layer in the mountain tundras of the Northern Urals]. In: Olschwang VN, Bogacheva IA, Nikolajeva NV (eds) Uspekhi entomologii na Urale. Aerokosmoekologiya, Ekaterinburg, pp 130–134

Ermakov AI (1998) Ekologo-faunisticheskij obzor zhuzhelits (Coleoptera, Carabidae) gornyh tundr massiva Denezhkin Kamen' [Ecological-faunistic review of carabids

(Coleoptera, Carabidae) of mountain tundras of Denezhkin Kamen' massif]. In: Mikhailova IN, Golovachev IB (eds) Sovremennye problemy populyatsionnoi, istoricheskoi i prikladnoi ekologii. Yekaterinburg Publ, Yekaterinburg, pp 53-58

Esben-Petersen P (1916) Ephemeridae. Scientific results of the Kuznetzov brothers' expedition to Polar Urals. Zap Imp Akad Nauk Ser 8 28:1-10

Esyunin SL, Efimik VE (1994) Raznoobrazie fauny paukov Urala: geograficheskaya izmenchivost' [Diversity of the spider fauna of the Urals: geographic variability]. Usp Sovr Biol 114:415-427 (English summary)

Esyunin SL, Efimik VE (1997) Fauna paukov Urala: istoriya izucheniya i nekotorye itogi [The spider fauna of the Urals: the history of investigations and some results]. In: Olschwang VN, Bogacheva IA, Nikolajeva NV (eds) Uspekhi entomologii na Urale. Aerokosmoekologiya, Ekaterinburg, pp 118-121

Fileva ON (1983) K izucheniyu mezofauny gorno-tundrovyh gruppirovok Severnogo Urala [On the investigation of the mezofauna of mountain tundra communities in the Northern Urals]. In: Bogacheva IA (ed) Fauna i ekologiya nasekomyh Urala. UNC AN SSSR, Sverdlovsk, pp 55-56

Fridolin VY (1936) Fauna Severnogo Urala kak zoogeograficheskaya edinica i biocenoticheskoe celoe [The fauna of the Northern Urals as a zoogeographic unit and as a biocoenotic entity]. Tr Lednikovyh Exp 4:245-270

Friese H (1918) Über die Bienen (Apidae) der Russischen Polar-Expedition 1900-1903 und einiger anteren arktischen Ausbenten. Zap Imp Akad Nauk Ser 8 28 (Separata):1-5

Gorbunov PY, Olschwang VN (1993) Fauna dnevnyh babochek Uralskogo Zapolar'ya [The diurnal butterfly fauna of the Transpolar Ural regions]. In: Utochkin AS (ed) Fauna i ekologiya nasekomyh Urala. Perm University Press, Perm, pp 19-34

Gorbunov PY, Olschwang VN (1997) Itogi izucheniya fauny dnevnyh babochek (Lepidoptera, Rhopalocera) Yuzhnogo, Srednego i Severnogo Urala [The results of investigation of diurnal butterflies (Lepidoptera, Rhopalocera) of the southern, central and northern Urals]. In: Olschwang VN, Bogacheva IA, Nikolajeva NV (eds) Uspekhi entomologii na Urale. Aerokosmoekologiya, Ekaterinburg, pp 88-98

Gorbunov PY, Olschwang VN, Lagunov AV, Migranov MG, Gabidullin AS (1992) Dnevnye babochki Yuzhnogo Urala [The diurnal butterflies of the southern Urals]. URO RAN, Ekaterinburg

Gorchakovskij PL (1975) Rastitelnyi mir vysokogornogo Urala [The vegetation of the high mountains of the Urals]. Nauka, Novosibirsk (English summary)

Grosser N (1985) Versuch einer Darstellung der Arten-kombinationen der Lepidopteren in aus gewahlten Vegetationseinheiten Baschkiriens (UdSSR). Faun Abh Mus Tierk Dresden 12:93-105

Hagvar S (1976) Altitudinal zonation of the invertebrate fauna on branches of birch (*Betula pubescens* Ehrl.). Norw J Entomol 23:61-73

Hodkinson ID, Bird J (1998) Host-specific insect herbivores as sensors of climate change in arctic and alpine environments. Arctic Alpine Res 30:78-83

Jacobson GG (1897) Materialia ad cognitionem faunae Chrysomelidarum provinciae Orenburgensis. Horae Soc Ent Ros 30:429-437

Kaisila J (1962) Immigration und Expansion der Lepidopteren in Finnland in den Jahren. Acta Ent Fen 18:452

Karavaev V (1916) Formicidae. Nauchnye resultaty ekspeditsii bratjev Kuznezovyh na Polyarny Ural [Formicidae. Scientific results of Kuznetzov brothers' expedition to Polar Urals]. Zap Imp Akad Nauk Ser 8 28 (Separata):1-4

Keskula T, Luig J (1996) On butterflies (Lepidoptera, Rhopalocera) collected in Polar Urals in July 1992. Lepid Inf 10:40-50

Keskula T, Viidalepp J, Miller R (1996) Moths (Heteroptera and Microlepidoptera) collected in Polar Urals in July 1992. Lepid Inf 10:51-54

Kharitonov AY (1976) Fauna strekoz (Insecta, Odonata) Urala i vostochnogo Priuralja [The dragon fly fauna of the Urals and eastern Cis-Urals]. In: Zolotarenko GS (ed) Fauna gelmintov i chlenistonogih Sibiri. Nauka, Novosibirsk, pp 157–161

Kirichenko AN (1916) Poluzhestkokrylye (Heteroptera, Hemiptera). Nauchnye resultaty ekspeditsii bratjev Kuznezovyh na Polyarny Ural [Heteroptera, Hemiptera. Scientific results of Kuznetzov brothers' expedition to polar Urals]. Zap Imp Akad Nauk Ser 8 28 (Separata):1–6

Klapalek F (1916) Vesnyanki. Nauchnye resultaty ekspeditsii bratjev Kuznezovyh na Polyarny Ural [Plecoptera. Scientific results of Kuznetzov brothers' expedition to the Polar Urals]. Ibid:6–12

Koponen S (1981) Outbreaks of *Dineura virididorsata* and *Eriocrania* (Lepidoptera) on mountain birch in northernmost Norway. Notulae Ent 61:41–44

Korshunov YP, Gorbunov PY (1995) Dnevne babochki aziatskoi chasti Rossii [Diurnal butterflies of the Asian part of Russia]. Ural University Press, Yekaterinburg

Korobejnikov YI (1991) Zhuzhelitsy gornyh tundr Urala [Carabidae of the mountain tundras of the Urals. In: Olschwang VN (ed) Ekologicheskiye gruppirovki zhuzhelits (Coleoptera, Carabidae) v estestvennyh i antropogennyh landshaftah Ural. UB AS USSR, Sverdlovsk, pp 51–60

Korobejnikov YI, Olschwang VN, Erokhin NG (1990) Geograficheskij analiz entomofauny gornyh lesov Yuzhnogo Urala v svyazi s istoriei ego razvitiya [Geographical analysis of entomofauna of the mountain forests of the Southern Urals in the context of its history]. In: Smirnov NG (ed) Istoricheskaya ekologiya zhivotnyh gor Yuzhnogo Urala. UB AS USSR. Sverdlovsk, pp 45–67 (English summary in a separate edition: Yekaterinburg 1992)

Kozyrev AV (1997) Rezultaty izucheniya zhuzhelits (Coleoptera, Carabidae) Urala I sopredel'nyh raionov [The results of the investigation of carabid-beetles (Coleoptera, Carabidae) of the Urals and adjacent regions]. In: Olschwang VN, Bogacheva IA, Nikolajeva NV (eds) Uspekhi entomologii na Urale. Aerokosmoekologiya, Ekaterinburg, pp 44–50

Kozyrev AV, Zinovjev EV, Korobejnikov YI, Malozemov AY (1993) Fauna zhuzhelits (Coleoptera, Carabidae) Pripolyarnogo Urala [Fauna of ground beetles (Coleoptera, Carabidae) of the Subpolar Urals]. Manuscript deposited in Veras-Eko, 18.02.1993, 10–03 N 232, Minsk

Kozyrev AV, Kozminyh VO, Esyunin SL (2000) Sostav lokal'nyh faun zhuzhelits (Coleoptera, Carabidae) Urala i Priural'ya [The composition of the local faunas of carabids (Coleoptera, Carabidae) of the Urals and Pre-Urals]. Perm University Herald Ser Biol 2:165–215

Kryzhanovskij OL (1969) Eshche o sostave i proiskhozhdenii alpijskih faun zhestkokrylyh [A contribution to the comparison and origin of alpine faunas of Coleoptera]. Zool Zh 48:1156–1165

Kuznezov NJ (1925) Some new Eastern and American elements in the fauna Lepidoptera of Polar Europe. Dokl Akad Nauk SSSR Ser A 1:119–122

Lopatin IK (1996) High altitude fauna of the Chrysomelidae of Central Asia: Biology and biogeography. In: Jolivet PHA, Cox ML (eds) Chrysomelidae biology, vol 3: general studies. SPB Academic Publishing, Amsterdam, pp 3–12

Makunina AA (1974) Landshafty Urala [The landscapes of the Urals]. Moscow State University Publishers, Moscow

Malozemov AY (1989) K izucheniyu dvukrylyh v gorah Pripolyarnogo Urala [On the investigation of flies in Subpolar Urals Mountains]. In: Vorobeichik EL (ed) Aktual'nye problemy ekologii: ekologicheskie sistemy v estestvennyh i antropogennyh usloviyah sredy. UB AS USSR, Sverdlovsk, pp 61–62

Malozemov AY (1992) O faune i ecologii nastoyashchih mukh (Diptera, Muscidae) vostochnogo makrosklona gor Pripolyarnogo Urala [On the fauna and ecology of muscid flies (Diptera, Muscidae) of the Eastern macroslope of Subpolar Urals]. In: Nikolaeva NV (ed) Nasekomye v estestvennyh i antropogennykh biogeocenozah Urala. Nauka, Ekaterinburg, pp 91–93

Malozemov AY, Stepanov LN (1990) Vertikal'naya struktura soobshchestv dvukrylyh (Diptera) v gorah Uralskoi Subarktiki [Vertical structure of the Diptera communities in the mountains of the Subarctic Urals]. In: Negrobov OP (ed) Problemy kadastra, ekologii i ohrany zhivotnogo mira Rossii. Materials of All-Russian Conference. Voronezh University Press, Voronezh, pp 65–66

Malozemov AY, Grichanov IY, Ovsyannikov EI (1997) K ekologii muh semeistva Dolichopodidae (Diptera) Severnogo Urala [To the ecology of flies of the family Dolichopodidae (Diptera) of the Northern Urals]. In: Zaitsev AF (ed) Mesto i rol' dvukrylyh nasekomyh v ekosistemah. Nauka, St Petersburg, pp 78–79

Mani MS (1968) Ecology and biogeography of high altitude insects. JNV Publishers, The Hague

Martynov AV (1916) Trichoptera. Nauchnye resultaty ekspeditsii bratjev Kuznezovyh na Polyarny Ural [Trichoptera. Scientific results of Kuznezov brothers' expedition to Polar Urals]. Zap Imp Akad Nauk Ser 8 28 (Separata):1–10

Marvin MY (1977) Zonal'noe raspredelenie mlekopitayushchih na zapadnyh sklonah Pripolyarnogo Urala [Altitudinal zonation of mammals on the Western slopes of Subpolar Urals]. In: Pokrovsky AV (ed) Ekologiya, metody izucheniya i organizaciya ohrany mlekopitayushchih gornyh oblastei. UNC AN SSSR, Sverdlovsk, pp 93–94

Medvedev LN (1996) Leaf beetles in the Arctic. In: Jolivet PHA, Cox ML (eds) Chrysomelidae biology, vol. 3. General studies. SPB Academic Publishing, Amsterdam, pp 57–62

Mikhailov YE (1997) Listoedy (Coleoptera, Chrysomelidae) Urala: istoriya i perspektivy izucheniya [Leaf beetles of the Urals (Coleoptera, Chrysomelidae): history and prospects of investigations]. In: Olschwang VN, Bogacheva IA, Nikolajeva NV (eds) Uspekhi entomologii na Urale. Aerokosmoekologiya, Ekaterinburg, pp 68–75

Mikhailov YE (2000) New distributional records of Chrysomelidae from the Urals and Western Siberia [on some "less interesting" faunistic regions]. Faun Abh Mus Tierk Dresden 22:23–37

Navas L (1914) Neuroptera. Scientific results of Kuznezov brothers' expedition to Polar Urals]. Zap Imp Akad Nauk Ser 8 28 (Separata):1–3

Nikolaeva NV (1990) Fauna krovososushchih komarov Polyarnogo Urala [Fauna of blood-sucking mosquitoes of the Polar Urals]. In: Utochkin AS (ed) Fauna i ekologiya nasekomyh Urala. Perm University Press, Perm, pp 61–73

Nupponen K, Bengtsson BÅ, Kaitila J-P, Junnilainen J, Olschwang VN (2000) The scythrridid fauna of the southern Ural Mountains, with description of fourteen new species (Lepidoptera: Scythridae). Ent Fen 11:5–34

Olenev AM (1965) Ural i Novaya Zemlya [The Urals and Novaya Zemlya]. Mysl', Moscow

Olschwang VN (1978) Zametki o komarah-dolgonozhkah roda Tipula v zonal'nyh i gornyh tundrah Urala i poluostrova Yamal [On the crane-flies from the genus *Tipula* in the zonal and mountain tundras of the Urals and the Yamal peninsula]. In: Vigorov YL (ed) Fauna, ecologiya i izmenchivost' zhivotnyh. UNC AN SSSR, Sverdlovsk, pp 15–16

Olschwang VN (1980) Nasekomye Polyarnogo Urala i Priobskoi lesotundry [Insects of the Polar Urals and forest tundra of the Ob area]. In: Danilov NN (ed) Fauna i ekologiya nasekomyh Priobskogo Severa. UNC AN SSSR, Sverdlovsk, pp 3–37

Olschwang VN (1981) Pochvennye bespozvonochnye v gornyh tundrah Urala [Soil invertebrates in the mountain tundra of the Urals]. In: Dolin VG (ed) Problemy pochvennoi zoologii. Abstracts of the VIIth All Union Conference. Kiev, pp 154–155

Olschwang VN (1992) Soobshchestva nasekomyh Urala kak naselenie ecotona [Insect communities of the Urals as ecotone communities]. In: Nikolajeva NV (ed)

Nasekomye v estestvennyh i antropogennykh biogeocenozah Urala. Nauka, Ekaterinburg, pp 109–110

Olschwang VN, Bogacheva IA (1990) Zhuki-dolgonosiki (Coleoptera, Curculionidae) Priobskogo Severa [Weevils (Coleoptera, Curculionidae) of the Northern Ob area]. Ent Obozr 69:332–341 (English summary)

Olschwang VN, Gorbunov PY (1996) The balance between west and east species of butterflies (Lepidoptera, Diurna) in north Europe, Urals and west Siberia. In: Xth European Congress of Lepidopterology. Programme and Absracts. Universidad Autonoma de Madrid, Madrid, p 28

Olschwang VN, Malozemov AY (1987) Naselenie khortobiontnyh chlenistonogih v gornoi tundre Yuzhnogo Urala [Communities of chortobiont Arthropods in the mountain tundra of Southern Urals]. In: Malozyomov YA (ed) Fauna i ekologiya nasekomyh Urala. Urals University Press, Sverdlovsk, pp 121–130

Ozenda P (1985) La vegetation de la chaine alpine dans l'espace montagnard europeen. Masson, Paris

Panova NK (1990) Istoria razvitiya rastitel'nosti gornoi chasti Yuzhnogo Urala v pozdnem pleistocene i golocene [History of development of vegetation of the mountain part of the Southern Urals in the late Pleistocene and Holocene]. In: Smirnov NG (ed) Istoricheskaya ekologiya zhivotnyh gor Yuzhnogo Urala. UB AS USSR. Sverdlovsk, pp 144–159 (English summary in separate edition: Yekaterinburg 1992)

Parmuzin YP (1985) Taiga SSSR [The taiga forests of the USSR]. Mysl', Moscow

Pavlov SI (1988) Redkie vidy zhukov-listoedov Kuibyshevskoi oblasti i prichiny sokrashcheniya ih chislennosti [Rare species of leaf-beetles of Kuibyshev region and causes of decrease of their number]. In: Gorelov MS (ed) Ohrana zhivotnyh v Srednem Povolzhje. KGPI Press, Kuibyshev, pp 51–56

Perel TS (1979) Rasprostranenie i zakonomernosti raspredeleniya dozhdevyh chervei fauny SSSR [Distribution and spread peculiarities of earthworms of the USSR]. Nauka, Moscow

Portenko LA (1958) Ocherk fauny pozvonochnyh Urala [Essay on the vertebrate fauna of the Urals]. In: Vinogradov BS (ed) Zhivotnyi mir SSSR, vol 5. Gognyie oblasti Evropeiskoi chasti SSSR. Izdatel'stvo Akademii nauk, Moskva Leningrad, pp 576–571

Putz A (1992) *Curimopsis uralensis* n. sp. – eine neue Art der Gattung Curimopsis GANGLBAUER, 1902 aus dem Polarnyi Uralgebirge (Coleoptera, Syncaliptidae). Entomol Nachr Ber 36:250–242

Riedel M (1919) Nematocera polineura. Scientific results of Kuznezov brothers' expedition to Polar Ural]. Zap Imp Akad Nauk Ser 8 28 (Separata):1–3

Schwartz SS, Pavlinin VN, Danilov NN (1951) Zhivotnyi mir Urala (Nazemnye pozvonochnye) [Animals of the Urals (Terrestrial vertebrates)]. Sverdlovskoje Oblastnoje Gosizdatelstvo, Sverdlovsk

Sedykh KF (1974) Zhivotnyi mir Komi ASSR [Animals of Komi Republic]. Nauka, Syktyvkar

Semenov-Tian-Shanskij AP (1937) Osnovnye cherty razvitiya alpijskih faun [The main traits of the alpine fauna developments]. Izv AN SSSR Otd Mat i est nauk:1211–1222

Tatarinov AG, Dolgin MM (1997) Rezultaty izucheniya dnevnyh babochek (Lepidoptera, Diurna) v severnyh raionah Urala [The results of investigations of diurnal butterflies in the northern regions of the Urals]. In: Olschwang VN, Bogacheva IA, Nikolajeva NV (eds) Uspekhi entomologii na Urale. Aerokosmoekologiya, Ekaterinburg, pp 104–107

Tatarinov AG, Dolgin MM (1999) Bulavousye cheshuekrylye [Butterflies]. Fauna Evropeiskogo Severo-Vostoka Rossii 7 1:1–183

Voronin AG, Esyunin SL (1990) Kompleksy zhuzhelits (Coleoptera, Carabidae) gor Srednego Urala [Complexes of carabids (Coleoptera, Carabidae) of the mountains of Central Urals]. In: Medvedev GS (ed) Uspehi entomologii v SSSR: Zhestkokrylye nasekomye. Materials of the X Congress of the All-Union Entomological Society. Zoological Institute, Leningrad, pp 31–32

Voronova LD (1973) Pochvoobitayushchie bespozvonochnye bezlecnyh uchastkov Ilmenskogo zapovednika i hrebta Taganai (Yuzhnyi Ural) [Soil invertebrates of open (non-wooded) areas of Ilmensky Nature Reserve and the Taganai Range (Southern Urals). In: Gilyarov MS (ed) Ekologiya pochvennyh bespozvonochnyh. Nauka, Moscow, pp 84–94

Warchalowski A (1993) Chrysomelidae. Stonkowate (Insecta: Coleoptera) 3. Fauna Polski 15:1–279

Zaizev FA (1952) K faune vodnyh zhestkokrylyh Polyarnogo Urala i Karskoi tundry [On the fauna of water beetles of the Polar Urals and the Karskaya tundra]. Entomol Obozr 33:226–232 (English summary)

Zhuravsky AV (1909) Rezultaty issledovanij "Pripolyarnogo" Zapecher'ya v 1907 i 1908 godah. [The results of investigations of the "Subpolar" Trans-Pechora in the years 1907 and 1908]. Izv Imp Russ Geogr Ob-va 45:197–231

15 The Diversity of High Altitude Arachnids (Araneae, Opiliones, Pseudoscorpiones) in the Alps

K. Thaler

15.1 Introduction

Arachnids are a major component in alpine assemblages, both for their diversity and abundance (Schmölzer 1962), spiders making up between 9–51 % of the epigeic macrofauna (Meyer 1980; Matthey et al. 1981; Dethier 1984). Arachnids have been broadly discussed in coenological studies dealing with the Central Alps (Franz 1943; Schmölzer 1962; Christandl-Peskoller and Janetschek 1976), the nival zone (Bäbler 1910; Handschin 1919; Janetschek 1993), and glacier forelands (Janetschek 1949). However, few studies have focused on the spiders alone, e.g. in the Ötztal Alps (Palmgren 1973; Puntscher 1980), Dolomites (Zingerle 1999a,b), Bavarian Alps (Muster 2001) and at Grisons, Munt La Schera (Dethier 1983), since the work of Heller and Dalla Torre (1882) on the distribution of arachnids (and other arthropods) across life zones and along a longitudinal transect from Mt. Baldo in Trentino to the northern Calcareous Alps in Tyrol and a pioneering comment on high alpine spiders in Switzerland by Lessert (1909). This paper focuses mainly on the Eastern Alps because the spider fauna of the SW Alps is lesser known (Maurer and Thaler 1988).

15.2 High Altitude Arachnids of the Alps: a Systematic Overview

Of the minor arachnid orders of Europe, none is found at high altitudes. An exceptionally high elevation record of palpigrades is based on one specimen, which was collected in 1922 in a giant cave above the treeline in Tennengebirge near Salzburg at 1810 m and now is lost. Eleven of 13 records of this order in Austria come from sites below 1000 m (Christian 1998). *Euscorpius germanus* C.L. Koch can be found at the treeline ecotone on the southern macroslope of the Alps up to an elevation of 2200 m (Hadzi 1931; Kofler 1977; Braunwalder 1996; Komposch et al. 2001). Mites (Acari) occur abundantly and are diverse in high mountain environments; however, they are not covered here (for information on Acari in the Eastern Alps, see Irk 1941; Willmann 1951; Schatz 1979; Janetschek 1993; Schmölzer 1993, 1999).

15.2.1 Opiliones and Pseudoscorpiones

Several harvestmen of mid-Europe (Stipperger 1928; Komposch and Gruber 1999), even Trogulidae (*Trogulus*), occur across two to three life zones, with the treeline ecotone as an upper limit. Few species are restricted to the high Alps, (Table 15.1), all of which belong to the Palpatores. Species of the other two suborders inhabit soil and leaf litter in old forests. Two widely distributed central European species have a vertical distribution range of almost 3000 m. The others apparently prefer the high alpine (*Dicranopalpus gasteinensis*, *Megabunus armatus*, *Mitopus glacialis*) or subalpine zone and are endemic to either the whole of the Alps or to one of their regions (*Mitostoma alpinum*, *Ischyropsalis* and *Megabunus* spp.). These Phalangiidae overwinter as juveniles, with the exception of *Mitopus morio*, which is a univoltine species, overwintering in the egg stage and is most remarkable for its ability to maintain its cycle across the whole range of vertical distribution. *Mitopus glacialis* migrates in its search for prey even to glacier surfaces (Steinböck 1931). *Megabunus lesserti* shows regional parthenogenesis in most of its range between Lower Austria, Styria, N Tyrol and the Bavarian Alps. Males have only been collected so far from a peripheral population in an unglaciated area in Lower Austria (Komposch 1998; Muster 2000).

The lesser-known high alpine false scorpions are similar in that they have few species, belonging to two hygrophilic families, Neobisiidae and Chthoniidae, which are abundantly present in forest soil and litter. *Neobisium carcinoides* is euryzonal and is widely distributed in extra-Mediterranean Europe. The other species have restricted ranges, which cover either most of the Alps (*N. jugorum*) or some subregions only. *Chernes montigenus* (Simon) (Chernetidae), a W-Alps putative myrmecophilous species (Schmarda 1995), occurs at the treeline ecotone in the Ötztal Alps and in the Dolomites. For *N. jugorum* breeding nests, which are as well camouflaged as in its lowland congeners, were observed as early as July (Janetschek 1948). At the treeline the life cycle of *N. carcinoides* lasts longer than in its lowland populations and there is only one generation per year (Meyer et al. 1985).

Table 15.1. High mountain Opiliones and Pseudoscorpiones in the Eastern Alps

Taxon	Altitude range (m)	Range and habitat
Nemastomatidae (11)		
Mitostoma alpinum (Hadzi 1931)	(1300)–2000	SE Calcareous Alps, N Alps: Hochschwab 2150 m (unpubl.), Schneeberg
M. chrysomelas (Hermann 1804)	400–2700	Mt. Festkogel 3035 m, Ötztal Alps, in subnival vegetation
Ischyropsalididae (7)		
Ischyropsalis helvetica Roewer 1916	2300, 2570	Rätikon, Samnaun A. (Breuss 1993)
I. kollari C.L. Koch 1839	1100–2600	East Alps, an isolated lowland population in a cave near Graz, Styria, at 350 m
Phalangiidae (34)		
Dicranopalpus gasteinensis (Doleschall 1852)	(1300) 2000–3280	Ötztal Alps, in open scree from subalpine to nival zone; Alps (Carpathians?)
Megabunus armatus (Kulczynski 1887)	(1000) 1850–3025	Dolomites, Sass Rigais, on rock faces in SE Calcareous Alps
M. lesserti Schenkel 1927	1100–2100	N Calcareous Alps, on rock faces
Mitopus morio (Fabricius 1779)	400–2700 (3000)	in high abundance in alpine grassland, locally at subnival sites
M. glacialis (Heer 1845)	1940–3450	Alps, among boulders, at rocky ridges
Pseudoscorpiones		
Chthonius jugorum Beier 1952	2000–2200	Dolomites
Neobisium carcinoides (Hermann 1804)	400–3000	
N. dolomiticum Beier 1952	2000–2400	Dolomites, Lechtal Alps
N. jugorum (L. Koch 1873)	1700–3200 (3520)	Alps, from France to Stubai Mts.
N. noricum Beier 1939	2500	Tauern range, known from type locality only

Data for harvestmen from Martens (1978), Komposch and Gruber (1999), for false scorpions from Beier (1963) and K. Thaler (pers. observ.). For harvestmen families the approximate number of species in the Eastern Alps is added in parentheses.

15.2.2 Araneae

Spiders are a very diverse order and many inhabit the high mountain environments of the Alps. There are about 90 species that regularly inhabit the Austrian Alps (Table 15.2). This number is ca. 13% of all the species in the area. Most spiders (70%) of the alpine zone occur there preferentially, with the rest being euryzonal, occurring from the lowlands to the upper alpine and even subnival zones. There are some additional species that casually invade the lower alpine zone from lower regions. Other occasionals include some aeronautic Linyphiidae and 'money spiders' (e.g. genera *Dicymbium*, *Erigone*, *Oedothorax*). They will not be considered further. High mountain spiders in the Alps are epigeal hunters – diurnal (Lycosidae, Salticidae) or nocturnal (Gnaphosidae, Clubionidae), ambush predators (Thomisidae, Philodromidae), and spinners of webs, mostly sheet webs (Linyphiidae), and some Araneidae, Agelenidae and Theridiidae. All the above belong to families rich in species. No orthognathous, haplogyne or cribellate spider is confined to the alpine zone in central Europe.

The high mountain spider fauna of the Eastern Alps includes ca. 30% species endemic to the Alps and a further 17% that are confined to the alpine mountain system. This is in contrast with spider communities from high latitudes, which consist mainly of species with large ranges (Otto and Svensson 1982). Three species occur in various mountains of the Holarctic region, but not in the Arctic. Another 20% show N-S disjunction of range (Reinig 1965). These differ in vertical distribution and include boreo-alpine spiders with a centre of distribution around the treeline (e.g. *Micaria aenea* Thorell [Gnaphosidae] and *Xysticus obscurus* Collett [Thomisidae]), arctic-alpine species confined to the subnival and nival zones (e.g. *Diplocephalus rostratus* Schenkel, *Erigone tirolensis* L. Koch [Erigoninae]), and others intermediate between these chorological extremes. Systematic affinities of high alpine endemic species are poorly investigated. *Sitticus longipes* (Canestrini) [Salticidae] apparently is an old element with distant relatives in the Nearctic (Proszynski 1983). Four species in the *Lepthyphantes annulatus* group (Linyphiidae) in European mountains may have developed from an eastern stem species, which invaded Europe in a cryocratic period. Retreat of population fragments to oreal refugia in mesocratic conditions and isolation was followed by speciation (Thaler et al. 1994).

15.3 Regional Distribution

Arachnids are not evenly distributed within the high Alps. Spiders and some minor orders exhibit also the well-known difference between the Central Alps, which were strongly glaciated in the Pleistocene, and the periphery

Table 15.2. High mountain spiders (Araneae) in the Eastern Alps

Family	Alps endemic	Alpine system endemic	Arctic-alpine	Palaearctic/holarctic	Euryzonal	Euryzonal, disperse	North Tyrol total	Switzerland total
Theridiidae	–	1	–	–	1+5	–	50	61
Linyphiidae	15	6	12	–	10?	3	275	335
Araneidae	1	–	–	–	0+3	–	34	45
Lycosidae	4	4	1	–	1	–	58	70
Agelenidae	2	–	–	–	–	–	19	31
Hahniidae	–	–	–	–	1	1	7	7
Dictynidae	–	1	–	–	–	–	11	13
Clubionidae	–	–	–	–	0+2	–	23	30
Gnaphosidae	4	1	2	–	2+2	2	51	77
Philodromidae	–	–	–	1	1+2	–	15	25
Thomisidae	1	1	1	1	1+4	–	35	40
Salticidae	1	1	1	1	1+3	2	45	70
Other families	–	–	–	–	–	–	60	118
Total	28	15	17	3	18+21	8	683	922

Alps endemic, species endemic to the Alps; alpine system endemic, endemic to the alpine system; arctic-alpine, north-south disjunction of range (distribution boreoalpine, arctic-alpine); euryzonal, species occurring from the lowlands to the upper alpine and even subnival zones, + separates number of additional species (not named), which invade regionally the lower alpine zone; euryzonal disperse, lowland distribution very scattered; palaearctic/holarctic, species occurring in other mountain systems, but not in arctic region; North Tyrol total, species numbers in N Tyrol (Thaler 1998); Switzerland total (Maurer and Hänggi 1990). A full species list can be found at: http://www.stir.ac.uk/departments/naturalsciences/DBMS/nagy/alpine_biodiversity/.

(Holdhaus 1954; Janetschek 1956). In the Central Alps, survival in the harsh environment of nunataks may have been the case for a few nival species. Conditions in the peripheral 'massifs de refuge' were more favourable, as witnessed by the broader array of life forms present. It is safe to assume that between these two extremes existed intermediate sites. Peripheral refugia existed mainly along the southern, but also along the northern and eastern borders of the Alps. Even in the Karwendel range (Northern Calcareous Alps), two endemic species occur, *Cryphoeca l. nigerrima* Thaler (Agelenidae) and *Lepthyphantes severus* Thaler 1990 (Linyphiidae). Endemic arachnids with small ranges now exist near the periphery (Thaler 1976, 1990; Maurer 1982, 1992; Zingerle 1999a; Muster 2000). Endemic spiders of this type mainly belong to Dysderidae, Leptonetidae, Linyphiidae, Nesticidae, Agelenidae, and Amaurobiidae. High alpine species among them are hitherto known in Linyphiidae (Fig. 15.1) and Agelenidae only. They live in rock fissures and among boulders, an environment that has remained stable in the face of climatic changes. In the SW Alps, the fauna is richer. The famous monotypic lycosid, *Vesubia jugorum* (Simon), probably a palaeoendemic relict (Maurer and Thaler 1988), is confined to the Maritime Alps, where it occurs on screes between 2000 and 2800 m.

For some species, the route of post-glacial re-immigration is indicated by the shape and limits of range. Limits both at the western (e.g. *Cryphoeca nivalis* Schenkel [Agelenidae], *Drassodes heeri* (Pavesi) [Gnaphosidae], *Sitticus longipes* [Salticidae]) and at the eastern border of the Ötztal/Stubai Alps (*Diplocephalus rostratus* [Erigoninae], *Neobisium jugorum* ([Pseudosc.]) suggest severe impacts of glacial events in the Central Alps of Austria and recolonisation from the west (Thaler 1988). Western limits indicate re-immigration from the broad unglaciated area at the eastern border of the Alps, e.g. *Ischyropsalis kollari* and *Megabunus lesserti* (Opiliones, see Martens 1978). The linyphiid *Lepthyphantes variabilis* (Kulczynski) apparently is common on screes in Trentino and Tyrol, but absent from the Alps of Switzerland and eastern Austria and it is thought that it spread from its southern refugium directly northwards (Thaler 1982). For some other high alpine spiders, the causes of distribution are unclear. Why are two conspicuous boulder-dwelling species (*Aculepeira carbonaria* (L. Koch), Araneidae, *Theridion petraeum* L. Koch, Theridiidae) and some others (Thaler 1998, 1999) absent from the Calcareous Alps? Why do some 'boreo-alpine' spiders occur in the Alps at a few isolated sites only, e.g. *Gnaphosa lapponum* (L. Koch), *Micaria rossica* Thorell [Gnaphosidae], *Xysticus obscurus* [Thomisidae], *Pardosa cincta* (Kulczynski) [Lycosidae], *Pellenes lapponicus* Sundevall [Salticidae]? Why is the crab spider *X. bonneti* Denis confined to a few subnival sites at 3000 m in the Central Alps, while it is present also in the Pyrenees, the Pirin and in the Ural Mts? Apparently, some high alpine Lycosidae are not distributed continuously in the Eastern Alps (Thaler and Buchar 1996; Buchar and Thaler 1997; Muster 2002).

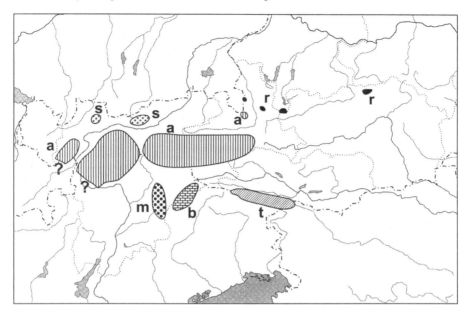

Fig. 15.1. Vicariance patterns of some high alpine *Lepthyphantes* species (Linyphiidae) confined to screes in the Eastern Alps (Thaler 1990; Muster 2000): *a L. armatus* Kulczynski 1905; *b L. brunneri* Thaler 1984; *m L. merretti* Millidge 1974; *r L. rupium* Thaler 1984; *s L. severus* Thaler 1990; *t L. triglavensis* Miller and Polenec 1975

15.4 Biology

15.4.1 Vertical Distribution

Spiders, similarly to other invertebrate groups, decrease in species number with an increase in altitude. The decrease is stepwise at the main ecotones (treeline and transition from the high alpine to the nival zone), and gradual within the alpine zone (Maurer and Hänggi 1990; Meyer and Thaler 1995). Figure 15.2 illustrates this using data from two transects, one at Mt. Festkogel, Ötztal Alps, where pitfalls were placed along an elevational gradient at distances of ca. 100 m (Puntscher 1980), and the other from the Tauern range, where a number of sites were investigated at three altitudes (Thaler 1989). The decrease in species numbers is paralleled by a decrease in activity-density values. Species numbers in alpine grassland in the Dolomites ranged between 20 and 30 (Zingerle 1999a,b), with a higher value at the treeline owing to microhabitat variation. The lowest values (ca. 5) were on nival screes. If three outlier sites are omitted, diversity values appear constant with a mean H' of 3.08 along the Festkogel transect despite the loss of species. Three spider commu-

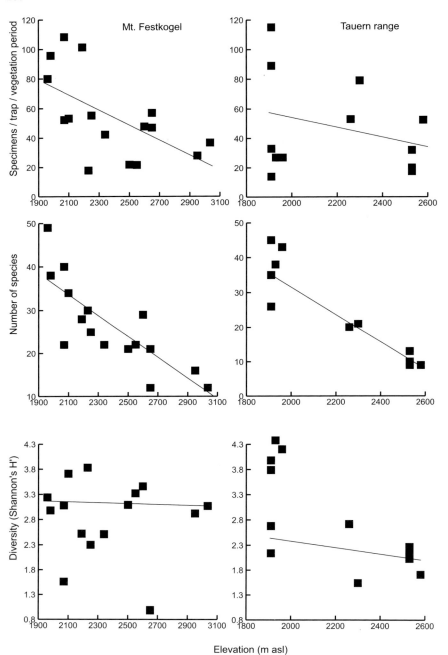

Fig. 15.2. Activity density (*top row*), species numbers (*middle row*) and diversity (*bottom row*) of spider communities along an elevational gradient in the Ötztal Alps (*left column*, Mt. Festkogel; Puntscher 1980) and in the Tauern range (*right column*; Thaler 1989). Trend lines: linear regression with values of <1.8 (Mt. Festkogel) and >3.3 (Tauern) excluded, see text

nities had low diversity at sites with late snow cover and on a ski piste. There is also some exchange of species along the gradient. The wolf spider *Pardosa oreophila* Simon, which is very common in the lower alpine zone, is replaced by *P. giebeli* (Pavesi) in the upper alpine zone. Mean diversity was H'=2.16 for the grassland sites in the Tauern range (four treeline ecotone sites with high diversity excepted). Treeline ecotone communities showed highest diversity (H'=3.4–3.9) in the Dolomites also (Zingerle 1999a,b). As vegetation cover changes with altitude it is difficult to separate the effects of altitude alone (Lomnicki 1963). The gradient of diversity was not investigated in the other main habitat types, screes and snowbeds. All these communities converge in the poor fauna of screes and boulder fields of the nival zone (Thaler 1981; Janetschek 1993).

In the Northern Calcareous Alps some species such as *Pardosa giebeli* [Lycosidae], *Gnaphosa petrobia* [Gnaphosidae], and *Hilaira montigena* L. Koch [Linyphiidae] are found at a lower elevation (ca. 2200 m) than in the Central Alps, where they occur in upper alpine and subnival communities.

15.4.2 Spatial Distribution

Arachnid species characteristic of the three main habitat types of the high alpine zone, grasslands, snowbeds and screes, were named first in the Tauern range (Franz 1943) and in the Zillertal Alps (Schmölzer 1962; Christandl-Peskoller and Janetschek 1976). Later investigations concentrated on grassland sites, with snowbeds considered by Puntscher (1980), Janetschek (1993) and screes by Zingerle (1999a,b). Arachnid diversity in snowbeds at 2650 m in the Ötztal Alps is very low; two erigonines dominate (*Erigone remota* L. Koch, *Mecynargus paetulus* (O.P.-Cambridge); Linyphiidae). Between nine and 30 species were found on screes between 1800–2200 m in the Dolomites, among them many influents from neighbouring habitats (Zingerle 1999a). Of course, the microhabitat requirements of arachnids are narrower, which is suggested by community tables and trellis diagrams in Puntscher (1980), Dethier (1983, 1984), Matthey et al. (1981), and Zingerle (1999a,b). For example, Czermak (1981) found that three Lycosidae in the Tauern range at 2300 m occupied different microhabitats: *Pardosa oreophila* was confined to the most sheltered sites, *P. mixta* (Kulczynski) to hollows with *Deschampsia* and *Luzula*, and *P. cincta* to ridges with *Carex curvula*.

In the nival zone, two main species assemblages exist. Exposed rock crevices of crests and boulders with lichens and mosses have a very poor fauna, the stenotopic species being *Diplocephalus rostratus* and *Lepthyphantes armatus* Kulczynski (Linyphiidae). The fauna is richer and intermediate to the subnival ecotone at favourable sites with cushion plants, where, for example, *Erigone tirolensis* and *Hilaira montigena* (Linyphiidae) occur (Thaler 1981; Janetschek 1993).

In glacier forelands, which have been ice-free since 1850, spider succession has been observed along a transect from the front of the glacier to climax grassland (Janetschek 1949; Kaufmann 2001). On bare ground close to the glacier, two euryzonal ripicolous species existed (*Janetschekia monodon* (O.P.-Cambridge) [Erigoninae], *Pardosa saturatior* [Lycosidae]), whereas nival Linyphiidae were absent.

15.4.3 Phenology, Winter Activity, Life Cycles

Epigeal locomotor activity of spiders in the alpine zone is mostly unimodal, with a peak in June/July soon after snow melt (Fig. 15.3). At the treeline ecotone, activity may be bimodal with a second peak in October, owing to the autumn activity of diplochronous Linyphiidae (Dethier 1984; Zingerle

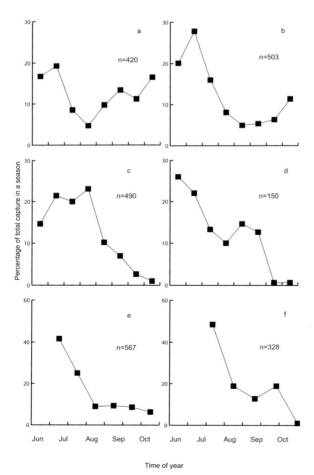

Fig. 15.3a–f. Annual locomotor activity of spiders at Mt. Patscherkofel in 1997, Innsbruck, 1990–2130 m (Ebenbichler 1998) and at Mt. Festkogel, Ötztal Alps, in 1999, 3035 m (K. Thaler, pers. observ.). **a, b** Sites at treeline ecotone ca. 2000 m; **c, d** grassland and *Loiseleuria* heath ca. 2130 m; **e** dwarf shrub 2130 m; **f** subnival zone 3035 m

1999b). In the Central Alps, high alpine spiders do not exhibit winter activity. Near the treeline at around 2000 m only few erratic specimens were caught in pitfall traps from November to March under snow and no peak of activity was found in any species (Puntscher 1980). In the subnival zone at 3000 m, in seven pitfall traps three spiders were trapped between 2 October and 1 November 1999, against 62 specimens between 30 August and 2 October (K. Thaler, pers. observ.). The pattern of total activity is a result of the specific activities involved (e.g. Puntscher 1980; Dethier 1983). Only two out of the five phenological types distinguished by Schaefer (1976) for the spiders of central Europe are present in the alpine zone, i.e. stenochronous species with a short adult life, maturing in early summer (most non-linyphiids), and diplochronous species with an extended adult period, maturing in late summer/autumn (most Linyphiidae, *Cryphoeca* [Agelenidae]). At the treeline some other stenochronous Linyphiidae can be found, maturing in autumn [*Bolyphantes luteolus* (Blackwall), *Centromerus pabulator* (O.P.-Cambridge)] and even in winter [*Macrargus carpenteri* (O.P.-Cambridge)]. Remarkably, epigeal *Pachygnatha* (Tetragnathidae) do not occur in alpine grasslands. Their absence is probably explained by a short growing season, which would limit the completion of their life cycles. All three species belonging to this genus in low regions are annual, with long-living adults and a short period of development in late spring/summer (Toft 1976). As can be inferred from pitfall catches, high alpine Gnaphosidae and Lycosidae overwinter as inadults at least twice, albeit exact data are few. According to Czermak (1981), the postembryonic development of *Pardosa mixta* (Lycosidae) lasts 3 years in grassland at 2400 m as opposed to 2 years at the treeline at 1900 m.

15.4.4 Biotic Interactions

In the high alpine environment, wind-blown allochthonous dipterans and aphids form an important additional food supply (Liston and Leslie 1982; Dethier 1984; Heiniger 1989). In Bernese Oberland, Switzerland, an average biomass of 310 mg (dry wt.) per week was deposited on a 1000 m^2 snowfield at 2320 m from March to the end of July. Specialist predators do not occur there. The highest elevation records for Mimetidae, which exclusively feed on web spiders, ant-feeding Gnaphosidae (*Callilepis*) and *Trogulus* (Opiliones) feeding on snails are from the treeline ecotone. Biotic interactions are scarce in an environment controlled by physical factors. In ant nests around the treeline, under stones and in anthills, four myrmecophilic spider genera occur, *Diastanillus*, *Evansia* (Erigoninae), *Syedra* (Linyphiinae), *Mastigusa* (Agelenidae), together with the false scorpion *Chernes montigenus*. Arachnid parasites and parasitoids have not yet been properly investigated in the high Alps. A nematode has been reported from 1750 m from the phalangiid *Mitopus morio* (Mermithida; Stipperger 1928) and at 3000 m from the abdomen of a subadult *Tiso aestivus*

(L. Koch) (Erigoninae; K. Thaler, pers. observ.). Acroceridae (Diptera), which exclusively develop in spiders, are present in the alpine zone, together with parasitic Hymenoptera (Terebrantes) and mites (Irk 1941).

15.4.5 Various Aspects

Most Lycosidae (Flatz 1987) and Salticidae are diurnal. Nocturnal activity may exist among alpine arachnids, as was demonstrated for the alpine zone of Colorado by pitfall trapping from 3600–4270 m (Schmoller 1971) and on Mount Rainier (Mann et al. 1980). Spiders may exhibit aeronautic behaviour in the Alps similarly to that in the Arctic (Braendegaard 1938). The ability for dispersal and colonisation of suitable habitats probably is an essential ecological trait for nival species. Parthenogenesis is known in one phalangiid only, *Megabunus lesserti* (see Muster 2000). Physiological properties have not been investigated in any euryzonal arachnid. Likewise, studies are lacking on cold tolerance, thermal melanism (Handschin 1919) or the function of pigment as a protection against radiation in high alpine arachnids. For an overview of adaptations of terrestrial arthropods to the alpine environment, see Somme (1989) and Somme and Block (1991).

15.5 Conclusions

1. The dominant arachnid orders in the Austrian Alps are spiders and mites (which are not considered here), both for their diversity and abundance. About 90 spider species (ca. 13 % of the species total in the area), nine harvestmen and five false scorpion species regularly inhabit the alpine zone. The most species-rich family is the Linyphiidae, followed by Lycosidae and Gnaphosidae; one or two species represent each of the Thomisidae, Salticidae, Araneidae, Theridiidae, Clubionidae, 'Agelenidae'. Endemism is considerable despite the impact of Pleistocene glaciations. As information about inventories and distribution of alpine species is still insufficient, further field work (which is certain to identify new species) is needed to clarify geographic patterns.
2. Among alpine arachnids, there are isolated species (e.g. *Sitticus longipes* [Salticidae], *Lepthyphantes armatus* [Linyphiidae]), others show vicariance patterns (e.g. *Cryphoeca* ['Agelenidae'], some species in the group of *L. mughi*), still others have a north-south disjunction of the range. There are a few euryzonal species with wide ranges, and others, with highly fragmented ranges, cover various mountain systems. Phylogenetic and phylogeographic studies are needed to understand the ways of evolution leading to these alpine species.

3. The biology and physiology of high altitude arachnids have been poorly investigated. Apparently, most alpine spiders are either stenochronous (Lycosidae), maturing soon after snow melt, or diplochronous, maturing in late summer/autumn, both with prolonged post-embryonic development. They do not show narrow biotic interactions, albeit at the treeline ecotone where some feeding specialists (on other spiders, ants, snails) and myrmecophilic species are present.

Acknowledgements. I am grateful to R. Maurer, E. Meyer, C. Muster, V. Zingerle, and in particular to Barbara Knoflach for information, co-operation and support. Sincere thanks are due to L. Nagy for linguistic improvement.

References

Bäbler E (1910) Die wirbellose, terrestrische Fauna der nivalen Region (Ein Beitrag zur Zoogeographie der Wirbellosen). Rev Suisse Zool 18:761–915 Pl 6
Beier M (1963) Ordnung Pseudoscorpionidea (Afterskorpione). Akademie-Verlag, Berlin
Braendegaard J (1938) Aeronautic spiders in the Arctic. Medd Gronland 119(5):1–9
Braunwalder ME (1996) (Un)heimliche Bewohner in den Bündner Südtälern. Terra Grischuna (Chur) 55(3):33–35
Breuss W (1993) Zum Vorkommen von *Ischyropsalis helvetica* Roewer in Graubünden und in Nordtirol (Samnaun-Gruppe) (Arachnida, Opiliones, Ischyropsalididae). Ber Nat Med Ver Innsbruck 80:251–255
Buchar J, Thaler K (1997) Die Wolfspinnen von Österreich 4 (Schluß): Gattung *Pardosa* max. p. (Arachnida, Araneae: Lycosidae) - Faunistisch-tiergeographische Übersicht. Carinthia II 187/107:515–539
Christandl-Peskoller H, Janetschek H (1976) Zur Faunistik und Zoozönotik der südlichen Zillertaler Hochalpen. Mit besonderer Berücksichtigung der Makrofauna. Veröff Univ Innsbruck 101 Alpin Biol Stud 7:1–134
Christian E (1998) *Eukoenenia austriaca* from the catacombs of St. Stephen's Cathedral in the centre of Vienna and the distribution of palpigrades in Austria (Arachnida: Palpigradida: Eukoeneniidae). Senckenbergiana Biol 77:241–245
Czermak B (1981) Autökologie und Populationsdynamik hochalpiner Araneen unter besonderer Berücksichtigung von Verteilung, Individuendichte und Biomasse in Grasheidebiotopen. Veröff Österr MaB Hochgebirgsprogramm Hohe Tauern 4:101–151
Dethier M (1983) Araignées et Opilions d'une pelouse alpine au Parc national suisse (Arachnoida: Opiliones, Aranei). Ber Nat Med Ver Innsbruck 70:67–91
Dethier M (1984) Etude des communautés d'arthropodes d'une pelouse alpine au Parc national suisse. Mitt Schweiz Entom Ges 57:317–334
Ebenbichler G (1998) Die epigäischen Spinnen des Patscherkofel bei Innsbruck (Waldgrenze und alpine Stufe). Diploma Thesis, University of Innsbruck
Flatz U (1987) Zur Tagesrhythmik epigäischer Webspinnen (Arachnida, Aranei) einer mesophilen Wiese des Innsbrucker Mittelgebirges (Rinn, 900 m, Nordtirol, Österreich). Ber Nat Med Ver Innsbruck 74:159–168

Franz H (1943) Die Landtierwelt der mittleren Hohen Tauern. Ein Beitrag zur tiergeographischen und -soziologischen Erforschung der Alpen. Denkschr Akad Wiss Wien Math Naturwiss Kl 107:1–552 Taf 1–14 Karte 1–11

Hadzi J (1931) Der Artbildungsprozess in der Gattung *Euscorpius* Thor. Arch Zool It 16:356–362

Handschin E (1919) Beiträge zur Kenntnis der wirbellosen terrestrischen Nivalfauna der schweizerischen Hochgebirge. Lüdin and Co, Liestal

Heiniger PH (1989) Arthropoden auf Schneefeldern und in schneefreien Habitaten im Jungfraugebiet (Berner Oberland, Schweiz). Mitt Schweiz Entom Ges 62:375–386

Heller C, Dalla Torre C (1882) Über die Verbreitung der Thierwelt im Tiroler Hochgebirge. II. Abtheilung. Sitz Ber Akad Wiss Wien (I) 86:8–53

Holdhaus K (1954) Die Spuren der Eiszeit in der Tierwelt Europas. Abh Zool Bot Ges Wien 18:1–493 Taf 1–52

Irk V (1941) Die terricolen Acari der Ötztaler und Stubaier Hochalpen. Veröff Mus Ferdinandeum (Innsbruck) 19:145–189

Janetschek H (1948) Zur Brutbiologie von *Neobisium jugorum* (L. Koch) (Arachnoidea, Pseudoscorpiones). Ann Naturhist Mus Wien 56:309–316

Janetschek H (1949) Tierische Successionen auf hochalpinem Neuland. Schlern-Schriften (Innsbruck) 67:1–215 Taf 1–7

Janetschek H (1956) Das Problem der inneralpinen Eiszeitüberdauerung durch Tiere (Ein Beitrag zur Geschichte der Nivalfauna). Österr Zool Z 6:421–506

Janetschek H (1993) Über Wirbellosen-Faunationen in Hochlagen der Zillertaler Alpen. Ber Nat Med Ver Innsbruck 80:121–165

Kaufmann R (2001) Invertebrate succession on an alpine glacier foreland. Ecology 82:2261–2278

Körner C (1999) Alpine plant life. Functional plant ecology of high mountain ecosystems. Springer, Berlin Heidelberg New York

Kofler A (1977) Zur Verbreitung des Deutschen Skorpions in Osttirol. Osttiroler Heimatblätter 45(1):3–4

Komposch C (1998) *Megabunus armatus* und *lesserti*, zwei endemische Weberknechte in den Alpen (Opiliones: Phalangiidae). Carinthia II 188/108:619–627

Komposch C, Gruber J (1999) Vertical distribution of harvestmen in the Eastern Alps (Arachnida: Opiliones). Bull Br Arachnol Soc 11:131–135

Komposch C, Scherabon B, Fet V (2001) Scorpions of Austria. In: Fet V, Selden PA (eds) Scorpions 2001. In Memoriam Gary A. Polis. British Arachnological Society, Burnham Beeches, pp 267–271

Lessert R de (1909) Notes sur la répartition géographique des Araignées en Suisse. Rev Suisse Zool 17:483–499

Liston AD, Leslie AD (1982) Insects from high-altitude summer snow in Austria, 1981. Mitt Entom Ges Basel 32:42–47

Lomnicki A (1963) The distribution and abundance of ground-surface-inhabiting arthropods above the timber line in the region of Zolta Turnia in the Tatra Mts. Acta Zool Cracov 8:183–250 Pl 3–5

Mann DH, Edwards JD, Gara RI (1980) Diel activity patterns in snowfield foraging invertebrates on Mount Rainier, Washington, USA. Arctic Alpine Res 12:359–368

Martens J (1978) Weberknechte, Opiliones (Spinnentiere, Arachnida). Tierwelt Deutschlands 64. Fischer, Jena

Matthey W, Dethier M, Galland P, Lienhard C, Rohrer N, Schiess T (1981) Etude écologique et biocénotique d'une pelouse alpine au Parc national suisse. Bull Ecol 12:339–354

Maurer R (1982) Zur Kenntnis der Gattung *Coelotes* (Araneae, Agelenidae) in Alpenländern. I. Die Arten aus dem Gebiet der Schweiz. Evolution der *pastor*-Gruppe. Rev Suisse Zool 89:313–336

Maurer R (1992) Zur Gattung *Cybaeus* im Alpenraum (Araneae: Agelenidae, Cybaeinae) - Beschreibung von *C. montanus* n.sp. und *C. intermedius* n. sp. Rev Suisse Zool 99:147-162

Maurer R, Hänggi A (1990) Katalog der schweizerischen Spinnen. Documenta Faunistica Helvetiae 12:1-412

Maurer R, Thaler K (1988) Über bemerkenswerte Spinnen des Parc National du Mercantour (F) und seiner Umgebung (Arachnida: Araneae). Rev Suisse Zool 95:329-352

Meyer E (1980) Ökologische Untersuchungen an Wirbellosen des zentralalpinen Hochgebirges (Obergurgl, Tirol). 4. Aktivitätsdichte, Abundanz und Biomasse der Makrofauna. Veröff Univ Innsbruck 125 Alpin Biol Stud 13:1-53

Meyer E, Thaler K (1995) Animal diversity at high altitudes in the Austrian Central Alps. In: Chapin FS III, Körner C (eds) Arctic and alpine biodiversity. Patterns, causes and ecosystem consequences. Springer, Berlin Heidelberg New York, pp 97-108

Meyer E, Wäger H, Thaler K (1985) Struktur und jahreszeitliche Dynamik von *Neobisium*-Populationen in zwei Höhenstufen in Nordtirol (Österreich) (Arachnida: Pseudoscorpiones). Rev Ecol Biol Sol 22:221-232

Muster C (2000) Arachnological evidence of glacial refugia in the Bavarian Alps. Ekologia (Bratislava) 19 Suppl 3:181-192

Muster C (2001) Biogeographie von Spinnentieren der mittleren Nordalpen (Arachnida: Araneae, Opiliones, Pseudoscorpiones). Verh Naturwiss Ver Hamburg NF 39:5-196

Muster C (2002) Substitution patterns in congeneric arachnid species in the northern Alps. Div Distrib 8:107-121

Otto C, Svensson BS (1982) Structure of communities of ground-living spiders along altitudinal gradients. Holarctic Ecol 5:35-47

Palmgren P (1973) Beiträge zur Kenntnis der Spinnenfauna der Ostalpen. Comment Biol Helsinki 71:1-52

Platnick NI (1997) Advances in spider taxonomy 1992-1995. New York Entomological Society and The American Museum of Natural History, New York

Proszynski J (1983) Tracing the history of a genus from its geographical range by the example of *Sitticus* (Arachnida: Araneae: Salticidae). Verh Naturwiss Ver Hamburg NF 26:161-179

Puntscher S (1980) Ökologische Untersuchungen an Wirbellosen des zentralalpinen Hochgebirges (Obergurgl, Tirol). 5. Verteilung und Jahresrhythmik von Spinnen. Veröff Univ Innsbruck 129 Alpin Biol Stud 14:1-106

Reinig WF (1965) Die Verbreitungsgeschichte zweier für die Apenninen neuer boreoalpiner Hummelarten mit einem Versuch der Gliederung boreoalpiner Verbreitungsformen. Zool Jb Syst 92:103-142

Schaefer M (1976) Experimentelle Untersuchungen zum Jahreszyklus und zur Überwinterung von Spinnen (Araneida). Zool Jb Syst 103:127-289

Schatz H (1979) Ökologische Untersuchungen an Wirbellosen des zentralalpinen Hochgebirges (Obergurgl, Tirol) 2. Phänologie und Zönotik von Oribatiden (Acari). Veröff Univ Innsbruck 117 Alpin Biol Stud 10:15-120

Schmarda T (1995) Beiträge zur Kenntnis der Pseudoskorpione von Tirol und Vorarlberg: Faunistik; taxonomische Charakterisierung; Aktivitätsdynamik. Diploma thesis, University of Innsbruck

Schmölzer K (1962) Die Kleintierwelt der Nunatakker als Zeugen einer Eiszeitüberdauerung. Ein Beitrag zum Problem der Prä- und Interglazialrelikte auf alpinen Nunatakkern. Mitt Zool Mus Berlin 38:171-400

Schmölzer K (1993) Die hochalpinen Landmilben der östlichen Brennerberge (Acarina terrestria). Veröff Mus Ferdinandeum (Innsbruck) 73:47-67

Schmölzer K (1999) Prä- und interglaziale Elemente in der Acarofauna der Alpen. Carinthia II 189/109:573-602

Schmoller R (1971) Nocturnal arthropods in the alpine tundra of Colorado. Arct Alpine Res 3:345–352
Somme L (1989) Adaptations of terrestrial arthropods to the alpine environment. Biol Rev 64:367–407
Somme L, Block W (1991) Adaptations to alpine and polar environments in insects and other terrestrial arthropods. In: Lee RE Jr, Denlinger DL (eds) Insects at low temperature. Chapman and Hall, New York, pp 318–359
Steinböck O (1931) Zur Lebensweise einiger Tiere des Ewigschneegebietes. Z Morph Ökol Tiere 20:707–718
Stipperger H (1928) Biologie und Verbreitung der Opilioniden Nordtirols. Arb Zool Inst Univ Innsbruck 3:17–79
Thaler K (1976) Endemiten und arktoalpine Arten in der Spinnenfauna der Ostalpen (Arachnida: Araneae). Ent Germ 3:135–141
Thaler K (1981) Neue Arachniden-Funde in der nivalen Stufe der Zentralalpen Nordtirols (Österreich) (Aranei, Opiliones, Pseudoscorpiones). Ber Nat Med Ver Innsbruck 68:99–105
Thaler K (1982) Weitere wenig bekannte *Leptyphantes*-Arten der Alpen (Arachnida: Aranei, Linyphiidae). Rev Suisse Zool 89:395–417
Thaler K (1988) Arealformen in der nivalen Spinnenfauna der Ostalpen (Arachnida, Aranei). Zool Anz 220:233–244
Thaler K (1989) Epigäische Spinnen und Weberknechte (Arachnida: Aranei, Opiliones) im Bereich des Höhentransektes Glocknerstrasse-Südabschnitt (Kärnten, Österreich). Veröff Österr MaB Programm 13:201–215
Thaler K (1990) *Lepthyphantes severus* n.sp., eine Reliktart der Nördlichen Kalkalpen westlich des Inn (Österreich) (Arachnida: Aranei, Linyphiidae). Zool Anz 224:257–262
Thaler K (1995) Beiträge zur Spinnenfauna von Nordtirol. 5. Linyphiidae 1: Linyphiinae (sensu Wiehle) (Arachnida: Araneida). Ber Nat Med Ver Innsbruck 82:153–190
Thaler K (1998) Die Spinnen von Nordtirol (Arachnida, Araneae): Faunistische Synopsis. Veröff Mus Ferdinandeum (Innsbruck) 78:37–58
Thaler K (1999) Beiträge zur Spinnenfauna von Nordtirol. 6. Linyphiidae 2: Erigoninae (sensu Wiehle) (Arachnida: Araneae). Veröff Mus Ferdinandeum (Innsbruck) 79:215–264
Thaler K, Buchar J (1996) Die Wolfspinnen von Österreich 3: Gattungen *Aulonia, Pardosa* (p.p.), *Pirata, Xerolycosa* (Arachnida, Araneae: Lycosidae) - Faunistisch-tiergeographische Übersicht. Carinthia II 186/106:393–410
Thaler K, van Helsdingen P, Deltshev C (1994) Vikariante Verbreitung im Artenkomplex von *Lepthyphantes annulatus* in Europa und ihre Deutung (Araneae, Linyphiidae). Zool Anz 232:111–127
Toft S (1976) Life-histories of spiders in a Danish beech wood. Nat Jutlandica 19:5–40
Willmann C (1951) Die hochalpine Milbenfauna der mittleren Hohen Tauern insbesondere des Großglockner-Gebietes (Acari). Bonner Zool Beitr 2:141–176
Zingerle V (1999a) Spider and harvestman communities along a glaciation transect in the Italian Dolomites. J Arachnol 27:222–228
Zingerle V (1999b) Epigäische Spinnen und Weberknechte im Naturpark Sextner Dolomiten und am Sellajoch (Südtirol, Italien) (Araneae, Opiliones). Ber Nat Med Ver Innsbruck 86:165–200

16 Patterns of Butterfly Diversity Above the Timberline in the Italian Alps and Apennines

L. Tontini, S. Castellano, S. Bonelli and E. Balletto

16.1 Introduction

The Italian butterfly fauna comprises 275 native species (Balletto et al. 1995), of which 106 occur in the Italian Alps (25 strictly alpine) and 64 in the Apennines (10 alpine). They form loose assemblages, apparently exhibiting low inter-specific competition for either space or other resources (Gilbert and Singer 1973; Gilbert 1984; Balletto et al. 1985; Porter et al. 1992). Density-independent processes (Den Boer 1998) generally determine their population sizes. The assemblages are made up of various combinations of stenotopic species (characterised along gradients) and broadly eurychorous or, sometimes, migratory species. Strict stenotopy, where it occurs, is generally a consequence of adult behaviour rather than of larval biology; the genus *Maculinea* (see Thomas 1994 for a review) affords a rare example in Europe. Even rarer are cases where stenotopy derives from strict larval monophagy, such as that observed for some species feeding on *Vaccinium*, or on some leguminous plants (Balletto et al. 1985, and literature cited therein).

Most butterfly assemblages are inextricably associated with vegetation types along a succession gradient (e.g. from open xerophilous or mesophilous meadows via scrub to secondary forest regrowth). This seral relationship, however, breaks down above the treeline, where grasslands dominate, and ecological conditions are characterised by a series of unique physical and biotic parameters (Ozenda 1985; Nagy 1998). These grasslands are well represented both in the Italian Alps and in the Apennines, and are generally less affected by human activities than vegetation at lower elevations. In alpine grasslands, various levels of livestock grazing represent the main potential source of human impact.

Butterfly assemblages overlap broadly with plant communities (see Balletto et al. 1982a,b; Balletto and Casale 1991). There are about 10 butterfly species that occur consistently and exclusively in as many alpine vegetation formations. For a butterfly assemblage to be associated with a given plant community, the area covered by the plant community should be large enough to accommodate one or more discrete butterfly populations (in practice, this area is ≥1 ha).

We report the species richness, population densities, assemblage structure, and similarities among assemblages in relation to latitude and altitude in the Italian Alps and the Apennines. We also report how grazing intensity may impact butterfly assemblage species richness and abundance.

16.2 Materials and Methods

16.2.1 Assemblage Description

Data collected between 1977 and 1998 in the Italian Alps and Apennines, between the timberline and the alpine-nival ecotone (where appropriate), were analysed. The data set comprised 76 butterfly communities sampled in the Alps and 62 in the Apennines, between 1350 and 2700 m in the Alps, and between 1400 and 2600 m in the Apennines. In the Apennines, in particular, the timberline has been lowered by man in many places to increase the area for pastures. Butterfly assemblages were sampled as uniformly as possible in *Nardus stricta*-dominated grasslands (Alps and Apennines), in acid *Festuca violacea* grasslands, calcareous *Sesleria apennina–Carex kitaibeliana*, and *Sesleria–Bromus* grasslands (Apennines) and in *Rhododendron–Vaccinium* dwarf-shrub, *Thlaspi* alpine screes, calcareous *Carex firma* grasslands, and

Table 16.1. Overall butterfly species richness in various vegetation types in relation to latitude and altitude (multiple regression analyses)

Vegetation type	Complete		Stepwise			
	Altitude b	Latitude b	Altitude b	P	Latitude b	P
Festuca halleri grassland	−0.005	−1.557	−	−	−	−
Rhododendron-Vaccinium dwarf-shrub	0.026	−4.020	−	−	−4.187	0.083
Thlaspi scree	−0.002	−0.928	−	−	−0.952	0.001
Nardus stricta grassland	−0.004	−0.079	−0.005	0.036	−	−
Carex firma grassland	−0.003	−2.438	−	−	−2.379	0.001
Festuca violacea grassland	−0.003	−0.815	−0.004	0.072	−	−
Sesleria-Carex grassland	−0.014	−8.814	−	−	−6.181	0.005
Sesleria-Bromus grassland	−0.017	−5.070	−	−	−	−

Acid grasslands are *Festuca halleri* (Alps) and *F. violacea* (Apennines).

Festuca halleri acid grasslands (Alps, Table 16.1). Each of these formations was provisionally considered to contain a distinct butterfly assemblage type. Species composition and mean population densities (specimens ha^{-1}) were recorded by the multiple transect method (Pollard 1977).

Species richness, population density, assemblage structure and similarity among assemblages (Sørensen's similarity index) were related to geographical (latitude, mountain range, distance) and ecological (altitude, habitat type) factors. All data were analysed by multiple regression analysis, covariance analysis (SYSTAT, Wilkinson 1988), and the multiple Mantel test (Manly 1991) on the matrix of Euclidean distances among communities, using 5000 randomisations. All analyses were undertaken for the full data set; separate, identical analyses were undertaken for species that occur exclusively above the timberline. Raw data on species richness and mean population densities are available from the corresponding author (E. Balletto).

16.2.2 Grazing Impacts

Grazing impacts were studied in 96 sites, 56 of which were in the Italian Alps and 40 in the Apennines. Since most grasslands are grazed by cattle or sheep (or both) and only a few are regarded as near-ungrazed (Balletto et al. 1982a,b), we distinguished two groups of biotopes: lightly grazed ($n=69$) and heavily grazed ($n=27$). The latter were often, although not always, characterised by a dense cover of *Nardus stricta* (cf. Giacomini and Pignatti 1955), a plant known to be the larval food of almost exclusively eurychorous species. Analyses of variance (ANOVA; with Tukey's HSD test) were used to separate the effects of grazing on butterfly assemblage species richness and total density.

Although the relationship between diversity and abundance in species assemblages is known to be log-normal (Preston 1962a,b), under conditions of environmental stress they can often shift from this pattern (e.g. Belaoussoff and Kevan 1998). To test whether grazing affected the log-normal relationships between diversity and abundance, covariance analyses were carried out.

16.3 Results

16.3.1 Richness and Density of Species

In total, 119 butterfly species were found to form relatively discrete and constant populations above the timberline. The numbers varied widely with habitat and the average species richness per assemblage was eight (±4.11 SD; $n=138$). The average density was 29 (±23.7 SD; $n=138$) individuals ha^{-1}.

16.3.1.1 Species Richness along Latitude and Altitude

Species richness (total in all habitats, total above the timberline, and strictly alpine species only) decreased with latitude (Fig. 16.1), especially above the timberline ($\chi^2=27.2$; d.f.=8; $P<0.001$). This north-to-south decrease in overall species richness along the Apennine peninsula, known as the peninsular effect, reflects re-colonisation patterns following Pleistocene glacial events (whereby progressively fewer species extended southwards).

Within individual assemblages, species richness showed different patterns of variation at different hierarchical levels. Multiple regression analyses showed that, overall, as altitude increased, species richness decreased in butterfly assemblages (one-tailed sign test: $P=0.035$) and latitude (one-tailed sign test: $P=0.004$). Step-wise regressions for individual habitats showed a statistically significant decrease in species richness with latitude for screes, calcicolous grasslands, and for *Sesleria–Carex* grasslands, and with altitude for *Nardus*-dominated grasslands (Table 16.1). All factors considered were significant in a covariance analysis carried out with mountain range (F=6.1; d.f.=1, 123, $P<0.05$) and habitat type (F=7.5; $P<0.001$) as independent factors, and latitude (F=21; $P<0.001$) and altitude (F=5.8; $P<0.05$) as covariates.

We subdivided the species into rare (occurring in <10 assemblages) and common (occurring in ≥10 assemblages) butterflies in order to see if the trends were specific to either category of butterflies. The results showed that, above the timberline, common species tend to decrease in communities with latitude (b=–0.545; $P<0.001$), whereas rare species increase slightly (b=0.168; $P<0.05$; Fig. 16.2).

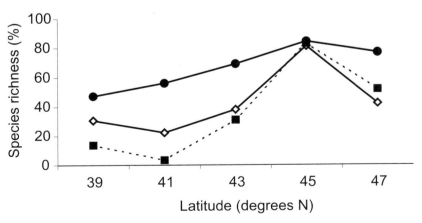

Fig. 16.1. Latitudinal variation of species richness observed in Italy considered separately for all species in all habitats (*filled circles*), all species observed above the timberline (*diamonds*) and strictly alpine species (*squares*) above the timberline

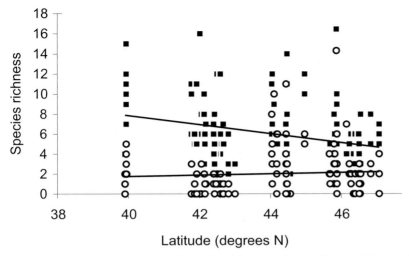

Fig. 16.2. Latitudinal variation of species richness in butterfly assemblages. Above the treeline, the number of common species (occurring in ten assemblages or more, *squares*) increases towards the south, whereas the number of rare species (occurring in less than ten assemblages, *circles*) tends to decrease

16.3.1.2 Population Densities and Assemblage Structure Variation

Population densities within communities were positively correlated with altitude in all but one habitat type, and were statistically significant for the *Festuca*-dominated and acid grasslands (Table 16.2). There was no clear pattern in relation to latitude. Covariance analysis showed that habitat type (ANCOVA: F=4.7; $P<0.001$) and altitude (ANCOVA: F=6.4; $P<0.05$), but not latitude, were the only observed factors responsible for patterns of variation. For assemblage structure (species richness, mean species densities) the results of the multiple Mantel test showed that habitat type ($P<0.001$) alone was correlated with the observed patterns of variation. The same test showed that all factors (habitat, geographic proximity, altitude, and mountain range, all at $P<0.001$) were statistically significant within communities.

16.3.2 Exclusively Alpine Species Richness and Densities

Twenty-five truly alpine species occurred in six of the eight habitat types considered (cf. Tables 15.1 and 15.3). All 25 species occurred in the Italian Alps, with nine of them present in the Apennines. Species richness increased with latitude in five of the six habitat types, although this was statistically significant only for *Nardus* grasslands (Table 16.3). Covariance analysis, however, showed that all factors, i.e. mountain range (F=4.7, d.f.=1, 123; $P<0.05$), habi-

Table 16.2. Mean population density for all species in relation to latitude and altitude (results of multiple regression analyses)

Vegetation type	Complete		Stepwise			
	Altitude b	Latitude b	Altitude b	P	Latitude b	P
Festuca halleri grassland	0.005	0.379	0.005	0.004	–	–
Rhododendron-Vaccinium dwarf-shrub	–0.024	–2.774	–	–	–	–
Thlaspi scree	0.001	0.286	–	–	–	–
Nardus stricta grassland	0.000	–0.080	–	–	–	–
Carex firma grassland	0.001	–0.113	–	–	–	–
Festuca violacea grassland	0.003	0.566	0.004	0.016	–	–
Sesleria-Carex grassland	0.012	2.493	–	–	–	–
Sesleria-Bromus grassland	0.001	0.116	–	–	–	–

Table 16.3. Richness of alpine species in relation to latitude and altitude (results of multiple regression analyses)

Vegetation type	Complete		Stepwise			
	Altitude b	Latitude b	Altitude b	P	Latitude b	P
Festuca halleri grassland	0.002	0.085	0.003	0.094	–	–
Rhododendron-Vaccinium dwarf-shrub	0.010	–0.426	0.010	0.105	–	–
Thlaspi scree	0.000	–0.098	–	–	–	–
Nardus stricta grassland	–0.001	0.187	–	–	0.155	0.028
Calcareous grassland	0.003	0.054	0.003	0.007	–	–
Acid grassland	0.001	0.405	–	–	0.544	0.101

tat type F=6.5, P<0.001), latitude (F=12.3, P≤0.001), and altitude (F=12.3, P<0.001), contributed to the observed pattern. Habitat type was the only statistically significant predictor of alpine species densities.

16.3.3 Grazing Impacts

Grazing could induce as much as a 55 % decrease in both species richness and total density. Heavily grazed *Nardus* grasslands had significantly fewer species and lower population densities than the other grasslands (Table 16.4), both in the Alps (ANOVA: F=23.676; d.f.=1, 53; P<0.001) and the Apennines (ANOVA: F=22.320; d.f.=1, 40; P<0.001).

In the Alps, butterfly species richness and total population density had similar values in calcareous and acid (excluding *Nardus*-type) grasslands (Tukey's HSD test 0.327 and 0.474), indicating a similar grazing impact. Considering the strictly alpine species only, there were no differences in butterfly species richness or population density between *Nardus* and calcareous grasslands (HSD=0.877 and 0.253, respectively). Acid grasslands supported more alpine species (HSD=0.001) and at higher densities (HSD=0.008) than the other two grassland types.

Similar results were found in the Apennines where, for all species, *Nardus* grasslands supported less species-rich butterfly communities than the calcareous *Sesleria* (HSD=0.001) or the acid *Festuca* (HSD=0.006) grasslands. Species densities did not differ among grassland types. The richness of strictly alpine species was higher in *Festuca* grasslands (HSD=0.001) than in *Nardus* grasslands. Alpine species richness and population densities did not differ between *Nardus* and *Sesleria* grasslands, probably because *Sesleria apennina* grasslands have very few strictly alpine species, whilst they virtually disappear from the heavily grazed Apennine grasslands. The *Festuca* and *Sesleria* grassland types did not differ in terms of butterfly species richness or in population densities.

Grazing impacts were independent of altitude (although in the Alps Tukey's test approached significance, F=3.0, P=0.057). A comparison between communities on calcareous and siliceous soils showed that the latter may have been studied at slightly higher elevations. Similarly, in the Apennines, plant communities dominated by *Sesleria apennina* tend to occur lower than those dominated by *Festuca violacea* (P=0.026).

Table 16.4. Mean butterfly species richness for all species (±standard deviation, SD) and mean population density (±SD), in heavily and lightly grazed grasslands above the timberline in the Alps and in the Apennines

Grassland type	Number of stands	Grazing impact	Altitude (m ±SD)	Species richness (individual ha^{-1} ±SD)	Population density (individual ha^{-1} ±SD)
Alps					
Nardus stricta	11	Heavy	1961 (177)	4 (2.3)a	10 (12.6)a
Carex firma	16	Light	1931 (239)	9 (3.6)b	37 (23.5)b
Festuca halleri	29	Light	2125 (325)	10 (4.1)b	40 (21.1)b
Total	56				
Apennines					
Nardus stricta	16	Heavy	1713 (126)	4.5 (1.8)a	11 (8.4)a
Sesleria apennina	13	Light	1760 (215)	10 (3.6)b	22 (8.0)b
Festuca violacea	11	Light	1959 (196)	8 (2.7)b	21 (10.3)b
Total	40				

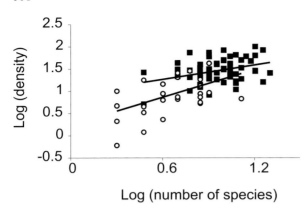

Fig. 16.3. Regression lines of $\log_{10}$-transformed cumulative data of species richness and abundance; *squares* lightly grazed grasslands; *circles* heavily grazed grasslands

A regression analysis of the $\log_{10}$-transformed data of species richness and abundance showed that they were log-normally distributed, both overall (F=89.0, P<0.001) and separately for the lightly grazed (F=8.1, P<0.01) and heavily grazed habitats (F=20.9, P<0.01). A covariance analysis, however, showed that the slopes of the two regression lines differed significantly (F=6.9; d.f.=1, 93; P<0.01; Fig. 16.3). Hence, although there was not a disruption of the log-normal distribution of species richness and species abundance, the impact of long-term heavy grazing was obvious. The existence of a log-normal distribution suggests that recovery of species richness after releasing grazing pressure is likely, although its exact nature is difficult to predict.

16.4 Conclusions

1. This study, which is heavily reliant on multivariate analyses, showed that habitat type is the most important factor influencing species composition and density of butterfly assemblages. All parameters of assemblage structure (species richness, mean species densities and their relative standard errors) were a function of habitat type rather than of geographical proximity or altitude. However, population parameters were also affected by latitude, the mountain range and distance between sample locations.
2. Within individual communities, overall species richness showed an inverse relationship with both latitude and altitude. Butterfly assemblages were more species-rich southwards (where they tended to include a number of non-specialist alpine species). Common species were progressively more abundant southwards, whereas rare species and strictly alpine species were more abundant northwards.
3. The general pattern of variation appeared to be strongly influenced by livestock grazing. This trend appeared to be fairly general and was not influ-

enced by substrate type, or elevation. Strictly alpine species, however, appeared less sensitive to the impacts of grazing than species also occurring below the timberline, probably because alpine specialist species are better adapted to grazing.
4. The slope of the log-normal distribution of total species richness and species abundance observed in lightly grazed biotopes was significantly different from that in heavily grazed ones. The log-normal distribution, however, did not become completely disrupted even in some of the most heavily affected grasslands. The recovery of species richness after the release of grazing pressure is possible and will depend on the metapopulation structure of the species involved and, in most cases, is likely to last many years with somewhat unpredictable outcomes.

References

Balletto E, Casale A (1991) Mediterranean insect conservation. In: Collins NM, Thomas JA (eds) The conservation of insects and their habitats. Academic Press, London, pp 121–142

Balletto E, Barberis G, Toso GG (1982a) Le comunità di Lepidotteri ropaloceri dei consorzi erbacei dell'Appennino. Quaderni sulla Struttura delle Zoocenosi Terrestri CNR Roma 2(II 1):77–143

Balletto E, Barberis G, Toso GG (1982b) Aspetti dell'ecologia dei Lepidotteri ropaloceri nei consorzi erbacei delle Alpi italiane. Quaderni sulla Struttura delle Zoocenosi terrestri CNR 2(II.2):11–96

Balletto E, Lattes A, Toso GG (1985) An ecological study of the Italian rhopalocera. In: Heath J (ed) Proceedings of the 3rd Congress of European Lepidopterology, Cambridge 1982. Societas Europaea Lepidopterologica, Karlsruhe, pp 7–22

Balletto E, Camporesi S, Cassulo L, Fiumi G, Karsholdt O, Zangheri S, (1995) Lepidoptera Cossoidea, Sesioidea, Zygaenoidea, Choreutoidea. In: Minelli A, Ruffo S, La Posta S (eds) Checklist delle specie della fauna italiana. 84. Calderini, Bologna e Ministero per l'Ambiente, Roma

Belaoussoff S, Kevan PG (1998) Research and application: toward an ecological approach for the assessment of ecosystem health. Ecosyst Health 4:4–8

Den Boer PJ (1998) The role of density-independent processes in the stabilisation of insect populations. In: Dempster JP, McLean IFG (eds) Insect populations in theory and in practice. Kluwer, Dordrecht, pp 53–80

Dennis RH (ed) (1992) The ecology of butterflies in Britain. Oxford University Press, Oxford

Giacomini V, Pignatti S (1955) Flora e vegetazione dell'alta Valle del Braulio, con speciale riferimento ai pascoli d'altitudine. Mems Soc Ital Sci Nat Mus Civ St Nat Milano 11:1–194

Gilbert LE (1984) The biology of butterfly communities. In: Vane-Wright RI, Ackery PR (eds) The biology of butterflies, Symposium of the Royal Entomological Society of London, no 11. Academic Press, London, pp 41–52

Gilbert LE, Singer MC (1973) Dispersal and gene flow in a butterfly species. Am Nat 107:58–72

Manly BFJ (1991) Randomisation and Montecarlo methods in biology. Chapman and Hall, New York
Nagy J (1998) European mountain biodiversity: a synthetic overview. Pirineos 151-152:7-41
Ozenda P (1985) La végétation de la chaîne alpine dans l'espace montagnard européen. Masson, Paris
Pollard E (1977) A method for assessing changes in the abundance of butterflies. Biol Conserv 12:115-134
Porter AH, Steel CA, Thomas JA (1992) Butterflies and communities. In: Dennis RH (ed) The ecology of butterflies in Britain. Oxford University Press, Oxford, pp 139-177
Preston FW (1962a) The canonical distribution of commonness and rarity. Part I. Ecology 43:185-215
Preston FW (1962b) The canonical distribution of commonness and rarity. Part II. Ecology 43:410-432
Thomas JA (1994) The ecology and conservation of *Maculinea arion* and other European species of large blue butterfly. In: Pullin AS (ed) Ecology and conservation of butterflies. Chapman and Hall, London, pp 180-197
Wilkinson L (1988) SYSTAT: the system for statistics. Systat Inc, Evanston

17 Diversity Patterns of Carabids in the Alps and the Apennines

P. Brandmayr, R. Pizzolotto, S. Scalercio, M.C. Algieri and T. Zetto

17.1 Introduction

Carabid beetles are widespread in all terrestrial environments of the world and it is thought that they originated from the tropics (Erwin 1979). They have complex evolutionary pathways; taxon-specific cul-de-sacs have been found in tropical forest canopies, in the tundra and in the mountains, where carabids are an important part of the animal diversity (Holdhaus 1954; Mani 1968). Their ability to colonise isolated environments is well documented (Darlington 1943; Lindroth 1949; Den Boer 1977; Den Boer et al. 1980). Carabid assemblages have been studied by several authors in European mountains; most importantly, in the Alps (Franz 1943, 1950; Janetschek 1949; Schmölzer 1962; Amiet 1967; Christandl-Peskoller and Janetschek 1976; Lang 1975; De Zordo 1979a,b; Chemini and Werth 1982; Focarile from 1973, 1975a,b, 1976a,b; Brandmayr and Zetto Brandmayr 1988). However, little is known about their biogeography, a fundamental of insect conservation, in the Mediterranean (Balletto and Casale 1991). In other European mountain chains, studies on carabids in the alpine zone are few, with the exception of Norway and Finnish Lapland (Forsskåhl 1972).

The lack of knowledge about the diversity patterns of invertebrates in Mediterranean high mountains is particularly worrisome, as global change is likely to first impact on the lowest mountains of the Mediterranean peninsulas. Sømme (1993) proposed some recommendations on monitoring and studies to produce realistic models in order to evaluate population changes in the boreal arctic and alpine fauna as a consequence of climate change. Solbreck (1993) stressed that the prediction of faunal dynamics in a changing climate required long-term studies and that all the predictions made from a northern European perspective involve only recent immigrants, which are perhaps "less sensitive to future climatic change..." and that "present knowledge of faunal patterns and populations gives us little guidance to predict the future faunal changes". Most recent literature is concentrated on animal responses to climate change in northern Europe (Hofgaard et al. 1999) and in the Alps (Guisan et al. 1995). Research on invertebrates has mostly been confined to coleopteran subfossil faunas from the beginning of the last glaciation (Foddai and Minelli 1994), while relationships between Mediterranean and Alpine coleopteran faunas have rarely been dealt with (Ponel et al. 1995).

The endemic-rich carabid faunas of southern European mountains have been described in terms of geographical distribution; however, there is a paucity of data on the carabid assemblages in the alpine belt, their long-term stability and responses to short-term climate oscillations. There is no data for single species population numbers using a standardised methodology (e.g. pitfall traps) comparable with other European ecosystems and habitats, and the ecology of the species and their life histories are little known.

Plant names in the following sections follow Pignatti (1984) and carbid names are after Magistretti (1965) or Vigna Taglianti (1993).

17.2 Overall Features of Alpine Carabid Assemblages in the Alps

Stenotopic carabid species of the open alpine habitats mostly belong to particular taxa, but there are well-known exceptions. In the genus *Carabus*, for instance, a group of larger predators, many so-called high alpine species are found in the subgenus *Orinocarabus* Kraatz. Many smaller predators are found in the genus *Nebria* (*Oreonebria*, *Alpaeus*) and in the nebriine genus *Leistus*, a collembola hunter, in the genus *Bembidion* (*Testedium* and *Testediolum*) and in the taxonomically related *Trechus*, in the genus *Pterostichus* sensu stricto and in the related *Oreophilus*. Phytophagous carabid species are not common; two subgenera of the seed-eating *Amara* (*Leirides* and *Leiromorpha*) live in large numbers in some dry alpine grass mats. There are many species from lower altitudes that are frequently found in alpine habitats, possibly because of their preference for lower temperatures or higher soil humidity.

From a geographical distribution point of view there are two main groups of species: widely distributed taxa, some with a clear boreo-alpine pattern, and more or less endemic ones. The alpine grassland endemics can either occur over large parts of the Alps chain, or they may be stenoendemic and restricted to certain sectors of the chain. The largest number of endemics in the Alps is found on the southern border of the chain, in the so-called 'massifs de refuge' of Holdhaus (1954).

The autecology of alpine carabids is not so thoroughly investigated as that of many central European species; however, habitat affinity studies indicate that they are true open land dwellers. All stenotopic species avoid shading by trees and dwarf-shrubs such as *Rhododendron* formations and their more or less acid, peaty soils. Alpine carabids avoid perturbed habitats such as anthropogenic grasslands in the treeline ecotone, and, for this reason, they have to be seen as useful indicators of habitat condition. Kühnelt (1968) proposed that a high alpine fauna would begin appearing around late snowbeds with snow melting typically in July, such as in the central Alps. In the Italian Pre-Alps, the activity of carabid populations begins in June (Brandmayr and Pizzolotto 1989), probably because of the more oceanic climate in the Italian Pre-Alps than in the central Alps, combined with the lower elevation of the former. The basic requirements for alpine carabids appear to be a grassland habitat (or bare, stony soils with scattered plant cushions), prolonged snow cover watering the soil from the end of May until July, good soil drainage, intense light and insolation.

The dispersal power of alpine carabids is normally low, similarly to many other mountain insects. Most carabid species have rudimentary hind-wings. However, especially on well-drained calcareous bedrocks, the brachypterous or wingless form is the dominant; on siliceous mountains about 30 % of the species are macropterous, i.e. able to fly (Brandmayr 1983, 1991; Brandmayr and Zetto Brandmayr 1988).

17.3 Species Recruitment in Alpine Habitats

Species recruitment is perhaps the most important issue to consider for any prediction concerning the impacts of climate change in alpine habitats. Mani (1968) and some other authors (e.g. Erwin 1979) discussed evolutionary species recruitment in high mountain areas. Their work, however, was mostly biogeographic and morphological and no published studies are available on the habitat preferences of extant carabid beetles. From a study by Brandmayr and Zetto Brandmayr (1988) across an altitudinal sequence of temperate deciduous forest, montane coniferous forest, and alpine (high mountain) biome in the Dolomites, it appeared that three main groups of carabids colonised alpine grasslands (Fig. 17.1). Firstly, the more or less stenotopic high alpine open ground dwellers, some of which, such as *Carabus creutzeri*, are thought to have colonised from the forest zone as recently as the last ice ages, as inferred from their more or less 'amphitopic' area-population structure (Brandmayr and Zetto Brandmayr 1979). The second group consists of species predominantly found in the coniferous forests, such as most *Oreophilus* species, which are also thought to be quaternary recruits to the alpine grasslands. Nonetheless, most *Oreophilus* species occur in the high montane forests and in the treeline ecotone, and it is possible to distinguish an *Oreophilus* assemblage belt both in the Alps (see, e.g. Vesubie Valley, Mar-

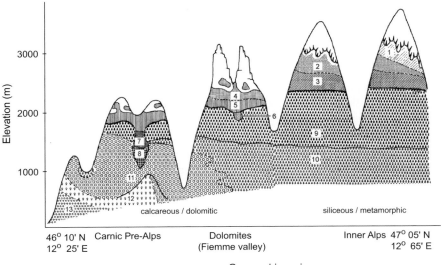

Fig. 17.1. Carabid species assemblages from the southern part of the Eastern Alps, simplified from Brandmayr and Zetto Brandmayr (1988). The assemblages are named by the most important characteristic or 'guide' species: *1 Nebria atrata*, subnival; *2 Nebria germari* and *Amara quenseli*, high alpine pioneer cushions; *3 Carabus alpestris* and *Cymindis vaporariorum*, siliceous alpine grasslands; *4 Nebria germari* and *Trechus dolomitanus*, high alpine pioneer cushions; *5 Carabus bertolinii* and *Amara alpestris*, calcareous alpine grasslands; *6 Pterostichus schaschli* and *Trichotichnus knauthi*, humid scree; *7 Carabus bertolinii* and *Pterostichus ziegleri* with *Pterostichus schaschli*, *Carex firma* sedge heath; *8 Amara uhligi* and *Cychrus angustatus*, endemic *Festuca spectabilis* grassland; *9 Leistus nitidus* and *Calathus micropterus*, upper montane and treeline ecotone *Picea abies* stands; *10 Carabus linnei* and *Argutor oblongopunctatus*, montane *Picea abies* forest; *11 Pterostichus metallicus* and *Molops tridentinus*, montane forest (*Fagus sylvatica* and *Abies alba*); *12 Abax ater* grouping, *Pinus sylvestris* stands and deciduous forest in southerly exposure; *13* thermophilic groupings in *Quercus pubescens* forest. The scheme is based on the characteristic landforms of the Alpago range near Pordenone, Carnic Pre-Alps, a generalised scheme of the Dolomites, and on the geomorphology of the Aurine Alps (Zillertaler Alpen) and the Hohe Tauern (the highest Austrian range)

itime Alps; Amiet 1967) as well in the Italian Apennines. The third group originates from the deciduous forests. This group, such as *Notiophilus biguttatus* and *Pterostichus metallicus*, is still actively colonising the most favourable sites in alpine grasslands in the Dolomites.

The stenotopic alpine open ground carabids largely dominate, especially on calcareous bedrock (e.g. 53–95% in the Dolomites). On siliceous rocks the numbers are generally lower and more variable: 13–21% in *Festuca halleri* grassland, 85% in well-conserved *Carex curvula* stands. Individuals of a single species, *Carabus* (*Orinocarabus*) *alpestris*, contributed 36% to samples from a *Festuca varia* stand, approximately 200 m from the *Carex curvula* stand described above.

17.4 Diversity Patterns in the Alps

The alpine assemblages are mostly composed of 6–12 ground beetles; the maximum so far observed was in the Vette di Feltre near Belluno, Venetian Pre-Alps, with a mean of about 15 species on five *Sesleria–Carex* sites (Brandmayr and Pizzolotto 1989). Because of the absence of strictly comparable data in wider regions, we discuss the regional patterns observed in the south-eastern Alps along the Prescudin-Carnic Pre-Alps to the Vette di Feltre and Dolomites sequence of about 70 km. Description, data and methods are presented in Brandmayr and Zetto Brandmayr (1988) and Brandmayr and Pizzolotto (1989).

On the whole, species diversity tended to decrease from the Pre-Alps to the inner chains. The well-structured acid humus-rich soils under *Sesleria–Carex* type vegetation had the highest species numbers (12–15) in the Vette di Feltre National Park (Busa delle Vette, about 24 km of the SE limit of the Alps). Vegetation dominated by *Alyssum ovirense* had an endemic carabid assemblage with *Pterostichus schaschli* and *Trichotichnus knauthi* as the main species. In the Prescudin Valley (18 km from the edge of the Southern Alps), the mean species numbers were between eight and nine. The low species richness may have in part been caused by the isolated nature of the valley. The most interesting faunulas in the Prescudin Valley are confined to calcareous scree vegetation dominated by *Festuca laxa* and *F. sieberi*. In the Dolomites between the Rolle Pass and the Fiemme Valley, about 50–65 km inside the Alps, calcareous/sandstone habitats were not richer in species than siliceous ones with *Festuca halleri* vegetation (Valles Pass). On the more Central Alpine Pala di Santa less humid typical *Carex curvula* acid soils had a lower mean of about five taxa per site.

There is a general trend of decrease in endemics from the Pre-Alps to the Dolomites. There are sometimes differences of the same magnitude between stands on calcareous skeletal soil and better developed, more humus- or silt-rich soils of the same sampling area. The species numbers for the habitat types (montane deciduous and coniferous forest, treeline ecotone, and all alpine habitats together) were weakly related to the percentage of endemics present. Five geographic areas are perhaps a too small sample for generalisation; however, it appears that endemic faunulas of the Southern Alps are especially well developed above the treeline on calcareous bedrock in the marginal chains, where there is a concentration of typical alpine species. Siliceous substrata with moist soils provide conditions similar to those of the tundra and are populated by widely distributed boreo-alpine elements such as *Nebria gyl-*

lenhali. Flightless endemics and scree specialists such as the flat-bodied *Pterostichus schaschli* are characteristic of calcareous lithosols. Larger data sets will make it possible to separate the influence of ecological factors and geographic location.

17.5 Carabid Assemblages of the Central Apennines

The alpine zone of the High Apennine of Lazio and Abruzzo, central Italy, highly contrasts with the surrounding Mediterranean bioclimate. The dominant vegetation from 1800–1900 m a.s.l. is discontinuous *Sesleria* grassland, in moister locations *Carex kitaibeliana* grasslands or *Festuca-Trifolium thalii* snowbeds. The carabid assemblages of two grasslands sampled on Mt. Terminillo had seven species in common and were composed of a mixture of xerophilous and Apennine orophilous elements (*Oreophilus morio samniticus**, *Zabrus orsinii**, *Nebria posthuma**) and European or Euro-Siberian xeromicrothermic species such as *Olisthopus sturmi*, *Harpalus xanthopus*

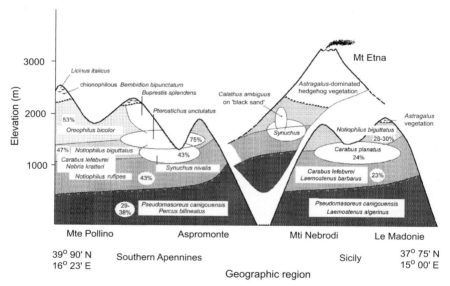

Fig. 17.2. Carabid beetle assemblages of the southern Apennines and the mountains of Sicily (Brandmayr et al. 1998). The alpine zone is confined to the highest elevations; on Mt. Pollino, a *Licinus italicus*-dominated assemblage occurs in natural grasslands; *Bembidion bipunctatum* assemblages occur in snowbeds and dolines. The upper montane and montane *Pinus leucodermis* forest has no ground beetle species; instead it is dominated by the buprestid beetle *Buprestis splendens*. There are no data for the ground beetles for the *Astragalus*-dominated shrub on Mt. Etna. *Percentage values* represent proportions of endemic species in each habitat type. *Ellipsoids* reflect a change in species assemblages in particular topo-climate conditions

winkleri and *Harpalus quadripunctatus* (regional endemic species or forms are marked by asterisks). The upper lying (2070 m) *Sesleria* stand (assumed to be primary) was dominated by a large and sedentary Molopine, *Percus dejeani**, found from Tuscany to Abruzzo, whilst in the other stand (1935 m), which is probably anthropogenic and derived from a beech forest, the dominant ground beetles were xerophilous (*Poecilus koyi*) or forest dwellers with an ability to spread into neighbouring grasslands (*Calathus fracassii*).

Figure 17.2 summarises the diversity of carabid species assemblages in elevation zones in the mountains of southern Italy. These southern assemblages are yet to be quantitatively described; however, preliminary data show few common species with the mountains of central Italy (P. Brandmayr et al., pers. observ.). The fraction of stenotopic alpine elements seems highly variable from site to site, both in the Alps and in the Apennines, and so is the species assemblage diversity (β- or inter-habitat diversity), a strong argument for the implementation of more targeted research projects.

17.6 Alpine Carabid Assemblages and Climate Change

Little is known about the diversity of carabids in the Alps and Apennines that could be used to predict local responses to a global climate change. Foddai and Minelli (1994) reported two carabid species, *Bembidion biguttatum* and *Amara alpina*, from a fully glaciated site in north-eastern Italy. The species are now extinct from other sites in Italy, but are still found in central and northern Europe. This indicates that changes in the last 10,000 years may have caused large shifts in the areal distribution of species. In the Apennines, at the Prato Spilla sequence, the changes inferred from sub-fossil beetle assemblages appeared to reflect the contemporary altitudinal zonation (Ponel and Lowe 1992). The impact of future changes seems less predictable because of the many collateral effects a temperature change may have. Undoubtedly, however, lower mountains could face the highest rates of extinction of many stenotopic chionophilous high alpine species. For example, at Mt. Terminillo, an increase of 135 m in altitude facilitated the appearance or a higher abundance of two stenotopic elements, *Nebria posthuma* and *Oreophilus morio*. As the summit of Mt. Terminillo is 2216 m, a shift of 1 °C could greatly reduce suitable habitats for *Nebria posthuma*, known from 15 mountains in the central Apennines.

17.7 Conclusions

1. Habitat conditions (bedrock, vegetation, and snow cover) and the faunal history of a mountain range appear to determine alpine diversity patterns of carabid beetles. The refugial massifs of the Pre-Alps with a history of less heavy glaciations than the Alps proper are generally more species rich than the Inner Alps. This is probably related to the survival of stenoendemic elements that behave ecologically as stenotopic alpine open ground dwellers.
2. Vegetation on shallow skeletal soils in the south-eastern Alps appear to have high alpine faunas richer in endemics than deeper, better developed soils. However, because of the lack of quantitative data from the Western Alps, no generalisation is attempted.
3. The southern carabid assemblages are yet to be quantified; however, preliminary data show few common species with the mountains of central Italy (P. Brandmayr et al., pers. observ.). *Sesleria* stands on Mt. Terminillo, Apennines, central Italy, are inhabited by an assemblage of 10–12 carabid species, a mixture of chionophilous and xerophilous species reflecting the upper Apennine landscapes, characterised by echinophytic vegetation (Pignatti 1979; Pignatti et al. 1980).
4. Ground beetle species diversity patterns in the alpine zone could be a suitable indicator of changes in habitat quality. Any such scheme should be designed so that vegetation formations along an altitudinal sequence span the upper montane–treeline ecotone–alpine zone to enable the understanding of species recruitment and habitat affinity.

Acknowledgements. We thank the Ministero per l'Università e la Ricerca Scientifica e Tecnologica, Italy, for funding the project entitled 'Descriptive and methodological aspects of studies on animal biodiversity in Italy'.

References

Amiet JL (1967) Les groupements des coléoptères terricoles de la haute vallée de la Vesubie (Alpes Maritimes). Mém Mus Hist Nat Paris S A Zool 46:124–213

Balletto E, Casale A (1991) Mediterranean insect conservation. In: Collins M, Thomas J (eds) The conservation of insects and their habitats. Academic Press, London, pp 121–142

Brandmayr P (1983) The main axes of the coenoclinal continuum from macroptery to brachyptery in carabid communities of the temperate zone. In: Brandmayr P, Den Boer PJ, Weber F (eds) Report 4th Symp Eur Carab, Haus Rothenberge, Westphalia, 24–26 September /1981. Publ Agric Univ, Wageningen, pp 147–169

Brandmayr P (1988) Zoocenosi e paesaggio I. Le Dolomiti – Val di Fiemme e Pale di S. Martino. Stud Trent Sci Nat Acta Biol 64 (Suppl):1–482

Brandmayr P (1991) The reduction of metathoracic alae and of dispersal power of carabid beetles along the evolutionary pathway into the mountains. In: Lanzavecchia G, Valvassori V (eds) Form and function in zoology. UZI selected Symp and Monographs, 5. Mucchi Editore, Modena, pp 363–378

Brandmayr P, Pizzolotto R (1989) Aspetti zoocenotici e biogeografici dei popolamenti a Coleotteri Carabidi nella fascia alpina delle Vette di Feltre (Belluno). Biogeogr NS 8:713–743

Brandmayr P, Zetto Brandmayr T (1979) Contribution to the ecology of an euryhypsic ground beetle of the Eastern Alps and Dinaric Karst, *Carabus creutzeri* Fabr. Zool Jahrb Syst 106:50–64

Brandmayr P, Zetto Brandmayr T (1986) Phenology of ground beetles and its ecological significance in some of the main habitat types of southern Europe. In: den Boer PJ (ed) Carabid beetles. G Fischer, Stuttgart, pp 195–220

Brandmayr P, Zetto Brandmayr T (1988) Comunità a Coleotteri Carabidi delle Dolomiti Sudorientali e delle Prealpi Carniche. Stud Trent Sci Nat Acta Biol 64 (Suppl):125–250

Brandmayr P, Cagnin M, Mingozzi T, Pizzolotto R, Scalercio S (1998) La biodiversità animale nel paesaggio del Mediterraneo, i rapporti con la componente vegetale ed il suo stato di conservazione. Atti 93 Congr Soc Bot Ital Arcavacata di Rende (CS) 1–3/10/98: 22–24

Chemini C, Werth F (1982) Censimenti di Carabidi in tre ambienti forestali di Magré e Favogna (Provincia di Bolzano) (Insecta: Coleoptera: Carabidae). Stud Trent Sci Nat Acta Biol 59:201–211

Christandl-Peskoller H, Janetschek H (1976) Zur Faunistik und Zoozönotik der südlichen Zillertaler Hochalpen. (Mit besonderer Berücksichtigung der Makrofauna). Veröff Univ Innsbruck 101. Alpin-Biol Stud 7:1–134

Darlington PJ (1943) Carabidae of mountains and islands: data on the evolution of isolated faunas, and on atrophy of wings. Ecol Monogr 13:37–61

Den Boer PJ (1977) Dispersal power and survival. Misc Pap Landbouwhogesch Wageningen 14:1–190

Den Boer PJ, van Huizen THP, den Boer-Daanje W, Aukema B, den Bieman CFM (1980) Wing polymorphism and dimorphism in ground beetles as stages in an evolutionary process (Coleoptera, Carabidae). Entomol Gen 6:107–134

De Zordo I (1979a) Phänologie von Carabiden im Hochgebirge Tirols (Obergurgl, Österreich) (Insecta: Coleoptera). Ber Nat Med Ver Innsbruck 66:73–83

De Zordo I (1979b) Ökologische Untersuchungen an Wirbellosen des zentralalpinen Hochgebirges (Obergurgl, Tirol). III. Lebenszyklen und Zönotik von Coleopteren. Veröff Univ Innsbruck 118. Alpin Biol Stud XI:1–131

Erwin TL (1979) Thoughts on the evolutionary history of ground beetles: hypothesis generated from comparative faunal analyses of lowland forest sites in temperate and tropical regions. In: Erwin TL, Ball G, Whitehead H (eds) Carabid beetles: their evolution, natural history and classification. Junk, The Hague, pp 539–592

Focarile A (1973) Sulla Coleotterofauna alticola del Gran San Bernardo (Versante Valdostano). Ann Fac Sc Agric Univ Stud Torino 9:51–118

Focarile A (1975a) Alcuni interessanti Coleotteri della Valle d'Aosta. Rev Valdot Hist Nat 29:8–52

Focarile A (1975b) Sulla Coleotterofauna alticola di cima Bonze m 2516 (Valle di Champorcher), del Monte Crabun m 2710 (Valle di Gressoney) e considerazioni sul popolamento prealpino nella Alpi Nord-Occidentali (Versante Italiano). Rev Valdot Hist Nat 29:53–105

Focarile A (1976a) Sulla Coleotterofauna alticola del Monte Barbeston m 2482 (Val Chelamy) e de Monte Nery m 3076 (Vald'Ayas). Rev Valdot Hist Nat 30:68–125

Focarile A (1976b) Sulla Coleotterofauna alticola della conca del Breuil (Valtournanche) e osservazioni sul popolamento pioniero delle zone di recente abbandono glaciale. Rev Valdot Hist Nat 30:126–168

Foddai D, Minelli A (1994) Fossil arthropods from a full-glacial site in northeastern Italy. Quat Res 41:336–342

Forsskåhl B (1972) The invertebrate fauna of the Kilpisjärvi area, Finnish Lapland: 9. Carabidae, with special notes on ecology and breeding biology. Acta Soc Fauna Flora Fenn 80:99–119

Franz H (1943) Die Landtierwelt der mittleren Hohen Tauern. Sitzungsber Akad Wiss Wien Math Nat Kl 107:1–552

Franz H (1950) Bodenzoologie als Grundlage der Bodenpflege. Akademie Verlag, Berlin

Franz H (1975) Die Bodenfauna der Erde in Biozonotischer Betrachtung I-II. Erdwiss. Forschung X. Verlag EF Steiner, Wiesbaden

Guisan A, Holten JI, Spichiger R, Tessier L (eds) (1995) Potential ecological impacts of climate change in the Alps and Fennoscandian Mountains. Second IPCC report, Working group II-C, Publ Hors-série n 8, Cons Jard Bot, Geneva

Hofgaard A, Ball JP, Danell K, Callaghan TV (eds) (1999) Animal responses to global change in the north. Ecol Bull Copenhagen 47:1–187

Holdhaus K (1954) Die Spuren der Eiszeit in der Tierwelt Europas. Abhandl Zool Bot Ges Wien 18:1–493

Holdhaus K, Lindroth CH (1939) Die europäischen Koleopteren mit boreoalpiner Verbreitung. Ann Naturhist Mus Wien 50:123–293

Janetschek H (1949) Tierische Sukzessionen auf hochalpinem Neuland. Nach Untersuchungen am Hintereis-, Niederjoch-und Gepatschferner. Ber Nat Med Ver Innsbruck 48/49:1–215

Kühnelt W (1968) Zur Ökologie der Schneerandfauna. Verh D Zool Ges Innsbruck, pp 707–721

Lang A (1975) Koleopterenfauna und-faunation in der alpinen Stufe der Stubaier Alpen (Kuhtai). Veroff Univ Innsbruck 99. Alpin Biol Stud 1:80

Lindroth CH (1949) Die Fennoskandischen Carabidae. Kungl Vetensk Vitterh Samh Handl SB4, 3 Allgemeiner Teil. Bröderna Lagerström Boktryckare, Stockholm

Magistretti M (1965) Coleoptera Cicindelidae, Carabidae. Fauna d'Italia, vol 8. Calderini, Bologna

Mani MS (1968) Ecology and biogeography of high altitude insects. Junk, The Hague

Pignatti S (1984) Flora d'Italia. Edagricole, Bologna

Pignatti S (1979) I piani della vegetazione in Italia. Giorn Bot Ital 113:411–428

Pignatti E, Pignatti S, Nimis P, Avanzini A (1980) Le vegetazione degli arbusti spinosi emisferici: contributo all'interpretazione della fasce di vegetazione delle alte montagne dell'Italia mediterranea. Quad Prog Finalizzati CNR AQ/1/71:1–130

Ponel P, Jay P, Lumaret JP (1995) Past and present changes in the coleopteran fauna since the end of the last glaciation: the case of the Western Alps and the Apennines. In: Guisan A, Holten JI, Spichiger R, Tessier L (eds) Potential ecological impacts of climate change in the Alps and Fennoscandian Mountains. Second IPCC report, Working group II-C, Publ Hors-Série n 8, Cons Jard Bot, Geneva, pp 159–172

Ponel P, Lowe JJ (1992) Coleopteran, pollen and radiocarbon evidence from the Prato Spilla 'D' succession, N. Italy. C R Acad Sci Paris 315(II):1425–1431

Schmölzer K (1962) Die Kleintierwelt der Nunatakker als Zeugen einer Eiszeit-Überdauerung. Mitt Zool Mus Berl 38:171–400

Solbreck C (1993) Predicting insect faunal dynamics in a changing climate – a northern European perspective. In: Holten JI, Paulsen G, Oechel WC (eds) Impacts of climatic change on natural ecosystems, with emphasis on boreal and arctic/alpine areas. NINA, Trondheim, pp 176–185

Sømme L (1993) Effects on boreal and arctic/alpine fauna. In: Holten JI, Paulsen G, Oechel WC (eds) Impacts of climatic change on natural ecosystems, with emphasis on boreal and arctic/alpine areas. NINA, Trondheim, pp 172–175

Thiele HU (1977) Carabid beetles in their environments. Zoophysiol Ecol 10. Springer, Berlin Heidelberg New York

Vigna Taglianti A (1993) Coleoptera Archostemata, Adephaga 1 (Carabidae). In: Minelli A, Ruffo S, La Posta S (eds) Checklist delle specie della fauna italiana, 44. Calderini Publ., Bologna

IV Vertebrates

18 Overview: Alpine Vertebrates – Some Challenges for Research and Conservation Management

D.B.A. THOMPSON

In his classic paper on arctic vertebrates, Chernov (1995) showed that summer temperature and the evolutionary 'age' of taxa had a bearing on the diversity of vertebrates in arctic areas. As a rule, vertebrate species richness declines across the globe as summer temperature decreases. Arguably, this is due as much to direct as indirect effects of temperature (e.g. Chernov 1989). Within groups such as birds, some of the more ancient taxa, such as Gaviiformes and Charadriiformes, tend to be more species-rich in the Arctic than do the younger taxa, notably the Passeriformes (also Chernov 1995). In some ways this is to be expected due to habitat differences – arctic and alpine areas have a structurally more simple habitat complex (compared with forest), and offer many fewer niches for passerines, in particular.

The five chapters in this section touch on the fundamentals of the diversity of alpine zone vertebrates, and in doing so raise a considerable number of questions for further research and conservation management.

18.1 Species Diversity Variation

Chapter 22 explores the diversity of alpine vertebrates in the Pyrenees and Sierra Nevada, Spain. The study shows, as many others have done, that species diversity declines with altitude. Three hypotheses are proposed to explain the importance of this phenomenon in relation to climate change impacts. Specifically, Chapter 22 poses the important hypothesis that vertebrate diversity may respond to climate change as a result of changes in environmental conditions, competition pressure, or changes in actual niches occupied by specialists. Whilst it is extremely difficult to separate these hypotheses, the results suggest that changes in niche characteristics due to climate change may be most important in determining any diversity response to climate change. This raises the question of whether the responses shown by specialist vertebrates are likely to differ substantially from those shown by generalists. Interestingly, this work bears out important parallels with Chernov's (1989, 1995) observations on diversity versus latitude gradients across taiga to arctic regions.

18.2 Population Cycling and Herbivores

One of the most tantalising questions in the textbooks on northern areas concerns the nature of cycling by rodent populations. There are hundreds of books and papers on this topic, yet still many research groups disagree on the key drivers of these cycles. As Jeffries and Bryant (1995) discuss, the composition of vegetation may have a key influence on vertebrate herbivore populations, and hence their tendency to cycle. In treeline ecosystems, and below, where woody and shrub-dominated vegetation persists, secondary compounds that deter herbivory may be so rich that herbivores are relatively uncommon. Hence, in grassy areas, and in graminoid/sedge-dominated plains of the arctic-alpine zone, herbivorous mammals predominate and cycle. Chapter 20 provides a fresh consideration of the population ecology and ecosystem impacts of the rodents of Europe's Alps (which have been less studied than the northern tundra and heaths in relation to rodent cycling; Stenseth and Ims 1993; CAFF 2002). The chapter considers the three rodent species of the alpine zone: snow vole (*Chionomys nivalis* Martins), common vole (*Microtus arvalis* Pallas), and alpine marmot (*Marmota marmota* L.). Populations of all of these are, to different degrees, importantly influenced by snow-lie and the pattern of snowmelt (see also Chap. 6). They probably have significant influences on the vegetation, although this has only been quantified in detail for the alpine marmot. Potential climate change influences are considered, and the indications are that these will lead to fragmentation of the alpine marmot populations. The authors conclude, however, that we still know little about the relative importance of vegetation and climate (or, strictly speaking, weather patterns) in regulating rodent populations. Jeffries and Bryant (1995) suggest that global warming, which will give rise to more shrub dominance at the expense of grasses, will lead to reductions in herbivore populations. It is striking that there is still a dearth of clear information on the impacts of rodents on alpine vegetation and predator populations. One is struck by the amount of detailed information we have for some of the arctic rodents in Europe. Perhaps there would be some merit in comparing many of the lowland rodent populations (which fluctuate greatly) with the alpine ones (which appear more stable, despite the greater variability in weather conditions).

We appear to know a lot more about the larger herbivores in alpine regions. Chapter 23 points out that domestic grazers have probably influenced alpine grasslands since 5000 B.P., though studies point to some impacts much earlier than that (albeit more locally, e.g. Thompson et al. 1995; CAFF 2002). Many observations of grazing impacts come from comparing exclosures with unenclosed areas. These demonstrate neatly that plant species diversity, abundance, structure, biomass and necromass are all markedly influenced by grazing. Yet two sorts of questions arise from such comparisons. First, there is likely to be a substantial difference between impacts on vegetation of no grazers (as is the case for exclosures) and low levels of grazing. Second, there is the issue of whether it is the physical action of grazing, trampling and/or dung/urine deposition that singly or in combination drive the observed changes in vegetation under grazing. Chapter 23 considers that dung/urine deposition is possibly more important in alpine areas. The authors make a plea for more detailed reviews of long-term exclosures, and for more detailed experimental studies. We consider this to be particularly important given the likely changes in temperature, nitrogen and ozone levels in the mountain regions (e.g. Hallanaro and Pylvänäinen 2001; CAFF 2002; Moorland Working Group 2002).

Chapter 21 considers the mountain ungulates of the Alps and Pyrenees. The authors demonstrate some of the profound impacts some of these ungulates have had on vegetation, and their role in ecosystem processes. They raise interesting questions about interactions between some of the ungulates, and about their potential influence on the distribution and abundance of predators. One of the important challenges raised within this study is the need for multi-faceted models to predict changes in the species structure, sex/age composition, and spatial distribution of assemblages in response to land management changes (e.g. Clutton-Brock et al. 2002). Research and modelling to address this need to be mindful of the relative scales at which different ungulates operate, and the fact that in many alpine regions human influences have already conditioned some responses more than others. Intriguingly, there is also the possibility that recent, growing numbers of some predators (notably wolf, *Canus lepus*) may cause ungulates to move out of habitats traditionally occupied (particularly at the lower edge of the alpine zone) and may give rise to increased levels of niche separation between some species.

18.3 Habitat Use

The first chapter in this section deals with an outlier area of Europe's alpine region, in the Scottish Highlands (Chap. 19). Here, in three study areas, the authors have quantified variability in habitat use by birds nesting in alpine habitats. This is the only paper we have on birds, which arguably reflects the greater research interest in mammals than birds in alpine regions (though not in the arctic). This paper highlights just how complex relationships are between invertebrates, habitats and birds. Indeed, the shifts in habitat use as the breeding cycle develops show how finely tuned to the environment predators have to be. One of the most interesting questions to emerge here is the extent to which just a few invertebrate species (notably *Tipula montana* Curtis) may drive differences in the numbers, distribution, habitat use and breeding ecology of some of the birds. For some species, microhabitats can be viewed as keystones: in the case of Euarasian golden plovers (*Pluvialis apricaria*), agricultural pastures can be viewed as such for some moorland and even tundra populations (e.g. Byrkjedal and Thompson 1998).

18.4 Conclusions

The five chapters in this section represent some fascinating aspects of vertebrate ecology in alpine areas in Europe. Nevertheless, some very basic questions still remain unanswered. The most fundamental of these is what variable (or complex of variables) governs the presence or absence of particular species, and why some species should be alpine zone specialists while others are more widely distributed? Chernov's (1989, 1995) observations point to physiological or phylogenetic explanations, but these are not clear. To what extent do climate conditions, habitat or food availability limit the presence of some species in this region? With growing concerns about vertebrate population conservation, one needs to look more at land management. Detailed practices have been devised for many of the herbivores, but are poorly developed for the rodents and the birds; this may need to change in the future as concerns grow for the conservation status of some of the small mammals and birds of the the alpine and arctic zone (see Hallanaro and Pylvänäinen 2001; CAFF 2002).

There appears to be considerable scope for developing geographical comparisons between the different alpine regions of Europe in order to determine rules to explain herbivore–vegetation associations, predator–prey relationships, and assemblage functions. Investigation of these issues has been given a significant stimulus by the papers reported here. It is noteworthy, of course, how little we seem to know about fish, and predator–prey relationships in alpine compared with arctic regions – something which should be addressed by fresh research.

Finally, the striking predominance of some vertebrate taxa in arctic rather than alpine areas is noteworthy – among birds, the Charadriiformes and, to a lesser extent, the Anseriformes are good examples. This may betray myriad adaptations to habitats, prey or some other feature of the environment not yet evident to researchers. With so many globally threatened species occurring in arctic/alpine areas (e.g. CAFF 2002 lists threatened arctic terrestrial species comprising 7 mammals, 12 birds and 10 fish), it will be important to determine precisely which factors account for variation in the composition and abundance of species in different alpine ecosystems – particularly those most sensitive to global warming.

Acknowledgements. I am grateful to Laszlo Nagy, Phil Shaw, Phil Whitfield and Paul Robertson for comments.

References

Byrkjedal I, Thompson DBA (1998) Tundra plovers: the Eurasian, Pacific, and American golden plovers and grey plover. Poyser, London

Chernov II (1989) Heat conditions and Arctic biota. Ekologiia 2:44–57

Chernov II (1995) Diversity of the Arctic terrestrial fauna. In: Chapin FS, Koerner C (eds) Arctic and alpine biodiversity. Springer, Berlin Heidelberg New York, pp 81–96

Clutton-Brock TH, Coulson TN, Milner-Gulland EJ, Thomson D, Armstrong HM (2002) Sex differences in emigration and mortality affect optimal management of deer populations. Nature 415:633–637

Conservation of Arctic Flora and Fauna/CAFF (2002) Arctic flora and fauna: status of conservation. Edita, Helsinki

Hallanaro E-L, Pylvänäinen M (2001) Nature in northern Europe – biodiversity in a changing environment. Nordic Council of Ministers, Copenhagen

Jeffries RL, Bryant JS (1995) The plant-veretbrate herbivore interface in Arctic ecosystems. In: Chapin FS, Koerner C (eds) Arctic and alpine biodiversity. Springer, Berlin Heidelberg New York, pp 271–281

Moorland Working Group (2002) Scotland's moorland – the nature of change. Scottish Natural Heritage, Battleby

Stenseth NC, Ims RA (1993) Population dynamics of lemmings: temporal and spatial variation – an introduction. In: Stenseth NC, Ims RA (eds) The biology of lemmings. Academic Press, London, pp 61–96

Thompson, DBA, Hester, AJ, Usher, MB (eds) (1995) Heaths and moorland: cultural landscapes. The Stationery Office, Edinburgh

19 Breeding Bird Assemblages and Habitat Use of Alpine Areas in Scotland

D.B.A. THOMPSON, D.P. WHITFIELD, H. GALBRAITH, K. DUNCAN, R.D. SMITH, S. MURRAY and S. HOLT

19.1 Introduction

'Where a species is adapted to a particular habitat, its distribution limits will tend to be circumscribed by whatever conditions restrict the occurrence of that habitat, but these conditions may vary over the total range ... Probably most of our upland species show a similar breeding biology in Britain to that elsewhere, and their habitats here thus tend to be scaled down equivalents of those over their much larger European, Eurasian or global ranges' (Ratcliffe 1990).

Despite several recent studies of British upland birds (e.g. Fuller 1982; Stroud et al. 1987; Ratcliffe and Thompson 1988; Haworth and Thompson 1990; Ratcliffe 1990; Thompson et al. 1995; Fielding 1999), there is still a considerable gap in our knowledge of upland breeding birds. Several 'azonal' species breed above the former treeline in Britain (in the alpine zone, sensu Horsfield and Thompson 1997), but only four 'specialist' species breed only there: ptarmigan, dotterel, purple sandpiper and snow bunting (scientific names are given in the Appendix). British populations of these four species are outliers of the Arctic Eurasian avifauna.

The alpine zone covers approximately 4% of Britain's land surface, ranging from above 650 m in the central Highlands down to above around 350 m in the north-west (Thompson and Brown 1992); its inaccessibility and harshness largely accounts for the scarcity of ecological information on its birds. However, the expansion of ski developments, wider realisation of sheep (*Ovisaries*) and deer (*Cervus* spp.) impacts on vegetation, greater recreational uses, concerns about global warming and local increases in acidic (notably nitrogen) deposition have highlighted the need to study this zone and its wildlife in more detail (e.g. Thompson et al. 1987; Thompson and Brown 1992; Galbraith et al. 1993b; Thompson et al. 2001; Whitfield 2002).

This chapter presents a quantitative comparison of bird assemblages on three mountain plateaux distinguishable by climate, geology and vegetation. It examines differences in species richness and abundance in relation to available habitat, and the use made of these habitats for foraging and nesting. The work provides a means of predicting how bird numbers may be affected by further habitat change, and complements autecological studies made over the past 30 years (e.g. Watson 1965; Nethersole-Thompson 1973, 1994; Thompson and Whitfield 1993; Thompson et al. 1995; Byrkjedal and Thompson 1998; Watson et al. 2000; Holt et al. 2002).

19.2 Study Area and Methods

Observations were made on two alpine plateaux in the central Scottish Highlands (sites A and B). Additional observations were made on a third plateau in the eastern Highlands (site C).

Site A is situated between approximately 1000 and 1200 m above sea level (a.s.l.), and comprises 8.2 km^2 of flat or gently sloping terrain. The solid geology is granite, producing an infertile, skeletal soil. The climate is severe with frequent gales and prolonged periods of precipitation (which regularly falls as snow as late as mid-June). Low cloud frequently blankets the site and snow lies locally for around 6–8 months each year, with more or less continuous cover between December and April. By mid-June snow cover is normally down to about 10–20 % (Table 19.1).

Site B comprises 4.0 km^2 of plateau between 800 and 950 m a.s.l. The underlying rock type is Moinian schist, which is base-rich in comparison with the granite site A. The weather is less severe than on the higher site A: temperatures are higher, winds less strong and low cloud less frequent. Snow cover may be continuous during the winter, but is generally down to less than 10 % by early May, and absent by June. The more fertile soil and less severe weather give rise to almost continuous plant cover (except on exposed, ablated ridges). Boulders are sparse except on a few steeper, eroding slopes.

Site C comprises 2.7 km^2 of plateau on Dalradian schist between about 900 and 1070 m a.s.l. The soil is the most base-rich of the three study areas. The climate is less severe than site A, but similar to that of site B.

The plant communities are readily distinguished in Britain's alpine areas (e.g. Thompson and Brown 1992). The relative cover for the major plant communities on the three sites is shown in Table 19.1. These data were obtained by visiting 50 randomly placed points on each plateau, recording the plant community at each, and then expressing the recorded percent frequency of each community as percent total in the study area. The estimated total areas of each community were then obtained from this last figure and the area of the site. We used detailed vegetation maps for each of the sites (Thompson and

Table 19.1. Habitat composition of the three study areas

Site	Vegetation (habitat) type[a]	Random sites (%)	Estimated area (km^2)
Site A	*Nardus* grassland	33	2.71
	Vaccinium–Empetrum heath	22	1.80
	Racomitrium heath	6	0.49
	Juncus trifidus heath	35	2.87
	Boulders	2	0.16
	Flush	2	0.16
Site B	Bog	34	1.36
	Nardus grassland	22	0.88
	Vaccinium–Empetrum heath	10	0.40
	Racomitrium heath	26	1.04
	Dwarf *Calluna* heath	8	0.32
Site C	Bog	8	0.21
	Nardus grassland	16	0.43
	Vaccinium–Empetrum heath	12	0.32
	Carex–Racomitrium heath	62	1.67
	Juncus trifidus heath	2	0.05

[a] Vegetation type refers to the dominant constant species or physical features (e.g. Thompson and Brown 1992).

Brown 1992; Brown et al. 1993) to confirm that there were no anomalies in Table 19.1.

Bird censusing and territory mapping were based on the 'Common Birds Census' technique (Williamson and Holmes 1964). Regular visits were made to all three sites between early May and August (the number of visits made annually to each site varied but was not less than 60). The locations of nesting/chick-rearing birds were distinguished by repeated observations of displaying or singing birds (a minimum of three such registrations in an area constituted a territory), and by finding nests and broods. Nine territorial species (dunlin, golden plover, ptarmigan, red grouse, wheatear, meadow pipit, skylark, snow bunting and dipper) were mapped in this way. Dotterel are non-territorial (Scottish Natural Heritage, unpubl.; Kålås and Byrkjedal 1984; Owens et al. 1994) and were surveyed by finding nests or broods. Additional information on dotterel movements was obtained from observations of colour-ringed adults. Display, nesting and brood-rearing habitats were recorded using the habitat categories in Table 19.1.

It is unlikely that every single territory and nesting pair were located using these methods. However, since the methods were the same on each site, and the sites were broadly similar in terms of detectability of birds, we are confident of the comparability of results between our sites (Galbraith et al. 1993b; Owens et al. 1994; Whitfield 2002).

19.3 Results

19.3.1 The Breeding Bird Assemblages

Both breeding densities and species richness were generally higher on sites B and C than on site A (Table 19.2). On sites B and C there was also a higher number of individuals of species typically associated with lower altitudes, and breeding densities of alpine species common to all three sites were higher than on A. However, site A held more specialist alpine species, including some which were not found on the other sites (i.e. snow bunting, purple sandpiper, and snowy owl, which summered but did not breed). Ptarmigan and passerine numbers were higher in 1988 at site A but not at site B; at site B, dunlin and golden plover numbers were markedly lower in 1988.

Table 19.2. Number of pairs and densities (pairs km^{-2}) of birds breeding on the three study sites in 1987 and 1988[a]

Species	Site A		Site B		Site C
	1987	1988	1987	1988	1988
Dotterel	18–22 (2.4)	16–29 (2.2)	28 (7.0)	29 (7.2)	7 (2.6)
Dunlin	–[a]	–	16 (4.0)	9 (2.2)	5 (1.8)
Golden plover	–	–	13 (3.2)	8 (2.0)	2 (0.7)
Purple sandpiper	3 (0.4)	2 (0.2)	–	–	–
Ptarmigan	21 (2.6)	37 (4.5)	23 (5.7)	18 (4.5)	16 (6.1)
Red grouse	–	–	3 (0.7)	3 (0.7)	–
Wheatear	6 (0.7)	11 (1.3)	4 (1.0)	3 (0.7)	4 (1.5)
Meadow pipit	6 (0.7)	14 (1.7)	10 (2.5)	12 (3.0)	8 (3.0)
Skylark	–	–	9 (2.2)	5 (1.2)	3 (1.1)
Snow bunting	6–9 (1.0)	14 (1.7)	–	–	–
Dipper	–	–	–	1 (0.2)	–
Total	(7.8)	(11.6)	(26.3)	(21.7)	(16.8)

[a] – Denotes not recorded.

19.3.2 Habitat Use

The breeding densities of birds in each of the vegetation types varied across the three sites (Table 19.3). Overall densities were lower on bog and *Nardus* grassland and higher on dwarf-shrub, *Juncus trifidus* (site A) and moss-dominated heaths.

Within these broad density patterns there are marked inter-specific differences in habitat use:

1. *Passerines*. Whereas meadow pipits on site B selected dwarf-shrub heaths for their nesting territories, skylarks preferred bog (Table 19.4). On site C data were too few for statistical analysis but again meadow pipits showed an apparent preference for dwarf-shrubs. On site A meadow pipits showed a significant preference for dwarf-shrub heaths and avoidance of mossy habitats, whereas wheatears preferred *Juncus trifidus* and dwarf-shrub heath and avoided *Nardus* grassland (significance levels given in Table 19.4). Ten of the 17 wheatear territories found on site A were on steep, rocky ground and four were on more level, but boulder-strewn ground. Similarly, on sites B and C wheatears were associated with steep, eroding ground.
Snow buntings on site A avoided *Nardus* grassland and selected *Juncus trifidus* and dwarf-shrub heath. Seven of the 20 territories were centred on the cliffs of corrie head-walls at the periphery of the plateau. Although these birds nested on the cliffs or associated talus slopes, they regularly fed on the plateau. Another six territories were on more level boulder fields on the plateau.
2. *Waders (Charadrii) and ptarmigan*. We found subtle differences in habitat use by dotterel between the incubation and chick-rearing phases of breeding (Table 19.5). On sites B and C, dotterel preferred to nest and raise their broods on *Racomitrium* heath (and on prostrate *Calluna* heath on site B), and generally avoided *Nardus* grassland and bog, particularly during chick rearing (Table 19.5). On site A, where *Racomitrium* heath was scarce, most dotterel nested and reared their broods on *Juncus trifidus* heath, though birds still showed a preference for *Racomitrium* heath; the avoidance of

Table 19.3. Overall breeding bird densities (pairs km^{-2}) in the main vegetation types of each of the three study sites

	Bog	*Nardus* grassland	Dwarf-shrub heath	Moss heath	*Juncus trifidus* heath
Site A	–	5.5	23.9	32.6	28.2
Site B	17.6	10.8	27.0	26.0	–
Site C	2.6	7.9	34.5	7.3	–

Table 19.4. Observed (and expected[a]) numbers of breeding territories of passerines and waders in the main vegetation types across the three study areas

Species	Site	Bog	*Nardus* grassland	Dwarf-shrub heath	Moss heath	*Juncus trifidus* heath	Chi-square value[b]
Meadow pipit	A	—	1 (6.3)	15 (4.2)	1 (1.1)	2 (6.6)	29.2**
	B	6 (10.2)	3 (6.6)	17 (5.4)	4 (7.8)	—	30.5***
	C	0 (0.6)	0 (1.3)	7 (1.0)	1 (5.0)	0	—
Skylark	A	—	0	0	0	—	
	B	11 (5.4)	1 (3.5)	3 (2.9)	1 (4.2)	—	9.1*
	C	0	0	3	0	—	
Wheatear	A	—	0 (5.6)	6 (3.7)	1 (1.0)	10 (5.9)	9.6*
	B	0	0	0	0	—	+
	C	0	0	0	0	—	+
Snow bunting		—	1 (6.6)	5 (4.4)	2 (1.2)	12 (7.0)	8.7*
Golden plover	B	17 (8.4)	4 (5.7)	3 (4.7)	2 (5.7)	—	10.8*
	C	0	0	0	2	—	42.7***
Dunlin	B	4 (8.5)	19 (5.5)	0 (4.5)	2 (6.5)	—	

[a] The expected number is calculated for territories as if these were distributed at random.
[b] *$P<0.05$; **$P<0.01$; ***$P<0.001$; Chi-squared (X^2) test comparing, for each species on a given site, observed and expected territory according to vegetation types. + denotes that all wheatear territories were associated with steep, naturally eroding ground.

Nardus grassland was marked. Interestingly, dotterel with broods showed more significant preferences for particular vegetation types than when with nests, with significant preferences for *Juncus trifidus* heath on site A and highly significant preferences for moss-dominated heaths on sites B and C.

Dunlin were scarce on site C and absent from site A. On site B there was a strong preference for *Nardus* grassland as a nesting habitat, with *Nardus* comprising 22% of the available area but containing 78% of nesting territories (Table 19.4). *Nardus* grassland and bog were the preferred chick-rearing habitats ($\chi^2=12.3$, $P<0.001$). Bog held only 11% of dunlin nests but 40% of brood registrations, indicating that broods moved from *Nardus* to bog after hatching. On site B, golden plover showed a preference for nesting in bog (Table 19.4). Brood observations were too few to allow statistical analysis.

There was no tendency for ptarmigan to choose particular vegetation types for nesting territories on sites A or B. However, after hatching, the broods on site B were found more often in dwarf-shrub heaths, whilst broods on site A were reared on *Juncus trifidus* heath.

Table 19.5. Habitat use by breeding dotterel during the incubation and chick-rearing phases[a]

Main vegetation types	Study areas					
	Site A		Site B		Site C	
	Incubation	Chick Rearing	Incubation	Chick Rearing	Incubation	Chick Rearing
Racomitrium lanuginosum heath	+***	+	+***	+***	+	+***
Nardus grassland	–***	–***	0	–***	+	–*
Vaccinium-Empetrum heath	0	–**	0	0	0	0
Carex flushes	0	–**	0	0	NP	NP
Calluna-Eriophorum bog	NP	NP	–*	–***	0	0
Prostrate *Calluna* heath	NP	NP	0	+	NP	NP
Juncus trifidus heath	+*	+*	NP	NP	NP	NP
Liverwort snowbed	0	0	NP	NP	NP	NP

[a] Habitats are preferred (+) or avoided (–). On each site birds have a choice of four or six vegetation types.
– denotes no preference; NP denotes not present. *$P<0.05$; **$P<0.01$; ***$P<0.001$; χ^2 test comparing observed and (randomly) expected habitats (see Table 19.4).

19.4 Discussion

19.4.1 Breeding Bird Assemblages

The breeding bird assemblages on the three study sites differed in population densities and species composition. On site A, densities were generally low, but there were more alpine species there (four out of six species). Some of these alpine birds breed on other British mountains, but their preponderance on site A signifies that the site is more closely allied to higher latitude areas (e.g. in Scandinavia) than to many other British sites.

The presence of more arctic-alpine specialists on site A may be due to a combination of severe, unpredictable weather and infertile soils producing an environment akin to arctic areas. The absence from there of some species found more commonly on lower altitude sites may be due to an absence of suitable habitat, rather than any climatic constraints. For example, the absence of dunlin and golden plover is probably due to these species showing a strong association with bog, which was absent from site A.

On sites B and C, there were fewer arctic-alpine species (two), and higher densities of typically lower-ground birds. Lower altitudes, less severe weather and more fertile soils result in less harsh environments more suitable for a wider range of upland birds. Thus, unlike site A, the breeding bird communities of sites B and C show closer affinities to those found on blanket bog and dry heath in sub-montane areas than to more arctic habitats. It is possible, however, that the relative smallness of sites B and C compared with A contributed to the absence of snow bunting and purple sandpiper.

Given that only sparse populations of waders were hitherto believed to nest in the alpine zone, the densities of breeding waders found on site B were unexpectedly high (11–14 pairs km^{-2}). The densities of golden plover and dunlin (3.2 and 4.0 pairs km^{-2}) were greater than the means of a range of densities recorded by Stroud et al. (1987) on the Caithness peat bogs (2.4 and 1.8 pairs km^{-2}) and by Byrkjedal and Thompson (1998) for many other parts of the northern hemisphere. The dotterel breeding density is high in comparison with other British sites (Galbraith et al. 1983b), and exceeds those found in Scandinavia (Kålås and Byrkjedal 1984). Indeed, the recorded densities in these Scottish sites are comparable to the highest densities recorded anywhere else (around 9–10 pairs km^{-2} in arctic Russia; Cramp and Simmons 1983). Ptarmigan densities on sites B and C were higher than found on more northern sites (e.g. West Greenland, Baffin Island and central Alaska), and as high as recorded on Iceland (Watson 1965; Watson et al. 2000).

19.4.2 Habitat Use

Ptarmigan showed no marked nest site preferences on any study area. However, they led their young to dwarf-shrub heath (dwarf shrubs are the main food plants of adults and older chicks; Cramp and Simmons 1980; Watson et al. 2000) on sites B and C, and to *Juncus trifidus* heath on site A (where there was only a limited area of dwarf-shrub heath available).

Waders showed marked habitat preferences on all three sites, and a degree of habitat partitioning was evident on site B; dotterel nested and reared their broods on *Racomitrium* heaths and usually avoided *Nardus* grassland and bog; dunlin nested on the former and led their young to the latter. Unlike the dotterel, golden plovers showed a strong preference for blanket bog and were less strongly associated with the more exposed habitats occupied by dotterel. The differential habitat use of the three commonest wader species found in this study may be a means of avoiding competition, or may simply reflect different preferences.

Differing nest site requirements in relation to vegetation structure also help explain the non-overlapping distributions of the waders on site B. On sites B and C, dotterel nested mainly in short, open vegetation (moss heaths) and avoided the dense, tussocky *Nardus* grassland. On site A, however, they did nest in *Nardus* grassland. This plant community on site A differs from its equivalents on sites B and C in that it is sparser, shorter and has more unvegetated patches (more closely resembling a moss-dominated or *Juncus* heath in horizontal and vertical structure). In contrast, dunlin, which build concealed nests in grass tussocks, preferred the *Nardus* grassland on sites B and C.

Snow buntings and wheatears preferred mainly rocky ground for nesting territories. Both nest under rocks, and their territory habitat use is, presumably, a reflection of this. Meadow pipits conceal their nests among vegetation, which explains their strong association with the comparatively tall dwarf-shrub heath vegetation.

19.4.3 Consequences of Habitat Modification

This study has shown that Scotland's alpine environment supports important breeding bird assemblages, with each species showing differences in habitat preferences. Any factor that modifies these habitats is likely to impact on the birds.

Heavy grazing, acidification, and heavy recreational use are all known to modify alpine habitats (e.g. Thompson et al. 1987; Thompson and Brown 1992; Thompson et al. 2001). Heavy grazing pressure is thought to have already led to the replacement of moss heaths with grassy heaths in large areas of the English, Welsh and southern Scottish uplands, with consequent losses of breeding birds (Thompson et al. 1987; Thompson and Brown 1992). Among the rare or restricted alpine birds, heavy grazing and acidic deposition, leading to such vegetation changes, may be expected to reduce dotterel and ptarmigan numbers. Returning to Ratcliffe's (1990) opening quote, we hope that these results can be of use to studies of habitat use and site differences for arctic-alpine species in other parts of their breeding range.

19.5 Conclusions

1. This chapter describes associations between breeding birds and their habitats in alpine areas in Scotland, and considers some implications of habitat change.
2. Densities of birds were low on alpine bog and *Nardus stricta* grasslands, and high on heaths dominated by dwarf-shrubs (*Erica* spp., *Vaccinium* spp., *Calluna vulgaris*), *Juncus trifidus* and bryophytes.
3. Habitat preferences varied across sites, but the following general preferences were found for: (i) alpine bog (skylark, golden plover and dunlin); (ii) *Nardus stricta* grassland (dunlin for nest sites); (iii) dwarf-shrub heaths (meadow pipit); and (iv) moss-dominated heaths and open fellfields (dotterel).
4. Dotterel, dunlin and ptarmigan exhibited strong habitat preferences during nesting and brood-rearing periods.
5. The British alpine breeding bird assemblage is a distinctive collection of species which is not found at such high densities or consisting of such a wide range of species in comparable oceanic habitats elsewhere in the world, and in other alpine areas in Europe.

Acknowledgements. We thank Stuart Rae and Chris Thomas for assistance in the field, Derek Ratcliffe, Adam Watson, John Atle Kålås, Ian Owens, David Horsfield, Michael B Usher, Suzanne Weinberg, John Hazlett, Janette Munneke, an anonymous referee and Laszlo Nagy for discussion or comments. The work was funded by the Nature Conservancy Council and Scottish Natural Heritage as part of the Montane Ecology project.

Appendix

Scientific names of species mentioned in the text.
Dipper, *Cinclus cinclus*
Dotterel, *Charadrius morinellus*
Dunlin, *Calidris alpina*
Golden plover, *Pluvialis apricaria*
Meadow pipit, *Anthus pratensis*
Ptarmigan, *Lagopus mutus*
Purple sandpiper, *Calidris maritima*
Red grouse, *Lagopus lagopus*
Skylark, *Alauda arvensis*
Snow Bunting, *Plectrophenax nivalis*
Snowy owl, *Nyctaea scandica*
Wheatear, *Oenanthe oenanthe*

References

Brown A, Thompson DBA (1992) Biodiversity in montane Britain: habitat variation, vegetation diversity and some objectives for conservation. Biodiv Conserv 1:179–208

Brown A, Birks HJB, Thompson DBA (1993) A new bio-geographical classification of the Scottish uplands. Vegetation-environment relationships. J Ecol 81:231–251

Byrkjedal I, Thompson DBA (1998) Tundra plovers. Academic Press, London

Cramp S, Simmons KEL (eds) (1980) The Birds of the western Palearctic, vol II. Oxford University Press, Oxford

Cramp S, Simmons KEL (eds) (1983) The Birds of the western Palearctic, vol III. Oxford University Press, Oxford

Fielding AH (1999) Ecological applications for machine learning methods. Kluwer, Boston

Fuller RJ (1982) Bird habitats in Britain. Poyser, Calton

Galbraith H, Duncan K, Murray S, Smith R, Whitfield DP, Thompson DBA (1993a) Diet and habitat use of the dotterel (*Charadrius morinillus*) in Scotland. Ibis 135:148–155

Galbraith H, Murray S, Rae S, Whitfield DP, Thompson DBA (1993b) Numbers and distribution of dotterel (*Charadrius morinellus*) breeding in Great Britain. Bird Study 40:161–169

Gibbons DW, Reid JAB, Chapman RA (ed) (1993) The new atlas of breeding birds in Britain and Ireland, 1988–1991. Poyser, London

Haworth PF, Thompson DBA (1990) Factors associated with the distribution of upland birds in the south Pennines, England. J Appl Ecol 27:562–577

Holt SD, Whitfield DP, Duncan K, Rae S, Smith RD (2002) Mass loss in incubating Eurasian dotterel: adaptation or constant? J Avian Biol 33:219–224

Horsfield D, Thompson DBA (1997) The uplands: guidance on terminology regarding altitudinal zonation and related terms. Information and Advisory Note 26. Scottish Natural Heritage, Battleby

Kålås JA, Byrkjedal I (1984) Breeding chronology and mating system of the Eurasian Dotterel. The Auk 101:838–847

McVean DN, Ratcliffe DA (1962) Plant communities of the Scottish Highlands. HMSO, London

Nethersole-Thompson D (1973) The dotterel. Collins, London

Nethersole-Thompson D (1994) The snow bunting. Reprint. Peregrine Books, Leeds

Nethersole-Thompson D, Watson A (1981) The Cairngorms. The Melven Press, Perth

Owens IPF, Burke T, Thompson DBA (1994) Extraordinary sex roles in the Eurasian Dotterel: female mating arenas, female-female competition and female mate choice. Am Nat 144:76–100

Ratcliffe DA (1990) Bird life of mountain and upland. Cambridge University Press, Cambridge

Ratcliffe DA, Thompson DBA (1988) The British uplands: their ecological character and international significance. In: Usher MB, Thompson DBA (eds) Ecological change in the uplands. Blackwell, Oxford, pp 9–36

Stroud DA, Reed TM, Pienkowski MW, Lindsay RA (1987) Birds, bogs and forestry: the peatlands of Caithness and Sutherland. Nature Conservancy Council, Peterborough

Thompson DBA, Brown A (1992) Biodiversity in montane Britain: habitat variation, vegetation diversity and some objectives for conservation. Biodiv Conserv 1:179–209

Thompson DBA, Whitfield DP (1993) Research on mountain birds and their habitats. Scottish Birds 17:1–8

Thompson DBA, Galbraith H, Horsfield DH (1987) Ecology and resources of Britain's mountain plateaux: conflicts and land use issues. In: Bell M, Bunce RG (eds) Agriculture and conservation in the hills and uplands. Institute of Terrestrial Ecology, Merlewood, pp 22–31

Thompson DBA, Hester AJ, Usher MB (eds) (1995) Heaths and Moorland: cultural landscapes. HMSO, Edinburgh

Thompson DBA, Gordon JE, Horsfield D (2001) Mountain landscapes: are these natural, relicts or artefacts. In: Gordon JE, Lees K (eds) The earth heritage of Scotland. The Stationary Office, Edinburgh, pp 105–119

Watson A (1965) A population study of ptarmigan in Scotland. J Appl Ecol 34:135–172

Watson A (1979) Bird and mammal numbers in relation to human impact at ski lifts on Scottish hills. J Appl Ecol 16:753–764

Watson A, Moss R, Rothery P (2000) Weather and synchrony in ten-year population cycles of rock ptarmigan and red grouse in Scotland. Ecology 81:2126–2136

Whitfield DP (2002) The status of breeding dotterel *Charadrius morinellus* in Britain in 1999. Bird Study 49:237–249

Williamson K, Holmes RC (1964) Methods and preliminary results of the common birds census. Bird Study 11:240–256

20 Rodents in the European Alps: Population Ecology and Potential Impacts on Ecosystems

D. ALLAINÉ and N.G. YOCCOZ

20.1 Introduction

Rodents have been extensively studied in most ecosystems of the world, and particularly so in northern regions (e.g. Hansson and Henttonen 1988; Stenseth and Ims 1993a; Reid et al. 1997). In the European Alps, however, relatively little is known about the population ecology of alpine rodent species, their role in shaping alpine ecosystems, and their interaction with vegetation growth and diversity. We review here some of the recent studies made on the three species of rodents living above the treeline in the European Alps: the alpine marmot (*Marmota marmota* L.), the snow vole (*Chionomys nivalis* Martins), and the common vole (*Microtus arvalis* Pallas). In particular, we discuss: (1) whether or not life history traits of alpine European rodents differ from those found in closely related species occurring in other alpine, northern or temperate environments; (2) the population dynamics and demography of these species and their potential consequences on ecosystem functioning; and (3) the consequences of changes in management practices and climate for these species.

20.2 Is There an 'Alpine Rodent'?

Alpine environments are rather heterogeneous because they exist across a broad band of latitudes and in very different climatic regimes (Barry 1992). The Alps are typical of temperate mountain environments, in general, with a deep winter snow cover and relatively mild temperatures, compared to arctic or continental areas. There is a degree of geographical variation, with Mediterranean influences in the south, and continental ones in the eastern or central parts of the range (Haeberli and Beniston 1998). One of the main differences from arctic regions is the absence of permafrost, except for glaciers (Haeberli and Beniston 1998) and sheet ice formation (caused by freezing rains, which are rare in the Alps). Ice cover on the ground has important consequences for herbivores, by limiting access to food resources, and in particular small mammals (Yoccoz and Ims 1999). The Alps have been heavily influenced by human activities for many thousands of years and the vegetation patterns, as well as the distribution of many mammal species, reflect this fact (Breitenmoser 1998; Chap. 21). The alpine zone has been used for grazing by cattle and sheep, and the location of the treeline reflects an interaction between local climatic factors and human influences.

Three species of rodents permanently inhabit the alpine zone: the alpine marmot, the common vole and the snow vole. Hibernation is the main factor besides body size distinguishing marmots from voles, and we show below that it determines most of the life-history characteristics as well as potential impacts on the ecosystem of the marmots.

While the genus *Marmota* covers the entire holarctic region, the spatial distribution of the alpine marmot (*M. marmota marmota*) is restricted exclusively to the European Alps (with a subspecies, *M. marmota latirostris* in the high Tatras, Slovakia; Kratochvil 1961). Following introduction or reintroduction events, some marmot populations are also found in the Pyrenees and in the Massif Central (France), in Slovenia and in the Romanian Carpathians.

Marmots occupy a wide range of open habitats that experience severe winter conditions (Barash 1989). Some species, such as the woodchuck (*Marmota monax*) in the New World or the bobak marmot (*Marmota bobac*) in Eurasia, inhabit flat fields or steppe (Formozov 1966), but most marmot species are found in mountain habitats. The alpine marmot is typically found in montane, treeline ecotone and alpine meadows at elevations from 800 to 3200 m (Forter 1975), but mostly between 2300 and 2700 m (Grimod et al. 1991). They prefer to occupy territories on southerly slopes with outcropping rocks (Allainé et al. 1994).

Marmot species present adaptations to the harshness of their environment. All 14 marmot species are hibernators; joint hibernation is observed in species living in the harshest environments, while solitary hibernation has been observed in *Marmota monax* and in *M. flaviventris*. Social thermoregulation has been described in the alpine marmot (Arnold 1988) where, during periodic 'arousals', yearlings and adults actively warm juveniles and allow them to save fat reserves (Arnold 1990). This social thermoregulation seems to increase the juvenile survival whenever subordinate males are present in the hibernaculum (Allainé et al. 2000). It has been proposed that sociality has evolved in those species facing the most severe conditions (Barash 1974a) possibly because of the need for social thermoregulation. The alpine marmot is one of the most highly social marmot species (Michener 1983). Its basic social unit is the family group that consists of a resident pair, subordinate adults, yearlings, and juveniles of the year (Perrin et al. 1993). All family members share a common home range and use the same burrows. Within-group interactions are mainly cohesive (Perrin et al. 1993). Social thermoregulation may have also implied a complete reproductive skew among females of family groups (Allainé 2000) and thus led predominantly to the social monogamy observed in this species (Goossens et al. 1998) and in other highly social marmot species (Blumstein and Armitage 1999). It has been shown that where short growing season is limiting, hibernating species (Clark 1970) and particularly marmots (Graziani et al., unpubl.) show an accelerated ontogeny compared to non-hibernating species of the same size.

Marmots are the largest true hibernators, and the alpine marmot is a medium-sized marmot (Armitage 1999). Sexual dimorphism in size exists both among adults (Farand, unpubl.) and juveniles (Allainé et al. 1998), but is relatively small. The adult size is reached when individuals are 2 years old.

Only two species of small rodents are permanently found above the treeline: the snow vole and the common vole. Other vole species such as the fossorial form of the water vole (*Arvicola terrestris* Scherman) or the subterranean voles (*Pitymys* spp.) are more typical of the treeline ecotone but can become abundant in the alpine zone in peak years (N.G. Yoccoz, pers. observ.). Wood mice, and in particular the alpine wood mouse (*Apodemus alpicola* Heinrich), as well as the bank vole (*Clethrionomys glareolus* Schreber), are found in the alpine zone wherever there is sufficient cover, and in particular bushes (*Rhododendron, Juniperus*; Yoccoz 1992; Reutter et al. 1999; N.G. Yoccoz, pers. observ.). The snow vole is found on screes from the sea level along the Mediterranean coast to above the snow line in the Alps (Le Louarn and Janeau 1975; Jones and Carter 1980; Kratochvil 1981; Krystufek and Kovacic 1989; Claude 1995; Yoccoz and Ims 1999). The common vole has a disjunct distribution; the species being absent from the treeline ecotone, and the populations found in the alpine zone have been claimed to belong to a different (sub)species, *Microtus a. incertus* (Le Louarn et al. 1970; see Meylan 1995). The common vole inhabits alpine meadows with sufficient grass cover. There is therefore very limited overlap in habitat use between the two species, and it is unlikely there is any direct competition between them.

The snow vole is a medium-sized vole, closely related to *Microtus*, showing specific adaptations to life in rocky habitats (long tail, claws and vibrissae; e.g. Frank 1953). There is no evidence for a change in body size (adults ca. 50 g) with altitude, and sexual dimorphism in body size is not large (males being about 25% heavier than the females; unpubl. data). The alpine common vole is relatively small (males 25–30 g, females 20–25 g), and appears to be somewhat smaller than its lowland conspecific. However, the large fluctuations in body size observed in lowland populations (e.g. Moravec and Vlasak 1988) make any comparison difficult. Both species show a rather typical social system for *Microtus* voles, with a polygynous mating system, and no evidence for strict territoriality (N.G. Yoccoz, unpubl. data). Kin females in both species seem to establish overlapping home ranges.

The three species of alpine rodents do not therefore differ in obvious ways from related species found in lowland or northern habitats. However, as we will emphasise below, they differ extensively in aspects of their population ecology.

20.3 Population Dynamics and Demography

The evolution of sociality in marmots has been well studied but, surprisingly, the dynamics of their populations is still poorly understood. Information on survival rates is available only for the yellow-bellied marmot (Schwartz et al. 1998), the alpine marmot (Farand et al. 2002), and, to a lesser extent, for the Vancouver Island marmot (Bryant 1996). The two former species seem to have stable populations but the declining demographic trend of the Vancouver Island marmot is worrying (less than 300 individuals remain). Compared with the yellow-bellied marmot, the alpine marmot shows a similar mean litter size at emergence from the natal burrow (about 4.1 in both cases, Van Vuren and Armitage 1994; Allainé et al. 2000), but a delayed age of dispersal (yearling age in the yellow-bellied, Armitage 1991; 2-year-old or more in the alpine marmot, Arnold 1990) and, consequently, a delayed age at first reproduction. In harsh environments, delaying dispersal may be a life-history tactic to avoid too high a mortality in the first year of life (Armitage 1981). The frequency of female reproduction also seems to decrease with increases in the harshness of the environment. Thus, female woodchucks and yellow-bellied marmots reproduce more regularly than alpine marmot females (which frequently skip reproduction). The more extreme case is observed in the hoary marmot where females reproduce biennially (Barash 1974b). Adult (0.65–0.7) and juvenile (0.54–0.6) survival rates are high in both species but somewhat higher in the alpine marmot. Climatic factors such as the duration of snow cover in the yellow-bellied marmot (Van Vuren and Armitage 1991) and the early onset of freezing in the alpine marmot (Farand et al. 2002) affect adult survival. In the alpine marmot, when freezing occurs before snowfall, the survival of adults is reduced (Farand et al. 2002), probably because of the steep decrease in temperature within the hibernaculum. In our study sites, most avalanches occur before marmot emergence. So, avalanches may potentially affect marmot survival by precluding emergence or access to food (D. Allainé, pers. observ.), particularly for territories located on avalanche corridors. Predation may cause significant mortality among juvenile yellow-bellied marmots (Schwartz et al. 1998), while infanticide may be the main source of juvenile summer mortality among alpine marmots (Coulon et al. 1995). However, the survival rates of juvenile alpine marmots are remarkably stable (coefficient of variation less than 15 %; Farand et al., submitted) probably as a consequence of social thermoregulation during winter. Little is known about the population dynamics and demography of voles and other alpine small rodents in other alpine ranges (e.g. in North America). The population dynamics of small rodents has, however, been extensively studied in other environments, mainly because of their well-known multiannual fluctuations in abundance (e.g. Stenseth 1999; Stenseth and Ims 1993b). These fluctuations are usually of a larger amplitude and a periodic component in northern, seasonal environments, and the same pattern could be expected in the alpine, seasonal environment. The two species of alpine rodents seem, however, to have stable populations, with very little variability in density between years, and relatively high densities in favourable habitats (Yoccoz and Ims 1999). The water vole, however, fluctuates cyclically in the treeline ecotone (Saucy 1994). The stability observed is based on the study of only a few populations and may be argued not to be a general phenomenon. However, the large differences observed in the demography of the two alpine species compared with their conspecifics or congeners indicate that stability is indeed a trademark of alpine small rodents. In fact both the snow and the common voles have small litter sizes, delayed age at first reproduction, and high survival, particularly in the winter period (Yoccoz and Ims 1999). A similar trend for a low turnover is found in the populations of the bank vole occurring in the treeline ecotone (Yoccoz and Mesnager 1998). It is not entirely clear what the ultimate causes for this pattern are, but the protective snow cover in the winter and the absence or scarcity of weasels in the (sub)alpine zone seem to be reasonable candidates (Yoccoz and Mesnager 1998; Yoccoz and Ims 1999). The snow vole may represent an extreme case because its rocky habitat provides environmental stability and protection against avian and mammalian predators. In fact, other species inhabiting similar rocky habitats such as the pika (a lagomorph, *Ochotona princeps* Richardson) have demographic traits very similar to the snow vole (Smith 1988; see also Millar and Teferi 1993 for the deer mouse *Peromyscus maniculatus* Wagner). The demography of other alpine voles such as *Alticola* spp. in central Asia or alpine populations of *Microtus montanus* Peale in North America would clearly be worth studying to show if this is a general characteristic of alpine small rodents.

20.4 Alpine Rodents and Plant Communities

Rodents interact with vegetation through their consumption of plants and their burrowing activities. Small rodents, being active all year round, are likely to have their greatest impact on standing vegetation during the winter. Their movements may be restricted in the subnival space and they may locally overgraze the vegetation. The absence of permafrost in most of the alpine zone, and the fact that the soil is usually not frozen in the winter because of the thick snow cover, result in important burrowing activities in marmots and common voles. The size difference between voles and marmots leads also to differences in food selection and consumption.

The burrowing activities of marmots seem to modify the vegetation around burrows by changing relative species cover, introducing species otherwise not characteristic, and by nutrient enrichment through dung deposition (Zimina and Zlotin 1980). For example, graminoids such as *Alopecurus alpinus* and *Poa arctica* may become the dominant species near the main burrow systems of *Marmota camtchatica* at the expense of tundra shrubs (*Dryas punctata, Cassiopa tetragona*), mosses and lichens which are dominant in the absence of marmots (Semenov et al. 2001). In the Olympic marmot *M. olympus*, its presence leads to an increase in unpalatable and ruderal species at the expense of graminoid and palatable species (Del Moral 1984). An example of plant species not characteristic of the current habitat appearing near burrow systems are *Polygonum laxmannii, Potentilla gelida* and *Rhodiola rosea* that have been found in the vicinity of main burrow systems in *Marmota camtchatica*. In the Alps, the high plant diversity makes more difficult the occurrence of plant species not characteristic of alpine meadows after marmot settlements. Digging and dunging may facilitate colonisation by nitrophilic plant species. For example, *Myosotis suaveolens* and *Delphinium chamissonis* are found near burrows of *Marmota camtchatica* (Semenov et al. 2001) in the tundra and *Poa alpina, P. pratensis,* and *Phlenum alpinum* are observed near alpine marmot burrows in alpine meadows (Y. Semenov, pers. comm.). These three kinds of impact have also been observed in *Marmota olympus* (Del Moral 1984) and *M. bobac* (Zimina et Zlotin 1980).

Grazing by marmots is also likely to affect plant communities; however, this is poorly understood, mainly because the diet of the marmot species is poorly known. The yellow-bellied marmot (*M. flaviventris*) is considered a generalist herbivore (Frase and Armitage 1989), whilst the Olympic marmot (*M. olympus*) was found to consume relatively more palatable species (Del Moral 1984). The alpine marmot was also reported to be selective by consuming a relatively small number of plant species in regards to the number available (Massemin et al. 1996).

The impact of snow voles is largely unknown, except for some local impact such as on alpine buttercups (*Ranuculus glacialis*) in high alpine areas (Diemer 1996). The role of the common vole may be substantially larger, because their use of habitat in the winter is more restricted (because of snow conditions), and they may therefore affect plant communities locally in a way similar to what has been demonstrated in northern areas (cf. Virtanen et al. 1997).

Overall, the presence of marmots can lead to local higher plant diversity by favouring the establishment of rare species even if plant species richness is locally reduced in the vicinity of burrow systems. The local impacts of voles on the plant communities should be studied by conducting long-term exclosure experiments.

20.5 Consequences of Changes in the Management of Alpine Ecosystems and in Climate

Marmots are influenced by human activities, directly through hunting and indirectly through the management of their habitats. Hunting pressure is less important than at the beginning of the twentieth century and has declined substantially in recent years (Wieser 1990). The continuation of agricultural practices in high elevation hay meadows (at least in France) is, however, often incompatible with the development of tourism activities and with the presence of marmots. Because of their burrowing activities and the stones they extract from the soil, farmers usually remove them from hay meadows. Marmots are also an important asset for tourism, as they often become relatively tame and are a symbol of the alpine fauna.

The alpine marmot prefers to settle in southern slopes (Allainé et al. 1994) and can be considered as thermophilic (Müller-Using 1954). However, marmots' fur absorbs solar radiation (Hayes 1976), which can be important on sunny days at high elevations. This may explain why alpine marmots are not active in the middle of the day and avoid elevations lower than 800 m (Türk and Arnold 1988). The alpine marmot is well adapted to periglacial conditions (Zimina and Gerasimov 1973) and lived in western Europe in the lowlands during the late Pleistocene (Chaline 1972). The warmer climate that followed after this period may have forced marmot populations to move up into their present alpine habitat. During this process, marmot populations may have suffered a severe genetic bottleneck leading to a low variation in structural genes among populations in Austria (Preleuthner et al. 1995).

Global climate change may lead to an elevation of ambient temperature in the alpine zone, and consequently an upward shift in vegetation zones (Ozenda and Borel 1991; Körner 1999). This is likely to reduce the range of alpine marmots, and increase the fragmentation of their populations. In particular, changes in climatic patterns such as increase in summer temperature and drought (Haeberli and Beniston 1998) may potentially have the largest impact on marmots, by reducing food availability and quality, and, in turn, reducing winter survival. Another potential consequence of global climate change is an increase in temporal variability of climate (Mearns et al. 1997). For alpine rodents, winter snow conditions are of particular importance in providing a stable environment. Unpredictability in climate is likely to have negative effects on them.

20.6 Needs for Further Work

The population dynamics and demography of alpine rodents are becoming better known through long-term studies based on capture-mark-recapture methods. Work has so far concentrated on the effects of different climatic parameters for marmots and voles, including, for voles, comparisons between alpine and arctic environments. One main conclusion is that alpine vole populations appear to be stable, in contrast to the widely fluctuating populations found in northern environments. There is a lack of comprehensive knowledge, however, regarding (1) how vegetation and climate regulate populations, (2) the role of small rodents in influencing the dynamics of plant communities, and (3) their indirect impact on other mammalian species through, for example, shared predators (golden eagle *Aquila chrysaetos* L., fox *Vulpes vulpes* L., and stoat *Mustela erminea* L.). Future work focusing on community and ecosystem dynamics in the alpine zone is needed to be able to better predict the consequences of changes in human management practices and climate.

20.7 Conclusions

Alpine rodents (common vole, snow vole and alpine marmot) may play an important role in the ecosystems through direct impacts on vegetation (herbivory) or as an important source of food for predators (stoats, red fox, golden eagle). This role is a direct function of the densities of these species, and how densities fluctuate between years. Recent studies using modern capture-mark-recapture statistical approaches have led to a better understanding of the factors determining the demography and population dynamics of these species. In particular, the weather-related factors driving reproductive and survival rates in an alpine marmot population have been identified as important. Early freezing and late snow melt negatively affect adult survival while juvenile survival is low during wet summers. Vole populations appear to be remarkably stable, compared with arctic or lowlands regions.

Acknowledgements. The studies on marmots were supported by the Parc National des Ecrins and Parc National de la Vanoise. The Ministère de l'Environnement (SRETIE), the CNRS and the Région Rhône-Alpes supported initial field studies. NGY thanks the Parc National des Ecrins, the National Geographic Society and the Club Alpin Francais for additional financial support.

References

Allainé D (2000) Sociality, mating system, and reproductive skew in marmots: evidence and hypotheses. Behav Proc 51:21-34

Allainé D, Rodrigue I, Le Berre M, Ramousse R (1994) Habitat preferences of alpine marmots, *Marmota marmota*. Can J Zool 72:2193-2198

Allainé D, Graziani L, Coulon J (1998) Postweaning mass gain in juvenile alpine marmots *Marmota marmota*. Oecologia 113:370-376

Allainé D, Brondex F, Graziani L, Coulon J, Till-Bottraud I (2000) Male-biased sex ratio in litters of alpine marmots supports the helper repayment hypothesis. Behav Ecol 11:507-514

Armitage KB (1981) Sociality as a life-history tactic of ground squirrels. Oecologia 48:36-49

Armitage KB (1991) Social and population dynamics of yellow-bellied marmots: results from long-term research. Annu Rev Ecol Syst 22:379-407

Armitage KB (1999) Evolution of sociality in marmots. J Mammal 80:1-10

Arnold W (1988) Social thermoregulation during hibernation in alpine marmots (*Marmota marmota*). J Comp Phys B 158:151-156

Arnold W (1990) The evolution of marmot sociality: 1. Why disperse late? Behav Ecol Sociobiol 27:229-237

Barash DP (1974a) The evolution of marmot societies: a general theory. Science 185:415-420

Barash DP (1974b) The social behavior of the hoary marmot (*Marmota caligata*). Anim Behav 22:256-261

Barash DP (1989) Marmots. Social behavior and ecology. Stanford University Press, Stanford

Barry RG (1992) Mountain weather and climate. Routledge, London

Blumstein DT, Armitage KB (1999) Cooperative breeding in marmots. Oikos 84:369-382

Breitenmoser U (1998) Large predators in the Alps: the fall and rise of man's competitors. Biol Conserv 83:279-289

Bryant AA (1996) Reproduction and persistence of Vancouver Island marmots (*Marmota vancouverensis*) in natural and logged habitats. Can J Zool 74:678-687

Chaline J (1972) Les rongeurs du Pléistocène moyen et supérieur de France (Systématique- biostratigraphie- paléoclimatologie). Cahiers de paléontologie. CNRS, Paris

Clark TW (1970) Early growth, development and behaviour of the Richardon's ground squirrel (*Spermophilus richardsonii elegans*). Am Midl Nat 80:197-205

Claude C (1995) *Chionomys nivalis* (Martins, 1842). In: Hausser J (ed) Mammifères de la Suisse. Birkhäuser, Basel, pp 339-343

Coulon J, Graziani L, Allainé D, Bel MC, Poudéroux S (1995) Infanticide in the alpine marmot (*Marmota marmota*). Ethol Ecol Evol 7:191-194

Del Moral R (1984) The impact of the Olympic marmot on subalpine vegetation structure. Am J Bot 71:1228-1236

Diemer M (1996) The incidence of herbivory in high-elevation populations *of Ranunculus glacialis*: a re-evaluation of stress-tolerance in alpine environments. Oikos 75:486-492

Farand E, Allainé D, Coulon J (2002) Variation in survival rates for the alpine marmot (*Marmota marmota*): effects of sex, age, year, and climatic factors. Can J Zool 80:342-349

Formozov AN (1966) Adaptive modifications of behavior in mammals of the Eurasian steppes. J Mammal 47:208-223

Forter D (1975) Zur Ökologie und verbreitungsgeechichte des alpenmurmeltiers im Berner Oberland. PhD Thesis, Universität Bern, Bern

Frank F (1953) Beiträge zur Biologie, insbesondere Jugendentwicklung der Schneemaus. Z Tierpsychol 11:1-9

Frase BA, Armitage KB (1989) Yellow-bellied marmots are generalist herbivores. Ethol Ecol Evol 1:353-366

Goossens B, Graziani L, Waits LP, Farand S, Magnolon S, Coulon J, Bel MC, Taberlet P, Allainé D (1998) Extra-pair paternity in the monogamous alpine marmot revealed by nuclear DNA microsatellite analysis. Behav Ecol Sociobiol 43:281-288

Grimod I, Bassano B, Tarellov V (1991) La marmotta (*Marmota marmota*) in valle d'Aosta. Ecologia e distribuzione. Regione autonoma della valle d'Aosta. Museo regionale di scienze naturali di Saint-pierre, Aosta

Haeberli W, Beniston M (1998) Climate change and its impacts on glaciers and permafrost in the Alps. Ambio 27:258-265

Hansson L, Henttonen H (1988) Rodent dynamics as community processes. Trends Ecol Evol 3:195-200

Hayes SR (1976) Daily activity and body temperature of the southern woodchuck, *Marmota monax monax*, in northwestern Arkansas. J Mammal 57:291-298

Jones JK, Carter DC (1980) The snow vole, *Microtus nivalis*, in the lowlands of western Yugoslavia. J Mammal 61:572

Körner C (1999) Alpine plant life. Functional plant ecology of high mountain ecosystems. Springer, Berlin Heidelberg New York

Kratochvil J (1961) *Marmota marmota latirostris* Ssp nova. Folia Zool 10:289-304

Kratochvil J (1981) *Chionomys nivalis* (Arvicolidae, Rodentia). Acta Sc Nat Brno 15:1-62

Krystufek B, Kovacic D (1989) Vertical distribution of the snow vole *Microtus nivalis* (Martins, 1842) in northwestern Yugoslavia. Z Säugetier 54:153-156

Le Louarn H, Janeau G (1975) Répartition et biologie du Campagnol des neiges *Microtus nivalis* Martins dans la région de Briancon. Mammalia 39:589-604

Le Louarn H, Spitz F, Grolleau G (1970) Le campagnol des champs *Microtus arvalis* Pallas dans le Brianconnais. Ann Zool Ecol Anim 2:423-426

Massemin S, Gibault C, Ramousse R, Butet A (1996) Premières données sur le régime alimentaire de la marmotte alpine (*Marmota marmota*) en France. Mammalia 60:351-361

Mearns LO, Rosenzweig C, Goldberg R (1997) Mean and variance change in climate scenarios: methods, agricultural applications, and measures of uncertainty. Clim Change 35:367-396

Meylan A (1995) *Microtus arvalis* (Pallas, 1778). In: Hausser J (ed) Mammifères de la Suisse. Birkhäuser, Basel, pp 328-333

Michener GR (1983) Kin identification, matriarchies and the evolution of sociality in ground-dwelling squirrels. In: Eisenberg JF, Kleiman, DG (eds) Advances in the study of mammalian behaviour. Spec Publ No 7, American Society of Mammalogy, Shippensburg, pp 528-572

Millar JS, Teferi T (1993) Winter survival in northern *Peromyscus maniculatus*. Can J Zool 71:125-129

Moravec J, Vlasak P (1988) Weight structure in a population of *Microtus arvalis* during the population cycle. Vestn Cesk Spol Zool 52:192-203

Müller-Using D (1954) Beiträge zur Ökologie der *Marmota m. marmota* (L.). Z Säugetier 19:166-177

Ozenda P, Borel JL (1991) Les conséquences écologiques possibles des changements climatiques dans l'arc alpin. Rapport Futuralp 1. ICALPE, Le Bourget-de-Lac

Perrin C, Allainé D, Le Berre M (1993) Sociospatial organization and activity distribution of the alpine marmot *Marmota marmota*: preliminary results. Ethology 93:21-30

Preleuthner M, Pinsker W, Kruckenhauser L, Miller WJ, Prosl H (1995) Alpine marmots in Austria. The present population structure as a result of the postglacial distribution history. Acta Theriol Suppl 3:87–100

Reid DG, Krebs CJ, Kenney SJ (1997) Patterns of predation on noncyclic lemmings. Ecol Monogr 67:89–108

Reutter BA, Hausser J, Vogel P (1999) Discriminant analysis of skull morphometric characters in *Apodemus sylvaticus*, *A. flavicollis*, and *A. alpicola* (Mammalia; Rodentia) from the Alps. Acta Theriol 44:299–308

Saucy F (1994) Density dependence in time series of the fossorial form of the water vole, *Arvicola terrestris*. Oikos 71:381–392

Schwartz OA, Armitage KB, Van Vuren D (1998) A 32-year demography of yellow-bellied marmots (*Marmota flaviventris*). J Zool 246:337–346

Semenov Y, Ramousse R, Le Berre M, Tutukarov Y (2001) Impact of the black-capped marmot (*Marmota camtschatica bungei*) on floristic diversity of arctic tundra in northern Siberia. Arctic Antarctic Alpine Res 33:204–210

Smith AT (1988) Patterns of Pika (genus *Ochotona*) life history variation. In: Boyce MS (ed) Evolution of life histories of mammals. Yale University Press, New Haven, pp 233–256

Smith AT, Weston ML (1990) *Ochotona princeps*. Mammal Species 352:1–8

Stenseth NC (1999) Population cycles in voles and lemmings: density dependence and phase dependence in a stochastic world. Oikos 87:427–461

Stenseth NC, Ims RA (eds) (1993a) The biology of lemmings. Academic Press, London

Stenseth NC, Ims RA (1993b) Population dynamics of lemmings: temporal and spatial variation – an introduction. In: Stenseth NC, Ims RA (eds) The biology of lemmings. Academic Press, London, pp 61–96

Swihart RK (1991) Influence of *Marmota monax* on vegetation in hayfields. J Mammal 72:791–795

Türk A, Arnold W (1988) Thermoregulation as a limit to habitat use in alpine marmots (*Marmota marmota*). Oecologia 76:544–548

Van Vuren D, Armitage KB (1991) Duration of snow cover and its influence on life-history variation in yellow-bellied marmots. Can J Zool 69:1755–1758

Van Vuren D, Armitage KB (1994) Survival of dispersing and philopatric yellow-bellied marmots: what is the cost of dispersal? Oikos 69:179–181

Virtanen R, Henttonen H, Laine K (1997) Lemming grazing and structure of a snow bed plant community – a long-term experiment at Kilpisjärvi, Finnish Lapland. Oikos 79:155–166

Wieser R (1990) Les marmottes. Payot, Lausanne

Yoccoz NG (1992) Présence de mulot (*Apodemus alpicola* ou *flavicollis*) en milieu alpin. Mammalia 56:488–491

Yoccoz NG, Ims RA (1999) Demography of small mammal in cold regions: the importance of environmental variability. Ecol Bull 47:137–144

Yoccoz NG, Mesnager S (1998) Are alpine bank voles larger and more sexually dimorphic because adults survive better? Oikos 82:85–98

Zimina RP, Gerasimov IP (1973) The periglacial expansion of marmots (*Marmota*) in middle Europe during late Pleistocene. J Mammal 54:327–340

Zimina RP, Zlotin RI (1980) Biotsenotitcheskoe znatchenie. V kn.: Surki, biotsenotitcheskoe i praktitcheskoe znatchenie (Biocenotic importance in marmots. Biocenotic and practical importance). Nauka, Moscow (in Russian)

21 Large Herbivores in European Alpine Ecosystems: Current Status and Challenges for the Future

A. LOISON, C. TOÏGO and J.-M. GAILLARD

21.1 Introduction

The mountain ecosystems of Europe are facing rapid changes in land use including an increasing use of mountain areas for tourism and transport, and a decrease in the extent of small-scale agriculture such as transhumance (Breitenmoser 1998; Fig. 21.1). All of these changes may have an impact on the landscape structure and on the flora and fauna. As most National Parks and other protected areas are in mountain areas, the protection of ecosystems with their plant and wildlife has to be undertaken in the face of an increasing human pressure. During the last decades of the twentieth century, most of the large herbivores increased in number as a direct result of human actions (such as the reintroduction of ibex, *Capra ibex* L.), or by natural changes in the size and dynamics of populations (Figs. 21.2–21.4). The introduction/reintroduction of an abundant and diverse fauna is perceived as mainly positive, especially by conservationists, tourists and hunters. However, it may be asked how mountain areas will react to the concomitant action of natural and human-made processes, especially in relation to specific areas of interest such as plant–herbivore relationships, interactions between wild and domestic herbivores, the role of large predators, the impact of wild herbivores on the landscape and the potential impacts of climate change on the population dynamics of ungulates.

The potential conflicts between herbivore populations and their environment are exemplified by the overgrazing of lichen heaths in the Scandes by semi-domestic reindeer under intensive management (Väisänen 1998; Hallanaro and Pylvänäinen 2001). In the Alps and Pyrenees, the impacts of high densities of ungulates on vegetation dynamics and landscape processes are as yet to be considered in an integrated manner. An understanding of interactions between ungulate populations and landscape processes is required for predictive models on the functioning of mountain ecosystems. We review the

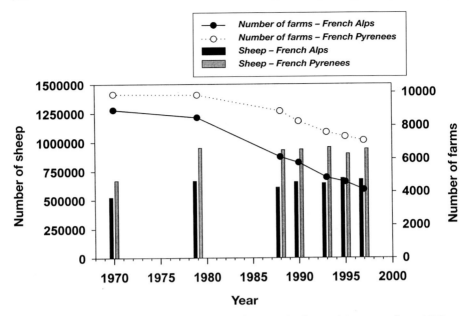

Fig. 21.1. Numbers of sheep and farmers in the French Alps and Pyrenees from 1970 to 1997

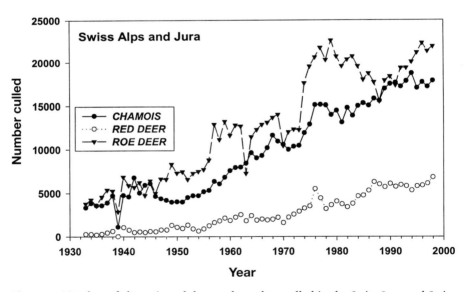

Fig. 21.2. Number of chamois, red deer and roe deer culled in the Swiss Jura and Swiss Alps between 1933 and 1998. Source: Swiss Wildlife Information Service

Large Herbivores in European Alpine Ecosystems

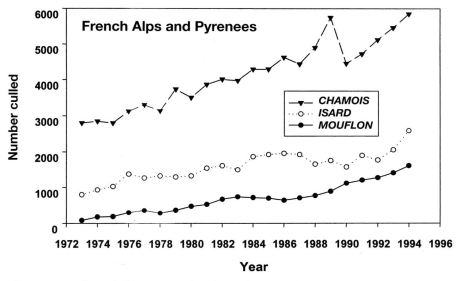

Fig. 21.3. Number of chamois, isard and mouflon culled in the French Alps and Pyrenees between 1973 and 1995. Source: Office National de la Chasse/Fédération National des Chasseurs – Réseau de correspondants 'Cervidés-sanglier'

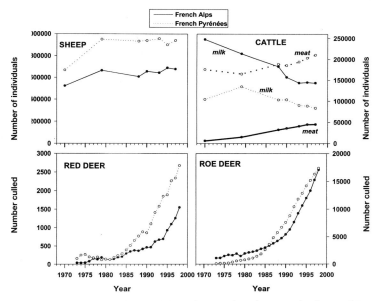

Fig. 21.4. Number of sheep and cattle raised in the French Alps and Pyrenees and number of red deer and roe deer harvested each year in the French Alps and Pyrenees since 1970. *Solid lines* are for the Alps and *dotted lines* for the Pyrenees. Source of culled numbers: Office National de la Chasse/Fédération National des Chasseurs – Réseau de correspondants 'Cervidés-sanglier'

common features of mountain ungulates in the Alps and Pyrenees with an emphasis on their population biology and dynamics, to summarise the current knowledge on the interactions between wild ungulates and their habitats, as well as with domestic ungulates and predators. Finally, we point out some future research needs.

21.2 The Species Concerned and Their Distribution in Europe

The reindeer (*Rangifer tarandus* L.) is the only large mountain herbivore in Fennoscandia; in central and southern Europe, there are five alpine specialist species, the Alpine ibex (*Capra ibex*), the Pyrenean ibex (*Capra pyrenaica* Schinz), the chamois (*Rupicapra rupicapra* L.), the isard (*Rupicapra pyrenaica* Bonaparte), and the mouflon (*Ovis gmelini* Pallas), the latter introduced from the Mediterranean islands during the nineteenth century. The two species in each of the genera *Rupicapra* and *Capra* are ecologically similar; however, they have distinct geographic distributions and their genetic and morphometric characteristics warrant their recognition as distinct species within each genus (Nascetti 1985; Catusse et al. 1996). Because of the ecological similarities between the species within the genera we apply the generic term chamois to both *Rupicapra* species, and ibex to the *Capra* spp. Reindeer are not considered further in this review. Mountain ungulates, after a low for centuries (Breitenmoser 1998), all have increased in numbers and distribution since the 1950s except for the Pyrenean ibex, the long-term survival of which remains uncertain (Catusse et al. 1996).

21.3 Ecological Features

21.3.1 Morphology and Physiology

Mountain ungulates have body weights ranging from 20 (isard) to 80 kg (ibex), with a dimorphism (male body weight/female body weight) ranging from 1.1 for chamois to 2.0 for ibex (Loison et al. 1999a). One common feature of mountain ungulates is the presence of horns in both sexes, which are larger in males than females.

Morphological adaptations in the above species to high mountain environments (steep slopes, snow in the winter and pronounced seasonality in climate) are subtle: all have relatively short and solid legs and moult between winter and summer. The species showing the most specific features are the chamois with its palmate hoof, improving walking on snow, and the ibex with its hoof clearly adapted to climbing. In addition, the chamois has a larger heart and higher number of red blood cells than other similar sized ungulates (Catusse et al. 1996).

21.3.2 Habitat

The specialised nature of mountain ungulates has been explained by their dependency on an escape terrain as an anti-predator strategy (Perez-Barbeira and Nores 1994; Kohlmann et al. 1996; Villaret et al. 1997) rather than on habitats providing specific types of food. Chamois appear to thrive in forests (e.g. Garcia Gonzales and Cuartas 1996; Herrero et al. 1996), as well as in alpine meadows (Lovari and Cosentino 1986; Catusse et al. 1996). For ibex, the main habitat characteristic appears to be steep terrain and cliffs (Wiersema 1984; Giacometti 1991).

21.3.3 Feeding Ecology

In terms of feeding type, mountain ungulates (sensu Hoffman 1989) range from grazers like mouflon (Cransac et al. 1997b) to intermediate feeders such as the ibex (Klansek et al. 1997). Chamois seem to prefer grass when available (Garcia-Gonzales and Cuartas 1996; Perez-Barbeira et al. 1997). However, all these species exhibit a wide flexibility in their feeding regimes depending on season and habitat (Klansek et al. 1997; Perez-Barberia et al. 1997). Both chamois and mouflon, for example, inhabit areas where grasslands are not the most abundant type of vegetation, and, therefore, an important part of their diet comes from browsing of dwarf shrubs (e.g. Garcia-Gonzales and Cuartas 1996). Inter-specific competition among ibex, chamois and mouflon could occur, but has not been studied extensively (Schröder and Kofler 1984).

21.3.4 Social and Mating System

All mountain herbivore species are social, sexually segregated (ibex: Villaret et al. 1997; mouflon: Petit et al. 1997; Cransac et al. 1998; chamois: Perez-Barberia et al. 1997) and polygynous (Loison et al. 1999a). Although mating systems may vary at the intra-specific level according to habitat structure, density and sex ratio (Lott 1991), all four species mainly reproduce on the basis of roaring male or a 'male tending' mating system (Clutton-Brock 1989).

The common scheme comprises labile female groups with young of the year of both sexes, and male groups of usually smaller size. Important variations around this general scheme occur depending on season, population density, habitat patchiness, vegetation quality and snow cover (Fandos et al. 1992; Perez-Barberia and Nores 1994; Pépin et al. 1997; Cransac et al. 1998).

21.3.5 Spatial Features

The spatial distribution of high mountain ungulates with respect to altitude and aspect varies according to season, snow pattern, temperature gradients, and vegetation (Cransac and Hewison 1997; Pépin et al. 1997; Perez-Barbeira et al. 1997; Michallet et al. 1999). No large-scale migrations similar to that observed for reindeer in Scandinavia are observed in mountain ungulates in the Alps and Pyrenees. The seasonal shift in spatial use is either altitudinal or a reduction in size of the home range (Loison 1995; Herrero et al. 1996). Dispersal patterns seem to follow the expected schemes for polygynous mammalian species, with males dispersing more than females, especially at the immature stages (e.g. chamois: Levet et al. 1995; Loison et al. 1999b). Natural colonisation has been observed in introduced ibex and chamois populations (chamois: Caughley 1963; Levet et al. 1995, ibex: Giacometti 1991), as opposed to mouflon, which seem to stay in areas where they have been released. Preliminary data on chamois and isard indicated no density-dependent dispersal (Levet et al. 1995; Loison et al. 1999b). Data on mouflon (Dubois et al. 1993; Petit et al. 1997) and isard (Levet et al. 1995) suggest that males may use sites for rutting where they are born, while remaining in a distant home range outside the rutting season.

21.4 Dynamics in Space and Time

21.4.1 Demographic Patterns

Mountain ungulates share demographic patterns of other large herbivores (Gaillard et al. 1998), i.e. a variable recruitment and a relatively stable and high adult survival rate. The age at first reproduction varies between 2 and 5 years of age depending on population density and climate (e.g. Bauer 1985 for chamois, Giacometti and Ratti 1994 for ibex). The age at first reproduction seems to occur later in males than females, even though sexual maturity is reached around 3 years of age.

Mountain ungulates are strict followers, i.e. new-born follow their mothers shortly after birth (chamois: Richard-Hansen and Campan 1992; Ruckstuhl and Ingold. 1999; ibex: Couturier 1962; mouflon: Réale et al. 1999). New-borns are preyed upon by golden eagles (*Aquila chrysaetos* L., on chamois: Locati

1990, and on ibex: Haller 1996) and red fox (*Vulpes vulpes* L.). Survival rates during the summer are nevertheless close to 100 % (for chamois: Loison 1995; for ibex: Toïgo 1998), probably because females can seek out nutritious feed by changing altitude, while kid winter survival is usually lower and more variable (Loison 1995).

The survival rates of adult males and females are generally very high in the absence of predation (chamois: Loison et al. 1994, 1999a,b; ibex: Toïgo et al. 1997; Girard et al. 1999; mouflon: Cransac et al. 1997a), although a decrease in the survival rates occurs with age, both for males and females (Loison et al. 1999d).

21.4.2 Density Dependence and Effects of Climate

The population dynamics of mountain ungulates follows the general patterns observed for large herbivores (Sæther 1997; Gaillard et al. 1998). Reintroduction of mountain ungulates does not appear to lead to boom-and-crash dynamics (e.g. Levet et al. 1995), but as the number of studies of reintroduced populations remain scarce, we do not know yet whether this is a general pattern for mountain ungulates in continental Europe. At high densities, mountain ungulates tend to have a delayed age at first reproduction (e.g. Bauer 1985 in chamois) and follow classical patterns of density dependence (Hanks 1981; Fowler 1987) with decreasing body weight, and fecundity rate (Storch 1989; Loison 1995). No evidence of density dependence in adult survival rate has been documented so far. Most of the evidence of density dependence in demographic parameters is based on the comparison of populations rather than on long-term studies of the same population at different densities. More long-term studies are therefore required for a better understanding of the density-dependent patterns of mountain ungulates. The existence of possible time lags in the effects of density, as suggested by Cappuro et al. (1997), in particular needs further investigation.

Cohort effects have been shown on chamois survival rates (Loison et al. 1999 c) and ibex horn growth and, thereby, probably condition (Toïgo et al. 1999), but not on the population dynamics of any of the species.

Snow depth and duration are probably key features for understanding the ecology of high mountain herbivore communities. However, few studies have documented the relationship between demographic parameters and climatic variables. Loison et al. (1999 c) reported no direct negative effect of snow cover on the survival rate of chamois, but rather, a 1-year delayed positive effect, probably mediated by the effect of snow on the duration of the growing season of the vegetation. Spring temperatures also affect survival rates during the following winter for chamois, in a positive or negative way depending on site location (Loison et al. 1999 c). As for density, the consequences of climate variables in mountain environments may be delayed (Loison et al. 1999 c) and long-term studies are necessary to highlight such demographic mechanisms. The dynamics of mountain ungulates may be affected by predicted changes in climate such as a decrease in precipitation, increase in the frequency of extreme events, and a variable snow cover at altitudes below 1700–2400 m (Martin et al. 1994; Beniston 1997).

21.5 The Role of Large Herbivores on the Mountain Landscape and Plant Communities

The mountain landscape has been shaped by several hundred years of human use for agriculture, grazing by domestic ungulates, and the collection of firewood (Breitenmoser 1998; Olsson et al. 2000). The forested areas have thereby been depleted, which has led to the opening and extension of alpine meadows (the timberline was lowered by 200 to 300 m in Switzerland already by the thirteenth century, see references in Breitenmoser 1998). Between 1960 and 2000, the number of sheep and cattle has either increased or remained unchanged in most mountain areas in France and Switzerland (Breitenmoser 1998), whilst the number of farmers has declined (Figs. 21.1 and 21.4). This has led to a concentration in grazing pressure in some areas and a decrease in others. When not used by sheep, meadows in the treeline ecotone tend to be invaded by shrubs and tall grasses. An increasing population of wild ungulates may compensate for the decrease in grazing/browsing pressure from domestic ones. However, they may differ in their feeding selectivity and in the seasonal pattern of their impact on the vegetation (Augustine and McNaughton 1998), as they could damage commercially important trees in winter (e.g. Tomiczek 1992; Abderhalden and Buchli 1999; Buchli and Abderhalden 1999).

Ungulates may increase or decrease species diversity, depending on their density and selectivity (Hobbs 1996). There is no general conclusion emerging from the studies in the Alps (Ammer 1996; Kienast et al. 1999). The conditions in the Alps and Pyrenees, with small-scale heterogeneity of the landscape inhabited by ungulates using small home ranges, appear to be amenable to the further study of herbivore–vegetation interactions.

21.6 Changes in Mountain Ecosystems and New Challenges for Large Herbivores

21.6.1 Interaction with Domestic Ungulates, Diseases and Parasites

The most abundant domestic ungulates, sheep and cattle, are grazers and any competition for food is therefore expected to be highest first with mouflon or ibex and then chamois. It is not known if the quality of feeding regimes of wild ungulates is affected by the presence of domestic ungulates. However, the latter exclude wild ungulates from areas that they would otherwise use (Herrero et al. 1996).

The interaction between domestic and wild ungulates that has generated the most concern is the transmission of diseases and common parasites (Durant and Gauthier 1996). Diseases such as keratoconjunctivitis are suspected to be spread from sheep and cattle (Giacometti et al. 1998) and might have a large impact on the demography of chamois, ibex and mouflon (Loison et al. 1996; Cransac et al. 1997a). The same is true for parasites (Gulland et al. 1993; Paterson et al. 1998), even if their role has so far been overlooked in studies of mountain ungulate population dynamics.

21.6.2 The Increase of Lowland Ungulates and Inter-Specific Competition

There has been a recent increase in the numbers of roe deer and red deer (Figs 21.2 and 21.4) in habitats where chamois and mouflon occur (Mattiello et al. 1997). The roe deer differs in its diet from high mountain ungulates (Schröder and Schröder 1984) but the red deer may have greater potential of overlap (Schröder and Schröder 1984; Matiello et al. 1997; Homolka and Matous 1999) during some periods of the year.

21.6.3 The Return of Large Carnivores

In most mountain areas, large carnivores such as the wolf (*Canis lupus* L.), brown bear (*Ursus arctos* L.) and lynx (*Lynx lynx* L.) have been drastically reduced or exterminated during the past centuries (Guidali et al. 1990; Breitenmoser 1998). As a result of concern about their conservation the populations of lynx and wolves have recently increased (Guidali et al. 1990; Breitenmoser 1998). From the few studies available it appears that in areas where all herbivore species are present, wolves first affect mouflon and then chamois populations, while the ibex is virtually unaffected (Merrigi and Lovari 1996; Poulle et al. 1997; Jobin 1998). At the individual level, increased flight distances and the use of habitat closer to refuge or escape areas were observed, mostly for chamois and mouflon, because ibex occupies, independently of the presence of predators, only the steepest terrain. Lynx, being a forest dweller, have a more local impact than wolves. Although the distribution of the lynx overlaps with that of the chamois in the northern Alps and Jura Mountains, chamois are seldom the only large herbivore species available to lynx (Jobin 1998; Weber and Weissbrodt 1999).

The presence of carnivores may reveal more clearly the life-history traits of ungulates that evolved under predator constraints (Byers 1997). Species with a 'follower' type of young rearings, such as ibex and chamois, rely on escape terrain and vigilance behaviour to guard against predators (Ruckstuhl and Ingold 1999; Toïgo 1999). The trade-off between high-quality food and escape terrain has been shown in these species even without the presence of predators (Alados 1985; Perez Barberia and Nores 1994; Villaret et al. 1997; Cransac et al. 1998), but may have been relaxed after many years with low risks (e.g. the colonisation of forests by chamois). We can therefore expect predators to push ungulate species out of habitats that they have colonised in the absence of large predators. The year-long use of forests by chamois, for instance, may be stopped by the return of the lynx. As a general point, the dependence of mountain ungulates on escape terrain (Perez-Barberia and Nores 1994) and of lowland ungulates on cover when predators are present may lead to a niche separation and a decrease in competition.

Carnivores are returning to a multiprey ecosystem in the European Mountains. A given vulnerable prey species can be eliminated from multiple prey–predator systems without a decline in the total prey population (Dale et al. 1995; Messier 1995). It will be interesting to monitor the future of the highly diverse ungulate community if wolves and lynx are to become more common.

21.6.4 Habitat Fragmentation

The construction of new roads and an increase in mountain road use (by 280 and 330 % by heavy goods vehicles between 1984 and 1995 in the French Alps and in the Pyrenees, respectively; IFEN, unpubl.) has exacerbated habitat fragmentation. Fragmentation may prevent natural exchanges of ungulates between massifs. Locally, habitat mosaics are changing as a result of abandonment of agricultural use. These changes are unlikely to affect ibex whose habitat tends to be close to steep non-cultural landscapes, but could have consequences for chamois and mouflon.

21.6.5 Recreational Use of the Mountains and Disturbance

The number of visitors using mountains for outdoor activities has increased steeply since the 1950s (e.g. the number of tourist facilities in the French Alps rose by 20-fold between 1950 and 1997; IFEN, unpubl.). These activities have the potential to disturb wildlife and force a change in habitat use patterns. The condition of the animals and their reproductive success (e.g. disturbance of the rut activities) as well as their impact on the vegetation may also be affected. For example, paragliders may make chamois or ibex seek cover and feed in forested habitats (Ingold et al. 1993; Gander and Ingold 1997; Szemkus et al. 1998), although habituation to such disturbance may occur in the long term.

21.7 Conclusions and Future Research Needs

1. Large herbivores in mountainous areas may be having an increasing impact on ecosystem function and dynamics, through their interactions with plant communities, their role in shaping habitat, and their importance for large carnivores. Changes occur rapidly, and, as this review suggests, there are large gaps in our knowledge of mountain ecosystems.
2. The traditional research approach based on single-species studies is necessary. However, an integrated approach, using long-term studies (and experiments where applicable), should focus on (1) interspecific competition, (2) interaction between domestic and wild ungulates, (3) the role of predation, (4) the role of wild ungulates on plant communities, (5) the effect of space, landscape patterns and their changes on the dynamics of ungulates, (6) the evaluation of climatic variation on population dynamics, and (7) defining management strategies (e.g. through culling) to satisfy multi-user management objectives.

Acknowledgements. We thank the Office National de la Chasse et de la Faune Sauvage for long-term collaboration on ungulate population dynamics, Michel Catusse, Dominique Dubray, Patrick Duncan, Dominique Allainé, Erling Johan Solberg, Nigel Gilles Yoccoz and an anonymous referee for comments on an early version of this manuscript, and Jonathan Heygate for correction of the language.

References

Abderhalden W, Buchli C (1999) The influence of alpine ibex (*Capra ibex*) on the forest. Z Jagdwiss 45:17–26

Alados CL (1985) An analysis of vigilance in Spanish ibex (*Capra pyrenaica*). Z Tierpsychol 68:58–64

Ammer C (1996) Impact of ungulates on structure and dynamics of natural regeneration of mixed mountain forests in the Bavarian Alps. For Ecol Manage 88:43–53

Augustine DJ, McNaughton SJ (1998) Ungulate effects on the functional species composition of plant communities: herbivore selectivity and plant tolerance. J Wildl Manage 62:1165–1183

Bauer JJ (1985) Fecundity patterns of stable and colonising chamois populations of New Zealand and Europe. In: Lovari S (ed) The biology and management of mountain ungulates. Croom Helm, London, pp 154–165

Beniston M (1997) Variations of snow depth and duration in the Swiss Alps over the last 50 years: links to changes in large-scale climatic forcings. Clim Change 36:281–300

Breitenmoser U (1998) Large predators in the Alps: the fall and rise of man's competitor. Biol Conserv 83:279–289

Buchli C, Abderhalden W (1999) Impact of alpine ibex (*Capra ibex*) on alpine meadows. Z Jagdwiss 45:77–87

Byers JA (1997) American pronghorn. Social organisation and the ghosts of predator past. University of Chicago Press, Chicago

Capurro AF, Gatto M, Tosi G (1997) Delayed and inverse density-dependence in a chamois population of the Italian Alps. Ecography 20:37–47

Catusse M, Corti R, Cugnasse JM, Dubray D, Gibert P, Michallet J (1996) La grande faune de montagne. Hatier, Paris

Caughley GC (1963) Dispersal rates of several ungulates introduced to New Zealand. Nature 200:280–281

Clutton-Brock TH (1989) Mammalian mating systems. Proc R Soc Lond 236:339–372

Couturier MAJ (1962) Le bouquetin des Alpes. Arthaud, Grenoble

Crampe JP, Caens JC, Dumerc JL, Pepin D (1997) Body mass as an index of physical condition during winter in the Isard, *Rupicapra pyrenaica* (Artiodactyla, Bovidae). Mammalia 61:73–85

Cransac N, Hewison AJM (1997) Seasonal use and selection of habitat by mouflon (*Ovis gmelini*): comparison of the sexes. Behav Processes 41:57–67

Cransac M, Hewison AJR, Gaillard JM, Cugnasse JM, Maublanc ML (1997a) Patterns of mouflon (*Ovis gmelini*) survival under moderate environmental conditions: effects of sex, age, and epizootics. Can J Zool 75:1867–1875

Cransac N, Valet G, Cugnasse JM, Rech J (1997b) Seasonal diet of mouflon (*Ovis gmelini*): comparison of population sub-units and sex-age classes. Rev Ecol 52:21–36

Cransac N, Gerard JF, Maublanc ML, Pépin D (1998) An example of segregation between age and sex classes only weakly related to habitat use in mouflon Soay sheep (*Ovis gmelini*). J Zool 244:371–378

Dale BW, Adams LG, Bowyer RT (1995) Winter wolf predation in a multiple ungulate prey system, gates of the Arctic National Park, Alaska. In: Carbyn LN, Fritts SH, Seip DR (eds) Ecology and conservation of wolves in a changing world. Occasional Publication No 35. Canadian Circumpolar Institute, Edmonton, pp 223–230

Dubois M, Quenette PY, Bideau E, Magnac MP (1993) Seasonal range use by European mouflon rams in medium altitude mountains. Acta Theriol 38:185–198

Durant T, Gauthier D (1996) The helminths of the chamois *Rupicapra rupicapra*: host parasite environment relationships and wildlife management. Vie Milieu Paris 46:333–343

Fandos P, Aranda Y, Orueta, JF (1992) Size and group type in Spanish ibex (*Capra pyrenaica*). Etologia 2:65-70

Festa-Bianchet M, Jorgenson JT, King WJ, Smith KG, Wishart WD (1996) The development of sexual dimorphism: seasonal and lifetime mass changes in bighorn sheep. Can J Zool 74:330-342

Forsyth DM, Hickling GJ (1998) Increasing Himalayan tahr and decreasing chamois densities in eastern southern Alps, New Zealand: evidence for interspecific competition. Oecologia 113:377-382

Fowler CW (1987) A review of density dependence in populations of large mammals. In: Genoway HH (ed) Current mammalogy. Plenum, New York, pp 440-441

Gaillard JM, Festa-Bianchet M, Yoccoz NG (1998) Population dynamics of large herbivores: constant maintenance and variable recruitment. Trends Ecol Evol 13:58-63

Gander H, Ingold P (1997) Reactions of male alpine chamois *Rupicapra rupicapra* to hikers, joggers and mountain bikers. Biol Conserv 79:107-109

Garcia-Gonzales R, Cuartas P (1996) Trophic utilisation of a mountain/subalpine forest by chamois (*Rupicapra pyrenaica*) in the central Pyrénées. For Ecol Manage 88:15-23

Giacometti M (1991) Contributions to the settlements dynamics and actual distribution of alpine ibex (Capra i. Ibex) in the Alps. Z Jagdwiss 37:157-173

Giacometti M, Ratti P (1994) On the reproductive performance of the free-ranging alpine ibex population (*Capra ibex*) at Albris (Grisons, Switzerland). Z Saügetierkd 59:174-180

Giacometti M, Bassano B, Peracino V, Ratti P (1997) The constitution of the Alpine ibex (*Capra ibex*) in relation to sex, age, area of origin and season in Graubunden (Switzerland) and in the Parco Nazional Gran Paradiso (Italy). Z Jagdwiss 43:24-34

Giacometti M, Nicolet J, Frey J, Krawinkler M, Meier W, Welle M, Johansson KE, Degiorgis MP (1998) Susceptibility of alpine ibex to conjunctivitis caused by the inoculation of a sheep strain of *Mycoplasma conjunctive*. Vet Microbiol 61:279-288

Girard I, Toïgo C, Gaillard JM, Gauthier D, Martinot JP (1999) Survival patterns of alpine ibex (*Capra ibex ibex*) in the Vanoise National Park. Rev Ecol 54:235-251

Guidali F, Mingozzi T, Tosi G (1990) Historical and recent distributions of lynx (*Lynx lynx*) in north-western Italy, during the 19th and 20th centuries. Mammalia 54:587--596

Gulland FMD, Albon SD, Pemberton JM, Moorcroft PR, Clutton-Brock TH (1993) Parasite associated polymorphism in a cyclic ungulate population. Proc R Soc Lond 254:-7-13

Hallanaro E-L, Pylvänäinen M (2001) Nature in northern Europe - biodiversity in a changing environment. Nordic Council of Ministers, Copenhagen

Haller H (1996) Predation and accidents among alpine ibex *Capra ibex* in the Engadine. Z Jagdwiss 42:26-35

Hanks J (1981) Characterisation of population condition. In: Fowler CW, Smith TD (eds) Dynamics of large mammals populations. Wiley, New York, pp 47-73

Herrero J, Garin I, Garcia-Serrano A, Garcia-Gonzalez R (1996) Habitat use in a *Rupicapra pyrenaica pyrenaica* forest population. For Ecol Manage 88:25-29

Hobbs NT (1996) Modification of ecosystems by ungulates. J Wildl Manage 60:695-713

Hobbs NT, Baker DL, Bear GD, Bowden DC (1996) Ungulate grazing in sagebrush grassland: mechanisms of resource competition. Ecol Appl 6:200-217

Hoffmann RR (1989) Evolutionary steps of ecophysiological adaptations and diversification of ruminants: a comparative view of their digestive systems. Oecologia 78:443-457

Homolka M, Matous J (1999) Density and distribution of red deer and chamois in subalpine meadow habitats in the Jeseniky Mountains (Czech Republic). Folia Zool 48:1-10

Ingold I, Huber B, Neuhaus P, Mainini B, Marbacher H, Schnidrig-Petrig R, Zeller R (1993) Tourism and sport in the Alps – a serious problem for wildlife. Rev Suisse Zool 100:529–545

Jobin A (1998) Predation patterns by Eurasian lynx in the Swiss Jura Mountains. PhD Thesis, University of Bern

Kienast F, Fritschi J, Bisseger M, Abderhalden W (1999) Modelling successional patterns of high elevation forests under changing herbivore pressure – responses at the landscape level. For Ecol Manage 120:35–46

Klansek E, Vavra III, Onderscheka K (1997) The composition of the rumen contents among alpine ibex (*Capra ibex*) in relation to the seasons, age and browse supply. Z Jagdwiss 41:171–181

Kohlmann SG, Muller DM, Alkon PU (1996) Antipredator constraints on lactating Nubian ibexes. J Mammal 77:1122–1131

Levet M, Appolinaire J, Catusse M, Thion N (1995) Demographic data, spatial behaviour and dispersion of an isard (*Rupicapra p pyrenaica*) population in a stage of colonisation. Mammalia 59:489–500

Lincoln GA (1998) Reproductive seasonality and maturation throughout the complete life-cycle in the mouflon ram (*Ovis musimon*). Anim Reprod Sci 53:87–107

Locati M (1990) Female chamois defends kids from eagle attacks. Mammalia 54:155–156

Loison A (1995) Aspect inter et intra spécifiques de la dynamique des populations: l'exemple du chamois. PhD Thesis, University Lyon 1, Lyon

Loison A, Gaillard JM, Houssin H, Jullien JM (1994) New insights on survival of female chamois (*Rupicapra rupicapra*) from marked animals. Can J Zool 72:591–597

Loison A, Gaillard JM, Jullien JM (1996) Recovery of a chamois population after a keratoconjunctivitis epizootic: demographic patterns and management implications. J Wildl Manage 60:517–527

Loison A, Gaillard JM, Pélabon C, Yoccoz NG (1999a) What factors shape sexual size dimorphism in ruminants? Evol Ecol Res 1:611–633

Loison A, Jullien JM, Menaut P (1999b) Subpopulation structure and dispersal in two populations of chamois. J Mammal 80:620–632

Loison A, Jullien JM, Menaut P (1999c) Response of chamois survival to changes in global and local climate: contrasted examples in the Alps and Pyrénées. Ecol Bull 47:126–136

Loison A, Festa-Bianchet M, Gaillard JM, Jorgenson JT (1999d) Age-specific survival in five populations of ungulates: evidence for senescence. Ecology 80:2539–2554

Lott DF (1991) Intraspecific variation in the social systems of wild vertebrates. Cambridge University Press, Cambridge

Lovari S, Cosentino R (1986) Seasonal habitat selection and group size of the Abruzzo chamois (*Rupicapra pyrenaica ornata*). Boll Zool 53:73–78

Martin E, Brun E, Durand Y (1994) Sensitivity of the French Alps snow cover to the variation of climatic variables. Ann Geophys 12:469–477

Mattiello S, Bergami G, Redaelli W, Verga M, Crimella MC (1997) Ecology and behaviour of red deer (*Cervus elaphus*) in an alpine valley. Z Saugetierkd 62:129–133

Michallet J, Gaillard JM, Yoccoz NG, Toïgo C (1999) Sélection des quartiers d'hivernage par le chamois, *Rupicapra rupicapra* dans les massifs montagnards de l'Isère (France). Rev Ecol 54:351–363

Meriggi A, Lovari S (1996) A review of wolf predation in southern Europe: does the wolf prefer wild prey to livestock. J Appl Ecol 33:1561–1571

Merrigi A, Brangi A, Matteucci, Sacchi O (1996) The feeding habits of wolves in relation to large prey availability in northern Italy. Ecography 19:287–295

Messier F (1995) On the functional and numerical responses of wolves to changing prey density. In: Carbyn LN, Fritts SH, Seip DR (eds) Ecology and conservation of wolves in a changing world. Occasional Publication No 35. Canadian Circumpolar Institute, Edmonton, pp 187-197

Nascetti G (1985) Revision of Rupicapra genus. In: Lovari S (ed) The biology and management of mountain ungulates. Croom Helm, London. pp 57-62

Okarma H (1999) The trophic ecology of wolves and their predatory role in ungulate communities of forest ecosystems in Europe. Acta Theriol 40:335-386

Olsson EGA, Austrheim G, Grenne SN (2000) Landscape change patterns in mountains, land use and environment diversity, Mid-Norway 1960-1993. Landscape Ecol 15:155-170

Paterson S, Wilson K, Pemberton JM (1998). Major histocompatibility complex variation associated with juvenile survival and parasite resistance in a large unmanaged population (*Ovis aries*). Proc Natl Acad Sci USA 95:3741-3719

Pépin D, Faivre R, Menaut P (1996) Factors affecting the relationship between body mass and age in the isard. J Mammal 77:351-358

Pépin D, Joachim J, Ferrie E (1997) Variability of spring habitat selection by isards (*Rupicapra pyrenaica*). Can J Zool 75:1955-1965

Petit E, Aulagnier S, Bon R, Dubois M, Crouau Roy B (1997) Genetic structure of populations of the Mediterranean mouflon (*Ovis gmelini*). J Mammal 78:459-467

Perez-Barberia FJ, Nores C (1994) Seasonal variation in group size of cantabrian chamois in relation to terrain and food. Acta Theriol 39:295-305

Perez-Barbeira FJ, Olivan M, Osoro K, Nores C (1997) Sex, seasonal and spatial differences in the diet of Cantabrian chamois *Rupicapra pyrenaica parva*. Acta Theriol 42:37-46

Poulle ML, Carles L, Lequette B (1997) Significance of ungulates in the diet of recently settled wolves in the Mercantour Mountains (southeastern France). Rev Ecol 52:357-368

Réale D, Bousses P, Chapuis JL (1999) Nursing behaviour and mother lamb relationships in mouflon under fluctuation population densities. Behav Processes 47:81-94

Richard-Hansen C, Campan R (1992) Social environment of isard kids *Rupicapra pyrenaica* during their ontogeny. Z Saügetierkd 57:351-363

Ruckstuhl KE, Ingold P (1999) Aspects of mother-kid behavior in Alpine chamois, *Rupicapra rupicapra rupicapra*. Z Saugetierkd 64:76-84

Schaller GB (1977) Mountain monarchs. University of Chicago Press, Chicago

Schröder J, Kofler H (1984) Coexistence and competitive exclusion between ibex *Capra ibex* and chamois *Rupicapra rupicapra*. Acta Zool Fenn 172:87-88

Schröder J, Schröder W (1984) Niche breadth and overlap in red deer *Cervus elaphus*, roe deer *Capreolus capreolus* and chamois *Rupicapra rupicapra*. Acta Zool Fenn 172:85-86

Storch I (1989) Condition in chamois population under different harvest levels in Bavaria. J Wildl Manage 53:925-928

Stüwe M, Nievergelt B (1991) Recovery of alpine ibex from near extinction: the result of effective protection, captive breeding and reintroductions. Appl Anim Behav Sci 29:279-387

Szemkus B, Ingold P, Pfister U (1998) Behaviour of alpine ibex (*Capra ibex ibex*) under the influence of paragliders and other air traffic. Z Saügetierkd 63:84-89

Sæther BE (1997) Environmental stochasticity and density dependence in population dynamics of large herbivores: a search for mechanisms. Trends Ecol Evol 12:143-149

Tomiczek H (1992) Browsing, fraying and stripping damage caused by ibex. Z Jagdwiss 38:63-67

Toïgo C (1998) Strategies biodemographiques et selection sexuelle chez le bouquetin des Alpes (Capra ibex ibex). PhD Thesis, University of Lyon 1, Lyon

Toïgo C (1999) Vigilance behavior in lactating female alpine ibex. Can J Zool 77:1060-1063

Toïgo C, Gaillard JM, Michallet J (1996) La taille des groupes: un bio-indicateur de l'effectif des populations de bouquetin des Alpes (*Capra ibex ibex*). Mammalia 60:463–472

Toïgo C, Gaillard JM, Michallet J (1997) Adult survival of the sexually dimorphic alpine ibex (*Capra ibex ibex*). Can J Zool 75:75–79

Toïgo C, Gaillard JM, Michallet J (1999) Cohort affects growth of males but not females in alpine ibex (*Capra ibex ibex*). J Mammal 80:1021–1027

Väisänen RA (1998) Current research trends in mountain biodiversity in NW Europe. Pirineos 151–152:131–156

Villaret JC, Bon R, Rivet A (1997) Sexual segregation of habitat by the alpine ibex in the French Alps. J Mammal 78:1273–1281

Weber JM, Weissbrodt, M (1999) Feeding habits of the Eurasian lynx in the Swiss Jura Mountains determined by faecal analysis. Acta Theriol 44:333–336

Wiersema G (1984) Seasonal use and quality assessment of ibex habitat. Acta Zool Fenn 172:89–90

Williams JS (1999) Compensatory reproduction and dispersal in an introduced mountain goat population in central Montana. Wildl Soc Bull 27:1019–1024

22 Diversity of Alpine Vertebrates in the Pyrenees and Sierra Nevada, Spain

J.P. Martínez Rica

22.1 Introduction

There are over 250 mountains surpassing 2000 m in 25 ranges of the five main mountain systems of the Iberian Peninsula, the Pyrenees, the Cantabrian, Iberic, Central and Betic Systems. The Pyrenees and the Sierra Nevada are the most important for their alpine zone. They are representative of the other three systems and, in addition, they include the two highest mountains of the peninsula (Aneto 3404 m and Mulhacén 3483 m) that are also the highest in Europe outside the Alps and the Caucasus. These mountain systems are contrasting because the Pyrenees in the north of Spain have an alpine zone, similar to those of other European mountains belonging to the euro-Siberian domain, while the Sierra Nevada in the south belongs to the Mediterranean domain, and the zone above the climatic treeline is referred to as cryoro-Mediterranean (Rivas Martinez et al. 1987). All the remaining mountain ranges of the Iberian Peninsula may be considered intermediate between these.

In the Pyrenees, the alpine zone is on average above 2300 m on southern slopes and above 2000 m on northern ones, descending to 1800 m in some localities in the Northern Pyrenees. True alpine areas (above 2300 m) are found in about 36 isolated places. In the Sierra Nevada, the cryoro-Mediterranean zone is above 2600 m (Esteve and Prieto 1971) and consists of a narrow range of peaks of about 30 km long running from SW to NE. The two chains differ in floristic and vegetation diversity. The number of Pyrenean plant species is about 3500 (800 in the alpine), as opposed to 2100 (500 in the cryoro-Mediterranean) in the Sierra Nevada (Blanca et al. 1998). Conversely, the percentage of endemic species is much higher in the isolated high mountains of the Sierra Nevada (ca. 35%) than in the Pyrenees (16%, Sese et al. 1998).

22.2 Alpine Vertebrate Species

As in other habitats, vertebrates are less suitable than invertebrates in defining alpine communities. However, being the best-known species, they are frequently chosen as representatives of the alpine fauna and used to evaluate alpine biodiversity. In the Pyrenees, about 50 species have been recorded frequently from the mountains above 2000 m; 40 of them often from above 2300 m. Six species frequently occur above 3000 m. The 40 species (for full list, see http://www.stir.ac.uk/departments/naturalsciences/DBMS/nagy/alpine_biodiversity/) occurring above 2300 m can be considered to make up the alpine communities; however, not all of them can be called alpine as many of them are found in lower vegetation zones also (Table 22.1). The Sierra Nevada, owing to the small extension of its cryoro-Mediterranean habitats, has fewer species (Table 22.1). About 30 can be found more or less frequently, although many of these have their optimum at lower altitudes.

Vertebrate alpine communities in both the Pyrenees and the Sierra Nevada consist of few truly alpine species. In the Pyrenees, 15 species occur above a mean altitude of 2000 m; six species, three of them small lizards, (which may be conspecific), can be called truly alpine, i.e. having their mean distribution above 2300 m. In the Sierra Nevada, only the alpine accentor (*Prunella collaris* Scopoli) and the European snow vole (*Chionomys nivalis* Martins) are restricted to the uppermost zone. The Spanish ibex (*Capra pyrenaica* Schinz), an introduced species in the Sierra Nevada, can be seen often near the summits, but lives everywhere above the forest level, or even within the forest. The smooth snake (*Coronella austriaca*) is found also above the forests, and its populations in the Sierra Nevada are the highest in Europe. The Iberian wall lizard (*Podarcis hispanica* Steindachner) occurs from sea level to 3398 m

Table 22.1. The number of vertebrates found above 2300 m in the Pyrenees and above 2600 m in the Sierra Nevada. The mean altitude distribution of the species considered ranges between 800 and 3050 m a.s.l., with most of them well below the treeline (for full list of species and elevations, see http://www.stir.ac.uk/departments/naturalsciences/DBMS/nagy/alpine_biodiversity/). Figures in parentheses show the number of species whose mean distribution is in the alpine zone

	Pyrenees (>2300 m)	Sierra Nevada (>2600 m)
Fish	2	1
Amphibians	5	1
Reptiles	6 (3)	5
Birds	15 (3)	17 (1)
Mammals	12	8 (2)

Table 22.2. Alpine amphibian and reptile species on European and North African mountains. The upper group includes species mostly restricted to alpine or alpinoid habitats while the bottom group includes species inhabiting the alpine zone but also found lower elevation. Data from Pleguezuelos and Martinez Rica (1997), Bons and Geniez (1996) and Bannikov et al. (1977)

Pyrenees	Sierra Nevada	Alps	Caucasus	Atlas
Lacerta agilis	Coronella austriaca	Salamandra atra	Lacerta valentini	Quedenfeldtia trachyblepharus
Lacerta aranica		Salamandra lanzai	Vipera dinniki	Lacerta andreanszky
Lacerta aurelioi				Vipera monticola
Lacerta bonnali				
Alytes obstetricans	Bufo bufo	Triturus alpestris	Triturus vittatus	Coronella girondica
Triturus helveticus	Podarcis hispanica	Bufo bufo	Mertensiella caucasica	
Bufo bufo		Rana temporaria	Rana macrocnemis	
Vipera aspis		Lacerta vivipara	Lacerta caucasica	
Podarcis muralis		Podarcis muralis	Lacerta raddei	
Euproctus asper		Vipera berus	Lacerta saxicola	
		Vipera aspis	Vipera ursinii	

(Pleguezuelos 1986), which is the highest record in Europe for any reptilian or low vertebrate species.

The numbers of high mountain amphibian and reptile species in the Alps (>2000 m), Caucasus (>2600 m) and the Atlas Mts. (>2800 m) are shown in Table 22.2. Using these values the number of alpine species in the Alps is nine, almost the same number as in the Pyrenees; all of these, however, are also found at lower altitudes, even at sea level, and perhaps only two species of salamanders, *Salamandra atra* Laurenti and *Salamandra lanzai* Nascetti, Andreone, Capula, Bullini, usually live above 1000 m a.s.l. In the Caucasus, the number of alpine amphibians and reptiles is also nine species; however, the cut-off limit may have been chosen too strictly. The number of alpine species in the Atlas chain, a range isolated from the European mountains, is only four; none of these are amphibians. The Sierra Nevada has the least species, probably caused by the small size of the range. The Atlas chain has one more species but is also poor. The Alps, Pyrenees and the Caucasus sites look comparable, perhaps because of a similar history within the European continent. Alpine zoocoenoses including herpetocoenoses comprise mountain endemics and widespread species that are able to adapt to the conditions of alpine habitats. The difference between the rates of plant and animal species endemism may be explained by the mobility and the capability of the animals to colonise distant areas as opposed to plants, which are poorly dispersed, compared with most animals. Hence, the high proportion of endemic plants in such relatively small mountain ranges as the Sierra Nevada.

22.3 Species Richness vs. Diversity Indices

Until now, we have dealt with specific richness only, a crude but widespread measure of diversity. Species-richness comparisons are made on a unit area basis. Whilst it is rather a semiquantitative measure it has been shown to correlate with other measures of diversity (Magurran 1988) and high species richness is often indicative of high ecological diversity.

Even for well-known groups such as vertebrates, and for well-studied mountains such as the Pyrenees and Sierra Nevada, there are few data on the populations of the different species. Without them, quantitative measures of diversity, such as the Shannon-Weaver or Brillouin index, cannot be calculated. Moreover, population sizes of animals change within and between years, and hence the estimates of local diversity (α-diversity) change as well. For general patterns that apply over an entire mountain range, a measure of diversity with a geographic component (γ-diversity) is required. Based on bird, reptile and amphibian distribution maps covering the Pyrenees and Sierra Nevada (Pleguezuelos 1992; Pleguezuelos and Martinez Rica 1997;

Purroy 1997), we calculated Shannon indices. Overall values were over 90 % of the maximum diversity possible. In separate analyses for the three groups, evenness values for amphibians and reptiles (83 and 84 % of the maximum feasible) were consistently lower than that for birds (96–98 %). If this result reflects reality, it may lead us to conclude that birds are poor indicators of γ-diversity, probably owing to their greater mobility than that of the others. The high values were an artefact because distribution maps data are ultimately presence/absence data. We propose that when no true abundance data are available the only measure that should be used is specific richness, given the biases introduced.

22.4 Altitude Gradients and Biodiversity

Altitude, and its surrogate mean temperature, which when water supply is adequate drives mean plant production, is thought to be the main variable controlling species richness in mountains. Species richness generally decreases with altitude in a more or less continuous manner with outliers occurring at ecotones, where the contact of two different zones produces an increase in taxonomic diversity. However, a general function that describes the dependency of species diversity on altitude is not easy to establish, as probably each mountain, or at least each mountain system, has its own speciality. Such a function may change over the year and perhaps from year to year.

The relationship between altitude and species richness may be used to generate or test hypotheses about the determinants of species richness along altitudinal gradients. The mathematical functions best describing this relationship are usually complex. The type of the relationship (peak, transition or monotonic), however, is more important than the exact mathematical function. For example, a monotonic function would imply a more or less regular decrease in the number of vertebrate species with altitude from sea level. For a transition function, the decrease may begin at the beginning of the montane belt. For a peak function one would expect a maximum in the montane zone, where both lowland and mountain species can be found, with the number decreasing slightly at lower altitudes and more sharply at higher altitudes. Findings from an altitudinal gradient study can then test the following hypotheses:
1. If biodiversity largely depends on climatic and environmental variables, climatic change would have a direct effect on alpine biodiversity.
2. If, on the contrary, diversity depends on ecological interactions between species and communities, such as competition, these interactions would continue to operate in spite of climatic changes, and biodiversity might be little affected.

3. Yet, if specific richness depends on other forms of diversity, such as the variety of ecological niches available, then the climatic changes could affect that variety and through it specific richness.

Mountain vertebrates from the Andes and from other tropical mountains have been well studied. However, little is known for temperate mountains. Studies of vertebrate species diversity in mountains have often tested the predictions emanating from the theory of insular biogeography (Villeumier 1972; Wohlgemuth 1993), a rather inadequate approach, because of the limited isolation of mountain areas and of the rather more complex nature of the speciation, extinction and migration processes in the mountains than on islands. The number of studies specifically dealing with elevation gradients is fewer and mostly confined to birds and flowering plants. Causal factors invoked for a decrease in species richness included the size of the mountain and the variety of habitats (e.g. Terborgh 1977; Graham 1990; Sfenthourakis 1992), environmental factors such as habitat constriction (Huey 1978), temperature (Balent et al 1988), or variables linked to temperature, such as primary production (Scheiner and Rey-Benayas 1994) or soil freezing and thawing (Grabherr et al. 1995).

In the Pyrenees and the Sierra Nevada, the historical biogeographical factors which may have contributed to lowering species numbers at high altitudes include the size of the area above treeline, an important factor accounting for the fewer species found in the Sierra Nevada than in the Pyrenees or the other chains, where the available area above the treeline is much smaller. The degree of isolation may also be considered for the higher percentage of endemic plants in the Sierra Nevada than in the Pyrenees. For example, the sedentary nature of flowering plants resulted in isolation during climatic oscillations and in an on-site evolution in the Sierra Nevada. In the Pyrenees, instead, connection with European or Iberian populations was not completely severed and the much larger size of the residual populations caused a slower rate of speciation. This does not hold, however, for vertebrates with a higher resource requirement than plants. Thus, in a small mountain system such as the Sierra Nevada, most vertebrate populations are likely to have become extinct at such maxima, without in-site adaptation and speciation. There is only one and doubtful endemic vertebrate in the Sierra Nevada, the local population of the European snow vole (*Chionomys nivalis*), which is different from the subspecies of other Spanish mountains; however, the differences might not warrant a subspecies status (Nadachovski 1991). Pyrenean vertebrates include many more endemic taxa – six species (Iberian mole vole, *Microtus pyrenaicus* de Sélys-Longchamps; three lacertoid lizards, *Lacerta bonnali* Lantz, *L. aranica* Arribas, and *L. aurelioi* Arribas; the Pyrenean brook newt, *Euproctus asper* Dugès; and the frog, *Rana pyrenaica* Serra-Cobo) and at least 12 subspecies, not counting forms.

The pattern for vertebrates in the Pyrenees and Sierra Nevada seems to be a sigmoidal function This is especially clear in Sierra Nevada, where vegetation belts are more or less concentric; in the Pyrenees, where mountains and peaks are located in a less regular way, it is not so evident. There are abrupt drops in species richness at altitudes that mark the beginning of the upper montane and subalpine belts (1600 and 2100 m in the Pyrenees), and the upper montane and oro-Mediterranean at 2000 and 2400 m in the Sierra Nevada. The drop in species richness above an ecotone, where vegetation cover is simpler than that at the ecotone, may be explained by a niche reduction. This feature points to habitat restriction, and environmental constraints, as the main factor controlling species richness in mountains.

The general pattern of variation along elevational gradients for lower vertebrates in the Pyrenees and the Sierra Nevada involves a decrease in richness with elevation and some local peaks, mainly at the location of ecotones between different vegetation belts (Fig. 22.1). Above these peaks, from the montane belt upward, there are sharp decreases in specific richness. For plants, sharp decreases were reported at 2300 m (alpine level) and 2850 m (nival level) in the Alps (Grabherr et al. 1995) and much less sharp changes at 1200 m (montane level) and 2000 m (oro-Mediterranean level) for Mt. Hermon (Wilson and Shmida 1984).

The number of species depends on the number of available habitats, which is related to the available area, a well-known premise of the theory of island biogeography. Notwithstanding the criticisms levelled at that theory, it is illuminating to use available area to predict the number of species. The decrease of available area with altitude for the Pyrenean and Sierra Nevada vertebrates

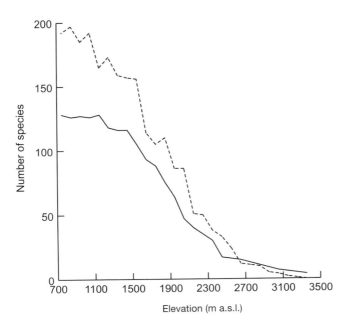

Fig. 22.1. Relationship between altitude and the number of vertebrate species in the Pyrenees (*dashed line*) and the Sierra Nevada (*solid line*)

is similar to the decrease in species numbers, but not for all groups (Martinez Rica and Reine 1988). There may be other factors involved such as the lack of predation or restricted food supplies; however, these are difficult to test.

The available area shows a peak at about 900–1000 m (probably caused by largely excluding lowland areas), which is not so clearly shown in the curves of specific richness. Above 1000 m, the available area decreases exponentially whilst species numbers decline linearly. While the decrease of the available area is a clear component of habitat restriction, it is not a sufficient explanation for the pattern of variation in the number of species along elevation gradients, at least in the mountains in the Pyrenees and the Sierra Nevada. Within the same type of habitat, e.g. alpine or nival zones, the function relating species and altitude has been reported to be exponential. In the Alps, the number of species of plants increases exponentially from 4450 to 3000 m (Grabherr et al. 1995). A similar, but less marked, increase has been reported for Mt. Hermon, from 2800 to 1200 m (Wilson and Shmida 1984). This may suggest a general pattern for plants at least at the upper levels of temperate mountains. However, vertebrates, especially birds and mammals, can move freely across altitudinal zones and are unlikely to conform to such a pattern.

22.5 Conclusions

1. This paper discusses three hypotheses related to the factors underlying the pattern of taxonomic diversity along altitude gradients. According to our data, all selected functions show a good fit to what is reported, being slightly better for the Sierra Nevada, but differences are so small that a much larger set of data would be necessary to test the stated hypotheses.
2. Of special importance would be the data corresponding to lowland areas, below 700 m, not considered here. If these data show a lower specific richness, then the third hypothesis would be favoured, and habitat restriction (and hence environmental constraints) would appear to be the main factor controlling biodiversity in the upper belts of temperate mountains.

References

Balent G, Genard M, Lescourret F (1988) Analysis of the distribution patterns of breeding birds in southern France (Midi-Pyrenees). Acta Oecol Oecol Gen 9:247–264

Bannikov AG, Darevskii IS, Ischenko VG, Rustamov AK, Scherbak NN (1977) Opredelitel zemiovodnikh i presmikayuschikhsya fauni S.S.S.R. Ed. Prosveschenie, Moskva

Blanca G, Cueto M, Martinez-Lirola M, Molero-Mesa J (1998) Threatened vascular flora of Sierra Nevada (southern Spain). Biol Conserv 85:269–285

Bons J, Geniez PH (1996) Amphibiens et reptiles du Maroc (Sahara Occidental compris). Atlas biogéographique. Asociación Herpetológica Española, Barcelona
Esteve F, Prieto P (1971) Vegetación y Flora Nevadense. In: Ferrer M (ed) Sierra Nevada. Anel, Granada, pp 393–403
Grabherr G, Gottfried M, Gruber A, Pauli H (1995) Patterns and current changes in alpine plant diversity. In: Chapin SF II, Körner C (eds) Arctic and alpine diversity. Springer, Berlin Heidelberg New York, pp 167–181
Graham GL (1990) Bats versus birds: comparison among Peruvian flying vertebrate faunas along an elevational gradient. J Biogeogr 17:657–668
Huey RB (1978) Latitudinal pattern of between-altitude faunal similarity: mountains might be higher in the tropics. Am Nat 112:225–229
Magurran A (1988) Ecological diversity and its measurement. Princeton University Press, Princeton
Martinez-Rica JP, Reine A (1988) Altitudinal distribution of amphibians and reptiles in the Spanish Pyrenees. Pirineos 131:57–82
Nadachowski A (1991) Systematics, geographic variation, and evolution of snow voles (Chionomys) based on dental characters. Acta Theriol 36:1–45
Pleguezuelos JM (1986) Distribución altitudinal de los reptiles en las Sierras Béticas Orientales. Rev Esp Herpetol 1:65–83
Pleguezuelos JM (1992) Avifauna nidificante de las Sierras Béticas Orientales y depresiones de Guadix, Baza y Granada. Su Cartografiado. Ed. Universidad de Granada, Granada
Pleguezuelos JM, Martinez Rica JP (eds) (1997) Distribución y biogeografía de los anfibios y reptiles de España y Portugal. Universidad de Granada and Asociación Herpetológica Española, Granada
Purroy FJ (ed) (1997) Atlas de las aves de España: 1975–1995. Lynx, Barcelona
Rivas Martinez S, Gandullo JM, Serrada R, Allué JL, Montero de Burgos JL, González Rebollar JL (1987) Memoria del Mapa de las Series de Vegetación de España 1:400000. Serie Técnica. Ministerio de Agricultura, Madrid
Scheiner SM, Rey-Benayas JM (1994) Global pattern of plant diversity. Evol Ecol 8:331–347
Sese JA, Ferrandez JV, Villar L (1998) La flora alpina de los Pirineos: Un patrimonio singular. In Villar L (ed) Espacios naturales protegidos del Pirineo: Ecología y cartografía. Publ. Del Consejo de Protección de la Naturaleza de Aragón. Serie Conservación, Zaragoza, pp 55–74
Sfenthourakis S (1992) Altitudinal effect on species richness of Oniscidea (Crustacea; Isopoda) on three mountains in Greece. Global Ecol Biogeogr Lett 2:157–164
Terborgh J (1977) Bird species diversity on an Andean elevational gradient. Ecology 58:1007–1019
Villeumier F (1972) Bird species diversity in Patagonia (temperate South America). Am Nat 106:266–271
Wilson MV, Shmida A (1984) Measuring beta diversity with presence-absence data. J Ecol 72:1055–1064
Wohlgemuth T (1993) Der Verbreitungsatlas der Farn- und Blütenpflanzen der Schweiz (Welten und Sutter 1982) auf EDV: Die Artenzahlen und ihre Abhängigkeit von verschiedenen Faktoren. Bot Helvetica 103:55–71

23 The Impacts of Vertebrate Grazers on Vegetation in European High Mountains

B. ERSCHBAMER, R. VIRTANEN and L. NAGY

23.1 Introduction

The complex nature of herbivore–plant relationships has an extensive literature (e.g. Crawley 1997) and grazers play an important role in many ecosystems (Grime 1979; McNaughton 1983; Milchunas et al. 1988; Huntly 1991; Augustine and McNaughton 1998; Mulder 1999; Ritchie and Olff 1999). Grazing in the alpine zone of Europe has a long history and has caused large-scale vegetation changes especially in the treeline ecotone (e.g. Hallanaro and Pylvänäinen 2001). In this chapter, we give an overview of how mammalian grazers impact on the species composition, species richness, and canopy structure of the vegetation in the alpine zone and the treeline ecotone in Europe's high mountains. However, the impacts of grazers in European alpine areas have been little investigated and there are controversial opinions about the general ecological significance of grazing in alpine environments. We highlight some problems deserving more attention in future. Plant names follow Adler et al. (1994); the nomenclature for animals is after Schaefer (1994).

23.2 The Historical Perspective

It has been estimated, using pollen evidence, that domestic grazers have affected the alpine grasslands of the inner Oetz Valley, Tyrol, Austria (2240–2760 m a.s.l.) since about 5000 years B.P. (Vorren et al. 1993; Bortenschlager 1993, 2000, 2001). Pollen records have shown that grazing indicators such as *Ligusticum mutellina*, species of the genera *Lotus* and *Plantago*, species of the Rosaceae, and unpalatable (poisonous) plants such as *Gentiana*, *Rhinanthus*, and species of the Ranunculaceae increased from the Neolithic period (4500–6000 years B.P.) onwards (Bortenschlager 1999).

In the Central Alps, the timberline was lowered by 200 to 400 m during the Middle Ages as a result of burning and clear-cutting (Friedel 1967; Breitenmoser 1998). Human impacts have been more pronounced in the zone below the potential timberline (1800–2100 m a.s.l.) than higher up (Patzelt et al. 1997). It is thought that sustained grazing allowed *Nardus stricta*, a species largely avoided by domestic grazers, to form extensive stands in the zone below the potential timberline, in the treeline ecotone and in the lower alpine zone (e.g. Schaminée and Meertens 1992).

Local nutrient enrichment through the deposition of faeces and urine over hundreds of years has resulted in the development of anthropo-zoogenic vegetation of the *Poa alpina*-, *Alchemilla–Poa supina*-, and *Rumex*-dominated communities (Ellmauer and Mucina 1993). Some of the *Poa alpina* communities are found at altitudes over 2300 m (Heiselmayer 1982; Ellmauer and Mucina 1993). In addition, pasture weeds, such as *Alchemilla* spp. or *Cirsium spinosissimum*, are distributed far outside their natural range up to the subnival zone. This suggests that grazing impacts have reached high altitudes in the Alps.

In the Highlands of Scotland, long-term heavy grazing by sheep is thought to have caused diverse plant communities to converge and form some of the structurally rather homogeneous anthropogenic grassland types (McVean and Ratcliffe 1962; Rodwell 1992; Thompson and Brown 1992). It is thought that the treeline has been heavily depressed by grazing and periodic burning (Crawford 1989). Grazing has also been recognised as one of the essential factors influencing the forests in northernmost Fennoscandia (Oksanen et al. 1995). The forest structure of the treeline areas in Fennoscandia is heavily influenced by reindeer (*Rangifer tarandus* L.) and sheep (Oksanen et al. 1995; Hofgaard 1997; Suominen and Olofsson 2000).

23.3 The Grazers and Their Diet

In the Alps and Pyrenees, wild (chamois *Capra rupicapra*, red deer *Cervus elaphus*, roe deer *Capreolus capreolus*, ibex *Capra ibex*, and mouflon *Ovis ammon*) and domestic ungulates (cattle *Bos*, goat *Capra*, horse *Equus*, and sheep *Ovis*) and rodents (hare *Lepus europaeus* and *L. timidus*, marmot *Marmota marmota*, and voles Microtinae) are the most important mammalian grazers. Selectivity in large grazers such as cattle, red deer, and sheep varies (Spatz 1994; Armstrong 1996); however, all feed preferentially on graminoids (Poaceae, Cyperaceae) and, when not available, on dwarf shrubs. Hofmann and Schnorr (1982) and van Soest (1994) distinguished grass/roughage eaters (cattle, mouflon, and sheep), an intermediate type (chamois, goat, ibex, and red deer) and concentrate selectors (roe deer) according to feeding and rumination frequency. However, there is a shortage of knowledge about their preferred forage plants with the exception of those of chamois, ibex and sheep (Spatz 1994; Catusse et al. 1996).

Ibex preferentially eat grasses (up to 85% *Anthoxanthum odoratum*, *Festuca ovina*, *F. spadicea*, *F. violacea*, *Nardus stricta*, *Sesleria* sp., *Trisetum flavescens*) with the rest being forbs (*Achillea millefolium*, *Anthyllis*, *Astragalus*, *Gentiana*, *Onobrychis*, *Plantago alpina*, *Potentilla*, *Primula*, *Saxifraga* and *Senecio*) and dwarf shrubs (*Arctostaphylos uva-ursi*, *Daphne mezereum*, *Empetrum*, *Juniperus*, *Loiseleuria procumbens*, *Rhododenderon ferrugineum*, *Vaccinium myrtillus*, *V. uliginosum*, *V. vitis-idaea*; Aymerich 1992; Catusse et al. 1996). Chamois prefer species of Poaceae and Fabaceae, and dwarf shrubs (Dunant 1977; Lambert and Bathgate 1977; Tataruch 1982; Ferrari and Rossi 1985; Perle and Hamr 1985; Bruno and Lovaci 1989). The diet of the sheep in the lowlands consists predominantly of grasses (e.g. *Anthoxanthum odoratum*, *Festuca* spp., and *Poa* spp.), legumes and herbs (Jans and Troxler 1990). In the alpine zone, in addition to Poaceae, they also eat sedges (Erschbamer 1994). In the absence of herbs, sheep will browse less palatable dwarf shrubs (Armstrong 1996). Horses are biters and graze short (Armstrong 1996). The number of horses in the Central Alps has risen recently, contributing an additional pressure on the treeline ecotone and on alpine environments (B. Erschbamer, pers. observ.). Heavy impacts due to over-grazing and trampling are to be expected.

In the Scandes, the most important grazers are the rodents (mountain hare, lemming *Lemmus lemmus* L., and voles) and the reindeer. Lemmings feed on mosses and graminoids (Tast 1991; H. Henttonen, pers. comm.); most of the voles feed on monocots and dicots (Tast 1974; N. G. Yoccoz and H. Henttonen, pers. comm.); the grey-sided voles (*Clethrionomys rufocanus* Sund.) prefer *Vaccinium myrtillus* and other dicots (Kalela 1957; H. Henttonen, pers. comm.). Reindeer mostly feed on lichens and to a lesser extent on grasses, herbs, mushrooms and shrubs (Ö. Danell, pers. comm.).

23.4 Grazing Impacts

Grazers can create, maintain, or reduce environmental heterogeneity and thereby influence species and community diversity. For example, small mammals and invertebrates are usually most numerous in areas with tall, not heavily grazed vegetation (MacDonald et al. 1998; Chap. 16).

Whether grazing impacts are manifested at the micro-, small- or patch-scale, or on the landscape-scale is dependent on the kinds of grazers present, their relative density, feeding habits, ranging behaviour and range use, and on the productivity of the vegetation types being grazed (Pignatti Wikus 1987; Armstrong et al. 1997a,b; MacDonald et al 1998; Austrheim and Eriksson 2001). Micro- and patch-scale impacts are largely a result of food preferences (e.g. selective grazing of flowers and fruits) and grazing habits of grazers (e.g. differences in minimum sward height). Landscape-scale changes are usually related to management and range use by large grazers such as cattle, red deer and reindeer. In addition to offtake, grazers create heterogeneity at the micro- and patch-scale by dung and urine deposition, seed dispersal, trampling and by pulling up plants.

Grazing-mediated heterogeneity is an important factor controlling the diversity, abundance and distribution of species in the canopy (Harper 1977; Crawley 1983, 1997; McNaughton 1983, 1985; Olff and Ritchie 1998). It can influence interspecific interactions (Knapp 1967), productivity and the availability of resources including soil nutrients (Batzli 1975, 1978; McKendrick et al. 1980). Sustained grazing may change the composition of functional types that make up the vegetation. The dominance of grasses may be enhanced together with species that have high concentrations of chemical defence substances (Coley et al. 1985; Oksanen 1990; Sundriyal 1992; Henry and Svoboda 1994; Wilson 1994; Klötzli 1996). Low stature species may establish in gaps in vegetation kept short by grazing. According to Oksanen (1990) and Oksanen and Ranta (1992), most Poaceae and Cyperaceae of the Scandinavian tundra are 'grazing-adapted plants'. Grasses have basal meristems, i.e. they can regrow after grazing. Defoliation and clipping experiments suggest that they respond positively to grazing. It has been shown that the growth of old tillers and production of new tillers and leaves is enhanced (Chapin 1980; Archer and Tieszen 1980; Kotanen and Jefferies 1987; Bazely and Jefferies 1989; Bock et al. 1995; Beaulieu et al. 1996; Rainer and Erschbamer 1996).

23.5 The Effects of Grazing Exclusion on the Vegetation

As grazers can have an array of effects, it is illuminating to examine what happens to vegetation in their absence. The exclusion of grazers by fencing certain areas for long periods is an efficient way to determine the degree of grazing impacts. We review below such exclusion studies.

23.5.1 Biomass and Species Composition

The responses of plant communities in grazing exclusion experiments largely support the predictions of the ecosystem exploitation hypothesis of Oksanen et al. (1981) and Oksanen (1990), who proposed that intense grazing favours low-stature plant forms, plants with secondary metabolites and grasses. In the absence of grazing, tall species, which are superior competitors for light, were found to be at an advantage (Oksanen 1988, 1990; Moen and Oksanen 1998; Virtanen 1998). In exclosure experiments the absence of grazing has led to an increase of highly competitive grasses (Lye and Lauritzen 1975; Dietl 1982; Moen and Oksanen 1998; Virtanen 1998, 2000) and competitive and tall herbs

(Oksanen 1990; Oksanen and Moen 1994; Moen and Oksanen 1998). A concurrent decrease or loss of weak competitors appears to be the norm (Dietl 1982; Schneiter 1997; Virtanen et al. 1997; Miller et al. 1999; Schütz et al. 2000). The persistence of some weak competitors such as the annual *Gentiana nivalis*, or low-statured herb *Crepis multicaulis* Ledeb. – the only population of the latter species became extinct from northern Norway probably as a result of protective exclosure fencing from sheep (Alm and Often 1997) – appears to depend on recurrent perturbation creating gaps for establishment.

Such responses on the functional type level depend on the initial plant community structure (Stüssi 1970; Dietl 1982), grazing pressure at the beginning of the exclusion (Andersson and Jonasson 1986; Moen et al. 1993; R. Virtanen, unpubl.), the productivity of the vegetation, environmental conditions (Moen and Oksanen 1998) and the time elapsed. For example, in snowbeds, together with graminoids, Polytrichaceae increased in the absence of lemmings (Virtanen 2000) whilst, in an alpine pasture, dwarf shrubs made a recovery when sheep were excluded (Dietl 1982). Virtanen (2000) demonstrated the time dependency of changes in the absence of grazing. After 5 years, only small changes in biomass were detected whereas highly significant biomass increases were found after 15 years (Fig. 23.1). Conversely, Dietl

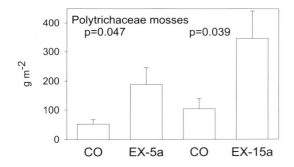

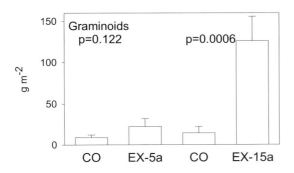

Fig. 23.1. Biomass fractions (g dry weight m^{-2} and standard error) of the Polytrichaceae mosses and the graminoids after 5 (*EX-5a*) and 15 (*EX-15a*) years of grazing exclosure as compared with controls (*CO*). Significance levels as p values from two-tailed t-tests are given for tests of differences in biomass between the exclosures and the appropriate controls (Virtanen 2000; reproduced with permission from Oikos)

(1982) found significant changes after 3–4 years, probably reflecting the capacity of the vegetation in the treeline ecotone to recover more rapidly than in alpine snowbeds. It is also possible that the initial grazing pressure by sheep on the treeline vegetation was higher than that of lemmings on snowbeds. A lack of response to cessation of grazing pressure such as that reported after 6 years by Schneiter (1997) for a *Carex curvula* grassland, Bidmer, Switzerland, by García-González et al. (1998) for *Bromus-* and *Nardus-*dominated grasslands in the Spanish Pyrenees, or that by Camenisch and Schütz (2000) for a *Crepis–Festuca nigrescens* grassland at the treeline ecotone in the Swiss National Park, may reflect an initial low grazing pressure. In another example, Cernusca (1989) reported that moderate grazing by sheep and young cattle had no visible effects on alpine grasslands (*Carex curvula* type, Hohe Tauern, Austria) in comparison with 10-year-old exclusion plots. Cernusca (1989) attributed this finding to the fact that 70 % of the living biomass of the grassland was between 0–1.5 cm above the ground. As sheep graze to about 3 cm and cattle to about 6 cm, a minor offtake has to be assumed. Grazing exclusion experiments have also provided insight into the temporal variation of offtake when population densities of grazers fluctuate. During outbreaks of Norwegian lemmings, Moen et al. (1993) showed that most plant groups are grazed. Effects on plant biomass were greatest when lemming population density was high (Andersson and Jonasson 1986).

23.5.2 Dead Standing Plant Material, Litter, and Soil Microbes

The amount of standing dead material and litter can be influenced by the exclusion of grazing. Considerable temporal and spatial variation may be recognised in short-term exclusions and the differences between exclosures and controls may not be statistically significant for each year (Fig. 23.2). In the long term, a significant accumulation of dead material and plant litter in the absence of grazing was shown by Virtanen (2000; Fig. 23.3). This is consistent with general findings at lower altitudes where plant litter accumulates after the cessation of grazing (Coughenour 1991; Knapp and Seastedt 1986; Henry and Svoboda 1994; Hofstede et al. 1995).

The effect of cattle exclusion on soil microbes was reported for anthropogenic montane grasslands (1600 m a.s.l.) in the Northern Calcareous Alps, Austria (Insam et al. 1996). The changes in microbial biomass, basal respiration and the influence on the metabolic quotient after 9 years exclusions were small and were mainly attributed to changes of litter quality. Körner (1999) estimated that cattle grazing and grasshoppers each contributed about 20 % to nutrient recycling in an alpine sedge-dominated grassland. The highly concentrated patchy nature of cattle dung and urine deposition is likely to cause small-scale heterogeneity in microbial processes. Stark (2002) has found marked large-scale variation in microbial processes in the northern Scandes.

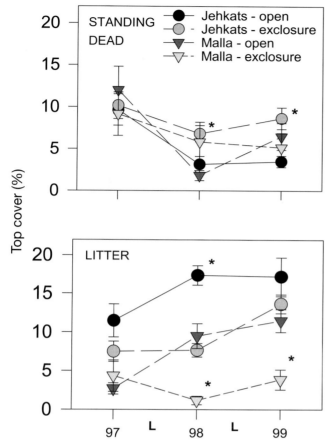

Fig. 23.2. Effects of grazing exclusion (exclosure) on snowbed vegetation (open = unfenced controls) in two mountain sites of NW Finnish Lapland during a population peak of the Norwegian lemmings (*Lemmus lemmus*) in 1997 and 1998. L shows the observed winter grazing at the site in the two winters. The sites: Malla (middle oroarctic zone, 890 m a.s.l.) is a late snowbed site characterised by *Ranunculus glacialis* and *Luzula arcuata* ssp. *confusa*, and bryophytes. Jehkats (middle oroarctic zone, 860 m a.s.l.) is a moderate snowbed site with *Anthoxanthum odoratum* ssp. *alpinum*, *Carex bigelowii*, and *Salix herbacea*. The *asterisks* indicate where the values within the exclosures differ from those of the unfenced control areas (t-test, $p < 0.05$)

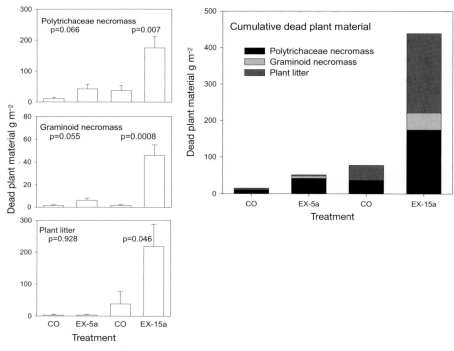

Fig. 23.3. Fractions of dead plant material (g dry weight m^{-2} and standard error) and their accumulation after 5 (*EX-5a*) and 15 (*EX-15a*) years of grazing exclosure as compared with controls (*CO*). Significance levels as in Fig. 23.1 (Virtanen 2000; reproduced with permission from Oikos)

23.6 Evidence from Outside Exclosures

23.6.1 Canopy Structure, Growth Form, Flowering and Seed Production

Sheep grazing has a pronounced influence on the canopy structure of grasslands below the potential timberline (Cernusca and Nachuzrisvili 1983; Cernusca 1989). In the Alps, the canopy height of meadows was found to be twice that of pastures (Cernusca 1989). In the Causasus (2050 m a.s.l.) the height of a pasture canopy was 2–3 cm whereas a nearby meadow was 35–40 cm before mowing (Körner 1980; Cernusca and Nachuzrisvili 1983). In the pasture, 80 % of the phytomass was of *Trifolium ambiguum*, *Alchemilla retinervis*, *Ranunculus oreophilus*, and *Plantago caucasica* as opposed to the meadow, where grasses accounted for more than 50 % of the canopy (Körner 1980).

Grasslands of the climax type of the Alps (such as the *Carex curvula* or the *Sesleria albicans–Carex sempervirens* types) seem to be stable because they are composed of a network of clonal plants (tussock grasses, sedges, deep-rooted rosette plants). Damage caused by trampling of the vegetation mat can easily be repaired by clonal growth, thereby preventing soil erosion (Klötzli 1977, 1990, 1991). In the absence of heavy mechanical perturbations, such grasslands do not show strong fluctuations in their dominant species and grazing may even enhance the production of new leaves and shoots of the dominant sedges (Erschbamer 1994; Rainer and Erschbamer 1996).

The impacts of grazing on flower and seed production have been little investigated. Järvinen (1987) and Diemer (1996) showed that *Ranunculus glacialis* inflorescences and flowering stalks were removed by voles. Galen (1990) pointed out that losses of up to 80 % of the annual seed production by *Polemonium viscosum* caused by predation might limit the distribution of this species. Forty percent of flower stalks of the rosette plant *Lychnis alpina* were reported lost through herbivory, presumably by voles and hares in the eastern Grampians, Scotland (Nagy and Proctor 1996).

23.6.2 Succession After the Abandonment of Pasturing

Changes in the distribution of sheep grazing pressure over the alpine landscape (Chap. 21) may lead to changes in vegetation composition and structure. For example, dwarf shrubs (*Arctostaphylos uva-ursi*, *Calluna vulgaris*, *Juniperus communis* ssp. *alpina*, *Rhododendron ferrugineum*, and *Vaccinium* spp.) or *Alnus viridis* scrub may invade abandoned pastures depending on elevation and nutrient availability (Cernusca 1978; Spatz 1980, 1983, 1994; Zoller et al. 1984; Spatz and Springer 1987; Spatz et al. 1993, Tappeiner et al. 1998; Tasser et al. 1999a,b). Forest regeneration on anthropogenic montane grasslands has been estimated to take over 100 years (Cernusca 1989).

In the Swiss National Park, it has been suggested that grazing by wild ungulates (red deer, chamois) has little influence on forest regeneration. However, an increase in wild ungulates was highly correlated with an increase in tree seedlings and saplings. Especially larch and mountain pine benefited from the gaps created by trampling of red deer (Krüsi and Moser 2000). Comparing 4-year exclosures and controls, Camenisch and Schütz (2000) found no differences in tree regeneration.

23.6.3 Nutrient Enrichment and Community Changes

Locally, and to a variable extent, nutrient enrichment from the urine and faeces of grazers may substantially change plant community composition (Ellmauer and Mucina 1993). Resting places of wild and domestic animals are in particular affected. The *Sesleria–Carex sempervirens* grasslands of calcareous soils or the *Carex curvula* grasslands of siliceous soils may be transformed to species-poor communities of *Poa alpina* or *Cirsium spinosissimum* in the Alps. For example, the *Sesleria–Carex sempervirens* community (1800–2400 m) of the Austrian Alps was reported to have changed on several sites into a *Deschampsia cespitosa–Poa alpina* community (Heiselmayer 1981, 1982, Medicus 1981; Gander 1984). Such changes have also been reported from the Dolomites (Oberhammer 1979; Steinmair 1999). In the Italian Alps, the alpine *Festuca halleri* grassland was changed to a *Festuca halleri–Alchemilla* type (Giacomini and Pignatti 1955), which has an increased abundance of nitrophilous species (*Alchemilla vulgaris, Capsella bursa-pastoris, Poa pratensis, Rumex arifolius, R. alpinus*).

High alpine scree vegetation (*Draba hoppeana* communities) may also be transformed into a *Poa alpina* community due to heavy fertilisation by grazers (Karrer 1980; Ellmauer and Mucina 1993). One of the highest alpine stands of this community is known from the Piza Sassal Masone at 3038 m a.s.l. (Braun 1913). Heavily dunged sites in cattle resting areas on montane pastures and around mountain huts in the Alps are typically dominated by tall-herb communities with more or less monotypic stands of *Rumex alpinus, Urtica dioica* or *Aconitum* spp. (Karner and Mucina 1993). Even after cessation of the agricultural use, these communities remain stable for long periods. In the Swiss National Park where former tall-herb communities have been heavily grazed by red deer, an increase in species number was detected (Achermann et al. 2000). The dominance of the tall-growing species was broken due to deer grazing and short-growing species invaded the stands (Achermann et al. 2000).

The deposition of dung in alpine grasslands can be regarded as a severe perturbation, creating special microhabitats where regeneration processes may continue for over 100 years (Schneiter 1997). In the short term, species richness may slightly increase on these deposition areas as they have similar colonisation dynamics to canopy gaps, favouring weak competitors and short-lived species. However, in the long term, competitive species tend to form monodominant stands on such nutrient-rich sites. In drier soil conditions, the deposition of urine and dung may create bare patches. However, regeneration of these patches and the effects of mechanical damage caused by grazer trampling have been poorly analysed in alpine environments.

23.7 Grazing in the Alpine Zone: Ecological Problems and Prospect

No consensus exists on the importance of grazing in alpine environments and even general ecological theories provide contrasting views. Alpine ecosystems have low productivity because of low temperatures, short growing seasons and low growth rates (Billings 1974; Körner and Larcher 1988).

Accordingly, Grime (1979) proposed that alpine environments were dominated by stress-tolerant species. As a result of a high degree of resistance to herbivores (evolved through natural selection), the impacts of grazing should be small on these species. This conforms to the model of Coley et al. (1985), which predicts that plant defences are well developed in low productive environments. By examining trends across a wide range of ecosystems, Cebrian and Duarte (1994) found that biomass removal (expressed as percentage plant biomass consumed by herbivores) was lower in low productivity ecosystems compared with ecosystems with high plant production (in absolute terms). The hypothesis put forward by them implies that plants of alpine environments should not experience high levels of herbivory. By contrast, Oksanen et al. (1981) and Oksanen and Oksanen (2000) suggested that herbivory impact could be enhanced in moderately unproductive habitats in the absence of large predators. The models developed by Oksanen (1990) predict that different conditions favour alternative plant strategies (in terms of resistance, avoidance and tolerance) to cope with grazing pressure in low productive environments. According to this theory, in the least productive habitats such as upper alpine or subnival zones, endothermic folivores are predicted to be absent and therefore plant biomass is at carrying capacity. More studies are needed to assess the role of plant defences and importance of herbivory in the alpine zone.

The role of grazing for the maintenance of plant diversity in the alpine environments is inadequately understood. From studies conducted elsewhere we can expect grazing to influence populations in several ways. Grazing may affect the persistence of certain types of life histories (Higgins et al. 2000), plant recruitment by increasing seed production (Lennartson et al. 1997), seed dispersal (Malo and Suárez 1996; Welch et al. 2000) and other aspects of plant population and community dynamics (e.g. Turnbull et al. 2000). The effects of grazing on plant coexistence (species richness) can depend on several mechanisms. On the local site scale, grazer-induced patchiness (i.e. spatio-temporal heterogeneity) has implications for species richness. It may be increased in a patchy environment (e.g. distribution of excrement patches; Schneiter 1997); however, theoretically, there is a point where too fine a scale patchiness results in an overall lower species richness (Palmer 1992). Models developed by Pacala and Crawley (1992) made several predictions depending on a species' frequency, the trade-off between its palatability and competitive ability, and the feeding behaviour of the herbivore(s). They showed that it is not self-evident that intermediate levels of grazing always favour plant coexistence. Proulx and Mazumder (1998) hypothesised that grazing decreases species richness in nutrient-poor environments while in nutrient-rich environments grazing increases plant species richness. Studies from the Swiss National Park, where grazing by wild ungulates increased species richness in tall-herb and tall-grass communities (Achermann et al. 2000) and less so on more nutrient-poor stands (Schütz et al. 2000), do not support the findings of

Proulx and Mazumder (1998). These recent studies demonstrate that alpine environments being characterised by high variation in productivity combined with different levels of grazing offer an excellent model system to further test general productivity–diversity hypotheses (Tilman and Pacala 1993; Tokeshi 1999; Waide et al. 1999). In addition, the presence of multiple grazers and their different feeding behaviour provide an interesting test of the effects of different interaction types and their compensatory and additive effects (Ritchie and Olff 1999). Considering the diet of mammalian grazers in European alpine environments, could additive effects prevail since polyphagous grazers use more or less the same functional groups? In Fennoscandian mountains, the presence of microtines and larger grazers may introduce compensatory effects. These relationships between grazers and their main food are inadequately known. Similarly, feedback mechanisms between plants and grazers as suggested for grasslands with long evolutionary histories of grazing (Milchunas et al. 1988) have been little studied in alpine ecosystems. It remains to be shown whether the environment–plant–grazer interactions lead to alternative states of alpine plant communities, similar to those known from arctic environments (Zimov et al. 1995; Mulder 1999).

23.8 Conclusions

1. The effects of excluding grazing depend on the initial composition of the community, the environmental conditions, the kind of grazers, the number and seasonality of offtake by grazers, the length of exclusion time, and the competitive interactions between the species.
2. Exclosure experiments in the Scandes show that long-term grazing markedly influences species diversity, abundance of functional groups, biomass and necromass.
3. In the Alps, heavy grazing impacts have been shown to modify the species composition of plant communities to different degrees. Dung deposition and nutrient enrichment have been considered to be more influential factors than biomass removal by grazing. Locally, severe community changes with highly decreased species diversity have occurred.
4. Within exclosures single species have declined or disappeared within a few years due to dense plant cover resulting from the cessation of grazing. The most vulnerable species are those with annual life cycles, and species that require canopy gaps for recruitment.
5. In many cases, the growth of Poaceae and Cyperaceae may be enhanced by grazing due to their ability to re-grow from basal meristems. However, highly competitive grasses are also favoured by grazing exclusion, due to their efficient clonal growth strategy.
6. Highly palatable species in subalpine and alpine grasslands may be con-

sumed at an early stage in their life history or have their reproduction heavily restricted by removal of flowers or seed heads.
7. High stocking rates in combination with abundant wild grazers are able to prevent the regrowth of trees in the treeline ecotone. Over-grazing and over-manuring have caused significant changes to plant communities. In the Central Alps, due to enhanced horse grazing, pronounced effects in alpine environments have to be expected locally.
8. More detailed analyses and long-term exclusion studies are required. Realistic spatial scales combined with manipulative experiments are recommended with varying densities and selections of grazers.

Acknowledgements. We thank A. Britton and F. Klötzli for their comments.

References

Achermann G, Schütz M, Krüsi BO, Wildi O (2000) Tall-herb communities in the Swiss National Park: long-term development of the vegetation. In: Schütz M, Krüsi BO, Edwards PJ (eds) Succession research in the Swiss National Park, vol 89. Nat Park Forsch Schweiz, Zernez, pp 67–88
Adler W, Oswald K, Fischer R (1994) Exkursionsflora von Österreich. Ulmer, Stuttgart
Alm T, Often A (1997) Species conservation and local people in E Finnmark, Norway. Plant Talk 11:30–31
Andersson M, Jonasson S (1986) Rodent cycles in relation to food resources on an alpine heath. Oikos 46:93–106
Archer A, Tieszen LL (1980) Growth and physiological responses of tundra plants to defoliation. Arct Alp Res 12:531–532
Armstrong HM (1996) The grazing behaviour of large herbivores in the uplands. Scottish Natural Heritage Information and Advisory Note 47. Scottish Natural Heritage, Battleby
Armstrong HM, Gordon, IJ, Grant SA, Hutchings NJ, Milne JA, Sibbald AR (1997a) A model of the grazing of hill vegetation by sheep in the UK. I. The prediction of vegetation biomass. J Appl Ecol 34:166–185
Armstrong HM, Gordon IJ, Hutchings NJ, Illius AW, Milne JA, Sibbald AR (1997b) A model of the grazing of hill vegetation by sheep in the UK. II. The prediction of offtake by sheep. J Appl Ecol 34:186–207
Augustine DJ, McNaughton SJ (1998) Ungulate effects on the functional species composition of plant communities: herbivore selectivity and plant tolerance. J Wildl Manage 62:1165–1183
Austrheim G, Eriksson O (2001) Plant species diversity and grazing in the Scandinavian mountains – patterns and processes at different scales. Ecography 24:683–695
Aymerich M (1992) Ejemplos originales de gestion de fauna de Espana con especial referencia a las especies amenazadas. Colloque International SNICEF Suppl 21 de FORESTIER. Bordeaux, 29–30 Sept 1992, pp 207–216
Batzli GO (1975) The influence of grazers on tundra vegetation and soils. In: Proceedings of the Circumpolar conference in Northern Ecology, Ottawa, 1975. Natural Research Council of Canada, Ottawa, pp I-215–I-225

Batzli GO (1978) The role of herbivores in mineral cycling. In: Adriano DC, Brisbin IL (eds) Environmental chemistry and cycling processes. Proceedings of Symposium, Augusta, Georgia, 28 April–1 May 1976, United States Dept Energy, Conf 76–0429, pp 95–112

Bazely DR, Jefferies RL (1989) Leaf and shoot demography of an arctic stoloniferous grass, *Puccinellia phryganodes*, in response to grazing. J Ecol 77:811–822

Beaulieu J, Gauthier G, Rochefort L (1996) The growth response of graminoid plants to goose grazing in a high arctic environment. J Ecol 84:905–914

Belsky AJ (1986) Does herbivory benefit plants? A review of the evidence. Am Nat 127:870–892

Billings WD (1974) Arctic and alpine vegetation: plant adaptations to cold summer climates. In: Ives JD, Barry RG (eds) Arctic and alpine environments. Methuen, London, pp 403–443

Bock JH, Jolls CL, Lewis AC (1995) The effects of grazing on alpine vegetation: a comparison of the central Caucasus, Republic of Georgia, with the Colorado Rocky Mountains, USA. Arctic Alp Res 27:130–136

Bortenschlager S (1993) Das höchst gelegene Moor der Ostalpen 'Moor am Rofenberg' 2760 m. Diss Bot 196:329–334

Bortenschlager S (1999) Die Umwelt des Mannes aus dem Eis und sein Einfluß darauf. In: Fowler B, Gaber O, DeStefano GF, DeStefano GF (eds) Die Gletschermumie aus der Kupferzeit. Schriften des Südt Archäologiemuseums. Folio Verlag, Wien, pp 81–96

Bortenschlager S (2000) The Iceman's environment. In: Bortenschlager S, Oeggl K (eds) The Iceman and his natural environment. Springer, Berlin Heidelberg New York, pp 11–24

Bortenschlager S (2001) Human influence on high altitude vegetation at the time of the Iceman. In: Goodman DK, Clarke RT (eds) Proceedings of the IX International Palynological Congress Houston, Texas, USA, 1996. American Association of Stratigraphic Palynologists Foundation, Texas, pp 517–525

Braun J (1913) Die Vegetationsverhältnisse der Schneestufe in den Rätisch-Lepontischen Alpen. Neue Denkschr Schweiz Naturforsch Ges Zürich 48:1–339

Breitenmoser U (1998) Large predators in the Alps: the fall and rise of man's competitor. Biol Conserv 83:279–289

Bruno E, Lovaci S (1989) Foraging behaviour of adult female Apennine chamois in relation to seasonal variation in food supply. Acta Theriol 34 37:513–523

Camenisch M, Schütz M (2000) Temporal and spatial variability of the vegetation in a four-year exclosure experiment in Val Trupchun (Swiss National Park). In: Schütz M, Krüsi BO, Edwards PJ (eds) Succession research in the Swiss National Park, vol 89. Nat Park Forsch Schweiz, Zernez, pp 165–188

Catusse M, Corti R, Cugnasse J-M, Dubray D, Gilbert P, Michalet J (1996) La Grande Faune de Montagne. Hatier, Paris

Cebrian J, Duarte CM (1994) The dependence of herbivory on growth rate in natural plant communities. Funct Ecol 8:518–525

Cernusca A (1978) Ökologische Veränderungen im Bereich aufgelassener Almen. In: Cernusca A (ed) Ökologische Analysen von Almflächen im Gasteiner Tal. Veröff Österr MaB-Hochgebirgsprogr Hohe Tauern 2. Univ Verlag Wagner, Innsbruck, pp 7–27

Cernusca A (1989) Ökosystemforschung in den österreichischen Zentralalpen (Hohe Tauern). In: Cernusca A (ed) Struktur und Funktion von Graslandökosystemen im Nationalpark Hohe Tauern. Veröff österr MaB-Progr 13. Univ Verlag Wagner, Innsbruck, pp 549–568

Cernusca A, Nachuzrisvili G (1983) Untersuchung der ökologischen Auswirkungen intensiver Schafbeweidung im Zentral-Kaukasus. Verh Ges Ökol 10:183–192

Chapin FS III (1980) Nutrient allocation and responses to defoliation in tundra plants. Arctic Alp Res 12:553-563

Coley PD, Bryant JP, Chapin F III (1985) Resource availability and plant antiherbivore defense. Science 230:895-899

Coughenour MB (1991) Biomass and nitrogen responses to grazing of upland steppe on Yellowstone's northern winter range. J Appl Ecol 28:71-82

Crawford RMM (1989) Studies in plant survival. Ecological case histories of plant adaptation to adversity. Blackwell, Oxford

Crawley MJ (1983) Herbivory: the dynamics of animal-plant interactions. University of California Press, Berkeley, CA

Crawley MJ (1988) Herbivores and plant population dynamics. In: Davy AJ, Hutchings MJ, Watkinson AR (eds) Plant population ecology. 28th Symp BES. Blackwell, Oxford, pp 367-392

Crawley MJ (1997) Plant-herbivore dynamics. In: Crawley MJ (ed) Plant ecology. Blackwell, Oxford, pp 401-474

Diemer M (1996) The incidence of herbivory in high-elevation populations of *Ranunculus glacialis*: a re-evaluation of stress-tolerance in alpine environments. Oikos 75:486-492

Dietl W (1982) Schafweiden im Alpsteingebiet (Ostschweizer Kalkalpen). Ber Geobot Inst ETH Stift Rübel 49:108-117

Dunant F (1977) Le régime alimentaire du Chamois des Alpes (Rupicapra rupicapra rupicapra L.): contribution personnelle et synthèse les domnées actuelles sur les plantes brontées. Rev Suisse Zool 84:884-903

Ellmauer T, Mucina L (1993) Molinio-Arrhenatheretea. In: Mucina L, Grabherr G, Ellmauer T (eds) Die Pflanzengesellschaften Österreichs. Fischer, Jena, pp 297-385

Erschbamer B (1994) Populationsdynamik der Krummseggen (*Carex curvula* ssp. *rosae*, *Carex curvula* ssp. *curvula*). Phytocoenologia 24:579-596

Ferrari C, Rossi G (1985) Preliminary observations on the summer diet of the Abruzzo chamois (*Rupicapra rupicapra ornata* Neum.). In: Lovari S (ed) The biology and management of mountain ungulates. Proc 4th Int Conf on Chamois and other Mountain Ungulates. Pescasseroli (Italy), 17-19 June 1983. Croom Helm, London, pp 85-92

Friedel H (1967) Verlauf der alpinen Waldgrenze im Rahmen anliegender Gebirgsgelände. Mitt Forstl Bundesversuchsanst Wien 75:81-172

Galen C (1990) Limits to the distributions of alpine tundra plants: herbivores and the alpine skypilot, *Polemonium viscosum*. Oikos 59:355-358

Gander M (1984) Die alpine Vegetation des hinteren Defereggentales (Osttirol). Diplomarbeit, University of Innsbruck, Innsbruck

Garcia-Gonzalez R, Gomez-Garcia D, Aldezabal A (1998) Resultados de 6 anos de esclusion del pastoreo sobre la estructura de comunidades del Bromion erecti y Nardion strictae en el P.N. de Ordesa y Monte Perdido. Actas de la 38° Reunion Cientifica de la Sociedad Espanola para el Estudio de los Pastos, 2-5 Junio 1998, Soria

Giacomini V, Pignatti S (1955) Flora e vegetazione dell'Alta Valle del Braulio. Fondazione per i problemi montani dell'arco alpino Milano 12:1-194

Grime JP (1979) Plant strategies and vegetation processes. John Wiley, Chichester

Hallanaro E-L, Pylvänäinen M (2001) Nature in northern Europe - biodiversity in a changing environment. Nordic Council of Ministers, Copenhagen

Harper JL (1977) Population biology of plants. Academic Press, London

Heiselmayer P (1981) Die Vegetationskarte als Grundlage für ökologische Kartierungen. Angew Pflanzensoz 26:59-73

Heiselmayer P (1982) Die Pflanzengesellschaften des Tappenkars (Radstädter Tauern). Stapfia 10:161-202

Henry GHR, Svoboda J (1994) Comparisons of grazed and non-grazed high-arctic sedge meadows. In: Svoboda J, Freedman B (eds) Ecology of a polar oasis. Captus Press, New York, pp 193-194

Higgins S, Pickett STA, Bond WJ (2000) Predicting extinction risks for plants: environmental stochasticity can save declining populations. Trends Ecol Evol 15:516-520

Hofgaard A (1997) Inter-relationships between treeline position, species diversity, land use and climate change in the central Scandes Mountains of Norway. Global Ecol Biogeogr Lett 6:419-429

Hofmann RR, Schnorr B (1982) Die funktionelle Morphologie des Wiederkäuer-Magens. Enke, Stuttgart

Hofstede RGM, Modragon Castillo MX, Rocha Osorio CM (1995) Biomass of grazed, burned, and undisturbed paramo grasslands, Colombia. I. Aboveground vegetation. Arc Alp Res 27:1-12

Huntly N (1991) Herbivores and the dynamics of communities and ecosystems. Annu Rev Ecol Syst 22:477-503

Insam H, Rangger A, Henrich M, Hitzl W (1996) The effect of grazing on soil microbial biomass and community on alpine pastures. Phyton 36:205-216

Jans F, Troxler J (1990) Weidenutzung und Landschaftspflege an Trockenstandorten mit Mutterkühen oder Schafen. I: Tierische Leistung. Landwirtschaft Schweiz 3:311-314

Järvinen A (1987) Microtine cycles and plant production: what is cause and effect? Oikos 49:352-357

Kalela O (1957) Regulation of reproductive rate in subarctic populations of the vole Cletrionomys rufocanus (Sund.). Ann Acad Sci Fenn (A IV) 34:1-60

Karner P, Mucina L (1993) Mulgedio-Aconitetea. In: Grabherr G, Mucina L (eds) Die Pflanzengesellschaften Österreichs. Fischer, Jena, pp 468-496

Karrer G (1980) Die Vegetation im Einzugsgebiet des Grantenbaches südwestlich des Hochtores (Hohe Tauern). Veröff Österr MaB-Hochgebirgspr Hohe Tauern Innsbruck 3:35-67

Kaufmann R, Spitaler T (2000) Preliminary exclosure experiments in a glacier foreland. ESF Alpnet News 2:33-34

Klötzli F (1977) Wild und Vieh im Gebirgsgrasland Aethiopiens. In: Tüxen R (ed) Vegetation und fauna. Ber Symp IVV, Rinteln, pp 499-512

Klötzli F (1990) African mountain grasslands in their global context, with an overview of puna as an orobiome. In: Winiger M, Wiesman U, Rheker J (eds) Mount Kenya area. Differentiation and dynamics of a tropical mountain ecosystem. Proc Int Workshop 1989, Geogr Inst Univ Bern, Afr Stud, Ser A8, Bern, pp 75-81

Klötzli F (1991) Zum Einfluß von Straßenböschungensaussaaten auf die umliegende naturnähere Vegetation am Beispiel des Schweizer Nationalparks. Lauf Sem Beitr 3:114-123

Klötzli F (1996) Verbiss. In: Brunold C, Rüegsegger A, Brändle R (eds) (1996) Stress bei Pflanzen. UTB, Haupt, Bern, pp 295-307

Knapp AK, Seastedt TR (1986) Detritus accumulation limits productivity of tallgrass prairie. BioScience 36:662-668

Knapp R (1967) Experimentelle Soziologie und gegenseitige Beeinflussung der Pflanzen, 2. Aufl. Ulmer, Stuttgart

Körner C (1980) Ökologische Untersuchungen an Schafweiden im Zentralkaukasus. Der Alm- und Bergbauer 30/5:2-8

Körner C (1999) Alpine plant life. Springer, Berlin Heidelberg New York

Körner C, Larcher W (1988) Plant life in cold climates. In: Long SF, Woodward FI (eds) Plants and temperature. CBL, Cambridge, pp 25-57

Kotanen P, Jefferies RL (1987) The leaf and shoot demography of grazed and ungrazed plants of *Carex subspathacea*. J Ecol 75:961-975

Krüsi BO, Moser B (2000) Impacts of snow and ungulates on the successional development of a mountain pine forest in the Swiss National Park (Munt la Schera). In: Schütz M, Krüsi BO, Edwards PJ (eds) Succession research in the Swiss National Park, vol 89. Nat Park Forsch Schweiz, Zernez, pp 131–164

Lambert RC, Bathgate JL (1977) Determination of the plane of nutrition of chamois. Proc N Z Ecol Soc 24:48–56

Lennartson T, Tuomi J, Nilsson P (1997) Evidence for an evolutionary history of overcompensation in the grassland biennial *Gentianella campestris*, Gentianaceae. Am Nat 149:1147–1155

Lye KA, Lauritzen EM (1975) Effect of grazing in alpine vegetation on Hardangervidda, south Norway. Norw J Bot 22:7–13

MacDonald A, Stevens P, Armstrong H, Immirzi P, Reynolds P (1998) A guide to upland habitats. Surveying land management impacts, vols 1 and 2. Scottish Natural Heritage, Battleby

Malo JE, Suárez F (1996) New insights into pasture diversity: the consequence of seed dispersal in herbivore dung. Biodiv Lett 3:54–57

McKendrick JD, Batzli GO, Everett KR, Swanson JC (1980) Some effects of mammalian herbivores and fertilization on tundra soils and vegetation. Arctic Alp Res 12:565–578

McNaughton SJ (1985) Ecology of a grazing ecosystem: the Serengeti. Ecol Monogr 55:259–294

McNaughton SJ (1983) Compensatory plant growth as a response to herbivory. Oikos 40:329–336

McVean DN, Ratcliffe DA (1962) Plant communities of the Scottish Highlands. Nature Conservancy Monograph No 1. HMSO, Edinburgh

Medicus R (1981) Die Vegetationsverhältnisse des Hollersbachtales Pinzgau – Salzburg. Dissertation, University of Salzburg, Salzburg

Milchunas DG, Sala OE, Lauenroth WK (1988) A generalized model of the effects of grazing by large herbivores on grassland community structure. Am Nat 132:87–106

Miller GR, Geddes C, Mardon DK (1999) Response of the alpine gentian *Gentiana nivalis* L. to protection from grazing by sheep. Biol Conserv 87:311–318

Moen J, Oksanen J (1998) Long-term exclusion of folivorous mammals in two arctic-alpine plant communities: a test of the hypothesis of exploitation ecosystems. Oikos 82:333–346

Moen J, Lundberg PA, Oksanen L (1993) Lemming grazing on snowbed vegetation during a population peak, northern Norway. Arctic Alp Res 25:130–135

Mulder CPH (1999) Vertebrate herbivores and plants in the Arctic and subarctic: effects on individuals, populations, communities and ecosystems. Persp Plant Ecol Evol Syst 2:29–55

Nagy L, Proctor J (1996) The demography of *Lychnis alpina* L. on the Meikle Kilrannoch ultramafic site. Bot J Scotl 48:155–166

Oberhammer M (1979) Die Vegetation der alpinen Stufe in den östlichen Pragser Dolomiten. Dissertation, University of Innsbruck, Innsbruck

Oksanen L (1988) Ecosystem organization: mutualism and cybernetics or plain Darwinian struggle for existence? Am Nat 131:424–444

Oksanen L (1990) Predation, herbivory, and plant strategies along gradients of primary production. In: Grace JB, Tilman D (eds) Perspectives on plant competition. Academic Press, San Diego, pp 445–474

Oksanen L, Moen J (1994) Species-specific plant responses to exclusion of grazers in three Fennoscandian tundra habitats. Ecoscience 1:31–39

Oksanen L, Oksanen T (2000) The logic and realism of the hypothesis of exploitation ecosystems. Am Nat 155:703–723

Oksanen L, Ranta E (1992) Plant strategies along vegetational gradients on the mountains of Iddonjarga – a test of two theories. J Veg Sci 3:175–186

Oksanen L, Fretwell S, Arruda J, Niemelä P (1981) Exploitation ecosystems in gradients of primary productivity. Am Nat 118:240–261

Oksanen L, Moen J, Helle T (1995) Timberline patterns in northernmost Fennoscandia. Acta Bot Fenn 153:93–105

Olff H, Ritchie ME (1998) Effects of herbivores on grassland plant diversity. Trends Ecol Evol 13:261–265

Pacala SW, Crawley MJ (1992) Herbivores and plant diversity. Am Nat 140:243–260

Palmer MW (1992) The coexistence of species in fractal landscape. Am Nat 139:375–397

Patzelt G, Kofler W, Wahlmüller B (1997) Die Ötztalstudie – Entwicklung der Landnutzung. In: Oeggl K, Patzelt G, Schäfer D (eds) Alpine Vorzeit in Tirol. Begleitheft zur Ausstellung. Univ Innsbruck, Innsbruck, pp 46–62

Perle A, Hamr J (1985) Food habits of Chamois (*Rupicapra rupicapra* L.) in northern Tyrol. In: Lovari S (ed) The biology and management of mountain ungulates. Proc 4th Int Conf on Chamois and other Mountain Ungulates, Pescasseroli (Italy), 17–19 June 1983. Croom Helm, London, pp 77–84

Pignatti Wikus E (1987) Alpine grasslands and the effect of grazing. In: Miyawaki A, Bogenrieder A, Okuda S, White J (eds) Vegetation ecology and creation of new environments. Tokai University Press, Tokyo, pp 225–234

Proulx M, Mazumder A (1998) Reversal of grazing impact on plant species richness in nutrient-poor vs. nutrient-rich ecosystems. Ecology 79:2581–2592

Rainer K, Erschbamer B (1996) Das Regenerationsverhalten von vier alpinen Pflanzenarten. Verh Ges Ökol 26:565–568

Ritchie ME, Olff H (1999) Herbivore diversity and plant dynamics: compensatory and additive effects. In: Olff H, Brown VK, Drent RH (eds) Herbivores: between plants and predators. The 38th Symposium of the British Ecological Society. Blackwell, Oxford, pp 175–204

Rodwell JS (1992) British plant communities, vol 3. Grasslands and montane communities. Cambridge University Press, Cambridge

Schaefer M (1994) Brohmer. Fauna von Deutschland. 19. Aufl. Quelle and Meyer, Heidelberg

Schaminée JHP, Meertens MH (1992) The influence of human activities on the vegetation of the subalpine zone of the Monts fu Forez (Massif Central, France). Preslia 64:327–342

Schneiter S (1997) Die Reaktion eines alpinen Rasens auf das Aussetzen der Beweidung. Diplomarbeit, University of Basel, Basel

Schütz M, Wolgemuth T, Krüsi BO, Achermann G, Grämiger H (2000) Influence of increasing grazing pressure on species richness in subalpine grassland in the Swiss National Park. In: Schütz M, Krüsi BO, Edwards PJ (eds) Succession research in the Swiss National Park, vol 89. Nat Park Forsch Schweiz, Zernez, pp 39–65

Spatz G (1980) Succession patterns on mountain pastures. Vegetatio 43:39–41

Spatz G (1983) Vegetationsveränderung auf Almen durch differenzierte Nutzung und deren Nachwirkung. Tuexenia 3:325–330

Spatz G (1994) Freiflächenpflege. Ulmer, Stuttgart

Spatz G, Springer S (1987) Vegetationsdynamik auf Almweiden im Alpenpark Berchtsgaden. In: Schubet R, Hillig W (eds) Erfassung und Bewertung anthropogener Vegetationsveränderungen, Teil 2. Wiss Beitr 25. Martin-Luther-Univ Halle-Wittenberg, Halle-Wittenberg, pp 62–74

Spatz G, Fricke T, Prock S (1993) Wirtschaftsbedingte Vegetationsmuster auf Almweiden der Hohen Tauern. Rev Géogr Alpine 3:83–93

Stark S (2002) Reindeer grazing and soil nutrient cycling in boreal and tundra ecosystems. Acta Univ Oul Ser A 382:1-31

Steinmair V (1999) Die Vegetation von unterschiedlich genutzten Almflächen auf der Plätzwiese (Dolomiten, Südtirol). Diplomarbeit, University of Innsbruck, Innsbruck

Stüssi B (1970) Naturbedingte Entwicklung subalpiner Weiderasen auf Alp La Schera im Schweizer Nationalpark während der Reservatsperiode 1939-1965. Lüdin AG, Liestal

Sundriyal RC (1992) Structure, productivity and energy flow in an alpine grassland in the Garhwal Himalaya. J Veg Sci 3:15-20

Suominen O, Olofsson J (2000) Impacts of semi-domesticated reindeer on structure of tundra and forest communities in Fennoscandia: a review. Ann Zool Fenn 37:233-249

Tappeiner U, Tasser E, Tappeiner G (1998) Modelling vegetation pattern using natural and anthropogenic influence factors: preliminary experience with a GIS based model applied to an alpine area. Ecol Model 113:225-237

Tasser E, Prock S, Mulser J (1999a) The impact of land-use on vegetation along the eastern alpine transect. In: Cernusca A, Tappeiner U, Bayfield N (eds) Land-use changes in European mountain ecosystems. Blackwell, Oxford, pp 235-246

Tasser E, Newesely C, Höller P, Cernusca A, Tappeiner U (1999b) Potential risks through land-use changes. In: Cernusca A, Tappeiner U, Bayfield N (eds) Land-use changes in European mountain ecosystems. Blackwell, Oxford, pp 218-224

Tast J (1974) The food and feeding habits of the root vole, *Microtus oeconomus*, in Finnish Lapland. Aquilo Ser Zool 15:25-32

Tast J (1991) Will the Norwegian lemming become endangered if climate becomes warmer? Arctic Alp Res 23:53-60

Tataruch F (1982) On the nutrition of chamois. Atti 'Simposio internazionale sulla cheratocongiuntivite infettiva del Camoscio', Vercelli-Varallo Sesia, 30 Nov-2 Dec, 1982. Tavola rotonda, pp 153-158

Thompson DBA, Brown A (1992) Biodiversity in montane Britain: habitat variation, vegetation diversity and some objectives for conservation. Biodiv Conserv 1:179-208

Tilman D, Pacala SW (1993) The maintenance of species richness in plant communities. In: Ricklefs RE, Schluter D (eds) Species diversity in ecological communities: historical and geographical perspectives. University of Chicago Press, Chicago, pp 13-25

Tokeshi M (1999) Species coexistence. Ecological and evolutionary perspectives. Blackwell, Oxford

Turnbull L, Crawley MJ, Rees M (2000) Are plant populations seed-limited? A review of seed sowing experiments. Oikos 88:225-238

Van Soest PJ (1994) Nutritional ecology of the ruminant, 2nd edn. Cornell University Press, Ithaca

Virtanen R (1998) Impact of grazing and neighbour removal on a heath plant community transplanted onto a snowbed site, NW Finnish Lapland. Oikos 81:359-367

Virtanen R (2000) Effects of grazing on above-ground biomass on a mountain snowbed, NW Finland. Oikos 90:295-300

Virtanen R, Henttonen H, Laine K (1997) Lemming grazing and structure of a snowbed plant community - a long-term experiment at Kilpisjärvi, Finnish Lapland. Oikos 79:155-166

Vorren K-D, Mörkved B, Bortenschlager S (1993) Human impact of the Holocene forest line in the central Alps. Veget Hist Archaeobot 2:145-156

Waide RB, Willig MR, Steiner CF, Mittelbach G, Gough L, Dodson SI, Juday GP, Parmenter R (1999) The relationship between productivity and species richness. Annu Rev Ecol Syst 30:257-300

Welch D, Scott D, Doyle S (2000) Studies on the paradox of seedling rarity in Vaccinium myrtillus L. in NE Scotland. Bot J Scotl 52:17-30

Wilson SD (1994) The contribution of grazing to plant diversity in alpine grassland and heath. Aust J Ecol 19:137–140

Zimov SA, Chuprynin VI, Oreshko AP, Chapin FS III, Reynolds JF, Chapin MC (1995) Steppe-tundra transition: a herbivore-driven biome shift at the end of the Pleistocene. Am Nat 146:765–794

Zoller H, Bischof N, Erhardt A, Kienzel (1984) Biocoenosen von Grenzertragsflächen und Brachland in den Berggebieten der Schweiz. Hinweise zur Sukzession, zum Naturschutzwert und zur Pflege. Phytocoenologia 12:373–395

V Long-Term Vegetation Dynamics

24 Alpine Vegetation Dynamics and Climate Change – a Synthesis of Long-Term Studies and Observations

G. Grabherr

24.1 Introduction

Vegetation dynamics – a complex process where the primary causal factor is often difficult to separate from other, often less important ones – require long-term investigations (Austin 1981; Zonnefeld 1988; Everson et al. 1990; Spellerberg 1991; Bakker et al. 1996; Pfadenhauer et al. 1996). Long-term studies using sample plots are few in the European alpine zone (e.g. Schütz et al. 2000); however, the long history of research has resulted in a large body of observational information. Within the scope of ALPNET we made use of this body of information and compared early and contemporary observations across Europe. These included repeat visual observations and detailed on-site comparisons by experts, including the use of old photographs. These methods unavoidably involved a certain degree of spatial inaccuracy. This approach of repeat observations complements published accounts about alpine vegetation dynamics in relation to climate warming (see Beniston and Tol 1998) in the European high mountains (Braun-Blanquet 1955, 1957; Shiyatov 1983; Thompson and Brown 1992; Grabherr et al. 1994, 2001; Pauli et al. 1996).

This contribution draws together the results of the examples presented as case studies in this section and includes the author's personal observations and pieces of anecdotal evidence presented during the 3 years of ALPNET (see also ESF-ALPNET News 1 & 2; http://stir.ac.uk/departments/naturalsciences/DBMS/nagy/alpine_biodiversity/).

24.2 Treeline Ecotone Trees and Forest Line Dynamics

The comparison of old and new photographs provided evidence for an altitudinal advance of the forest line (but not of the treeline as defined in Chap. 1) in the southern and northern Ural Mountains during the twentieth century (Chap. 26). On gentle slopes with a well-developed soil, the forest line nowadays is found 60–80 m higher than at the beginning of the twentieth century; in less favourable habitats, the advance has been estimated to be 20–40 m uphill. This advance is mainly caused by a filling process (see Grabherr et al. 1995). The increased cover of closed forest in the upper reaches of the Urals has reduced the area of open habitats (moss heaths and meadows) by an estimated 10–30%.

The observed change at the forest line was not a continuous process. In spruce (*Picea obovata*) narrower tree rings were formed during the cold phase between the 1950s and 1970s and some formerly established birch stands died. Dying birch stands during colder periods in the last century were also reported from the birch treeline in the Scandes (Kullmann 1979, 1990; Holtmeier 1994). At the treeline in the Central Alps, Arolla pine (*Pinus cembra*) was affected in the 1970s by a sequence of cold summers and severe late winter frosts, which caused drought (Bortenschlager 1977). Young individuals at the uppermost reaches died. However, these stands have since recovered.

Did this advance of the forest line exceed those during the Holocene? What might be the overall potential for treeline changes? The dramatic retreat of glaciers in the Alps during recent decades has left large moraines exposed (Barry 1994; Häberli 1994). In the Eastern Alps, old blocks of timber have been found in moraines up to 2600 m, which exceeds today's potential treeline of about 2400–2500 m. According to radiocarbon data, the trees date back to warm phases in the Holocene (G. Patzelt and F. Nikolussi, pers. comm.), suggesting that tall trees could have grown at high elevations several times in the past. However, these findings also confirmed that after the establishment of the mountain forests in the late glacial, tall trees never grew at elevations higher than 200 m above recent treelines (e.g. Patzelt and Bortenschlager 1973). During the last 1000 years (except in very recent times), the treeline might have been quite stable. Evidence was found for this in the *Pinus sylvestris* L. treeline at Creag Fhiaclach in the Cairngorms, Scotland (J. Nagy et al., unpubl.).

Comparisons of photographs, taken 25 years apart, of a ski run at the treeline in Obergurgl, Tyrol, showed a complete vegetation cover (mainly dwarf shrubs and lichens) where vegetation had previously been completely removed and no re-seeding was applied. Seedlings and young trees of *Pinus cembra*, the dominant species at the treeline in the Central Alps, were abundant. Thus, succession on open soils, and even in subsoils, appears to be a relatively rapid process in treeline environments. Species of late successional communities such as *Pinus cembra*, *Picea abies*, *Larix decidua* and associated dwarf shrubs, forbs, and lichens appear rapidly. Such a rapid establishment of tree stands is also well known from moraines of glaciers in the Alps (e.g. Aletsch, Morteratsch, Gepatsch) that reached the treeline during the last maximum of glacier advance. These moraines were colonised by trees after the retreat of the ice in the middle of the nineteenth century. It appears, therefore, that climate warming in the past could induce a rapid plant establishment on moraines or on other pioneer sites (e.g. screes, avalanche fields).

The rapid appearance of tree species on glacier moraines or on open ground is in contrast with their poor establishment in undisturbed heath or grassland. For example, anemochorous species such as *Picea abies* or *Larix decidua* are propagated over long distances, yet they rarely establish in closed alpine vegetation (Kuoch 1965; Kuoch and Amiet 1970; Ellenberg 1996, p. 352). These trees require open ground for germination and therefore depend on gaps created by wild or domestic animals. In contrast, the seeds of the zoochorous *Pinus cembra* are propagated by the bird *Nucifraga caryocatactes*, or nutcracker. Germinating seeds from a cache made by a nutcracker have been recently found at 3103 m (Stockkogel, Tyrolean Alps), which is at least 600 m above the uppermost *Pinus cembra* stand (M. Gottfried and H. Pauli, pers. comm.). We can assume that such seed caches are scattered throughout the alpine landscapes where *Pinus cembra* and the bird co-occur. Climate warming will enhance, and probably has already enhanced, germination success. Survival rates may therefore increase, giving rise to higher densities of the species in its current altitude range and may also allow the establishment of the species at higher elevations.

The different life strategies of treeline tree species appear to cause different and individualistic responses to climate change. However, the potential to react to climate warming rapidly is generally high, mainly resulting in a filling process in the existing range. Movement of the tree species line in the strict sense to a higher elevation, however, depends on many other factors, such as the availability of substratum, disturbance and biotic interactions.

24.3 Vegetation Dynamics Within the Alpine Zone (Grasslands, Dwarf Shrubs, Snowbeds)

Most of the dominant species in alpine grasslands and dwarf-shrub heath are long-lived clonal perennials (Grabherr 1997; Körner 1999). Some of these clones are hundreds, if not thousands, of years old. The associated, less abundant species are also long-lived perennials; annuals or geophytes are negligible (except in Mediterranean mountains). Dynamic processes therefore appear at a very slow rate, and alpine grasslands and dwarf-shrub heath appear as rather stable vegetation formations (e.g. Thompson and Brown 1992).

Not surprisingly, no changes in species composition and only small changes in abundance have been recorded in the world's oldest permanent plots in alpine grasslands (G. Achermann and M. Schütz, unpubl.). These were established in an *Kobresia* (= *Elyna*)-dominated ridge grassland community in the Swiss National Park at Ofenpass in 1914. In 1941, further plots were established in a typical low alpine grassland (*Carex curvula-Nardus stricta*) and in 1950 (1956) in a limestone grassland (*Sesleria albicans-Carex sempervirens*). A comparison of the old records with recent data revealed no change that could be related to an increase in temperature during the twentieth century. Similarly, the exclusion of domestic livestock from the area has not changed the species composition. These findings appear generally applicable to alpine grassland communities.

Alpine vegetation in Scandinavia is also stable (Chap. 25). No change has been found in vascular plant species composition in most plant communities since 1920, whilst lichen cover declined in the majority of cases because of increases in reindeer grazing. There was an observable increase in grassland species in snowbed communities which might be related to the observed warming as well as to increased grazing (Chap. 25).

Repeat observations of a mosaic of *Nardus stricta* grassland and snowbeds at Mt. Hohe Muth near Obergurgl (Tyrol, 2600 m), a site well known to the author for more than 30 years, showed similar trends to that reported in Chapter 25. *Nardus stricta*, the dominant species in the grasslands of the lower alpine zone, has clearly established in snowbeds just at its upper altitude limit. Young individuals of *Nardus* can be distinctly recognised in the field as their size is related to age. *Nardus* starts as small tussock-like clusters of tillers that expand to a 'fairy' ring and later disintegrate into a clonal population densely intermixed with other *Nardus* tussocks. The size structure of a snowbed population of *Nardus stricta* consisting of 65 individual tussocks (>5 cm in diameter) was noted by the author during a short visit in summer 1999. The largest tussock was about 28 cm in diameter. Among the 65 individuals, 52 % were between 5–10 cm, 31 % 10–15 cm, and 17 % >15 cm. It is likely, therefore, that the warm years of the 1980s and 1990s have induced the invasion of the snowbed.

From the foregoing it is evident that climate change related dynamics are most obvious as an invasion process. Internal dynamics work at a very slow rate in these alpine communities. This is confirmed also by results from an old research field at the Hohe Muth area. The site is covered by a dense carpet of *Carex curvula*, associated lichens, other grasses, herbs, and a few moss species (Grabherr et al. 1978). *Carex curvula* determines the vegetation structure, both horizontally and vertically (Grabherr 1989). In 1976–1977, monoliths of the *Carex curvula* sward were removed leaving behind 5-cm-deep hollows of 20 × 30 cm or 30 × 50 cm. No vegetative propagules remained below or at the bottom of the hollows, which were covered by a thin humus layer, and the seed bank at 5-cm depth is known to be negligible. After examining 26 such hollows in 1999, little regeneration of *Carex curvula* was observable (Table 24.1). The other species, lichens in particular, had colonised the hollows with varying success. The grasses *Agrostis rupestris* and *Festcua halleri* had a relatively high cover, indicating their ability to rapidly colonise available sites. Using a DCA-ordination (Hill 1979) to compare the vegetation of the hollows with

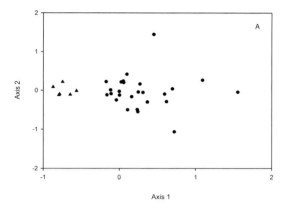

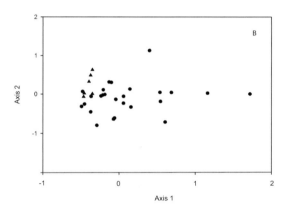

Fig. 24.1A, B. Ordination of the vegetation samples (see Table 24.1) from regenerating and unperturbed mature plots from Mt. Hohe Muth. Ordination method: DECORANA (Hill 1979); first and second axes displayed.
A Complete data set: *triangles* unperturbed mature swards of *Carex curvula* grassland; *circles* regenerating plots. **B** *Carex curvula* excluded from the data set. Note that mature swards show no clear distance from the regenerating plots when *Carex curvula* is removed from the data set

that in quadrats of the same size in nearby mature grassland, we found that *Carex curvula* alone discriminated the hollows from the mature stands (Fig. 24.1a). When *Carex curvula* was removed from the data set, some hollows closely resembled the mature grassland samples (Fig. 24.1b). Others showed little in common with mature grassland (Fig. 24.1b), indicating that recovery after perturbation may follow various pathways in alpine grasslands. A full recovery to the original mature state with well-established *Carex curvula* takes a very long time. Although the associated species have a higher colonisation success than *Carex curvula*, the latter has the largest influence on community processes by being the matrix species (for details, see Grabherr 1989). Its stabilising effect is thought to arise from reducing the internal species turnover to an extremely slow rate.

Table 24.1. Regeneration of a *Carex curvula* grassland, a characteristic type of alpine vegetation in the Central Alps, at Mt. Hohe Muth (Tyrol, 2600 m). The vegetation table represents the regenerating plots (right side; 009a–0034) and six plots, identical in size to those of the regeration plots, from undisturbed mature swards in the immediate vicinity of the regeneration plots (left side). Regeneration plots are revegetating after the removal of the total vegetation and top soil to a depth of 5 cm in 1996–97. The values are estimates of the percentage cover of species. Nomenclature after Adler et al. (1994)

Species	Mature swards						Regeneration plots						
	009b	011b	013b	022b	031b	032b	009a	0010	011a	0012	013a	0014	001
Carex curvula	50	50	60	60	40	30		+	+				+
Avenula versicolor	10	15	10	30	20	30	10	10	5	10	7	7	10
Festuca halleri	10	10	7	1	20	10	3	7	1	7	10	3	10
Veronica bellidioides		3	7	1	3	5	5	3	3	10	5	3	3
Euphrasia minima	+	+	1	+	1	1	+	1	1	+		+	+
Leucanthemopsis alpina		+	+	+	+	1	+	+	+	+	+	+	
Leontodon helveticus		+	+				+	1			+	+	+
Phyteuma hemisphaericum		+		+	+		+	+		+			
Agrostis rupestris		5								5			
Oreochloa disticha		10			5								
Gnaphalium supinum												+	+
Antennaria dioica												+	
Vaccinium vitis-idaea			+										
Minuartia sedoides			+					+					
Primula glutinosa													
Sibbaldia procumbens													
Anthoxantum odoratum													
Androsace obtusifolia								+					
Pulsatilla vernalis								+					
Homogyne alpina													
Cetraria islandica	40	10	5	40	20	30	+	3	1	5	10	10	1
Cetraria ericetorum	40	30	70	20	20	20	10	35	15	40	40	20	4
Cladonia arbuscula	10	20	10	20	20	20	3	5	10	5	10	3	1
Cladonia rangiferina	+	10	1		20	5		+			5	3	
Cladonia uncialis	+		1		1		+				+		
Crustose lichens		3						5		5		1	
Thamnolia vermicularis		1									3	1	
Stereocaulon alpinum							3	1	3				
Solorina crocea													
Polytrichum spp.	+	+		+	5	1	50	3	15	5	10	20	
Ptilidium sp.					5	1					5	5	
Small mosses							50	5	5	10		10	
Scleropodium purum					+								

Alpine Vegetation Dynamics and Climate Change

0017	0018	0019	0020	0021	022a	0023	0024	0025	0026	0027	0028	0029	0030	031a	032a	0033	0034
										+		1					
5	3	5	3	5	10	10	7	7	5	7	10	5	7	7	5	5	5
5	3	5	5	3	7	10	7	5	5	7	10	5	5	10	3	7	5
3	3	1	5	3	10	7	5	5	7	5	3	1	7	2	5	5	2
1	+		1	1		1	3	1	1		1		1			1	1
	+			1			+			+	1	+					+
	+	+					+			+				+			1
					1								1	1		1	
3							3				7			5			
	1						5				10				5		
									+		5						+
		1	15														
1	3																
					1												
														+			
3																	
	+																
																	1
20	20	10	3	20	25	10	5	10	10	20	10	50	30	20	15	15	10
30	25	10	20	25	30	25	50	40	15	25	10	10	30	15	10	15	10
30	5	10	3	15	30	10	10	40	10	25	5	15	20	15	15	5	2
3		2		1	1		3						1	1			
	1															2	
10		10	10			2	2			20	15		5	1	15	1	15	5
	1																
							3					2					
	1																
1	5	10	1	20	10	2	+	5	2	2	20	1	2	15	80	5	2
					5			+					2		1		
	2	5	5	20		2	5	5	+			+	1	5		5	20

24.4 Dynamic Processes at the Limits of Plant Life

The nival zone, where vascular plant growth reaches its limit, is around 3000 m in the Alps. Many records are available for over 100 such mountain tops. Scattered individuals of a few species are able to survive at 'safe' sites with a favourable microclimate and stable substrata. Pauli et al. (1999) demonstrated that, even at the limits of plant life, species were not distributed randomly. Their analysis of about 1000 1-m^2 samples from the subnival and nival zones of Mt. Schrankogel, Tyrol (3497 m), distinguished several community types, each of them with specific habitat requirements.

The establishment of new species in the subnival and nival zones is likely to be a strong indication of an environmental shift. The same holds true if populations of already existing species undergo a large expansion. It was this latter phenomenon that was documented for Piz Linard, the highest summit (3411 m) of the Silvretta Mountains in the Central Alps (Chap. 28). A detailed map of vascular plant distribution drawn in 1992 compared with that for 1947 showed that the populations of most of the nival plants increased and expanded to microsites not occupied 45 years earlier. On many summits, but not on Piz Linard, an increase in species richness was found (Grabherr et al. 1994, 2001; Pauli et al. 1996). The increase in species richness and abundance is ongoing, as could be shown for the top of Mt. Glungezer, Tyrol, when comparing recent floristic records (1999–2000) with those made in 1986 (Chap. 27).

No or small increases in species richness on high summits such as the Piz Linard are usually associated with conditions which largely restrict the arrival of propagules of species from lower elevations. These summits are gravelly and unstable with no rocky ridges to serve as migratory routes. The example of Piz Linard suggests that plants that are able to withstand the effects of cryoturbation in the nival zone are widely distributed at high elevations. Potential invaders from lower altitudes such as the species group of the alpine grasslands need safe migration routes. Exceptional events such as violent storms, however, may transport seeds, and sometimes vegetative parts, over long distances to the upper reaches of suitable habitats. A young *Carex curvula* population at Hohe Wilde (3480 m) not recorded 40 years previously may be taken as proof of this (Gottfried et al. 1994).

24.5 Conclusions

1. The observed climate warming during the twentieth century has affected alpine vegetation by increasing vascular plant species richness on nival mountain tops as well as through an increase in the population sizes of

long-established nival plant species. Grassland species have become established in snowbeds in the Alps and the Scandes. The forest line has moved to higher altitudes in pristine mountain areas of the Urals.
2. Alpine climax vegetation types (grasslands, dwarf-shrub heaths) are very stable communities. The internal vegetation dynamics caused by the biotic characteristics of the associated species operates at a very low rate. The climate warming of the last century has not been sufficient to stimulate these internal processes, and no change has been recorded.
3. The commonly assumed high sensitivity of alpine vegetation to climate change is largely due to invasion processes. These processes become most obvious in azonal vegetation types (see Grabherr et al. 1995) such as snowbeds or at the uppermost limits of plant life. Treeline tree establishment might be favoured predominantly on previously unvegetated substrata such as glacier moraines.
4. Alpine climax communities are vulnerable (sensitive and non-resilient) to perturbations. Once disturbed, alpine climax communities may need a very long time (>100 years) to regenerate.
5. Although climate change may not markedly affect zonal alpine vegetation, its impacts are important in specific habitats. By placing permanent monitoring sites at ecotones or in extreme sites (snow beds, ridges, mountain tops) we can establish the presence and nature of the dynamics of ecological change in a climatically changing environment As many alpine areas are little influenced directly by human activities, they offer good opportunities to study climate change in the absence of interference from land use impacts. Comparisons with areas that do have human activities offer scope for teasing apart the relative impacts of climate change and human-related disturbance.

References

Adler W, Oswald K, Fischer R (1994) Exkursionsflora von Österreich. Ulmer, Stuttgart
Austin MP (1981) Permanent quadrats: an interface for theory and practice. Vegetatio 46:1–10
Bakker JP, Olff H, Willems JH, Zobel M (1996) Why do we need permanent plots in the study of long-term vegetation dynamics? J Veg Sci 7:147–156
Barry RG (1994) Past and potential future changes in mountain environments. In: Beniston M (ed) Mountain environments in changing climate. Routledge, London, pp 3–34
Beniston M, Tol RSJ (1998) Europe. In: Watson RT, Zinyowera MC, Moss RH, Dokken DJ (eds) The regional impacts of climate change – an assessment of vulnerability. Special report of the IPCC working group II. Cambridge University Press, Cambridge pp 149–185
Bortenschlager S (1977) Ursachen und Ausmaß postglazialer Waldgrenzschwankungen in den Ostalpen. Erdwissenschaft Forsch 13:260–266

Braun-Blanquet J (1955) Die Vegetation des Piz Languard, ein Maßstab für Klimaänderungen. Svensk Bot Tidskrift 49:1-9

Braun-Blanquet J (1957) Ein Jahrhundert Florenwandel am Piz Linard (3414 m). Bull Jard Bot Bruxelles Vol Jubil W Robyns (Comm SIGMA 137):221-232

Ellenberg H (1996) Vegetation Mitteleuropas mit den Alpen. Ulmer, Stuttgart

ESF ALPNET News (1999) ALPNET, a European Science Foundation Scientific Network on alpine biodiversity (1997-2000) http://website.lineone.net/~mccassoc1/alpnet/alpnet1.html

Everson TM, Clarke GPY, Everson CS (1990) Precision in monitoring plant species composition in montane grasslands. Vegetatio 88: 135-141

Gottfried M, Pauli H, Grabherr G (1994) Die Alpen im "Treibhaus": Nachweise für das erwärmungsbedingte Höhersteigen der alpinen und nivalen Vegetation. Jahrb Ver Schutz Bergwelt 59:13-27

Grabherr G (1989) On community structure in high alpine grasslands. Vegetatio 83:223-227

Grabherr G (1997) The high mountain ecosystems of the Alps. In: Wielgolaski FE (ed) Polar and alpine tundra, ecosystems of the world, vol 3. Elsevier, Amsterdam, pp 97-121

Grabherr G, Mähr E, Reisigl H (1978) Nettoprimärproduktion in einem Krummseggenrasen (Caricetum curvulae) der Ötztaler Alpen, Tirol. Oecol Plant 13:227-251

Grabherr G, Gottfried M, Pauli H (1994) Climate effects on mountain plants. Nature 369:448

Grabherr G, Gottfried M, Gruber A, Pauli H (1995). Patterns and current changes in alpine plant diversity. In: Chapin FS III, Körner C (eds) Arctic and alpine biodiversity: pattern, causes, and ecosystem consequences, Springer, Berlin Heidelberg New York, pp 167-181

Grabherr G, Gottfried M, Pauli H (2001) Long-term monitoring of mountain peaks in the Alps. In: Burga C, Kratochwil A (eds) Biomonitoring. Task for vegetation science. Kluwer, Dordrecht, pp 153-177

Häberli W (1994) Accelerated glacier and permafrost changes in the Alps. In: Beniston M (ed) Mountain environments in changing climate. Routledge, London, pp 91-108

Hill MO (1979) DECORANA - a FORTRAN program for detrended correspondence analysis and reciprocal averaging. Cornell University, Ithaca, New York

Holtmeier F-K (1994) Ecological aspects of climatically caused timberline fluctuations. In: Beniston M (ed) Mountain environments in a changing climate. Routledge, London, pp 220-234

Körner C (1999). Alpine plant life. Springer, Berlin Heidelberg New York

Kullmann L (1979) Change and stability in the altitude of the birch tree-limit in the southern Swedish Scandes, 1915-1975. Acta Bot Suecica 65:1-121

Kullmann L (1990) Dynamics of altitudinal tree-limits in Sweden: a review. Norsk Geogr Tidsskrift 44:103-116

Kuoch R (1965) Der Samenanfall 1962/63 an der oberen Fichtenwaldgrenze im Sertigtal. Mitt Eidg Anst Forstl Vers 41:63-85

Kuoch R, Amiet R (1970) Die Verjüngung im Bereich der oberen Waldgrenze der Alpen unter Berücksichtigung von Vegetation und Ablegerbildung. Mitt Eidg Ans Forstl Vers 46:161-342

Patzelt G, Bortenschlager S (1973) Die postglazialen Gletscher- und Klimaschwankungen in der Venedigergruppe (Hohe Tauern, Ostalpen). Z Geomorphol Suppl 16:25-72

Pauli H, Gottfried M, Grabherr G (1996) Effects of climate change on mountain ecosystems - upward shifting of alpine plants. World Res Rev 8:382-390

Pauli H, Gottfried M, Grabherr G (1999) Vascular plant distribution pattern at the low temperature limits of plant life - the alpine-nival ecotone of Mt. Schrankogel (Tyrol, Austria). Phytocoenologia 29:297-325

Pfadenhauer J, Poschlod P, Buchwald R (1994) Überlegungen zu einem Konzept geobotanischer Dauerbeobachtungsflächen für Bayern. Ber Akad Nat Landschaft Lauf 10:41–60

Schütz M, Krüsi BO, Edwards PJ (eds) (2000) Succession research in the Swiss National Park. Nat Park Forsch Schweiz 89, Zernez

Shiyatov SG (1983) Experience in the use of old photographs for the study of changes of forest vegetation at the upper elevational limit of distribution. In: Gorchakovskii PL (ed) Floristic and geobotanical research in the Ural. Ural Scientific Centre of Academy of Science of USSR, Sverdlovsk, pp 76–109

Spellerberg IF (1991) Monitoring ecological change. Cambridge University Press, Cambridge

Thompson DBA, Brown A (1992) Biodiversity in montane Britain: habitat variation, vegetation diversity and some objectives for conservation. Biodiv Conserv 1:179–208

Zonneveld IS (1988) Monitoring vegetation and surveying dynamics. In: Küchler AW, Zonneveld IS (eds) Vegetation mapping. Kluwer, Dordrecht, pp 331–334

25 Long-Term Changes in Alpine Plant Communities in Norway and Finland

R. Virtanen, A. Eskelinen and E. Gaare

25.1 Introduction

High-mountain environments are regarded as susceptible to environmental change such as the global warming of climate (Grabherr et al. 1995; Körner 1999). It has been predicted that, in a warmer climate, lowland plants would migrate upwards and plant communities typically found in areas with late or moderate snow cover would be invaded by plant species characteristic of more favourable, lowland habitats (Holten and Carey 1992). Grabherr et al. (1994) presented evidence of upward migration of plants on mountain tops. Although migration may increase species richness in some habitats, elsewhere it may pose a threat. Some mountain plants are expected to decrease due to increased competition and changes in environmental conditions (Saetersdal and Birks 1997). Recent experiments have shown that temperature conditions strongly affect plant performance in alpine environments (Galen and Stanton 1993; Molau and Alatalo 1998; Totland and Nyléhn 1998); however, short-term experiments may give insufficient information on long-term vegetation changes.

The objective of this study was to analyse changes in plant species composition over 70 years at seven sites in two alpine areas in Finland and Norway. In particular, we compared species' abundances and composition reported in the literature (Nordhagen 1928; Lippmaa 1929) with those recorded in 1998 and 1999. We interpret the observed changes in relation to changes in reindeer grazing and temperature. An increasing trend of annual mean temperatures from the 1920s to the1990s has been observed near Kilpisjärvi, Finland (Hoofgaard 1999). Annual mean temperatures in the Norwegian mountains declined by 0–0.5 °C between 1925–1944 followed by an increase of ca. 1–2 °C between 1978–1999 (Grønås 1999; Jones et al. 1999; Kullman 2002). The duration of snow lie decreased during 1950–1990s by about 10 % (E. Gaare, unpubl. data). It has also been known for some time that reindeer grazing can have considerable impact on vegetation on mountains (Gaare 1968) and it has to be considered when interpreting detected vegetation responses.

25.2 Study Areas

Two mountain areas of the Scandes with detailed data on their alpine plant communities were re-sampled in 1998–1999, using identical methodology to the earlier sampling about 70 years earlier. The first area, Mt. Sylfjellet, central Norway (63°00'N, 12°10'E), was recorded by R. Nordhagen in the 1920s (Nordhagen 1928). Mount Sylan, the highest mountain of the area, reaches 1762 m a.s.l. and has a mountain birch (*Betula pubescens* subsp. *czerepanovii*) treeline at an altitude of 800 m a.s.l. Apart from the construction of a water reservoir in the 1960s between 706–729 m a.s.l., the mountains in the area have remained in a near-natural condition and human impact is still small. The reindeer population nearly doubled from about 2100 in the 1890s to 3400–4000 in the 1990s (H. Staaland, pers. comm.). Moreover, the seasonal pattern of grazing has changed: the formerly prevalent winter grazing has declined while summer grazing increased (T. Brandfjeld, pers. comm.). Nordhagen (1928) mentioned no sheep grazing and we observed no signs of sheep grazing in the area in the summer of 1999.

The second area was the Kilpisjärvi region, NW Finnish Lapland (69°03'N, 20°50'E), where T. Lippmaa studied the vegetation in the late 1920s (Lippmaa 1929). The highest mountains of the area reach above 1000 m a.s.l. The treeline formed by mountain birch lies at ca. 600 m a.s.l. In the Kilpisjärvi region, the density of the reindeer population has ranged from about 1.5 reindeer km^{-2} to 3.5 reindeer km^{-2} in 1991 (Kojola et al. 1993). There is little information available about the reindeer from the 1920s, and it is believed that their long-term average densities have remained relatively unchanged. For the *Sibbaldia procumbens–Trisetum spicatum* snowbed site changes in reindeer grazing are likely to have been small because the site is located in the Malla Nature Reserve where reindeer are not allowed to graze. Nevertheless, some stray animals have been sighted.

25.3 Data Collection and Analysis

We re-sampled the vegetation using, as much as possible, identical methods and numbers of vegetation quadrats to those used in the original studies.

At Mt. Sylfjellet, we relocated five of Nordhagen's (1928) sampling sites representing four community types (*Loiseleuria procumbens* wind-swept heath, *Vaccinium myrtillus* heath, *Deschampsia flexuosa–Anthoxanthum odoratum* snowbed community and *Kobresia simpliciuscula–Carex microglochin* alpine mire community). Following Nordhagen (1928), the cover of vascular plants, bryophytes and lichens was estimated using the Hult-Sernander scale.

At Kilpisjärvi, we relocated two of the sites sampled by Lippmaa (1929). The vegetation types included *Vaccinium myrtillus* heath and *Sibbaldia–Trisetum* snowbed. Following Lippmaa (1929), the shoots of individual plants were counted or their percentage cover class (75, 50, 40, 25, 13, 10, 6, 3 and <3) was estimated in each of ten 0.25-m^2 quadrats in both vegetation types. The quadrats were placed randomly by throwing a rod within the sampling area.

The nomenclature of bryophytes is according to Söderström et al. (1992), that of lichens follows Vitikainen et al. (1997), and that of vascular plants is after Hämet-Ahti et al. (1998).

We aimed to separate the significant changes in plants' abundances and occurrences between the 1920s and 1990s from those that could have been the result of chance. The tests can be regarded as planned a priori contrast analyses, where each data set is tested only once. For this reason, type I error levels do not need to be corrected (Sokal and Rohlf 1995). For plant species that occurred regularly in the quadrats, t-tests for independent samples were made. Before these tests, the original cover classes were transformed to corresponding percent cover values using geometric means of each cover class (Oksanen 1976). The comparison of cover changes was not suitable for more infrequent and subordinate species occurring in the lowest cover classes. The changes in species frequencies were tested by using chi-squared tests.

25.4 Changes in Species Composition and Abundance

25.4.1 Heath Vegetation

In wind-exposed *Loiseleuria* heath at Sylfjellet, reindeer lichens (*Cladina* spp.) and *Flavocetraria nivalis* declined (Table 25.1) and *Solorina crocea* disappeared. By contrast, the cover values of *Empetrum nigrum* ssp. *hermaphroditum*, *Salix herbacea* and bryophytes, notably *Gymnomitrion* spp. and *Racomitrium lanuginosum*, were higher in 1999 than in the 1920s. An annual vascular plant, *Euphrasia frigida*, present in the 1920s, was not recorded in 1999.

In the snow-protected *Vaccinium myrtillus* heath on Mt. Sylfjellet (Table 25.1), lichen species, mostly *Cladina* spp., declined or disappeared. By contrast, the cover of *Empetrum nigrum* and the bryophytes *Barbilophozia* spp. and *Pleurozium schreberi* were higher in 1999 than in the 1920s. *Calluna vulgaris*, not recorded in the 1920s, was present in one-third of the quadrats in 1999. At Kilpisjärvi, in the *Vaccinium myrtillus* heath reindeer lichens have declined,

Table 25.1. Changes in plant species abundance in heath vegetation at Mt. Sylfjellet and Kilpisjärvi in the 1920s and 1999. *Loiseleuria procumbens* heath $n=15$; *Vaccinium myrtillus* heath, Sylfjellet $n=25$; and *Vaccinium myrtillus* heath, Kilpisjärvi $n=10$. t-test and χ^2 results (* $p<0.5$, ** $p<0.01$, *** $p<0.005$)

	Loiseleuria procumbens heath (Sylene)			
	Mean cover (%)		Frequency	
	1920s	1999	1920s	1999
Decreased				
Arctostaphylos alpina	7.6	0.2***	0.8	0.0***
Loiseleuria procumbens	66.0	28.9***	1.0	1.0
Vaccinium myrtillus				
Vaccinium vitis-idaea	4.4	0.7***	1.0	0.2***
Salix herbacea				
Carex bigelowii				
Deschampsia flexuosa				
Diphasiastrum alpinum				
Euphrasia frigida	2.0	0.0	0.6	0.0**
Hieracium alpinum				
Trientalis europaea				
Polytrichum piliferum				
Alectoria nigricans	4.4	1.1***	0.8	0.3
Alectoria ochroleuca	12.2	2.2***	1.0	0.7
Cetraria ericetorum	12.0	3.1***	1.0	0.9
Cladina arbuscula (incl. *C. mitis*)	8.2	2.9***	1.0	0.9
Cladonia coccifera	3.3	1.3***	1.0	0.4**
Cladonia deformis (+*C. sulphurina*)				
Cladonia gracilis	3.4	2.2*	0.9	0.7
Cladonia pyxidata (+*C. chlorophaea*)	2.0	0.7*	0.6	0.2
Cladonia uncialis				
Flavocetraria cucullata	10.2	0.9***	0.9	0.3*
Flavocetraria nivalis	17.7	4.9***	1.0	0.9
Solorina crocea	2.6	0.0	0.8	0.0***
Stereocaulon spp.	8.0	0.7***	1.0	0.2***
Thamnolia vermicularis	7.0	3.3***	1.0	1.0
Increased				
Calluna vulgaris				
Empetrum nigrum ssp. *hermaphroditum*	11.8	19.9*	1.0	1.0
Salix herbacea	0.7	2.4***	0.2	0.7*
Vaccinium uliginosum				
Melampyrum pratense				
Andreaea rupestris	0.0	1.5	0.0	0.5*
Barbilophozia spp.				
Gymnomitrion spp.	1.2	10.7***	0.3	0.9***
Polytrichum juniperinum				
Pleurozium schreberi				
Ptilidium ciliare	0.0	1.3	0.0	0.4
Racomitrium lanuginosum	2.8	15.8***	0.7	0.9
Cetraria islandica				
Cetrariella delisei	0.0	1.3	0.0	0.4
Cladonia subfurcata				

Long-Term Changes in Alpine Plant Communities in Norway and Finland   415

Vaccinium myrtillus heath (Sylene)				Vaccinium myrtillus heath (Kilpisjärvi)			
Mean cover (%)		Frequency		Mean cover (%)		Frequency	
1920s	1999	1920s	1999	1920s	1999	1920s	1999
				40	26*	1.0	1.0
5.5	2.0***	0.7	0.6				
4.6	0.5**	0.4	0.2				
6.9	3.1***	1.0	0.6				
7.3	3.8***	1.0	0.9				
2.9	0.7*	0.6	0.2				
3.7	0.9***	0.9	0.3**				
2.2	0.9***	0.7	0.3				
				1.9	0	0.8	0***
7.8	1.4***	1.0	0.4**				
30.1	11.4***	1.0	1.0	1.8	0.3*	1.0	0.6
7.1	1.2***	1.0	0.4	3.9	1.2*	1.0	0.8
3.1	0.1***	0.9	0.1***				
15.2	3.6***	1.0	0.9				
11.3	1.8***	1.0	0.6	0.9	0.3**	0.9	0.7
3.1	0.8*	0.4	0.2				
0	0.9	0	0.3*				
4.5	27.4***	0.7	1.0				
1.4	10.3***	0.3	0.4				
0.1	0.9*	0.1	0.3				
4.4	9.6*	0.7	1.0				
				0.5	7.6*	0.8	0.6
0.7	4.8***	0.2	0.8***				
				0.7	2.7***	0.7	1
0.0	0.9	0	0.3*				

Table 25.2. Changes in plant species abundances in snowbed vegetation at Mt. Sylfjellet and Kilpisjärvi in the 1920s and in 1998/1999 (*Deschampsia flexuosa–Anthoxanthum odoratum* snowbed; $n=20$; *Trisetum spicatum–Sibbaldia procumbens* snowbed, $n=10$). t-test and χ^2 results (* $p<0.5$, ** $p<0.01$, *** $p<0.005$). Values in italics are numbers of shoots 0.25 m^{-2}

	Deschampsia flexuosa–Anthoxanthum odoratum snowbed				*Trisetum spicatum–Sibbaldia procumbens* snowbed			
	Mean cover (%)		Frequency		Mean cover (%)		Frequency	
	1920s	1999	1920s	1999	1920s	1998	1920s	1998
Decreased								
Salix herbacea					51	25.4**	1.0	1.0
Carex bigelowii	14.3	0.7***	1	0.2***				
Carex lachenalii	2.5	0	0.5	0**				
Carex vaginata	2.2	0	0.5	0**				
Gnaphalium supinum	4.7	1.5**	0.7	0.5				
Rumex acetosa	3.0	1.0***	0.9	0.3***				
Salix herbacea	27.4	13.9**	1	0.9				
Trientalis europaea	6.2	0.8***	0.7	0.3				
Barbilophozia spp.	8.6	5.0*	1	0.9				
Kiaeria starkei	2.9	0.6*	0.5	0.1				
Moerckia blyttii	3.4	0	0.5	0**				
Pleurocladula albescens					3.1	0	1.0	0***
Polytrichastrum alpinum	3.8	0.2***	0.6	0.1**				
Cetraria islandica (+ *C. ericetorum*)	5.4	1.8***	1	0.6**	1.8	0.2**	1.0	0.2**
Cetrariella delisei					0.7	0	0.7	0*
Cladonia coccifera					1.6	0	0.8	0**
Cladonia gracilis (+*C. ecmocyna*)	11.7	2.0***	1.0	0.6	1.3	0.2*	0.9	0.3
Cladonia bellidiflora	8.3	0	0.9	0***				
Increased								
Agrostis mertensii					0	15.6	0	1.0***
Alchemilla alpina	0.2	5.6***	0.1	0.7***				
Bartsia alpina	0	1.2	0	0.4*				
Bistorta vivipara	1.0	4.0***	0.3	1.0***				
Deschampsia cespitosa	0.2	2.6*	0.1	0.4				
Deschampsia flexuosa					*3.1*	*43.5***	*0.2*	*0.8*
Empetrum nigrum	0.3	6.3***	0.1	0.8***				
Euphrasia frigida	0	1.8	0	0.6***				

Table 25.2. (*Continued*)

	Deschampsia flexuosa–Anthoxanthum odoratum snowbed				Trisetum spicatum–Sibbaldia procumbens snowbed			
	Mean cover (%)		Frequency		Mean cover (%)		Frequency	
	1920s	1999	1920s	1999	1920s	1998	1920s	1998
Geranium sylvaticum	0	1.7	0	0.5**				
Luzula multiflora ssp. *frigida*	0	1.8	0	0.5**				
Phleum alpinum	0.8	2.1*	0.3	0.6				
Pyrola minor	0	1.5	0	0.5**				
Ranunculus acris	0	2.6	0	0.6***				
Sibbaldia procumbens	0.5	1.7*	0.2	0.5				
Solidago virgaurea	0.8	2.9**	0.3	0.7*				
Taraxacum sp.					0	1.4	0	0.6*
Vaccinium myrtillus	0.2	13.3***	0.2	0.8***				
Veronica alpina	0	2.3	0	0.7***	1.3	4.9*	0.4	0.8
Polytrichum juniperinum	0	1.5	0	0.5**				
Sanionia uncinata	0	2.3	0	0.6***				
Cladina arbuscula	0	1.5	0	0.5**				

similarly to the Sylfjellet site. However, the response was different, with *Vaccinium myrtillus* having a lower cover in 1999 than in the 1920s. Of the mosses, *Polytrichum piliferum*, regularly present in the 1920s, was not found in 1999.

25.4.2 Snowbed Vegetation

In the *Deschampsia flexuosa–Anthoxanthum odoratum* community on Mt. Sylfjellet, typical snowbed plants such as *Carex lachenalii*, *Gnaphalium supinum*, *Kiaeria starkei*, *Moerckia blyttii* and *Salix herbacea* declined between the 1920s and 1999 (Table 25.2) and herbaceous species of grasslands or less extreme snowbed sites, such as *Bistorta vivipara* and *Solidago virgaurea*, increased. At Kilpisjärvi, in the *Trisetum spicatum–Sibbaldia procumbens* snowbed community a similar trend was found. Snowbed plants (*Pleurocladula albescens*, *Salix herbacea*, and *Veronica alpina*) declined and grasses (*Agrostis mertensii* and *Deschampsia flexuosa*) and *Taraxacum* spp. (mainly *T. croceum*) increased. In both areas, dwarf-shrub species, e.g. *Vaccinium myrtillus*, normally absent from snowbed sites, had established.

25.4.3 Alpine Soligenous Mires

The species-rich *Kobresia simpliciuscula–Carex microglochin* mire communities on Mt. Sylfjellet were studied at low (860 m a.s.l., Storbekken) and high altitude (1200 m a.s.l., Storsola). Lichens, albeit not dominant in 1920s, showed a significantly lower abundance in 1999 compared with the 1920s (Table 25.3). At high altitude there was a decline in vascular plant species, including dwarf shrubs, small herbs, and rosette species. Only *Juncus biglumis* had a greater cover in 1999 than in the 1920s. On the lower altitude site, species typically found at high altitudes, e.g. *Carex bigelowii*, declined. Of the mosses, *Blindia acuta* and *Scorpidium cossoni* showed some increase.

Table 25.3. Changes in plant species abundance in the *Kobresia simpliciuscula–Carex microglochin* communities in Storsola (1205 m a.s.l.) and in Storbekken (860 m a.s.l.) at Mt. Sylfjellet in the 1920s and 1999 (in each site $n=10$). t-test and χ^2 results (* $p<0.5$, ** $p<0.01$, *** $p<0.005$)

	Storsola				Storbekken			
	Mean cover (%)		Frequency		Mean cover (%)		Frequency	
	1920s	1999	1920s	1999	1920s	1999	1920s	1999
Decreased								
Polygonum viviparum	10.6	3.3***	1	1	5.5	2.3**	1.0	0.7
Carex atrofusca	4.8	1.7*	0.8	0.5	5.0	2.0**	1	0.6
Carex bigelowii					6.1	0.7	0.9	0.2*
Carex capillaris					8.3	2.3***	1.0	0.7
Carex microglochin					28.3	13.7*	1.0	0.8
Carex rupestris	8.8	2.5*	0.6	0.6				
Carex vaginata	13.3	5.2**	1	0.7				
Diapensia lapponica	2.9	0	0.7	0*				
Empetrum nigrum	6.4	0.3***	1	0.1***				
Euphrasia frigida					3.0	1.0***	0.9	0.3
Festuca vivipara					6.6	2.0***	1.0	0.6
Juncus triglumis					3.9	1.3***	1.0	0.4
Pedicularis oederi	6.6	2.0***	1	0.6				
Pinguicula vulgaris	3.3	1.3***	1	0.4*	3.3	1.3***	1.0	0.4
Vaccinium uliginosum	2.9	0	0.7	0*	4.4	2.0*	1.0	0.6
Aulacomnium turgidum	8.3	1.0***	0.8	0.3				
Hypnum bambergeri					10.9	0	1.0	0
Hypnum hamulosum (+*H. callichroum*)	6.0	1.0*	0.7	0.3				
Ptilidium ciliare					2.3	0.7*	0.7	0.2
Racomitrium lanuginosum		6.1	0	1	0***			
Scorpidium cossoni (+*S. revolvens*)	5.1	1.7*	0.7	0.5				
Cetraria islandica	3.3	0.3***	1	0.1***	3.3	0.3	1.0	0.1***

Table 25.3. (*Continued*)

	Storsola				Storbekken			
	Mean cover (%)		Frequency		Mean cover (%)		Frequency	
	1920s	1999	1920s	1999	1920s	1999	1920s	1999
Cladina arbuscula Cladonia pyxidata	3.3	0.7***	1	0.2***	3.3	0.7	1.0	0.2***
Flavocetraria cucullata	3.3	1.3***	1	0.4*				
Flavocetraria nivalis	3.3	0.3***	1	0.1***	3.0	0	0.9	0***
Thamnolia vermicularis	2.6	0.7**	0.8	0.2				
Nostoc commune					9.6	0.3	1	0.1***
Increased								
Juncus biglumis	1.3	3.0*	0.4	0.9				
Blindia acuta	0	8.0	0	1***				
Hypnum bambergeri	0	2.3	0	0.7*				
Scorpidium cossoni					0	2.9	0	0.7*

25.5 Discussion

The long-term comparative data show a number of changes in the species composition and abundance of plant species in several sites of the Scandes. The observed vegetation changes can result from different factors and they need careful interpretation. Causal links between observational data and processes rely on weak inference. However, we recognise that information obtained from long-term observational data is important to complement an experimental approach to vegetation change, which often suffers from time-lag deficiencies and unrealistic designs.

A possible reason for the decline in lichen cover in heath vegetation at Mt. Sylfjellet is the increased summer grazing, as opposed to mainly winter grazing earlier. Summer grazing and trampling are known to reduce lichen cover (Gaare 1968; Grabherr 1982; Haapasaari 1988) and recovery is slow after heavy lichen depletion (Gaare 1998). However, Cornelissen et al. (2001) found that an increase in dwarf shrubs could lead to a decrease in lichens. In the heath site at Kilpisjärvi, the cover of lichens had declined which suggests that the levels of reindeer trampling and grazing have become more intense during the last decades.

The increased cover of heath dwarf shrubs at Mt. Sylfjellet could be a response to the observed climatic warming (Holten and Carey 1992). Meanwhile, the decline in the cover of the dominant dwarf shrub *Vaccinium myrtillus* at the Kilpisjärvi heath site is most likely to have resulted from grazing by the Norwegian lemming. Lemmings cut *V. myrtillus* shoots during their population peak in the winter of 1997–1998 (R. Virtanen, pers. obs.).

A general trend of the change in snowbed sites was that the abundance of grassland and heath species increased while snowbed plants declined. Such changes in snowbed vegetation are unlikely to be related to changes in reindeer grazing pressure. In fact, heavy reindeer grazing would favour narrow-leaved graminoids and prostrate plants, whereas taller herbs and shrubs would increase under light grazing (Gaare 1968; Oksanen and Moen 1994; Moen and Oksanen 1998; Virtanen 1998). Rather, the changes in vegetation composition suggest that the sites with formerly late-lying snow became snow-free earlier and that the vegetation had already responded to such a lengthening of the growing season. Unfortunately, we cannot relate the observed changes in snowbed vegetation to direct observations of snow duration in the two areas; however, they are in accordance with the observed increase in surface temperatures and shorter duration of snow cover, particularly in the mountains of southern Norway (see also Klanderud and Birks 2003).

The species-rich alpine mires at Mt. Sylfjellet with rare mountain plants did not show any strong signs of vegetation change over 70 years. The decline of lichens may reflect the observed changes in reindeer grazing discussed above. Current grazing regime may be responsible for the increase of the moss *Blindia acuta* and the graminoid *Juncus biglumis*. The rarer plants are still regularly found and their abundance changed only a little. These communities are found in small habitat patches where ground water seeps onto the ground. The conditions in these habitat patches have remained stable and therefore no marked change in plant cover can be detected. A potential threat to mires, particularly on lowland sites, is the invasion of some plants which by dominating the space can exclude low-stature mountain plants. It is noteworthy that *Molinia caerulea* hardly recorded in the 1920s now has a cover of ca. 5 %, suggesting that it may have increased in this site. Its further expansion has to be considered as a potential threat to these rich alpine mires.

25.6 Conclusions

1. Long-term changes in species composition, abundance and diversity of alpine plant communities were analysed by using data sets from central Norway and NW Finland sampled in the 1920s and 1990s. Comparisons were made for seven sites representing heath, snowbed and soligenous mire vegetation.
2. The cover and number of lichens declined in most communities probably owing to increased reindeer grazing and trampling. In heath, the cover of some bryophytes and dwarf shrubs increased on the Norwegian sites.

Herbs and grasses increased in snowbed communities, while some typical snowbed plants disappeared or decreased. The data for snowbeds suggests that these communities may have responded to changes in snow cover as a result of the observed warmer climate in the 1990s.
3. The species that showed a significant decline or increase may be considered for use in monitoring as indicators of environmental change.

Acknowledgements. The Finnish Research Council of Natural Resources and Environment and European Science Foundation through the ALPNET network financially supported the study. M. Zobel helped in interpreting Lippmaa's cover classes. H. Staaland gave useful information on reindeer population at the Mt. Sylfjellet area. L. Nagy and an anonymous referee improved the presentation.

References

Cornelissen JHC, Callaghan TV, Alatalo JM, Michelsen A, Graglia E, Hartley AE, Hik DS, Hobbie SE, Press MC, Robinson CH, Henry GHR, Shaver GR, Phoenix GK, Gwynn Jones D, Jonasson S, Chapin III FS, Molau U, Neill C, Lee JA, Melillo JM, Sveinbjörnsson B, Aerts R (2001) Global change and arctic ecosystems: is lichen decline a function of increases in vascular plant biomass? J Ecol 89:984–994

Gaare E (1968) A preliminary report on winter nutrition of wild reindeer in the southern Scandes, Norway. Symp Zool Soc Lond 21:109–115

Gaare E (1998) Lav – reinens viktigste vinterføde. Reindriftsnytt 1998(4):57–59

Galen C, Stanton ML (1993) Short-term responses of alpine buttercups to experimental manipulations of growing season length. Ecology 74:1052–1058

Grabherr G (1982) The impact of trampling by tourists on a high altitude grassland in the Tyrolean Alps, Austria. Vegetatio 48:209–217

Grabherr G, Gottfried M, Pauli H (1994) Climate effects on mountain plants. Nature 369:448

Grabherr G, Gottfried M, Gruber A, Pauli H (1995) Patterns and current changes in alpine plant diversity. In: Chapin FS III, Körner C (eds) Arctic and alpine biodiversity: patterns, causes and ecosystem consequences. Springer, Berlin Heidelberg New York, pp 167–181

Grønås S (1999) Klimavariasjoner i våre områder de siste tusen år. Naturen 123:299–311

Haapasaari M (1988) The oligotrophic heath vegetation of northern Fennoscandia and its zonation. Acta Bot Fenn 135:1–219

Hämet-Ahti L, Suominen J, Ulvinen T, Uotila P (eds) (1998) Retkeilykasvio (Field flora of Finland), 4th edn. Finnish Museum of Natural History, Botanical Museum, Helsinki

Holten JI, Carey PD (1992) Responses of climate change on natural terrestrial ecosystems in Norway. NINA Forskiningsrapport 29:1–59

Hoofgaard A (1999) The role of 'natural' landscapes influenced by man in predicting responses to climate change. Ecol Bull 47:160–167

Jones PD, New M, Parker DE, Martin S, Rigor IG (1999) Surface air temperatures and its changes over the past 150 years. Rev Geophys 37:173–199

Kalela O, Koponen T, Lind EA, Skarén U, Tast J (1961) Seasonal change of habitat in the Norwegian lemming, *Lemmus lemmus* (L.). Ann Acad Sci Fenn Ser A IV Biol 55:1–72

Klanderud K, Birks HJB (2003) Recent increases in species richness and shifts in altitudinal distributions of Norwegian mountain plants. Holocene 13:1–6

Körner C (1999) Alpine plant life. Springer, Berlin Heidelberg New York
Kojola I, Aikio P, Helle T (1993) Influences of natural food resources on reindeer husbandry in northern Lapland (in Finnish with an English abstract). Res Rep Res Inst Northern Finland 116:1–39
Kullman L (2002) Rapid recent range-margin rise of tree and shrub species in the Swedish Scandes. J Ecol 90:68–77
Lippmaa T (1929) Pflanzenökologische Untersuchungen aur Norwegisch- und Finnisch-Lappland. Acta Inst Hortic Bot Univ Tartuensis 2:1–146
Moen J, Oksanen L (1998) Long-term exclusion of folivorous mammals in two arctic-alpine plant communities: a test of the hypothesis of exploitation ecosystems. Oikos 82:333–346
Molau U, Alatalo J (1998) Responses of subarctic-alpine plant communities to simulated environmental change: biodiversity of bryophytes, lichens, and vascular plants. Ambio 27:322–329
Nordhagen R (1928) Die Vegetation und Flora des Sylenegebietes. I. Die Vegetation. Skr Nor Vidensk Akad Oslo. I. Mat Naturvidensk Kl 1927(1):1–612
Oksanen L (1976) On the use of the Scandinavian type class system in coverage estimation. Ann Bot Fenn 13:149–153
Oksanen L, Moen J (1994) Species-specific plant responses to exclusion of grazers in three Fennoscandian tundra habitats. Écoscience 1:31–39
Oksanen L, Virtanen R (1995) Topographic, altitudinal and regional patterns in continental and suboceanic heath vegetation of northern Fennoscandia. Acta Bot Fenn 153:1–80
Saetersdal M, Birks HJB (1997) A comparative ecological study of Norwegian mountain plants in relation to possible future climatic change. J Biogeogr 24:127–152
Söderström L, Hedenäs L, Hallingbäck T (1992) Checklista över Sveriges mossor. Myrinia 2:13–56
Sokal RR, Rohlf JF (1995) Biometry: the principles and practice of statistics in biological research, 3rd edn. WH Freeman, New York
Totland Ø, Nyléhn J (1998) Assessment of the effects of environmental change on the performance and density of *Bistorta vivipara*: the use of multivariate analysis and experimental manipulation. J Ecol 86:989–998
Virtanen R (1998) Impact of grazing and neighbour removal on a heath plant community transplanted onto a snowbed site, NW Finnish Lapland. Oikos 81:359–367
Vitikainen O, Ahti T, Kuusinen M, Lommi S, Ulvinen T (1997) Checklist of lichens and allied fungi of Finland. Norrlinia 6:1–123

26 Vegetation Dynamics at the Treeline Ecotone in the Ural Highlands, Russia

P.A. Moiseev and S.G. Shiyatov

26.1 Introduction

Copious seedling establishment and an increase in stem density just below the actual treeline have been reported for various mountain areas around the world: North America (Kearney 1982; Jakubos and Romme 1993; Little et al. 1994; Taylor 1995; Woodward et al. 1995; Rochefort and Peterson 1996; Lloyd 1997), northern Europe (Ågren et al. 1983; Kullman 1986, 1988) and New Zealand (Wardle and Coleman 1992). A few studies have reported seedling establishment above the treeline and an upward shift of the treeline in the Urals (Gorchakovsky and Shiyatov 1978), eastern Canada (Payette and Filion 1985), and Colorado (Shankman and Daly 1988).

It has been debated whether the observed changes in growth dynamics can be related to climate change or they are within the limits of normal dynamics of the treeline ecotone (Kullman 1988, 1989; Hättenschweiler and Körner 1995).

We investigated, by using historic fixed-point landscape photographs, changes in forest growth and possible treeline shifts in the Ural Mountains, Russia. The changes we report took place over 70 years in the South Urals and over 35 years in the Polar Urals and correlate with general warming trends in these regions. We visited mountain peaks in the South and Polar Urals and took over 500 repeat photographs between 1995 and 2000. We present four pairs; however, general conclusions are drawn on the basis of all of them.

26.2 Materials and Methods

26.2.1 Study Areas

Repeat landscape photographs were taken of the peaks of the Bolsoi Taganai Ridge and Iremel Massif (South Urals) and of the Rai-Iz Massif, Tchernaya, Slancevaya and Yar-Keu Mountains (Polar Urals). These two regions were selected to contrast the high-latitude and low-latitude provinces of the Urals. There are many old photographs available from these regions as numerous studies have been reported on their geomorphology, soils, vegetation, and on dendrochronology (Tyulina 1929, 1931; Igoshina 1964; Shiyatov 1965, 1983, 1986; Sharafutdinov 1983).

The Iremel Massif (54°32'N, 58°51'E) is the highest in the central part of the South Urals (1586 m a.s.l.). It has a rugged topography, large altitudinal amplitudes (up to 1100 m) and a complex geological structure. The amount of precipitation is over 800–1000 mm year^{-1}. Dense coniferous forests (*Abies sibirica*, *Picea abies* ssp. *obovata*) predominate in valleys and slopes up to 1250 m. There are fragments of a Scots pine (*Pinus sylvestris*) sub-belt between 500 and 700 m. Birch (*Betula pubescens*, *B. pendula*), aspen (*Populus tremula*) and larch *(Larix sibirica)* grow together with the other conifers. The treeline ecotone is situated between about 1200–1400 m, with open *Picea abies* ssp. *obovata* forests and tall herbaceous vegetation occupying in the lower part and crooked spruce and birch forests (*Betula pubescens* ssp. *tortuosa*), moss-rich short grass meadows, shrubs (*Juniperus communis* ssp. *alpina*, *Salix* spp.), and moss-grass heaths in the upper half. The alpine zone is above 1400 m. Moss-grass, moss-dwarf shrub and moss-lichen heath communities occupy the more gentle slopes, plateaus and terraces. Steep stony and rocky slopes are covered by lichens and lithophytic communities (Tsvetaev 1960; Gorchakovsky 1975). The vegetation of the treeline ecotone and the alpine zone is largely natural without such anthropogenic impacts as grazing, fires and logging.

The Rai-Iz Massif and Slancevaya Mountain are on the eastern macroslope of the Polar Ural Mountains (66°46'–66°55'N, 65°36'–65°49'E). They are separated from each other by the Sob River valley. The Rai-Iz Massif is a large ultramafic formation 10 km wide and 15 km long with a maximum elevation of up to 900–1100 m a.s.l. Crystalline shale makes up the Slancevaya Mountain, which is about 3 × 4 km in size and reaches 417 m a.s.l. This area is situated within the forest-tundra zone. On mountain slopes the treeline ecotone is of forest-tundra (or subgoltsy) character; above the treeline there is the alpine zone, lower alpine (mountain tundra) and high alpine (the Goltsy Desert; Gorchakovsky 1975). The treeline ecotone is found in deep valleys and slopes up to 200–300 m, with *Larix sibirica* and *Picea abies* ssp. *obovata–Larix sibirica* open forests dominating with patches of closed larch-spruce forests with

Betula pubescens ssp. *tortuosa*. A dense scrub of *Alnus viridis* ssp. *fruticosa* often covers the crystalline shale and gabbro slopes. Bushes of different *Salix* species (*S. glauca*, *S. lanata*, *S. lapponicum*, *S. phylicifolia*) usually occupy the wettest sites. The shrub layer is mainly represented by *Betula nana*, *Vaccinium uliginosum* and *Ledum decumbens*. The alpine zone between about 250 and 650 m is represented by different types of dwarf-shrub heath communities with numerous lichen and moss species. Snowbeds and well-irrigated sites are dominated by meadows. The Goltsy Desert, the approximate upper limit to higher plants, is above 650 m where the vegetation is represented by fragmentary moss-lichen communities with some scattered small herbaceous plants.

26.2.2 Long-Term Weather Data

Based on observational data between 1890–1999 from Zlatoust, South Urals (Fig. 26.1) and Salehard, Polar Urals (Fig. 26.2) weather stations, it was estimated that winters, on average, have become about 4 °C warmer in the Polar Ural area and 3 °C warmer in the Middle to South Urals. Summers for the same period showed smaller changes: 1.4 °C in the Polar Urals and 0.6 °C in the South Urals.

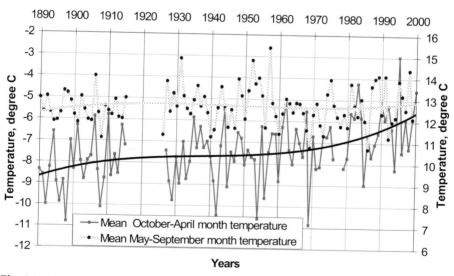

Fig. 26.1. May–September and October–April temperatures at Zlatoust, South Urals, Russia, between 1890 and 1999

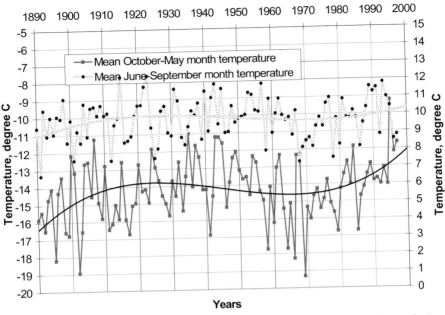

Fig. 26.2. June–September and October–May temperatures at Salehard, Polar Urals, Russia, between 1890 and 1997

26.2.3 Repeat Photography and Image Analysis

From landscape photographs it is possible to estimate many qualitative and quantitative parameters of past and present woody vegetation (such as stand composition and density, cover of tree and bush layers, growth form, height and diameter of trees, fructification). The comparison of old photographs with recent ones may give an indication of the degree of regeneration and mortality of trees, forest cover, and displacement of borders between forest and non-woody communities.

The baseline for comparisons was provided by old photographs taken by L.N. Tyulina in 1929–1931, K.N. Igoshina in 1956–1960 and by S.G. Shiyatov in 1960–1969. Repeat photographs and evaluation of vegetation changes were made in 1976 and between 1996–1999. The location of the points from where the original photographs were taken was made with the help of prints of the original photographs, which were also used to make an on-site evaluation of any vegetation change. The original camera positions were determined by seeking and overlapping the most noticeable, outstanding landmarks (rocks, hills, summits) of different line of horizon. On establishing a point we took a repeat photograph and estimated, calculated and measured different vegetation parameters depending on distance from the photo point to the objects.

Because the same objects were described from different photo points and therefore different distances, the data obtained were corrected and verified twice or more times. Occasional field truthings of visual estimates were made.

In addition to the field comparisons, comparisons of the scanned images of the original and repeat photographs were made by using Adobe Photoshop v5.2 software. We superimposed the two or more images and verified data obtained in the field. We contoured altitudinal lines of former and contemporary borders of closed and low open forests and calculated their altitudinal and horizontal displacement.

26.3 Four Examples of Photo Pairs

The comparison of current and past vegetation showed that trees have regenerated vigorously in low open forests and established in former sites dominated by meadows and moss heath communities over the last 70 years. This is quite clear from an analysis of the image pairs in Figs. 26.3–26.6.

The first pair was taken in 1929 and 1999 (Fig. 26.3) from above the upper limit of low open forest (1490 m) on the abrupt north-west-facing slope of Bolshoi Iremel Peak, South Urals. We estimated that the closed forest on the saddle in the middle of the photograph, behind the bouldery slope in the foreground, has advanced upslope from 1260–1270 to 1320–1330 m since 1929. *Picea abies* ssp. *obovata* elfin and low open forests became closed (from 10–30 % up to 50–80 %). Trees grew four to six times higher from 1–2 to 6–8 m and their stem forms changed from contorted to shrubby and upright forms (obtained from other photo points). In 1999, many new individuals of spruce were apparent on previously open ground up to 1380–1410 m. There was also an increase in the areas occupied by closed spruce forests on the south-facing slope of the western spur of Maly Iremel from 1220–1230 to 1280–1310 m a.s.l. (far mountain on extreme left-hand side of Fig. 26.3).

The second pair was taken in 1929 and 1999 (Fig. 26.4) from a small hill (1335 m a.s.l.) on the northern slope of Bolshoi Iremel (flat area with a rock in the foreground). Large differences can be observed in the vegetation on the intermountain saddle running from Bolshoi to the far mountain of Maly Iremel (in the second line) and on the south-west-facing slope of Maly Iremel (on the left-hand side in the background). In 1929, there were many gaps and the *Picea* forest had a cover of about 50 %. In 1999, closed (up to 90 % canopy cover) *Picea* or *Betula pubescens* ssp. *tortuosa* forest dominated. Forest cover on the south-west-facing gentle slope of Maly Iremel increased greatly. We estimated that there was an upward shift of the upper forest limit of about 60–80 m along 600–900 m of horizontal front: low open forest from 1280–1300 up to 1340–1360 m and closed forest from 1220–1230 up to 1300–1310 m. The upward movement of closed forest involved an increase from 20 % to 60–70 %

Fig. 26.3. Repeat photographs of the north-facing slope of Bolshoi Iremel Peak, Southern Urals, and the south-facing slope of the western spur of Maly Iremel, taken in 1929 and 1999. In the *foreground*, stony slope; *middle distance*, saddle and summit with cairn marking altitude (1335 m a.s.l.); *far distance*, western spur of Maly Iremel

Vegetation Dynamics at the Treeline Ecotone in the Ural Highlands, Russia

1929

1999

Fig. 26.4. Repeat photographs taken in 1929 and 1999 of the saddle between Bolshoi and Maly Iremel and of the northern extremity of the north-eastern spur of Bolshoi Iremel (*middle distance*) and of the gentle south-west-facing (*left*) and stony south-facing (*right*) slope of Maly Iremel (*far distance*). Pictures taken from a small horizontal platform (with a rock in the foreground) located on the northern slope of Bolshoi Iremel, Southern Urals

in cover in the formerly open forest. Meanwhile a concurrent increase in canopy closure was estimated from 50% to 80–90% in the closed forest. An upward shift of open and closed forest boundaries by up to 20–40 m along a 100–300 m horizontal stretch is also well visible on the southern stony slope of Maly Iremel (on the right-hand side in the background).

The third pair is of the south-eastern slope of the Rai-Iz Massif, the Polar Urals (in the background), taken in 1962 and 1997 (Fig. 26.5) from the north-

1962

1997

Fig. 26.5. Repeat photographs taken in 1962 and 1997 of the south-eastern slope of the Rai-Iz Massif, Polar Urals

ern slope of the hill, at an elevation of 300 m. There is a woodless valley bottom in the middle distance. The upper limit of *Larix sibirica* open forest is between 280–350 m. A large increase in canopy cover from an initial 20–30 % to 50–70 % occurred on the south-eastern slope between 1962 and 1997. At the same time, the average tree height increased from 8 m to 11–12 m. On gentle slopes of the Rai-Iz Massif the upper open forest communities boundary moved upwards to 20–40 m (obtained from other photo points). The cover and height of the alder layer have increased significantly and young individu-

1962

1999

Fig. 26.6. Repeat photographs taken in 1962 and 1999 of the north-eastern slope of Slancevaya Mountain (220 m a.s.l.), Polar Urals. *Larix sibirica* growing at its upper limit in the foreground and in the background

als and patches of *Alnus viridis* ssp. *fruticosa* occupied many previously open sites in 1997.

The fourth pair was taken of the Slancevaya Mountain, Polar Urals (ca. 220 m) in 1962 and 1999 (Fig. 26.6) from the north-eastern slope. The upper treeline forming species is *Larix sibirica* and the dark patches are dense *Alnus* scrub. The cover of larch stands increased at least two-fold between 1962 and 1997. Larches occupied many open sites and formed open stands at the upper former treeline.

26.4 Discussion

The comparison of historic landscape photographs is relatively rarely reported, as it is often difficult to determine the exact points from where the original photographs were taken (Kullman 1979; Shiyatov 1983). After examining more than 500 historic photographs from the Ural Highlands, we judged that the method was suitable for a study of vegetation dynamics because there are readily identifiable landmarks which make camera orientation relatively easy. Overall, there appeared to be abundant and vigorous regeneration present with few standing and fallen dead trees all over the Ural Highlands. There appeared to be a remarkable upward shift of the upper limit of open forest, especially on gentle slopes. *Picea abies* ssp. *obovata* (South Urals) and *Larix sibirica* (Polar Urals) have regenerated and propagated best and increased their relative cover. On gentle slopes with well-developed soil, upper limits of closed and open forests were estimated to be up to 60–80 m higher in the South Urals than 70 years earlier and by 20–40 m higher in the Polar Urals than 40 years earlier. A clear increase in forest canopy cover along the contours of up to 600–900 m over the last 70 years could be detected in the South Urals. In less favourable conditions on the steep boulder fields, or in damp hollows, an upward shift of up to 20–40 m altitudinal and 100–300 m of horizontal expansion were estimated in the South Urals.

Weather data has shown that at the end of the nineteenth and the beginning of the twentieth century, on average, the mean monthly temperature for May–September (warm season) was about 12.3 °C and October–April (cold season) was about −8.7 °C, indicating a cold climate in the South Urals (Fig. 26.1). In the Polar Urals the mean June–September temperature was about 8.9 °C and October–May temperature was −16.4 °C for the same period. From that period until ca. 1960, a general warming occurred. In the South Urals the increase for warm seasons was ca. 0.4 °C and for cold seasons ca. 0.9 °C. In the Polar Urals warming was more pronounced: 1.2 °C for warm seasons and 2.6 °C for cold seasons on average. In the 1960s and early 1970s, a slight summer cooling occurred, which was more marked in the Polar Urals. Finally, warm season temperatures increased up to about 12.9 °C in the South

Urals and up to about 10.3 °C in the Polar Urals by the late 1990s. In the Polar Urals, cooling by 0.9 °C was also apparent for cold seasons, followed by a period of sharp warming to about –12.1 °C at the present time. There were gradual changes to the cold seasons, i.e. a warming from –7.6 up to about –5.4 °C since 1930 to present in the South Urals.

The warm and relatively wet climate of the twentieth century appears to have resulted in an improvement in conditions for the growth and survival of woody plants at the upper limit of their distribution. This may explain the abundant tree regeneration and their establishment at the treeline ecotone and including the lower alpine zone, especially between 1960 and 2000. Expanding forest cover reduced the area of grasslands and alpine heath in the alpine zone by about 10–30 %.

26.5 Conclusions

1. Landscape-scale changes in montane forest growth in the Urals, Russia, were analysed by using repeat photography. The changes reported took place over 70 years in the South Urals and over 35 years in the Polar Urals and correlate with general warming trends in these regions.
2. There was an abundant regeneration and increase in the relative cover of *Picea abies* ssp. *obovata* (South Urals) and *Larix sibirica* (Polar Urals). The line of closed forest appeared to have advanced up to 60–80 m higher than it was 70 years earlier in the South Urals and by 20–40 m higher than 40 years earlier in the Polar Urals.
3. The closing of open forest and expansion appear to have reduced the area of non-woody vegetation in the treeline ecotone by about 10–30 %.

References

Ågren J, Isaksson L, Zackrisson O (1983) Natural age and size structure of *Pinus sylvestris* and *Picea abies* on a mire in the inland part of northern Sweden. Holarct Ecol 6:228–237

Briffa KR, Jones PD, Schweingruber FH, Shiyatov SG, Cook ER (1995) Unusual twentieth-century summer warmth in a 1000-year temperature record from Siberia. Nature 376:156–159

Gorchakovsky PL (1975) The vegetation of the Ural Highlands. Nauka, Moscow (in Russian)

Gorchakovsky PL, Shiyatov SG (1978) The upper forest limit in the mountains of the boreal zone of the USSR. Arct Alp Res 10:349–363

Hättenschweiler S, Körner C (1995) Responses to recent climate warming of *Pinus sylvetris* and *Pinus cembra* within their montane transition zone in the Swiss Alps. J Veg Sci 6:357–368

Houghton JT, Meira Filho LG, Callender BA, Harris N, Kattenberg A, Maskell K (eds) (1995) Climate change 1995 – the science of climate change. Contribution of Working Group I to the Second Assessment Report of the Intergovernmental Panel on Climate Change (IPCC). Cambridge University Press, Cambridge

Igoshina KN (1964) Vegetation of USSR and foreign countries: vegetation of the Urals. Researches of Botanical Institute of Academy of Science of USSR, Series 3, Issue 16. Academy of Science of USSR, Moscow (in Russian)

Innes JL (1991) High-altitude and high-latitude tree growth in relation to past, present and future climate change. Holocene 1:168–173

Jakubos B, Romme WH (1993) Invasion of subalpine meadows by lodgepole pine in Yellowstone National Park, Wyoming, USA. Arct Alp Res 25:382–390

Kearney MS (1982) Recent seedling establishment at timberline in Jasper National Park, Alberta. Can J Bot 60:2282–2287

Kelly PM, Jones PD, Sear SB, Cherry BSG, Tavakol RK (1982) Variations in surface air temperatures: Part 2. Arctic regions, 1881–1980. Mon Weather Rev 110:71–83

Kullman L (1979) Change and stability in the altitude of the birch tree-limit in the southern Swedish Scandes, 1915–1975. Acta Phytogeogr Suecica 65:1–121

Kullman L (1986) Recent tree-limit history of *Picea abies* in the southern Swedish Scandes. Can J For Res 16:761–771

Kullman L (1988) Subalpine *Picea abies* decline in the Swedish Scandes. Mountain Res Dev 8:33–42

Kullman L (1989) Recent retrogression of the forest-alpine tundra ecotone (*Betula pubescens* Ehrh. ssp. *tortuosa* (Ledeb.) Nyman) in the Scandes Mountains. Sweden. J Biogeogr 16:83–90

Lamb HH (1977) Climate. Present, past and future, vol 2. Climatic history and future. Methuen, London

Little RL, Peterson DL, Conquest LL (1994) Regeneration of subalpine fir (*Abies lasiocarpa*) following fire: effects of climate and other factors. Can J For Res 24:934–944

Lloyd AH (1997) Response of tree-line populations of foxtail pine (*Pinus balfouriana*) to climate variation over the last 1000 years. Can J For Res 27:936–942

Payette S, Filion L (1985) White spruce expansion at the tree line and recent climatic change. Can J For Res 15:241–251

Rochefort RM, Peterson DL (1996) Temporal and spatial distribution of trees in subalpine meadows of Mount Rainier National Park, Washington, USA. Arct Alp Res 28:52–59

Shankman D, Daly C (1988) Forest regeneration above tree limit depressed by fire in the Colorado Front Range. Bull Torr Bot Club 115:272–279

Sharafutdinov MI (1983) The mountain tundras of the Iremel Mountains, Southern Urals. In: Gorchakovsky PL (ed) Floristic and geobotanical research in the Urals. Ural Scientific Centre of Academy of Science of USSR, Sverdlovsk, pp 110–119 (in Russian)

Shiyatov SG (1965) Age structure and formation of larch open stands at the upper tree-line in the basin of Sob River (The Polar Ural Mountains). Geogr Dyn Veg Cover 42:81–96 (in Russian)

Shiyatov SG (1983) Experience in the use of old photographs for the study of changes of forest vegetation at its upper elevational limit of distribution. In: Gorchakovskii PL (ed) Floristic and geobotanical research in the Urals. Ural Scientific Centre of Academy of Science of USSR, Sverdlovsk, pp 76–109 (in Russian)

Shiyatov SG (1986) Dendrochronology of the upper timberline in the Urals. Nauka, Moscow (in Russian)

Taylor AH (1995) Forest expansion and climate change in the mountain hemlock (Tsuga mertensiana) zone, Lassen Volcanic National Park, California, USA. Arct Alp Res 27:207–216

Tsvetaev AA (1960) The Iremel Mountains of the south Urals. Geographical Society of the USSR, Bashkirian Division, Ufa (in Russian)

Tyulina LN (1929) From the highlands of the South Urals (Iremel). In: Phytosociological and phytogeographical essays. Novaya derevnya, Leningrad, pp 345–359 (in Russian)

Tyulina LN (1931) On the highland vegetation of the South Urals. Izvestia Vsesouznogo Geographicheskogo Obshestva 63:453–499 (in Russian)

Wardle P, Coleman MC (1992) Evidence for rising upper limits of four native New Zealand forest trees. N Z J Bot 30:303–314

Woodward A, Schreiner EG, Silsbee DG (1995) Climate, geography, and tree establishment in subalpine meadows of the Olympic Mountains, Washington, USA. Arct Alp Res 27:217–225

27 Recent Increases in Summit Flora Caused by Warming in the Alps

M. BAHN and CH. KÖRNER

27.1 Introduction

Plant species richness increased on many high-altitude mountain summits during the twentieth century. This has resulted from an up-slope migration of species, which has been attributed to climate warming (Braun-Blanquet 1957; Hofer 1992; Gottfried et al. 1994; Grabherr et al. 1994, 1995; Pauli et al. 1996; Chaps. 24 and 29). In 1986, Mount Glungezer, a high alpine summit in the Austrian Central Alps, was found to host 83 vascular plant species, a remarkably high number in an area of 4000 m² (Bahn and Körner 1987). The present study reports a recent re-recording of the site by the same authors after the warmest period on record in the Alps. The aim of the study was to assess if an upward migration of plant species had taken place and if any changes in species composition and abundance occurred between 1986 and 1999–2000 within the different plant communities.

27.2 The Mount Glungezer Summit

Mount Glungezer (2600 m a.s.l.) is a high alpine summit near Innsbruck, Austria. The area has mainly siliceous bedrock and poorly developed soils with a pH ranging from 4.6 to 6.6, with a median of 4.7. It has a small permanent snowfield. The length of the growing season is 60–120 days. The mean air temperature of the warmest month, July, is around 5 °C with frost possible every night. The vascular plant species composition and the vegetation of a 4000-m² area near the summit of Mount Glungezer were first assessed in 1986. The area is characterised by a highly variable topography, including both north- and south-facing slopes. The 83 higher plant species recorded comprise elements from the treeline ecotone, alpine, subnival and nival vegetation

Table 27.1. Types of vegetation and changes in species richness on Mount Glungezer, 1986 vs. 1999/2000

Type of vegetation	Subunits characterised by most abundant species	Area (%)	Exposure	No. of species 1986	No. of species 2000	New species 1999/2000
Fragmented alpine mats	(a) *Carex curvula, Oreochloa disticha, Juncus trifidus, Leontodon pyrenaicus* ssp. *helveticus, Geum montanum*	29.1	S	70	77 (+7)	*Alchemilla alpina, Alchemilla vulgaris* agg., *Calluna vulgaris, Gentiana punctata, Homogyne alpina, Vaccinium uliginosum, Veronica alpina*
	(b) *Salix herbacea, Carex curvula, Oreochloa disticha*	34.2	SW	50	55 (+5)	*Arabis alpina, Cirsium spinosissimum, Geum reptans, Oxyria digyna, Sempervivum montanum*
	(c) *Salix retusa, Oreochloa disticha, Ligusticum mutellina, Luzula alpinopilosa*	7.5	NE	34	40 (+6)	*Achillea moschata, Carex curvula, Gentiana alpina, Minuartia recurva, Poa alpina vivipara, Vaccinium uliginosum*
	(d) *Avenula versicolor, Agrostis rupestris, Poa alpina, Leontodon pyrenaicus, Phyteuma hemisphaericum*	0.2	S	22	22 (–)	–
Open scree vegetation	(e) *Saxifraga moschata*[a]*, Saxifraga bryoides, Cerastium uniflorum, Doronicum clusii*	5.2	NE	31	31 (–)	–
	(f) *Geum reptans, Cerastium uniflorum, Doronicum clusii, Saxifraga moschata, Saxifraga bryoides*	0.5	NE	24	25 (+1)	*Minuartia recurva*
	(g) *Oxyria digyna, Cerastium uniflorum, Doronicum clusii, Geum reptans, Saxifraga bryoides, Saxifraga androsacea*	2.3	N, NE	34	34 (–)	–
	(h) *Ranunculus glacialis*	1.6	SW	19	20 (+1)	*Euphrasia minima*
Boulder field vegetation	(i) *Arabis alpina, Cerastium uniflorum, Cardamine resedifolia, Linaria alpina*	13.1	N, S	17	19 (+2)	*Luzula alpinopilosa, Oxyria digyna*
	(j) *Luzula alpinopilosa, Oreochloa disticha, Cerastium uniflorum, Doronicum clusii*	4.6	N, SW	35	38 (+3)	*Carex curvula, Erigeron uniflorus, Vaccinium uliginosum*
Snowbed vegetation	(k) *Veronica alpina, Cerastium cerastoides, Gnaphalium supinum*	1.6	E	12	20 (+8)	*Carex curvula, Festuca halleri, Ligusticum mutellina, Oreochloa disticha, Oxyria digyna, Saxifraga androsacea, Saxifraga bryoides, Saxifraga moschata*

[a] Now largely *Salix retusa* dominated mats.

zones. Four types of fragmented alpine mats (dominated by *Avenula versicolor, Carex curvula, Salix herbacea,* and *Salix retusa*), four types of open vegetation on scree (dominated by *Geum reptans, Oxyria digyna, Ranunculus glacialis* and *Saxifraga moschata*), two types of boulder field vegetation (dominated by *Arabis alpina* and *Luzula alpinopilosa*) and a snowbed community (Table 27.1) were distinguished.

27.3 Floristic Changes Between 1986–2000

The summit area of Mount Glungezer was re-surveyed in the summers of 1999 and 2000 after a 13-year period, which included the warmest 10 years on record in central Europe. On the neighbouring Mount Patscherkofel summit, mean air temperatures have increased since registration started in 1967 (Fig. 27.1). Distinct vegetation changes were observed on Mount Glungezer. Five species from the treeline ecotone (*Alchemilla alpina, Alchemilla vulgaris* agg., *Calluna vulgaris, Gentiana punctata,* and *Vaccinium uliginosum*) had established in the fragmented grassland on the southern slope. All species recorded already in 1986 were refound, bringing the total number of vascular plant species to 88. The species richness of Mount Glungezer is remarkably higher than the mean of 55 species for altitudes of 2500–2800 m reported by Grabherr et al. (1995) for the Austrian Alps and is amongst the highest recorded for a summit area of a comparable altitude. The high number of species on Mount Glungezer can be explained by the diversity of habitats (including shaded northerly slopes, a snowbed and south-facing slopes), the moderate levels of soil pH, and by the fact that parts of Mount Glungezer were not glaciated during the ice age (Bahn and Körner 1987). The high degree of fragmentation within the habitats, which precludes the dominance of few

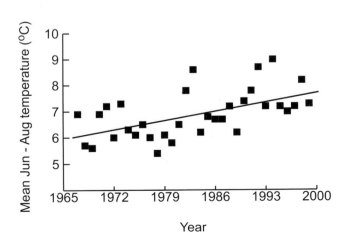

Fig. 27.1. Mean summer (June–August) air temperatures on Mount Patscherkofel (2246 m, 5 km distant from Mount Glungezer) between 1967 and 1999

competitive species, is also of importance (for general discussion of causes of alpine plant diversity, see Körner 1995). The fact that new species migrated from the treeline ecotone to Mount Glungezer is not really surprising, since, already in 1986, 16 typical treeline ecotone and alpine dwarf-shrub heath species were found in the summit area of Mount Glungezer (Bahn and Körner 1987). It is noteworthy that in 1999 most treeline ecotone dwarf-shrub species were found as established patches in the south exposed fragmented grass mats and were thus no longer confined to protected sites in the vicinity of large rocks, as was the case in 1986.

The area of the snowbed community expanded into formerly bare areas and had a distinct change in species composition, including species typical for shaded scree vegetation (e.g. *Oxyria digyna*) and alpine grass mats (e.g. *Carex curvula, Festuca halleri*) in 1999–2000. A filling of snowbeds with *Carex curvula* has been documented for many other sites in the Alps (Grabherr et al. 1995; Grabherr 1997). The actual area that remained covered by the snowfield throughout the vegetation period varied considerably from year to year, permitting plant growth on about twice the area in 1999 than in 2000.

The northerly *Saxifraga moschata* dominated scree of 1986 largely changed to fragmented *Salix retusa* mats, which resulted from an increase in abundance of the prostrate dwarf shrub and a concomitant stabilisation of the substrate.

During the re-recording, altogether 26 species were found in plant communities where they had not been present in 1986 (Table 27.1), *Carex curvula* and *Oxyria digyna* establishing best, together with the new species on site, *Vaccinium uliginosum*.

Ten of the 27 plots recorded in detail in 1986 could be relocated in 1999 and 2000, including all major vegetation units, except scree vegetation dominated by *Saxifraga moschata* or by *Oxyria digyna*. On the ten plots, a change in the abundance of 37 species (including the newly established ones) was recorded. The dynamics was most pronounced on the southern slope. Averaged over the re-surveyed plots, the most dynamic species included the prostrate dwarf shrub *Salix herbacea*, all viviparous species (*Festuca vivipara, Poa alpina* ssp. *vivipara, Polygonum viviparum*), three hemiparasitic species (*Bartsia alpina, Euphrasia minima, Pedicularis asplenifolia*), the graminoids *Carex curvula* and *Oreochloa disticha* and the herbs *Leucanthemopsis alpina* and *Oxyria digyna*. In most cases an increase in abundance was observed; in an alpine mat dominated by *Salix herbacea*, however, some species that were represented by very few individuals in 1986 had even disappeared from the plot.

27.4 Conclusions

From these results it can be concluded that:
1. on Mount Glungezer, in 13 years, distinct vegetation changes occurred involving (a) an up-slope migration of treeline ecotone species and (b) local changes of vegetation patterns;
2. these changes were most prominent on a south-facing slope and next to a retreating snowfield;
3. local increases in species occurrence and abundance were most pronounced for frequent graminoids, prostrate dwarf shrubs, and viviparous and hemiparasitic species.

Acknowledgements. We are grateful to Monika Weis (Zentralanstalt für Meteorologie und Geodynamik, Wetterdienststelle Innsbruck) for providing the climate data for Mount Patscherkofel, to Vincent Saxl for assisting with fieldwork, and to Laszlo Nagy and an anonymous referee for comments on the manuscript.

References

Bahn M, Körner C (1987) Vegetation und Phänologie der hochalpinen Gipfelflur des Glungezer in Tirol. Ber Naturwiss Med Ver Innsbruck 74:61–80
Braun-Blanquet J (1957) Ein Jahrhundert Florenwandel am Piz Linard (3414 m). Bull Jard Bot Brux 27:221–232
Gottfried M, Pauli H, Grabherr G (1994) Die Alpen im „Treibhaus": Nachweis für das erwärmungsbedingte Höhersteigen der alpinen und nivalen Vegetation. Jahrb Ver Schutz Bergwelt 59:13–27
Grabherr G (1997) The high-mountain ecosystems of the Alps. In: Wielgolaski FE (ed) Ecosystems of the world, vol 3. Polar and alpine tundra. Elsevier, Amsterdam, pp 97–121
Grabherr G, Gottfried M, Pauli H (1994) Climate effects on mountain plants. Nature 369:448
Grabherr G, Gottfried M, Gruber A, Pauli H (1995) Patterns and current changes in alpine plant diversity. In: Chapin FS III, Körner C (eds) Arctic and alpine biodiversity: patterns, causes and ecosystem consequences. Springer, Berlin Heidelberg New York, pp 167–181
Hofer HR (1992) Veränderungen in der Vegetation von 14 Gipfeln des Berninagebietes zwischen 1905–1985. Ber Geobot Inst ETH Stiftung Ruebel (Zürich) 58:39–54
Körner C (1995) Alpine plant diversity: a global survey and functional interpretations. In: Chapin FS III, Körner C (eds) Arctic and alpine biodiversity: patterns, causes and ecosystem consequences. Springer, Berlin Heidelberg New York, pp 45–62
Pauli H, Gottfried M, Grabherr G (1996) Effects of climate on mountain ecosystems – upward shifting of alpine plants. World Res Rev 8:382–390

28 The Piz Linard (3411 m), the Grisons, Switzerland – Europe's Oldest Mountain Vegetation Study Site

H. PAULI, M. GOTTFRIED and G. GRABHERR

28.1 Introduction

There are over 100 high summits in the Alps for which historical (>40 years) records on vascular plant species richness exist. The earliest summit record is from Piz Linard, a nival summit of 3411 m a.s.l. in SE Switzerland, from the year 1835 (Herr 1866). The majority of these summits exceed 3000 m, reaching the subnival belt above the closed alpine grassland, or the nival belt above the climatic snow line. A comparison of early (Braun-Blanquet 1957, 1955) and recent (Hofer 1992; Gottfried et al. 1994; Grabherr et al. 1994, 1995, 2001; Pauli et al. 1996) re-investigations provided evidence for an upward migration of vascular plants on many of these summits.

This paper focuses on vegetation changes on the summit of Piz Linard, determined by the longest observation series from high-mountain environments. We compared species richness among all seven records between 1835 and 1992 and species abundance between 1947 and 1992.

28.2 Method

The investigation area on Piz Linard refers to the uppermost 30 m elevation from the peak downwards. Complete lists of all vascular plant species occurring within this area were obtained from the literature for 1835, 1864, 1895, 1911, 1937, and 1947. In 1992, the authors visited the site and made an exhaustive species list. The qualitative species abundance scores of 1947 (i.e. common, intermediate abundance, rare, very rare) were used in 1992 for comparability (Table 28.1).

In 1947, a distribution map showing the locations of single individuals or groups of individuals of all plant species within the 30-m summit area was drawn by E. Campell (Braun-Blanquet 1957). According to Braun-Blanquet (1957), the map was made for future comparisons in good weather conditions allowing a detailed species sampling for several hours. Thus, we can consider these records as very reliable. Figure 28.1 shows a reproduction of the original map.

In 1992, the species mapping was repeated over 4.5 h, in good weather conditions on the summit. However, as it was not possible to determine each individual or group of individuals within a 1-day visit, as they were too numerous, we divided the summit area into sectors and scored species abundances in each sector (Fig. 28.1).

28.3 Results

28.3.1 Changes in Species Richness

In 1835, the sole species found was *Androsace alpina*. Since then, the number of species increased successively to 10 in 1937. Until 1992, no change in the species number was observed (Table 28.1).

Table 28.1. Changes in species richness within the uppermost 30 m elevation on Piz Linard (3411 m, SE Switzerland). *p* Species present; for the records from 1947 and 1992: *c* species common; *i* of intermediate abundance; *r* rare; *r!* very rare. The elevation of the uppermost occurrence is indicated. Data for 1835–1947 from Heer (1866), Braun (1913), and Braun-Blanquet (1957)

Species	1835	1864	1895	1911	1937	1947		1992	
Androsace alpina	p	p	p	p	p	c	3409	c	3407
Ranunculus glacialis		p	p	p	p	c	3409	c	3410
Saxifraga bryoides			p	p	p	c	3405	c	3408
Saxifraga oppositifolia			p	p	p	c	3407	c	3407
Poa laxa				p	p	c	3405	c	3409–3410
Draba fladnizensis				p	p	i	3405	i	3400
Gentiana bavarica				p	p	r	3387	r	3390–3395
Cerastium uniflorum				p	p	r	3397	r	3380–3382
Leucanthemopsis alpina		p			p	r	3402		
Saxifraga exarata					p	r!	3382		
Cardamine resedifolia								r!	3385–3395
Luzula spicata								r!	3380–3382

The Piz Linard – Europe's Oldest Mountain Vegetation Study Site 445

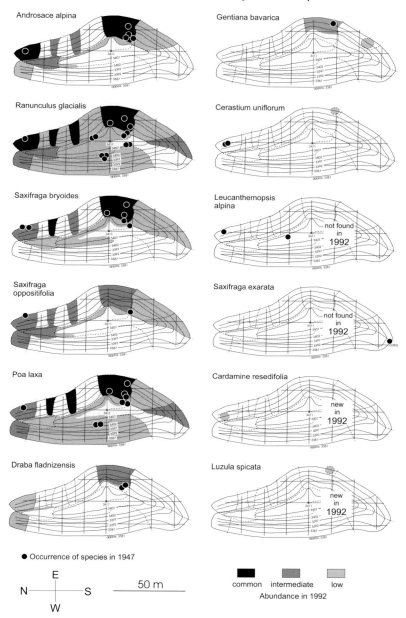

Fig. 28.1. The uppermost 30-m elevation of Piz Linard summit with the distribution of the vascular plant species which occurred in 1947 (after Braun-Blanquet 1957) and in 1992. In 1992, the summit area was divided into sectors, indicated by *broken lines*, according to aspect and geomorphologic structure. For each sector, species abundances are given as: common (*dark grey*) for common or scattered to common occurrences; intermediate (*medium grey*) for scattered occurrence; and low (*light grey*) for rare to scattered, locally scattered, rare, or very rare occurrences

28.3.2 Changes in Species Abundance

The comparison of species distribution in 1947 and 1992 (Fig. 28.1) showed that five species (*Androsace alpina, Ranunculus glacialis, Saxifraga bryoides, Saxifraga oppositifolia,* and *Poa laxa*) colonised formerly bare areas. Although these species were already well established in some spots in 1947 (cf. Table 28.1, Fig. 28.1), they expanded over large parts of the summit area until 1992. From those species found to be rare or of intermediate abundance in 1947, two (*Draba fladnizensis* and *Gentiana bavarica*) have possibly expanded their distribution slightly. *Cerastium uniflorum* was rare in 1992, like in 1947, but was found at another location. *Leucanthemopsis alpina* and *Saxifraga exarata,* two species which were rare in 1947, were not found in 1992 and two new species were recorded (*Cardamine resedifolia* and *Luzula spicata*), each in a single location.

28.4 Discussion

The increase in species richness on the summit of Piz Linard between 1835 and 1937 is clear evidence that high mountain plants migrate upwards. Braun-Blanquet (1957), based on evidence from the retreat of glaciers in SE Switzerland, with losses of 30 % of their surface area between 1895 and 1936–1944 (Zingg 1952), already suspected that a warmer climate was the main reason for the increase in the number of species at Piz Linard. This led him to propose mountain summits as indicator habitats of climatic change (Braun-Blanquet 1957).

Other causes that might have influenced the change of species richness on the summit of Piz Linard are largely negligible for the following reasons: Hill walkers and climbers may contribute to increasing species richness by incidentally dispersing seeds from lower elevations. On the other hand, species may be lost through perturbation caused by trampling. Mountain tourism, however, was not very common between 1835 and 1937 and the visit by Heer in 1835 is considered to be the first ascent to the summit. Even on the summit of Piz Julier, a popular tourist destination, species richness was found similar by comparing the 1900 and the 1992 records (Gottfried et al. 1994). It is possible that the area of potential habitats for plant establishment has been reduced by trampling impacts over the years.

Domestic mammals such as sheep or goats are unlikely to have grazed at that elevation. Seed dispersal by wild high mountain ungulates, such as *Capra ibex*, may be possible even at elevations over 3000 m. However, the ibex was near to extinction in the nineteenth century. Although the re-establishment of *Capra ibex* began in the first half of the twentieth century, substantial population increases occurred in the second half of the century (Giacometti 1991). Nutrient deposition from air pollution was likely to be of minor importance in high mountain areas, at least until 1937.

The constancy of species richness on Piz Linard after 1937 contrasted with other high summits of the Eastern Alps (e.g. Hohe Wilde, 3480 m; Stockkogel, 3109 m) where an increase in species richness was recorded between the 1950s and the 1990s (Gottfried et al. 1994; Grabherr et al. 1995, 2001; Pauli et al. 1996). Although no increase in species richness was observed on Piz Linard, five of the ten species recorded in 1937 and 1947 increased in abundance or grew in new locations in 1992. All these species, except *Androsace alpina*, reached the summit area after 1935 and were well established in 1947. After 1947, they continued to expand into bare habitats. The five species are common plants in the subnival belt (i.e. in the open plant assemblages above the alpine grassland belt) of the siliceous Central Alps (Pauli et al. 1999). This suggests that the summit environment of Piz Linard changed during the 167-year period of observation from harsh nival to subnival conditions, to which the five species are well adapted.

Therefore, the stagnation of species richness cannot be explained by a stagnation of the climate warming effect, but probably by a saturation of the subnival/nival species pool around 1940, followed by an increase of abundance. Potential invaders from the alpine species pool have not reached the summit to date. This may be owing to still too severe habitat conditions on the summit or, more likely, because of the absence of contacts between closed alpine grasslands and the summit area. On Mt. Hohe Wilde (3480 m), for example, where an increase in species richness from 10 to 19 was recorded between 1953 and 1992 (Grabherr et al. 1995), south-facing slopes down to the alpine grassland are more or less uninterrupted. Most of the new species on Mt. Hohe Wilde were plants of the alpine grassland belt.

28.5 Conclusions

1. Repeat observations on species richness on the summit of Piz Linard revealed the first evidence for mountain plants migrating upwards to higher altitudes (between 1835 and 1937). Evidence for climate warming in the Piz Linard area is available by the glacier retreats since the late nineteenth century. Therefore, climate warming is a likely cause for this upward establishment.
2. Seed dispersal by animals or humans, nutrient input by grazing mammals or air pollution can largely be excluded particularly with regard to species that reached the summit in the nineteenth and the early twentieth century.
3. The absence of further increase in species richness after 1937 may have been due to a saturation of the subnival/nival species pool. Species of the

subnival/nival species pool, however, expanded their populations, most likely supported by climate warming.
4. Potential invaders from the alpine species pool have not reached the summit so far. The upwards shift of the alpine plants is obviously dependent on appropriate migration corridors which are less present on Piz Linard compared with other high summits where alpine plants have already reached the summit during the last decades.

Acknowledgements. The study was financed by the Austrian Academy of Sciences as a contribution to the International Geosphere-Biosphere Programme.

References

Braun J (1913) Die Vegetationsverhältnisse der Schneestufe in den Rätisch-Lepontischen Alpen. Neue Denkschr Schweiz Naturforsch Ges 48:1-347
Braun-Blanquet J (1955) Die Vegetation des Piz Languard, ein Maßstab für Klimaänderungen. Svensk Bot Tidskrift 49/1-2:1-9
Braun-Blanquet J (1957) Ein Jahrhundert Florenwandel am Piz Linard (3414 m). Bull Jard Botan Bruxelles Vol Jubil W Robyns (Comm SIGMA 137):221-232
Giacometti M (1991) Beitrag zur Ansiedlungsdynamik und aktuellen Verbreitung des Alpensteinbockes (*Capra ibex* L.) im Alpenraum. Z Jagdwiss 37:157-173
Gottfried M, Pauli H, Grabherr G (1994) Die Alpen im „Treibhaus": Nachweise für das erwärmungsbedingte Höhersteigen der alpinen und nivalen Vegetation. Jahrb Ver Schutz Bergwelt München 59:13-27
Grabherr G, Gottfried M, Pauli H (1994) Climate effects on mountain plants. Nature 369:448
Grabherr G, Gottfried M, Gruber A, Pauli H (1995) Patterns and current changes in alpine plant diversity. In: Chapin FS, Körner C (eds) Arctic and alpine biodiversity: patterns, causes and ecosystem consequences. Springer, Berlin Heidelberg New York, pp 167-181
Grabherr G, Gottfried M, Pauli H (2001) Long-term monitoring of mountain peaks in the Alps. In: Burga C, Kratochwil A (eds) Biomonitoring. Tasks for vegetation science 35. Kluwer, Dordrecht, pp 153-177
Heer O (1866) Der Piz Linard. Jahrb Schweiz Alpin Club III Bern 457-471
Hofer HR (1992) Veränderungen in der Vegetation von 14 Gipfeln des Berninagebietes zwischen 1905 und 1985. Ber Geobot Inst ETH Zürich (Stiftung Rübel) 58:39-54
Pauli H, Gottfried M, Grabherr G (1996) Effects of climate change on mountain ecosystems – upward shifting of alpine plants. World Res Rev 8:382-390
Pauli H, Gottfried M, Grabherr G (1999) Vascular plant distribution patterns at the low-temperature limits of plant life – the alpine-nival ecotone of Mount Schrankogel (Tyrol, Austria). Phytocoenologia 29/3:297-325
Zingg T (1952) Gletscherbewegungen in den letzten 50 Jahren in Graubünden. Wasser- und Energiewirtschaft V-VII. Offiz. Organ des Schweizerischen Wasserwirtschafts- verbandes und des Schweizerischen Nationalkomitees für Große Talsperren, Zürich

VI Synthesis

29 Alpine Biodiversity in Space and Time: a Synthesis

L. Nagy, G. Grabherr, Ch. Körner and D.B.A. Thompson

29.1 Spatial Trends in Animal and Plant Species Richness

29.1.1 Latitudinal Trends

The area above the treeline in the European mountain ranges varies from ca. 1% (Corsica) to about 14% in the Caucasus (Table 29.1). The biota in the alpine zone across the different mountain systems is rather heterogeneous, as illustrated by a comparison of their alpine floras (Chap. 5). Mountain ranges such as the Alps and the Pyrenees, which are similar, contrast with ranges that have very little in common with the others (e.g. Sierra Nevada, the mountains of Crete). This heterogeneity is also evident when comparing treeline tree species in the different mountain systems (Table 29.2). An assessment of the climate in the alpine zone, based on soil temperatures at 10 cm below ground, indicated a mean growing season length of 155 (min. 105, max. 190) days year^{-1} across Europe (Chap. 2). Seasons tended to be slightly warmer in the south, particularly with regard to thermal sums (Chap. 2). However, no systematic differences in growing season length were observed across Europe, and absolute minima (mean –5 °C; range 0–15 °C) or maxima (mean +17 °C; range 12–19 °C) did not show any significant latitudinal trend, but depended on local snow cover and exposure.

Chapter 5 estimated the size of the total European alpine flora to be at about 2500 species and subspecies. This estimate is based on a 20% sample of the European alpine flora (species predominantly occurring above the treeline) and excludes the Caucasus. The final figure is likely to be higher (cf. Chap. 6) as work progresses in compiling the distribution atlas of the European flora. However, as it stands, this figure indicates that 20% of the European native vascular plants are confined to, or have their distribution centre, in the alpine zone. This figure highlights the importance of the alpine zone for species diversity. Clearly, this small (3%) land surface of Europe has a high importance for plants. About 10% of the species and subspecies that occur in the alpine zone are endemics confined to a single mountain or range. They,

Table 29.1. Summary table on plant species richness in the treeline ecotone and the alpine zone in the European high mountains

Range/ geographic region	Estimated proportion of area above tree line (%)	Taxonomic group	Species total for range/ geographic region	Species from tree line to highest peaks	Species total for alpine zone above the treeline ecotone	Species exclusive to the alpine and nival zones
Urals	?	Vascular plants	3000	525	350	165
		Bryophytes	–	–	–	–
		Lichens	–	–	–	–
Scandes	?	Vascular plants	1800 (Norway)	463	250	<200
		Bryophytes	1066 (Norway)	–	–	–
		Lichens	–	–	–	–
Scotland	2–4	Vascular plants	1117	ca. 220	–	58
		Bryophytes	928	–	–	–
		Lichens	1486	>700	ca. 275	–
Giant Mountains	10	Vascular plants	1226	467 (64 introduced)	–	–
		Bryophytes	–	–	–	–
		Lichens	–	–	–	–
Alps	7.5	Vascular plants	4530	ca. 2500	750–800	–
		Bryophytes	1100	–	–	–
		Lichens	2500	–	–	–
Italian Alps	40	Vascular plants	3264	1122	782	264
		Bryophytes	1032	439	–	–
		Lichens	–	–	–	–
SE Carpathians	11	Vascular plants	1715	672	130 (includes ssp.)	–
		Bryophytes	600	65	34	–
		Lichens	1875	218	–	–
Pyrenees	4	Vascular plants	3500 (includes ssp.)	803	506	22
		Bryophytes	530	150	–	–
		Lichens	1500	200	–	–
Caucasus	15	Vascular plants	6350	761 (Kazbegi)	447 (Kazbegi)	152 (Kazbegi)
		Bryophytes	900	–	–	–
		Lichens	1500	–	–	–
Apennines	5	Vascular plants	3091	728	495	190
		Bryophytes	497	150	–	–
		Lichens	–	–	–	–
Corsica	1	Vascular plants	2090	273 (>1700 m)	131 (includes ssp.)	11
		Bryophytes	ca. 600	–	–	–
		Lichens	ca. 1200	–	–	–
Greece	?	Vascular plants	5700 (native and fully naturalised)	1737 (>1800 m)	246 (>2400 m)	–

Data are largely from Chapters 3.1–3.10 and from various other sources.

Table 29.2. Species forming the timberline, treeline and closed arborescent vegetation in the European high mountains

Alps	*Larix decidua, Pinus cembra, Picea abies, Pinus mugo, Alnus viridis, Sorbus aucuparia*; in the south also *Fagus sylvatica* and *Abies alba*
Pyrenees	*Pinus uncinata*
Cantabrian Mts.	*Fagus sylvatica, Genista obtusiramea*
Sierra Nevada	*Quercus pubescens, Pinus sylvestris*
Apennines, Corsica, Sardinia	*Fagus sylvatica, Abies alba, Quercus pubescens, Pinus laricio, P. leucodermis, P. mugo, Alnus viridis* ssp. *suaveolens*
Balkan Peninsula	*Fagus sylvatica, F. moesiaca, Pinus heldreichii, P. peuce, P. mugo, P. sylvestris, Picea abies, Abies cephalonica, A. alba, Pinus nigra, Abies borisii-regis*
Carpathians	*Pinus cembra, P. mugo, Picea abies, Larix decidua*
Urals	*Picea abies* ssp. *obovata, Abies sibirica, Larix sibirica, Betula pubescens* spp. *czerepanowii*
Caucasus	*Acer trautvetteri, Betula litwinowii, B. medwedewii, Fagus orientalis, Picea orientalis, Pinus kochiana, Quercus macrantha*
Scandes	*Betula pubescens* spp. *czerepanowii*
Scotland	*Pinus sylvestris, Betula pubescens, Sorbus aucuparia*
Mediterranean islands	*Cupressus sempervirens, Acer sempervirens* (Crete), *Genista aethnensis* (Mt. Etna)

together with other narrow-range taxa, are most numerous in the Alps and Pyrenees, followed by the mountains of the Balkan Peninsula, Crete, Sierra Nevada, the Massif Central, Corsica and Central Apennines. Widely distributed species characterise the other ranges.

The percentage ratio of the flora exclusive to the alpine relative to the total for a single range (geographic area including low altitudes) ranges from 0.5 % (Corsica) to 16.6 % (Alps, Table 29.1). Figures such as these, however, do not take into account variability owing to the flora of northern Europe being much poorer than that of central and southern Europe. Taking all European mountains together, no clear relationship was found between vascular plant richness and the size (surface area) of mountains (Chap. 6). The correlation is poor because of low relative species richness in the extensive northern mountains in comparison with the relative high richness of the smaller mountain areas in the south. At the local (ecological) scale, calcareous habitats support more species than acidic ones (see also Chaps. 3.1–3.10), increasingly so from south to north. All habitats have less vascular plants species northwards whilst the number of cryptogams increases towards the north (Chap. 6), and there is no clear separation of the various exposure types (wind-blasted exposed ridges, intermediate snow-protected grasslands and snowbeds).

The vegetation of the Mediterranean high mountains south of ca. 41°N, i.e. the Pyrenees–Corsica–Central Apennines–Pindus and Olympus line, is rather dissimilar from that further north. Mediterranean high mountains have a flora and vegetation little influenced by glaciation. Although there is a period of snow lie, snowmelt is rather abrupt and little vegetation is associated with areas where seep water does not provide irrigation in the summer. High mountains along ca. 41°N have a dual character whereby the vegetation on the highly insolated S slopes is of a Mediterranean character whilst on the N slopes the flora and vegetation are more similar to those found on mountains in more northerly latitudes (Quézel 1967; Chaps. 3.6 and 3.9). Mid-latitude European high mountains, including the Alps, Pyrenees and the southern Carpathians, show high similarities in terms of flora (Chaps. 3.4 and 3.5) and vegetation (Chap. 3.4). The Highlands of Scotland, the Scandes and the Urals form another group of mountains bearing a tundra-like vegetation above the treeline with numerous circum-arctic species.

At the subregional scale in north-western Europe, glacial history was rather uniform, and patterns in species richness are related to macroclimate (Holten 1986, 1998) and ecological conditions (Birks 1996). At the local scale in the southern Scandes, an oceanic to drier inland macroclimate gradient has been shown to correlate with the distribution of certain groups of species, and overall species richness increases with decreasing oceanity (Chap. 7). Conversely, Brown et al. (1993) found that in Scotland vegetation diversity decreased from the western oceanic parts towards the less oceanic interior. The pattern reported in Chapter 7 might have been the result of the presence of species-rich calcareous mountains inland, rather than climate alone. Another example of climate effect at the local scale is that reported by Dirnböck et al. (2001), who postulated that the existence of some endemic plant communities at the relatively dry eastern edge of the Eastern Alps is related more to recent habitat conditions rather than to Pleistocene glacial history.

Information regarding cryptogams, especially on lichens, is incomplete. However, it appears that bryophytes and lichens tend to form a more important component of the vegetation with latitude (Chap. 6).

29.1.2 Altitude Trends

The species richness of vascular plants above the treeline shows a decreasing trend with altitude. Cumulative species numbers, i.e. the total flora above a certain altitude, decrease at a rate of 47 species 100 m^{-1} elevation ($y=1590-0.47x$; $P<0.001$; $R^2=0.61$; Fig. 29.1A). Examples are given for regions with a rather narrow alpine zone such as the Greek mountains (Chap. 3.10), the Pyrenees (Chap. 3.7), the Carpathians (Chap. 3.5) and for a single mountain (Mount Olympus, Strid and Tan 1992). This 47:100 ratio is based on local flora database records (e.g. Flora Hellenica database, Chap. 3.10), i.e. on the overall recorded presence of species in the region, rather than on studies comparing plot-based inventories. Nevertheless, it gives a graphic illustration of a decrease in species richness with altitude. This number is somewhat higher than the 30–45 suggested by Körner (2001), derived from a mix of studies using a variety of plot-based transects. The number of species recorded in ecological studies using sample plots in successive altitude belts appears to follow a decrease of 15 species 100 m^{-1} (y=484–0.15x; $P<0.001$; $R^2=0.78$; Fig. 29.1B). Estimates that have reported altitude patterns need careful interpretation. Generalisations from these studies appear to be somewhat vulnerable and misleading because the trend of a decreasing species richness with altitude depends on the scale at which the data were collected, both with regard to sample plot size, total extent and altitude range (Grabherr et al. 1995). The trend presented here may therefore only apply to studies which used 100-m altitude belts for sampling. From a wider sample including the Himalayas, it appears that the decline in species diversity is non-linear with an asymptotic tail at high elevations (Körner and Spehn 2002). The decline in species number reflects the reduction with elevation of the available land area (Körner 2000) and, accordingly, the species to area ratio changes little with altitude.

The relative importance of bryophytes and lichens increases with altitude, both in terms of number of taxa and biomass (Ellenberg 1996; Fryday 1997; Grabherr 1997; Körner 1999; Chap. 3.4). For example, Grabherr (1989) reported that in a *Carex curvula* grassland in the Alps the biomass of flowering plants was 61 g^{-2} (+346 g^{-2} litter) against 31 g^{-2} moss, and 283 g^{-2} lichen biomass.

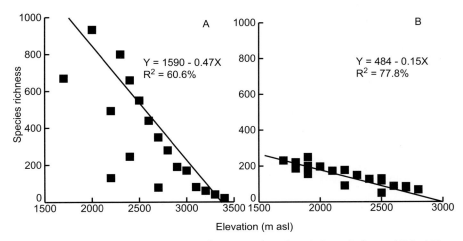

Fig. 29.1A, B. Vascular plant species richness vs. elevation. A Cumulative and B in 100-m altitude bands

Species richness of cryptogams seems to decrease with elevation in a similar fashion to that of vascular plants (Geissler and Velluti 1996), although locally may follow a different pattern (Chap. 8). Cryptogams often characterise the vegetation, e.g. bryophytes in the nival zone (Chap. 3.4) or in late snowbeds in the upper alpine zone, and may be the only evident organisms seen, e.g. open lichen communities on rocks (Ellenberg 1996; Fryday 1997; Chap. 3.5). Pitschmann and Reisigl (1954) reported 104 macrolichens from above 3000 m in the Ötztaler Alps and Ozenda (1985) suggested that the total number of lichen species including microlichens would be in excess of 200. The corresponding figures for vascular plants were in the range of 100–150 across a number of sites (Ozenda 1985) and 75 for bryophytes from Ötztal (Pitschmann and Reisigl 1954) and 30 for Belalp, Valais at 2700 m (Chap. 8).

29.1.3 Animals

Both invertebrates and vertebrates can be keystone species in many alpine regions. Some are vertebrate grazers, many are predators and several invertebrates are keystone prey. A series of single species studies of rodents and chamois underline the importance of snow as the most critical factor for determining the abundance of these animals. Detailed assessments for certain invertebrate groups reveal declining species richness with increasing elevation and latitude, but a number of peculiarities are noted. For instance, in butterflies (Lepidoptera), the overall species richness (across all altitudes) decreases from the Alps to the southern end of Italy, whereas, in the alpine communities, only the species richness increases from north to south. An additional regional phenomenon is noted in carabid beetles (Carabidae): the richness declines from the outer to the inner chains of the Alps. In the invertebrates as a whole, species richness decreases with latitude, reflecting direct or indirect effects of temperature (also Chernov 1995).

Both in the Alps as well as in the Urals, the endemism of beetles (Coleoptera) in some cases appears tightly linked to endemic plant species, the latter often found in scree habitats. Of the large invertebrate groups, such as the Opiliones and Arachnida, the alpine fauna represents 10–15% of the overall regional complement of the species of these taxa.

We do not know what fraction of species is generalist or specialist. Such a distinction is necessary with regard to food and habitat requirements. Butterflies, and possibly also spiders and carabids of alpine elevations, are predominantly food generalists, with only 10% having narrow (exclusive) habitat and food source requirements.

The ratio of alpine zonal to azonal species varies greatly between animal taxa. Generally, it is amongst the arachnids and beetles that we find the greatest degree of specialism, and therefore higher ratios of zonal to azonal species. Some birds clearly recognise the alpine boundary; mountain birds such as purple sandpiper *Calidris maritima* Brunnich, dotterel *Charadrius morinellus*, snow bunting *Plectrophenax nivalis* L. and ptarmigan *Lagopus mutus* rarely nest below the treeline, and Chapter 19 shows that these species have marked habitat preferences. It is possible that the invertebrate food supplies of these areas account for the high abundance of these birds in alpine regions; however, it does not explain their restriction to the zone. Area effects are also important – in the larger alpine ranges there is a disproportionately greater species richness of birds.

29.2 Temporal Changes

29.2.1 Environment

The main underlying causes of changes in species richness and abundance are both land use- and climate-related. Climatic trends such as the 'little ice age' and the warming since, especially the accelerated warming in some parts of Europe in recent decades, have been receiving increasing attention (Chaps. 24–28). Climate change or variability in climate may change habitats directly, e.g. through glaciation, deglaciation, or pH (e.g. Koinig et al. 1998, 2002). However, it is more likely to influence interspecific interactions and soil processes. Competition is important when changes in environmental conditions can permanently shift an existing dynamic balance among species.

The extent of arctic and alpine vegetation in north-western Europe has been decreasing over the last 17,000 years (Birks 1986). The vegetation itself has been characterised by high species richness, low vegetation stability, and rapid rates of change compared with reconstructions of earlier vegetation in north-western Europe and the Alps (Ammann 1995). Amman (1995), however, suggested that in the Alps for the past ca. 8000 years there has been a high stability above the treeline (which itself fluctuated within an altitudinal range of 150–200 m). Contemporary changes in some habitats in the alpine zone across Europe have been similar to those reported for the late Quaternary. Other vegetation types such as *Carex curvula* dominated alpine sedge grasslands in the Alps appear to have remained stable over the medium (ca. 25–50 years, Chap. 24) and the long term (at least 1000 years, Steinger et al. 1996).

The dynamic nature of the treeline ecotone has been demonstrated for the Alps (Chap. 24; Burga 1988). Treeline-forming trees in the Urals appear to have responded to higher temperatures by increasing their recruitment and growth (Chap. 26) within the formerly open ecotone and this, in some instances, has led to an advance of the altitudinal forest line in the Urals. However, vegetation dynamics at the treeline–alpine vegetation interface are not a simple function of temperature. Recruitment, in a species-specific manner, controls this interface. Perturbations such as fire, trampling by grazers or by small and medium-sized burrowing animals are important for creating available sites (Chap. 24).

Species richness has increased (Grabherr et al. 1994) and/or established populations of plants have expanded (Chaps. 27–28) on many nival summits in the Alps during the twentieth century. The altitudinal advance of plant species is related to dispersal traits, agents and geomorphology. In the absence of fruit production, some clonal and patch-forming plants such as *Carex curvula* and *Juncus trifidus* or cushion plants colonise frontally in small advances, whilst dispersal by seeds (fruits) results in a diffuse advance. Seed characters (e.g. size, the presence of appendices) determine dispersal by seed rain and influence further transportability by, e.g. wind or water. Because of gravitational processes, direct uphill dispersal by seed rain is negligible for most species. However, extreme weather events (e.g. storms) or zoochory may redistribute species well above their normal range (Chap. 24). This secondary redistribution of propagules is more effective over unbroken surfaces such as short grassland or rock as opposed to, for example, boulder fields and screes and has contributed to differences in changes in species richness in the subnival and nival zones.

Climate change may change wind patterns and the frequency of the occurrence of extreme events (e.g. Harrison 1997). This leads to changes in the uphill redistribution of propagules, which may establish in favourable years. The response to variation in growing season length (in terms of flowering, seed maturity and dispersal) has been found to vary with species and year (Alatalo and Totland 1997; Wagner and Mitterhofer 1998; Totland 1999; Totland and Eide 1999). In the subnival and nival zones, warm episodes may enable or enhance the sexual reproduction of some established plants of some species and thereby the expansion of their populations (Chap. 28).

29.2.2 Grazers

The impact of grazers on plant species richness and vegetation patterns can be quantified either by the exclusion or inclusion of grazers or by on-site observations which compare different time periods. The experimental exclusion of grazers (or, less commonly, their artificial introduction to enclosures) is a type of perturbation, and it may have different impacts in the short term from that in the long term (e.g. Virtanen 2000). The observed changes are a function of both the initial state of the ecosystem (biotic and abiotic) and the grazers and grazing pressures applied or released (Chap. 23). Time series observations are most often used at the landscape scale (e.g. McVean and Ratcliffe 1962; Thompson et al. 1995) and sometimes involve weak inferences. Changes which transform a community are especially apparent if the extent of losses, in terms of taxonomic and functional groups are unbalanced, e.g. dwarf-shrub heaths are converted to grassland by sustained heavy grazing pressure (Thompson et al. 1995).

Species richness under light grazing appears to be higher than when grazers are excluded (Körner 2000, 2001); heavy grazing tends to reduce species richness (Chap. 23). In addition to species richness, plant biomass and dead material in the long term are affected by grazing. Under incessant grazing pressure, species which require gaps for regeneration usually thrive (e.g. the therophyte *Gentiana nivalis* (Miller et al. 1999) and, in general, graminoids increase (Olofsson et al. 2001). Higher reproductive output after release from flower or seed predation may not compensate for the absence of competitive ability.

Prolonged heavy, long-term grazing, and associated with this the nutrient enrichment from faeces and urine, have influenced the species composition of plant communities, e.g. in the Caucasus (Chap. 3.8), Alps (Ellenberg 1996) and in the Scottish Highlands (Thompson and Brown 1992). At the landscape scale, grazers, in addition to forest and scrub clearing by man, have had a prominent impact on the treeline ecotone. Deforestation for grazing land has depressed treelines and expanded downwards the range of alpine vegetation. Locally, treeline tree species may take advantage of trampling mediated soil scarification – both of which enhance tree regeneration. However, browsing may control this, thus suppressing the growth of tree seedlings and saplings. Today, changes in the stocking densities of domestic grazers in the Alps and the Pyrenees (Chap. 21) may induce marked impacts on vegetation patterns at the landscape scale. Abandonment of high elevation pastures permits forest recovery whilst local overgrazing causes graminoids to spread at the expense of dwarf shrubs and bryophytes, notably on the richer soils.

Nevertheless, in the naturally graminoid-dominated alpine zone, vegetation changes caused by grazing are less prominent at the landscape scale. Locally, the largest changes are associated with sheltering areas, wallows, burrows (e.g. Chap. 20) and around feeding areas where species richness is usually much reduced and species composition and vegetation structure can be quite different from the original. In the Scandes, snowbed vegetation is much impacted by lemming grazing in years of outbreak (e.g. Virtanen 2000). Selective grazing in the alpine-nival zones may therefore have heavy impacts at the micro-scale by restricting the reproduction and dispersal of some species (e.g. Galen 1990; Nagy and Proctor 1996).

29.3 Outlook

ALPNET has provided the first continent-wide assessment of biological richness of high elevation biota. Many hypotheses and generalisations remain unproven and require further continent-wide comparisons. One of the virtues of ALPNET has been to stimulate cross-country and regional comparisons. The data on temperature profiles for different alpine regions provide a valuable snapshot of climate conditions across alpine regions: a common platform for predicting changes in growing season and gradations in vegetation types under various climate change scenarios.

Clearly, we need to do more to understand why some species are alpine specialists, whilst others are azonal, and of course many avoid alpine areas. The contrast between birds and mammals is interesting: many more birds are alpine specialists. We would like to see more work on this facet, looking at the animal–habitat–food–relationships, as well as at the different influences of temperature, precipitation and wind.

We have tried to draw out the anthropogenic influences on alpine areas – directly through grazing pressures by livestock, and less directly through global warming and acidic deposition. Interestingly, there is a dearth of historical (in the Holocene sense) data on alpine landscapes: too few alpine lakes provide palaeoecological data on vegetation changes. Looking forward, we still have some way to go in defining the elements of human influence that benefit alpine biodiversity. Undoubtedly, some herbivore grazing benefits many of the invertebrates.

Finally, we should touch on the end-users. The work of ALPNET shows that it is possible to have genuine cross-country collaboration to develop our knowledge of alpine ecosystems. As far as the implementation of the Habitats Directive (92/43/EEC) is concerned, we now have a much stronger basis for contexting the changing nature of alpine habitats. Each of the member states of the EC is monitoring their 'Special Areas for Conservation' to determine whether or not they remain in favourable condition. ALPNET provides an important context for this, not only in that it provides sources of baseline studies, but also indications of the scale and magnitude of likely changes in vegetation and animals in different parts of alpine Europe.

Above all else, however, ALPNET, and this volume in particular, crystallises the importance of alpine areas. These are special, fascinating and important parts of our planet. That we are able to describe so much about the variety, abundance, richness, populations and trends of so many species groups, across alpine regions and from the treeline upwards, is a massive tribute to the work of so many ecologists and naturalists. We need to build on this huge, collaborative venture to promote the importance of Europe's alpine areas, and to develop the collaborative and comparative approach needed now to develop a more holistic approach to understanding, conserving and managing these areas. It is hoped that this book will help maintain the momentum of the International Year of Mountains, 2002; this is our contribution to helping secure a healthier future for alpine ecosystems in Europe. The ALPNET initiative has already been extended worldwide by the establishment of a global mountain biodiversity assessment (GMBA) under the auspices of DIVERSITAS (Körner and Spehn 2002).

Acknowledgements. We thank R.M.M. Crawford and R. Virtanen for their comments.

References

Alatalo JM, Totland O (1997) Response to simulated climatic change in an alpine and subarctic pollen-risk strategist, *Silene acaulis*. Global Change Biol 3:74–79

Amman B (1995) Paleorecords of plant biodiversity in the Alps. In: Chapin FS III, Körner C (eds) Arctic and alpine biodiversity. Springer, Berlin Heidelberg New York, pp 137–149

Birks HJB (1986) Late-Quaternary biotic changes in terrestrial and lacustrine environments, with particular reference to north-west Europe. In: Bergland BE (ed) Handbook of Holocene palaeoecology and palaeohydrology. Wiley, New York, pp 3–65

Birks HJB (1996) Statistical approaches to interpreting diversity patterns in the Norwegian mountain flora. Ecography 19:332–340

Brown A, Horsfield D, Thompson DBA (1993) A new biogeographical classification of the Scottish Uplands. I. Descriptions of vegetation blocks and their spatial variation. J Ecol 81:207–229

Burga CA (1988) Swiss vegetation history during the last 18,000 years. New Phytol 110:581–602

Chernov II (1995) Diversity of the Arctic terrestrial fauna. In: Chapin FS III, Koerner C (eds) Arctic and alpine biodiversity. Springer, Berlin Heidelberg New York, pp 81–96

Dirnböck T, Dullinger S, Grabherr G (2001) A new grassland community in the Eastern Alps: evidence on the environmental distribution limits of endemic plant communities. Phytocoenologia 31:521–536

Ellenberg H (1996) Vegetation Mitteleuropas mit den Alpen, 6th edn. Ulmer, Stuttgart

Fryday A (1997) Montane lichens in Scotland. Bot J Scotl 49:367–374

Galen C (1990) Limits to the distributions of alpine tundra plants: herbivores and the alpine skypilot, *Polemonium viscosum*. Oikos 59:355–358

Geissler P, Velluti C (1996) L'écocline subalpin-alpin: approche par les bryophytes. Bull Murithienne 114:171–177

Grabherr G (1989) On community structure in high alpine grasslands. Vegetatio 83:223–227

Grabherr G (1997) The high mountain ecosystems of the Alps. In: Wielgolaski FE (ed) Ecosytems of the world, vol 3. Polar and alpine tundra. Elsevier, Amsterdam, pp 97–121

Grabherr G, Gottfried M, Pauli H (1994) Climate effects on mountain plants. Nature 369:448–44

Grabherr G, Gottfried M, Gruber A, Pauli H (1995) Patterns and current changes in alpine plant diversity. In: Chapin FS III, Körner C (eds) Arctic and alpine biodiversity. Springer, Berlin Heidelberg New York, pp 167–181

Harrison SJ (1997) Changes in the Scottish climate. Bot J Scotl 49:287–300

Holten JI (1986) Autecological and phytogeographical investigations along a coast–inland transect in central Norway. PhD Thesis. University of Trondheim, Trondheim

Holten JI (1998) Vertical distribution patterns of vascular plants in the Fennoscandian mountain range. Ecologie 29:129–138

Koinig KA, Schmidt R, SommarugaWograth S, Tessadri R, Psenner R (1998) Climate change as the primary cause for pH shifts in a high alpine lake. Water Air Soil Pollut 104:167–180

Koinig KA, Kamenik C, Schmidt R, Agustí-Panareda A, Appleby P, Lami A, Prazakova M, Rose N, Schnell ØA, Tessadri R, Thompson R, Psenner R (2002) Environmental changes in an alpine lake (Gossenköllesee, Austria) over the last two centuries – the influence of air temperature on biological parameters J Paleolimnol 28:147–160

Körner C (1995) Alpine plant diversity: a global survey and functional interpretations In: Chapin FS III, Körner C (eds) Arctic and alpine biodiversity. Springer, Berlin Heidelberg New York, pp 45–62

Körner C (1999) Alpine plant life. Springer, Berlin Heidelberg New York

Körner C (2000) The alpine life zone under global change. Gayana Bot 57:1–17

Körner C (2000) Why are there global gradients in species richness? Mountains may hold the answer. TREE 15:513

Körner C (2001) Alpine ecosystems In: Levin SA (ed) Encyclopaedia of biodivesity, vol 1. Academic Press, San Diego, pp 133–144

Körner C, Spehn EM (2002) Mountain biodiversity. A global assessment. Parthenon, London

McVean DN, Ratcliffe DA (1962) Plant communities of the Scottish Highlands. Nature Conservancy Monograph No. 1. The Stationery Office, Edinburgh

Miller GR, Geddes C, Mardon DK (1999) Response of the alpine gentian *Gentiana nivalis* L. to protection from grazing by sheep. Biol Conserv 87:311–318

Nagy L, Proctor J (1996) The demography of *Lychnis alpina* L. at the Meikle Kilrannoch Ultramafic outcrops, Angus, Scotland. Bot J Scotl 48:155–166

Olofsson J, Kitti H, Rautiainen P, Stark S, Oksanen L (2001) Effects of summer grazing by reindeer on composition of vegetation, productivity and nitrogen cycling. Ecography 24:13–24

Ozenda P (1985) La végétation de la chaîne alpine dans l'espace montagnard européen. Masson, Paris

Pitschmann H, Reisigl H (1954) Zur nivalen Mossflora der Ötztaler Alpen (Tirol). Rev Bryol 23:123–131

Quezel P (1967) La végétation des hauts sommets du Pinde et de L'Olympe de Thessalie. Vegetatio 14:127–228

Steinger T, Körner C, Schmid B (1996) Long-term persistence in a changing climate: DNA analysis suggests very old ages of clones of alpine *Carex curvula*. Oecologia 105:94–99

Strid A, Tan K (1992) Flora Hellenica and the threatened plants of Greece. Opera Bot 113:55–67

Thompson DBA, Brown A (1992) Biodiversity in montane Britain: habitat variation, vegetation diversity and some objectives for conservation. Biodiv Conserv 1:179–209

Thompson DBA, Macdonald AJ, Marsden JH, Galbraith CA (1995) Upland heather moorland in Great-Britain – a review of international importance, vegetation change and some objectives for nature conservation. Biol Conserv 71:163–178

Totland O (1999) Effects of temperature on performance and phenotypic selection on plant traits in alpine *Ranunculus acris*. Oecologia 120:242–251

Totland O, Eide W (1999) Environmentally-dependent pollen limitation on seed production in alpine *Ranunculus acris*. Ecoscience 6:173–179

Virtanen R (2000) Effects of grazing on above-ground biomass on a mountain snowbed, NW Finland. Oikos 90:295–300

Wagner J, Mitterhofer E (1998) Phenology, seed development, and reproductive success of an alpine population of *Gentianella germanica* in climatically varying years. Bot Acta 111:159–166

Subject Index

A

Abies alba 80, 108, 310, 455
Abisko 223
Acantholimon echinus 119
accuracy 14, 130, 210, 211, 215
– assessment 214, 215
– of vegetation maps 217
– producer's 216
– user's 216
Acerbia 247, 267
acid grassland 62, 70, 118, 119, 211, 298, 302
acidic deposition 327, 335
acidification 335
acidophilous 69-70, 81, 83, 163
activity density 287-288
Adenostyles alliaria 62, 120
– *briquetii* 109
Agelenidae 284-286, 291-292
Agrostis alpina 211, 213, 215
Agrotis 248
air pollution 51, 447, 448
airborne pollution 43
Alnus viridis 4, 10, 62, 66, 70, 101, 107-109, 118, 120, 385, 425, 432, 455
Alopecurus gerardii 119, 152
alpha-diversity 83, 222, 227
alpine 4
– bio-temperatures 26
– bioclimate 13
– carabid assemblages 313
– definition of 4
– fauna 244, 249, 309, 345, 368
– faunal type 242-243, 247
– flora 56, 58, 59-61, 87, 145, 453
– – size of, European 453
– grassland 16, 23, 27, 54, 55, 60, 68, 81, 108, 211-212, 322, 377, 382, 386, 401-405
– habitats 126, 309, 335, 369, 370
– ibex 354
– marmot 322, 339, 345
– mats 100, 438
– meadows 97-100
– mires 35, 413, 420
– permafrost 182
– rodents 338-346
– specialist species 458
– ungulates 351-361
– vegetation 9, 27, 34-35, 42, 53, 55, 62, 69, 81-82, 90-91, 98-99, 109, 119-120, 150-155, 322, 401, 459
– – recovery after perturbation 403
– vegetation dynamics 399-407, 460
– vertebrates 321, 367
– zone 4, 9-10, 13, 453
ALPNET 461-462
Alps 6, 15, 18, 21, 22, 24, 53–57, 61, 62, 74, 101, 102, 140, 150, 196–203, 205, 206, 235, 240, 243–247, 249, 281, 282, 287, 303, 307–311, 314, 339–341, 354, 356, 358, 360, 369, 370, 373, 374, 382, 384, 385, 400, 401, 437, 440, 453–457, 460, 461
– Austrian 241, 284, 292, 386, 439
– calcareous 212, 243, 281, 283, 286, 289
– – Austria 213
– eastern 283, 285
– central 404
– French 352, 353, 360, 361
– Italian 75-81, 234, 297-299, 301, 386, 454
– Ötztal 15, 18, 21, 24, 59, 150, 283, 288, 290
– Swiss 14, 16, 27, 53, 56, 59, 60, 352
altitude 129
– belts 4, 55, 57, 457
– distribution 368
– effect 259
– gradient 173, 180, 189, 198, 371
– range 4, 55, 173, 175, 176–178, 283, 401, 457

– trends   456
– zones   4, 33, 80, 96, 108, 188, 235
altitudinal advance of plant
    species   399, 460
altitudinal distribution   55, 57, 109, 201
altitudinal gradient   270, 271, 371
Amara   308, 310
ambush predators   284
amphibians   368-370
Androsace alpina   62, 444-447
– *vandellii*   90
anemo-orographic systems   50
Anemonastrum biarmiensis   272
animals   458
– species richness   453
– species diversity   262
annual locomotor activity   290
anthropogenic grassland   80, 81, 82, 378
Anthus pratensis   130, 223, 227
Antitype   248
Apennines   7, 15, 18, 21, 73-82, 150, 297,
    298, 303, 307, 312, 313, 454, 455
Apodemus alpicola   341
Aporia crataegi   269
Aporophila   248
Aquila chrysaetos   346
arachnid parasites   291
Arachnida   262
arachnids   235, 281
– spatial distribution   289
Araneae   281, 284, 285
Araneidae   285
arboreal zonobiomes   247
arctic   8
– -alpine   181, 243, 285, 322, 335
– -alpine flora   33, 40, 76, 145
– -alpine specialists   334
– -alpine species   49, 249, 266, 267, 334
Arctiidae   264
Arctostaphylos alpina   35, 91, 155
Argynnis aglaja   269
– *paphia*   269
Arvicola terrestris   341
Atlas Florae Europaeae   133
Austria   24, 195, 196, 199, 212, 235, 281,
    282, 345, 382, 437
avalanche   47, 49, 342
Aves   262

B

Baetic Cordillera   7, 200, 201
Balkan Peninsula   113, 119, 240, 244,
    246, 249, 455
bank vole   341
Belalp   186
Bellardiochloa variegata   80, 109, 119
Bembidion   308, 312
Berberis aetnensis   109
beta diversity   222
Betula nana   175
biogeographical connections   246
bioindicators   272
biomass   191, 270, 271, 291, 323, 382
– removal   43, 387
bird assemblages   327, 330
bird territory   130, 227
– mapping   329
Boloria   267
– *graeca*   244
Bolshoi Iremel   428
Bolsoi Taganai   424
Borderea   87
– *pyrenaica*   90
boreal   8
– species, increase   269
– -alpine   176
boreo-montane species   266, 267
Brassicaceae   59, 134, 135
breeding bird assemblages   333
brown bear   261, 359
Bruckenthalia spiculifolia   67, 120
bryophyte   33, 41, 42, 68, 77, 87, 158, 165,
    187, 189-190, 383, 454
– importance   457
– species richness   187
Buprestis   312
butterflies   243, 267
– assemblage   297, 298, 300
– diversity   297
– species richness   298, 302, 303
– – altitude   300
– – latitude   300

C

calcareous   39, 58, 88, 90, 91, 119, 152,
    154, 159, 163, 178, 311, 312
– alpine grasslands   302, 310
– scree vegetation   210
– soil substrata   35, 163
– secondary grasslands   119

Subject Index

calcicolous 162
– endemics 88
calcifuge 163
Calluna vulgaris 42, 155, 414, 439
Calophasia 248
Campanula pulla 202, 205
Canis lupus 359
canonical correspondence analysis 213
canopy structure 384
Cantabrian Mountains 244, 246, 455
Capra ibex 354, 378, 447
– *pyrenaica* 354, 368
– *rupicapra* 378
Capreolus capreolus 378
carabid beetle 235, 307
– altitude distribution 234
– assemblages 308, 310, 312
– classification 265
– stenotopic species 308
Carabidae 234, 243, 263
Carabus 309, 310
– *karpinskii* 266
Cardamine resedifolia 445
Carex atrofusca 175
– *bigelowii* 42, 44, 154
– *curvula* 16, 55, 62, 90, 119, 150, 152, 403, 404, 457
– *ferruginea* 62, 119, 210
– *firma* 62, 150, 152, 210, 213, 215, 310
– *microglochin* 413
– *rupestris* 154
– *sempervirens* 150, 210, 213, 215
Carpathians 6, 101, 152, 240, 243, 244, 246, 247, 265, 340, 454, 455, 456
– South-Eastern 65, 66, 67, 68, 69, 71
Cassiope hypnoides 35
– *tetragona* 154
cattle 340, 353, 358, 378
Caucasus 7, 93, 102, 244, 454, 455
centres of diversification-speciation 247, 249
Cerastium kazbek 98
– *soleirolii* 109
– *uniflorum* 445
Cercyonops caraganae 266
Cervus elaphus 43, 378
Cetraria islandica 154
chamaephytes 34, 40, 48, 59, 68, 79, 88, 89, 96, 108, 188
chamois 353, 354, 378
change in plant species abundance 414, 416, 418, 446

change in species composition 411, 413, 440
change in species richness 438, 444, 446
changing environment 34, 191, 407
Charadrii 331
Charissa 247, 249
Chenopodium bonus-henricus 91
Chilopoda 262
Chionomys nivalis 322, 339, 368
chorological spectrum 88, 200, 202
Chrysolina 265
– *hyperboreica* 266
– *septentrionalis* 272
Chrysomelidae 263
Cicerbita alpina 35
circumpolar species 270
Clethrionomys glareolus 261, 341
– *rufocanus* 261
climate 40, 66, 74, 75, 85, 94, 105, 114
– change 32, 43, 71, 83, 95, 195, 205, 237, 245, 268, 272, 307–309, 313, 321, 322, 345, 399, 401, 423, 459, 460
– – related dynamics 402
– data 24
– warming 35, 76, 204, 399, 401, 419, 437, 447
Clubionidae 285
coast–inland transect 174, 175, 179, 180
Cochlearia pyrenaica 90
Coenonympha 267
Coleoptera 262, 263
colonisation 162, 236, 243, 292, 386
– by trees 50
– by invading species 204, 344
– faunal 236, 243, 268, 356, 360,
– success 403
colour polymorphism 272
common stoat 261
common vole 322, 339
community 360
competition 335
– interspecific 355, 359
– pressure 204, 205, 321
competitive exclusion 165, 190, 191
conservation 82, 128, 146, 321, 462
Copiphana 248
Coronella austriaca 368
Corsica 7, 15, 18, 20, 21, 105, 152, 454, 455
Corvus corax 227
Cotoneaster integerrima 91
crane flies 267

Crepis multicaulis 381
cryophilic 202
– endemic species 204
cryoro-Mediterranean 10, 89, 108, 109, 196, 199, 200, 202, 367, 368
cryoturbation 9, 182, 406
cumulative species richness with elevation 190
Curculionidae 243, 263
cutworm-type larvae 248

D

dead plant material 382, 384
deforestation 461
demography 342
– rodents 342-343
– ungulates 356
density dependence 357
Deschampsia cespitosa 211
– *flexuosa* 416, 417
Dianthus alpinus 202
Diapensia lapponica 127
Dichagyris 248
Dicranopalpus gasteinensis 283
Dictynidae 285
Dinaric Alps 7, 113
Dinarids 9, 119-120, 152
– Outer and Inner 113
Diphasiastrum alpinum 144
dipper 329, 330
Diptera 262
disjunct distribution 268, 341
disjunct subspecies 243, 244, 249
disjunction of range 284, 292
distribution of plant species 53, 213, 222,
– vascular, Piz Linard 445
disturbance
– human 10, 361
– natural physical 47, 50, 203,
DIVERSITAS 462
diversity 70, 235, 288
– index 70, 370
– of habitats 439
– patterns 125, 235
– – of carabids 307
– scale 125, 128
– temporal 127
Dolomites 15, 18, 283
domestic and wild grazers 43
Donus opanassenkoi 266

dotterel 327, 330
Dovrefjell 15, 24, 174
Draba fladnizensis 445
– *loiseleurii* 109
– *sauteri* 202, 205
– spp. 134, 137
– *stellata* 202
drought 105, 163, 204
Dryas caucasica 98
– *octopetala* 35, 150, 152, 154
dunlin 329, 330, 332
dwarf-shrub heath 8, 35, 42, 49, 54, 62, 69, 81, 82, 91, 98, 120, 210, 332, 334, 401, 425

E

East Balkan Uplands 113
ecological classification 8
ecosystem
– alpine 34, 217, 272, 345, 351, 386, 462
– diversity pattern 222
– dynamics 346
edaphic endemism 117
Edraianthus graminifolius 119
Edraianthus sp. 152
effects of climate 357
effects of grazing 387
– exclusion 383
Elbrus 96
elevation gradient 187, 191, 205, 372, 374
– Alps 185
elevation range 97, 98, 185
elevation zones 4, 313
Elophos 247, 249
Empetrum nigrum ssp. *hermaphroditum* 35, 42, 54, 62, 98, 120, 150-155, 414
endemic 81, 95, 200, 285
– arachnids 286
– carabid assemblage 311
– chrysomelids 261
– cushion plants 10
– faunulas 311
– genera 58, 87, 95
– Lepidoptera 244
– Microlepidoptera 239
– microthermic species 65
– plant communities 456
– plants 88, 195, 370, 372

Subject Index

- species   50, 58, 67, 68, 76, 96, 100–102, 107, 109, 142, 198, 199, 202, 203, 205, 234, 245, 249, 266, 267, 286, 312, 313, 367
- spider   284, 286
- taxa   36, 201, 235, 372
- vascular plants   61
- vertebrate   372
endemism   58, 116
- areas of   242, 248
- of beetles   458
Entephria   249
environment   459
environmental change   217, 272, 411, 421
environmental conditions   411, 459
environmental constraints   373
environmental factors   222, 372
environmental gradients   164, 173, 180, 216
environmental heterogeneity   272, 379
environmental impacts   212
environmental response models   216, 217
Erebia   247, 249
- *ottomana*   245
- *rhodopensis*   244
eremic zonobiome   242
Erigoninae   290, 291
Eriophorum vaginatum   127
error matrix   215
euryzonal   282, 284, 285
- ripicolous   290
Euxoa   248
evaluation of vegetation changes   426
exclosures   44, 322, 323, 383, 385
- experiments   344, 380
- fencing   381
exclusion of grazing   33, 380, 382
exposure   109
extinction   40, 82, 205, 268, 313
extreme weather events   460

F
Fagus orientalis   97, 99, 117, 455
- *sylvatica*   10, 80, 455
faunal types   239, 242
feeding ecology   355
ferns   87
Festuca clementei   201
- *eskia*   90
- *hallerii*   55, 150, 310
- *ovina*   154, 175
- *pumila*   211, 213, 215
- *rubra*   211
- *sardoa*   109, 152
- *violacea*   62
filling   440
- process   400, 401
fine-scale relief   213
fine-scaled patterns   217
Finland   411
Finnish Lapland   152
flexible areas   142
flora   32, 40, 48, 56, 67, 76, 87, 95, 106, 116
floristic diversity   56, 81
floristic similarity   60, 146
flowering plants   34, 57, 68, 96, 116
forest line dynamics   399
forest canopy cover, increase   432
Formica truncorum   270
freeze-and-thaw cycles   47

G
gamma (regional) diversity   83, 187, 222
generalised additive model   156, 158, 159, 161
generalist species   458
Genista salzmannii ssp. *lobelioides*   109
Gentiana bavarica   445
- *nivalis*   381, 461
geographical information systems   221
geographical pattern   245, 247
geology   6-7, 31, 39, 65, 74, 85, 93
Geometridae   264
geomorphology   5-7, 39, 206
Geranium sylvaticum   35
Giant Mountains   11, 15, 18, 24, 47, 48, 454
GIS   130, 217, 218, 221, 225, 226
glacial refugia   235, 245, 249
glacial survival   32, 162
glacier forelands   281, 290
glacier moraines   401
Glacies   247, 249
GLORIA   195
GMBA   462
Gnaphosidae   285, 291
goat   43, 378
golden eagle   346
golden plover   329, 330, 332
graminoid   9, 59, 79, 322, 344, 378, 381, 461
grass and moss heaths   33, 424

grasslands   42, 69, 90, 109, 119
- calcareous   62
- type   303
- grazed   303
grazers   460
grazing   100, 118, 130, 212, 234, 298, 323, 335, 380, 424
- by domestic ungulates   358
- by marmots   344
- by sheep   378
- by wild ungulates   385, 387
- exclusion   380, 384
- experiments   226
- impact   299, 302, 303, 305, 322, 378, 379, 380, 385, 387
- indicators   377
- pressure   304, 305, 335, 358, 381, 382, 387, 460
- -adapted plants   380
Greece   7, 454
Greek mountains   138, 140
ground beetles   266
ground temperature   13, 14, 17, 18, 20-22, 26, 27
- minima and maxima   19
growing period   23
growing season   16, 115
- lenght   22, 76, 205
- temperatures   17
- - seasonal mean   22
growth form   79, 186, 188, 384
Gulo gulo   227, 261

H

habitat   455
- available   373
- conditions   456
- diversity   41, 195
- - sources of   3
- fragmentation   48, 360
- modification   335
- mosaics   360
- patches   420
- preferences   459
- restriction   373
- use   327, 330, 334
- patterns   361
- type   149
Hahniidae   285
hare   43, 378
heath   333
- vegetation   335, 413, 414

Hellenides   113, 119
hemicryptophytes   34, 40, 48, 60, 68, 79, 87, 96, 108, 188
Hemiptera   262
herbivore-vegetation interactions   271
herbivore–plant relationships   377
Hercynian middle-mountains   47
Hercynian mountains of Central Europe   6
Herniaria boissieri   202
hierarchial GIS   130
high altitude fauna, Urals   265
historical perspective   377
Hochschwab   196
- mountain range   210, 212, 213
- region   195, 197, 200, 202
Homoptera   262
Hormathophylla purpurea   202
horse   378, 379
hotspot areas   144
human impact   70, 95, 106, 378
humid scree   310
hunting pressure   345
hygric continentality index   75
Hymenoptera   262
- parasitic   292

I

ibex   354, 378, 379
- Pyrenean   354, 368
- Spanish   368
image analysis   209
immigration   191, 286
increased temperature   44
Insecta   262
internal species turnover   403
invading species   204
Invertebrata   262
invertebrate diversity   233, 259
Iremel Massif   424
irreplaceable areas   142
isard   353, 354
Ischyropsalididae   283
ITEX   222, 225, 226

J

Jaccard's similarity index   140, 145
Juncus trifidus   119, 152, 154, 330

K

Karkonosze 47
karst 7, 113, 212
Kazbegi 96
Kiaeria starkei 42, 43, 155, 416
Kilpisjärvi 15, 412, 414, 416
Kobresia myosuroides 90, 150, 152, 154, 175
– *simpliciuscula* 413, 418
Krkonoše 47
krummholz 16, 211

L

Lagopus mutus 43
Lagotis uralensis 272
land clearing 70
landforms 5
landscape scale 56, 128, 236, 460
large carnivores 359
large herbivores 351
large-scale 221
large-toothed redback vole 261
Larix decidua 54, 80, 400, 401
late snowbed 42, 49, 69, 383
latitudinal differentiation 162, 163
latitudinal trends 158, 162, 453
latitudinal variation 300, 301
Latnjajaure 221–226
leaf beetles 266
– oligophagous 272
– polyphagous 272
lemming 379
– grazing 461
Lemmus lemmus 379
Lepidoptera 235, 262, 263, 265
Lepthyphantes 287
– *brunneri* 287
– *merretti* 287
– *rupium* 287
– *severus* 287
– *triglavensis* 287
Lepus europaeus 378
– *timidus* 43, 378
Leucanthemopsis alpina 445
lichen 35, 41, 48, 158, 351, 454, 457
– heaths 41
Licinus 312
life cycles 290
life forms 34, 40, 48, 59, 60, 68, 79, 96, 108, 182, 186, 188
life zone classification 9
limiting factors 181

Linyphiidae 284, 285, 289, 290, 291
livestock grazing 297, 304
local species richness 149, 161, 163, 164
Loiseleuria procumbens 35, 54, 62, 91, 151, 155, 412, 414
long-term dynamics 195
long-term grazing 461
long-term observations 269
long-term studies 399
long-term vegetation changes 411
long-term weather data 425
longitudinal transect 281
Lophozia opacifolia 77
Lumbricidae 264
Luzula alpinopilosa 153
– *arcuata* 177
– *spicata* 445
Lycaenidae 264
Lychnis alpina 385
Lycosidae 285, 289, 290, 291, 292
Lynx lynx 359

M

Macedonia 21, 24, 27, 116
Macrolepidoptera 233, 239, 246
– phylogenetic patterns 247
macrolichens 162
macroslope 282
Mamisoni 96
Mammalia 262
marmot 378
Marmota 340
– *marmota* 322, 339, 378
'massifs de refuge' 245, 249, 286, 308
meadow pipit 329, 330, 332
Mediterranean high mountains 456
– environments 10
Mediterranean islands 455
Megabunus armatus 283
Megaloptera 262
Mercantour 15, 18, 21, 24, 55
Mermithida 291
meso-scale 221, 228
Metopoceras 248
microhabitats 191, 289, 324, 386
Microtinae 378
Microtus arvalis 322, 339
mid-latitude European high mountains 456
middle mountains 11
migration 270
– routes 203, 206, 406

Mimetidae 291
Minuartia aizoides 98
— *cerastifolia* 90
— *corcontica* 48
— spp. 136
mires 42, 69, 98, 109, 211
— soligenous 49, 62, 69, 120, 418
Mitopus morio 283
Mitostoma alpinum 283
modelling 217, 222, 226, 323
moisture availability gradients 181
monitoring of climate 272
Monte Perdido 85
moss heath 332, 335, 400
mouflon 353, 354, 378
Mount Etna 7, 15, 19, 312
Mount Glungezer 437, 438, 439
Mount Patscherkofel 439
Mount Pollino 312
Mount Schrankogel 406
Mount Sylan 412
Mount Sylfjellet 412, 414, 416, 418
Mount Terminillo 313
mountain hare 379
mountains of Europe 5
mouse-hare 261
multi-scale patterns 166
multi-summit approach 197
multiple transect method 299
Mustela erminea 261, 346

N

Nardus stricta 42, 62, 78, 90, 98, 119, 150, 175, 211, 402
— grassland 332-336
narrow range endemics 147
natural grasslands 81, 82
near-minimum area set 142, 143
Nebria 308, 310
— *nivalis* 266
Nemastomatidae 283
neo-endemic species 243
Neuroptera 262
nitrophilous species 344, 386
nival 57
— zone 102, 281, 406, 460
Noctuidae 264
Noctuinae 248
north-south gradient 203
north-western Europe 456
Norway 411

Notodontidae 264
Nucifraga caryocatactes 401
nunatak 32, 87, 243, 245, 286
nutrient deposition from air pollution 447
nutrient enrichment 43, 81, 344, 378, 385
Nymphalidae 264

O

Ochotona alpina hyperborea 261
— *princeps* 343
Oenanthe oenanthe 130, 223, 227
Oligochaeta 262
Omalotheca hoppeana 110, 152
— *supina* 90, 109, 153
ombrogenous mires 49, 69
Ompalophana 248
open cryptogamic communities 69
open habitats 400
Opiliones 281, 282, 283, 291
oreal fauna 242
oreal refugia 284
Oreochloa disticha 211
Oreophilus 308, 309
oro-Mediterranean 89, 108, 196, 199, 245, 373
orobiomes 239, 248
orographic isolation 202, 203, 205
Orthoptera 243, 262
Ortles-Cevedale mountain group 76
Otiorhynchus dubius 266
overwintering 282
Ovis ammon 378
— *gmelini* 354
Oxyria digyna 55, 62, 90, 109

P

palaearctic 285
paleo-endemics 117
palynological 66, 108
Papilionidae 264
parasites 359
Pardosa oreophila 289
parthenogenesis 282, 292
passerines 331
pastures 62, 70, 78, 81, 89, 211, 324, 384
— abandonment 385, 461
— canopy structure 384
— clearing 70, 289, 324

Subject Index

- improvement 212
- montane 386
- species composition 78, 211, 378, 386
- species richness 62, 81, 89
patterned ground 9
Pelagonides 113
Peleponnisos 152
per pixel modelling 213, 214
permanent plots 197, 401
Phalangiidae 283
phenology 290
Philodromidae 284, 285
Phyllodoce caerulea 35, 153
physiography 53, 191
Picea abies 9, 50, 54, 65, 310, 400, 401, 455
Pieridae 264
pika 343
Píndos 152
Pinus cembra 54, 80, 400, 401
- *leucodermis* 80
- *mugo* 4, 50, 62, 80, 120
- - scrub 215
- *sylvestris* 42
- *uncinata* 91
pisces 262
Pitymys 341
Piz Linard 445, 446
plant communities 48, 210, 381, 411, 440
- changes 385
- composition 385
- distribution 216
- diversity 228, 379
- dynamics 387
- structure 225
- similarities 60
- species richness 49
- types 35, 42, 49, 60-62, 69, 81-82, 90-91, 98-99, 109, 119-120, 212, 406
plant litter 382
plant species richness 437, 453, 454
- ecological trends 157
- geographical trends 157
plant taxa 32, 41, 68
plot-based transects 457
pluvial continentality index 75
Poa alpina 35
- *arctica* 175
- *flexuosa* 175, 177
- *laxa* 445
Pohlia andalusica 77
Polar Ural 152, 234, 424, 425, 426

Polemonium viscosum 385
Polygala chamaebuxus 80
Polytrichaceae 381, 384
Polytrichum sexangulare 77
pool exhaustion 189, 190
population density 298, 301-303
post-glacial 163
- re-immigration 286
- re-colonisation 32, 161, 300
potential invaders 406
Potentilla crantzii 154
predators 227, 261, 308, 346, 360
- abundance 323
- populations 322
- pressure 228
predator-prey relationships 324
Primula clusiana 202
proportional sampling model 166
prostrate dwarf-shrub heaths 39, 54, 80, 100, 333
Pseudoscorpiones 281, 282, 283
Pseudovesicaria digitata 98
ptarmigan 43, 327, 329, 330, 331
pteridophytes 77, 78
Pterostichus 310
Pulsatilla halleri 151
purple sandpiper 327, 330
Pyrenees 7, 15, 18, 21, 24, 85-88, 90, 91, 101, 108, 138, 140, 152, 240, 243, 244, 246, 323, 340, 354, 356, 358, 360, 367-370, 372-374, 382, 453-456, 461

Q
Quaternary glaciations 242, 249

R
Racomitrium lanuginosum 43, 44, 333, 414, 418
Rai-Iz Massif 424
Ramondia nathaliae 119
range-size rarity 140, 141, 145
Rangifer tarandus 261, 354
Ranunculus glacialis 177, 445
- *marschlinsii* 109
rare-quartile richness 141, 146
raven 227
recreational disturbance 361
recreational use 335
red deer 43, 327, 352-353, 359, 378
- grazing 385-386

red fox   261, 346
red grouse   329, 330
redback vole   261
refugia   49, 50, 243, 286
refugial massifs   314
regeneration of alpine vegetation   404
regional altitude ranges   173
regional distribution, Arachnids   284
regional mountain range richness   145
regional orobiomes   247
regional scale   165, 205
regional species pools   156
regional species richness   149, 161, 164
reindeer   354, 378, 379
– grazing   411, 412, 420
relief   9, 11, 47, 114, 174
remote sensing   209, 217, 221, 222
repeat landscape photographs   424, 426, 428, 429, 430, 431
repeat observations   402
reptile   369-370
Rhododendron ferrugineum   62, 77, 91
Rhododendron hirsutum   120, 211
Rhodope   152
Rila-Pirin   119
Rila-Rodopi Massif   7, 113
rock communities   62, 69, 81, 90
rock crevices   109
rock vegetation   119
rodents   261, 322, 339, 378, 379
– population dynamics   342
– population ecology   339
roe deer   352, 353, 359, 378
root zone temperature   13, 19, 21, 27, 28
Rubus chamaemorus   48
Rumex alpinus   386
Rupicapra pyrenaica   354, 378
– *rupicapra*   354, 378
Russia   264, 334, 423, 425, 426

S
Sagina pilifera   109
Salehard   426
Salix helvetica   62
– *herbacea*   35, 55, 62, 80, 90, 150, 152, 175
– *polaris*   35
– *reticulata*   35, 62, 90, 119, 151, 154
– *retusa*   119, 150
– spp.   134, 136

Salticidae   285, 292
Sardinia   455
saturation   190
– of the species pool   447
– sampling   189
Satyridae   264
Saxifraga aquatica   90
– *bryoides*   445
– *exarata*   445
– *nevadensis*   202
– *nivalis*   48
– *oppositifolia*   445
– *pedemontana* spp. *cervicornis*   109
– *pedemontana* ssp. *cymosa*   119
– *stellaris*   42, 62, 109, 120
scales   221
– continent   128
Scandes   6, 15, 31–36, 139, 140, 154, 173, 174, 176–178, 181, 182, 222, 227, 234, 236, 382, 400, 412, 419, 454–456
Scotland   15, 18-24, 39, 41, 43, 44, 154, 327, 400, 454, 455
Scottish Highlands   5, 6, 378, 456
scree   35, 42, 49, 82, 108, 109, 119, 215
– and rock communities   98
– communities   62, 69, 81, 90
– specialists   312
– vegetation   438
scrub   4, 35, 42, 49, 69, 81-82, 99, 120, 211, 248
Scythrididae   265
sedge heath   119, 310
seed bank   227
seed production   385
seed rain   227
Sesleria albicans   62, 150, 210, 211, 213, 215
– *apennina*   80
– *caerulea*   150
– *coerulans*   152
– *comosa*   152
Setina   249
Shar-Korab   119
Shar Planina   113, 117
Shara mountains   113, 115, 119-120, 152
sheep   33, 43, 353, 358, 378, 379, 447
– grazing   44, 78, 82, 384
– – pressure   385
shrews   261
Sibbaldia procumbens   109, 151, 153, 412, 416
– *semiglabra*   98

Sicily 312
Sierra Nevada 7, 15, 19, 22, 24, 154, 195, 196, 198-204, 206, 244-246, 249, 367-374, 453, 455
Silene acaulis 154
siliceous 54, 58, 81-82, 90-91, 151, 153, 155, 163
– alpine grasslands 310
similarity 189, 246, 249
skylark 329, 330, 332
Slancevaya Mountain 424, 431
small-scale 221
– heterogeneity 358
smooth snake 368
smoothing splines 157, 161
snow bunting 327, 329, 330, 332
snow cover 115, 157, 163, 181, 356-358
– duration 342
snow gradient 163
snow lie 118
snow line 100
snow vole 322, 339, 368
snow-protected 154, 159, 166
– heath 413
snowbed 35, 42, 49, 69, 81-82, 90, 98, 119, 151, 153, 155, 210, 223-224, 312, 383, 401-402, 416, 438
– communities 62, 417
– grasslands 109
– vegetation 438
snowmelt 322
soil 115
– Arthropods 271
– chemistry 228
– conditions 39, 386
– fauna 270
– microbes 382
– substratum 35, 149, 160, 162-163,
– temperature 13, 15, 16, 19, 20, 40, 105
Soldanella austriaca 202, 205
Sørensen's similarity index 186, 189, 299
Sorex spp. 261
spatial models 224
spatial patterns 128, 173, 217
spatial trends 453
specialist predators 291
speciation 76, 77, 116, 117, 247
species 80, 242, 268, 288
– abundance 80
– composition 36, 213, 215, 299, 304, 333, 380, 413, 419, 420, 437
– – of plant communities 461

– density 299
– diversity 227, 236, 261
– diversity, Ural 262
– pool 128, 163, 165, 166, 190, 191
– recruitment 309
– richness 179, 180, 186, 187-189, 199, 234, 298, 299, 301-303, 370, 460
– – of birds 459
– – of butterflies 236
– – of carabids 236, 270
– – of cryptogams 458
– – of spiders 236
– – patterns in Europe 149
– – to area relationship 163
– – vs. altitude 179
– turnover 192
– density 299
spiders 285, 290
– vertical distribution 287
– winter activity 290, 291
Spitsbergen 154
springs 109, 120
– and flushes 42, 49, 62, 69, 90, 98
Stara Planina 113, 114, 119, 246
stoat 346
sub-Mediterranean mountains 10
subnival 97
– nanocoenoses 98
– zone 33, 96, 101, 290
– – establishment of species 406
Sudetes 154
summer drought 196, 198
summit flora 437
summit habitats 195
Svaneti 96
Synchloe callidice 270

T
tall-herb vegetation 42, 62, 69, 91, 99, 101, 109, 201
Tatra Mountains 15, 139, 140
taxonomic diversity 59, 133, 186
taxonomic richness 78, 107, 135-139, 144
temperate mountains 9
temperature extremes 19
temperature–physiography hypothesis 191
temporal changes 459
temporal scale 128
Terebrantes 292
territory 332, 335

– size of, breeding birds   226
Theridiidae   285
thermal sums   23
Thlaspi alpinum   202
– *rotundifolia*   62
Thomisidae   284, 285
thorny cushion-scrub   248
Thysanoptera   262
tiger moths   267
timberline   3, 106, 236, 301
– species   455
Tipula   267, 324,
topographical variables   212
Torneträsk   15
tragacanthic vegetation   98
transects   185–190
tree regeneration   433
treeline   3, 4, 60
– ecotone   3, 35, 42, 49, 55, 65, 69, 81, 91, 97–99, 119, 186, 188, 210, 290, 310, 377, 399, 423, 454
– – scrub   62, 234
– forming species   455
– lowering of   89, 108, 118, 289, 358, 377, 461
Trisetum spicatum   412
Tundra Landscape Dynamics   222, 223
tundra-steppe species   266, 267

U

ungulates   354
– social and mating system   355
– use of habitat   360
– spatial distribution   356
up-slope migration   437, 441
upward migration   204
– of invaders   204
– of species   195, 411, 437, 443
upward shift of vegetation zones   345
Ural Mountains   259-261, 266, 267, 269, 400, 423, 432, 454, 455
– South   265, 425
Ursus arctos   261, 359
Urtica dioica   386

V

Vaccinium gaultheroides   80
– *myrtillus*   35, 42, 54, 91, 98, 151, 154, 412-414

– *uliginosum*   54, 62, 91, 151, 152
– *vitis-idaea*   77
Val d'Arpette   185, 187-190
Valais   15, 16, 18, 21, 23, 24, 27, 150, 185
vascular plant   91, 197-199, 201, 454
– diversity   185
– distribution, Piz Linard   445
– species richness   179-180, 188, 443, 455, 457
vegetation   89, 97, 117
– change   43, 335, 377, 419, 420, 439, 443
– – comparison of current and past   427
– cover   198, 210, 400
– diversity   123, 209, 217
– dynamics   399, 401, 423, 459
– maps   209, 212, 214, 216, 223, 225
– – errors   217
– – for biodiversity research   216
– types   35, 42, 49, 62, 69, 81-82, 90, 98, 109, 119-120, 150, 152, 154, 298, 302, 329, 331–333, 438
Veronica anagalis-aquatica   98
vertebrates   262, 368, 458
– grazers   377
– species   368, 373
– species richness   321
vicariance   287, 292
vicariant associations   68, 102
vicariant patterns   245, 249
vicariant species   76, 77, 243, 248
– sibling   244, 249
viviparous species   440, 441
voles   378-379
– subterranean   341
Vulpes vulpes   261, 346

W

waders   331
water vole   341
weevils   266
wheatear   329, 330, 332
wind exposure   212
wind-exposed heath   35, 413
winter snow conditions   345
winter temperatures   181
wolf   359
wolverine   227, 261
wood mouse   341
WORLDMAP software   133

X

xeromontane fauna 233, 248, 250
xeromontane faunal type 245, 249
xeromontane species 245, 249

Z

Zlatoust 425
zonobiomes 196, 197, 247, 259, 260
Zygaena exulans 235

Ecological Studies
Volumes published since 1997

Volume 129
Pelagic Nutrient Cycles: Herbivores as Sources and Sinks (1997)
T. Andersen

Volume 130
Vertical Food Web Interactions: Evolutionary Patterns and Driving Forces (1997)
K. Dettner, G. Bauer, and W. Völkl (Eds.)

Volume 131
The Structuring Role of Submerged Macrophytes in Lakes (1998)
E. Jeppesen et al. (Eds.)

Volume 132
Vegetation of the Tropical Pacific Islands (1998)
D. Mueller-Dombois and F.R. Fosberg

Volume 133
Aquatic Humic Substances: Ecology and Biogeochemistry (1998)
D.O. Hessen and L.J. Tranvik (Eds.)

Volume 134
Oxidant Air Pollution Impacts in the Montane Forests of Southern California (1999)
P.R. Miller and J.R. McBride (Eds.)

Volume 135
Predation in Vertebrate Communities: The Białowieża Primeval Forest as a Case Study (1998)
B. Jędrzejewska and W. Jędrzejewski

Volume 136
Landscape Disturbance and Biodiversity in Mediterranean-Type Ecosystems (1998)
P.W. Rundel, G. Montenegro, and F.M. Jaksic (Eds.)

Volume 137
Ecology of Mediterranean Evergreen Oak Forests (1999)
F. Rodà et al. (Eds.)

Volume 138
Fire, Climate Change and Carbon Cycling in the North American Boreal Forest (2000)
E.S. Kasischke and B. Stocks (Eds.)

Volume 139
Responses of Northern U.S. Forests to Environmental Change (2000)
R. Mickler, R.A. Birdsey, and J. Hom (Eds.)

Volume 140
Rainforest Ecosystems of East Kalimantan: El Niño, Drought, Fire and Human Impacts (2000)
E. Guhardja et al. (Eds.)

Volume 141
Activity Patterns in Small Mammals: An Ecological Approach (2000)
S. Halle and N.C. Stenseth (Eds.)

Volume 142
Carbon and Nitrogen Cycling in European Forest Ecosystems (2000)
E.-D. Schulze (Ed.)

Volume 143
Global Climate Change and Human Impacts on Forest Ecosystems: Postglacial Development, Present Situation and Future Trends in Central Europe (2001)
J. Puhe and B. Ulrich

Volume 144
Coastal Marine Ecosystems of Latin America (2001)
U. Seeliger and B. Kjerfve (Eds.)

Volume 145
Ecology and Evolution of the Freshwater Mussels Unionoida (2001)
G. Bauer and K. Wächtler (Eds.)

Volume 146
Inselbergs: Biotic Diversity of Isolated Rock Outcrops in Tropical and Temperate Regions (2000)
S. Porembski and W. Barthlott (Eds.)

Volume 147
Ecosystem Approaches to Landscape Management in Central Europe (2001)
J.D. Tenhunen, R. Lenz, and R. Hantschel (Eds.)

Volume 148
A Systems Analysis of the Baltic Sea (2001)
F.V. Wulff, L.A. Rahm, and P. Larsson (Eds.)

Volume 149
Banded Vegetation Patterning in Arid and Semiarid Environments (2001)
D. Tongway and J. Seghieri (Eds.)

Volume 150
Biological Soil Crusts: Structure, Function, and Management (2001)
J. Belnap and O.L. Lange (Eds.)

Volume 151
Ecological Comparisons of Sedimentary Shores (2001)
K. Reise (Ed.)

Volume 152
Global Biodiversity in a Changing Environment: Scenarios for the 21st Century (2001)
F.S. Chapin, O. Sala, and E. Huber-Sannwald (Eds.)

Volume 153
UV Radiation and Arctic Ecosystems (2002)
D.O. Hessen (Ed.)

Volume 154
Geoecology of Antarctic Ice-Free Coastal Landscapes (2002)
L. Beyer and M. Bölter (Eds.)

Volume 155
Conserving Biological Diversity in East African Forests: A Study of the Eastern Arc Mountains (2002)
W.D. Newmark

Volume 156
Urban Air Pollution and Forests: Resources at Risk in the Mexico City Air Basin (2002)
M.E. Fenn, L. I. de Bauer, and T. Hernández-Tejeda (Eds.)

Volume 157
Mycorrhizal Ecology (2002)
M.G.A. van der Heijden and I.R. Sanders (Eds.)

Volume 158
Diversity and Interaction in a Temperate Forest Community: Ogawa Forest Reserve of Japan (2002)
T. Nakashizuka and Y. Matsumoto (Eds.)

Volume 159
Big-Leaf Mahogany: Genetic Resources, Ecology and Management (2003)
A. E. Lugo, J. C. Figueroa Colón, and M. Alayón (Eds.)

Volume 160
Fire and Climatic Change in Temperate Ecosystems of the Western Americas (2003)
T. T. Veblen et al. (Eds.)

Volume 161
Competition and Coexistence (2002)
U. Sommer and B. Worm (Eds.)

Volume 162
How Landscapes Change: Human Disturbance and Ecosystem Fragmentation in the Americas (2003)
G.A. Bradshaw and P.A. Marquet (Eds.)

Volume 163
Fluxes of Carbon, Water and Energy of European Forests (2003)
R. Valentini (Ed.)

Volume 164
Herbivory of Leaf-Cutting Ants: A Case Study on *Atta colombica* in the Tropical Rainforest of Panama (2003)
R. Wirth, H. Herz, R.J. Ryel, W. Beyschlag, B. Hölldobler

Volume 165
Population Viability in Plants: Conservation, Management, and Modeling of Rare Plants (2003)
C.A Brigham, M.W. Schwartz (Eds.)

Volume 166
North American Temperate Deciduous Forest Responses to Changing Precipitation Regimes (2003)
P. Hanson and S.D. Wullschleger (Eds.)

Volume 167
Alpine Biodiversity in Europe (2003)
L. Nagy, G. Grabherr, Ch. Körner, D. Thompson (Eds.)

Volume 168
Root Ecology (2003)
H. de Kroon and E.J.W. Visser (Eds.)

Printing (Computer to Plate): Saladruck Berlin
Binding: Stürtz AG, Würzburg